The Scaffolding of Our Thoughts

Ancient Magic and Divination

VOLUME 13

The titles published in this series are listed at *brill.com/amd*

The Scaffolding of Our Thoughts

Essays on Assyriology and the History of Science in Honor of Francesca Rochberg

Edited by

C. Jay Crisostomo
Eduardo A. Escobar
Terri Tanaka
Niek Veldhuis

BRILL

LEIDEN | BOSTON

Library of Congress Cataloging-in-Publication Data

Names: Crisostomo, C. Jay, editor. | Escobar, Eduardo A., editor. | Tanaka, Terri, editor. | Veldhuis, Niek, editor. | Rochberg, Francesca, 1952- honouree.
Title: The scaffolding of our thoughts : essays on Assyriology and the history of science in honor of Francesca Rochberg / edited by C. Jay Crisostomo, Eduardo A. Escobar, Terri Tanaka, Niek Veldhuis.
Description: Leiden ; Boston : Brill, 2018. | Series: Ancient magic and divination ; VOLUME 13 | Includes index.
Identifiers: LCCN 2018001101 (print) | LCCN 2018008028 (ebook) | ISBN 9789004363380 (E-book) | ISBN 9789004363373 (hardback : alk. paper)
Subjects: LCSH: Science, Ancient. | Assyriology. | Civilization, Assyro-Babylonian.
Classification: LCC Q124.95 (ebook) | LCC Q124.95 .S33 2018 (print) | DDC 509.35--dc23
LC record available at https://lccn.loc.gov/2018001101

Typeface for the Latin, Greek, and Cyrillic scripts: "Brill". See and download: brill.com/brill-typeface.

ISSN 1566-7952
ISBN 978-90-04-36337-3 (hardback)
ISBN 978-90-04-36338-0 (e-book)

This book is printed on acid-free paper and produced in a sustainable manner.

Contents

Preface

> Now it gives our way of looking at things,
> and our researches, their form.
> Perhaps it was once disputed.
> But perhaps, for unthinkable ages,
> it has belonged to the
> *scaffolding* of our thoughts.
>
> L. WITTGENSTEIN, *On Certainty*

How did historical cultures engage with what we call science? What makes an explanation valid within a given culture? How do texts reflect the world (or worlds)? These questions would seem more appropriate for anthropology or philosophy than Assyriology. Yet they frame and characterize the work of Francesca Rochberg. Francesca Rochberg—Chessie to her friends, students, and colleagues—is an anthropologist of dead civilizations. Her research has focused on celestial divination in cuneiform cultures and what effect those cultures can have on challenging our own conceptions of scientific knowledge. In the 1980s and 1990s, this work took a comparative form and examined how Babylonian astral sciences could be juxtaposed alongside the Greek astronomical sources. Underlying these textual analyses were perennial questions of demarcation. What makes knowledge "science"? And what does science look like in pre-modern cultures wherein the "natural world" was not the object of inquiry? Especially in her latest monograph, *Before Nature*, one can find the anthropological inclination to understand different ways of thinking about the world, to engage with the unfamiliar through cuneiform texts.

We found it fitting, therefore, to title this Festschrift dedicated to Chessie Rochberg "The Scaffolding of our Thoughts"—a Wittgenstein quote that she is most fond of citing—for there are few about whom we can say have done more to examine, question, and uphold the scaffolding of Assyriological and ancient scientific inquiry. She improves us all by example, as scholars and individuals. And anyone who has had the good fortune to interact with Chessie would agree that she is in every way approachable. Until, of course, things get technical. In those instances, one should prepare to wrestle with an incisive mind, capable of challenging all pre-conceptions, even within itself.

Intellectual openness ranks among her most contagious qualities. And as it usually follows for great thinkers, her openness and curiosity for new inquiries couples with her unburdened ability to change her mind. "Canonicity" and "science," major thema tested in the waters of her research, have undergone significant revision in the last few decades. Indeed, any day now, we should expect a new take on the problem of demarcation (i.e., the problem of where we draw the boundary of "science"). Perhaps more than she realizes, Chessie is an influence on those around her. Before *Before Nature*, and even before *The Heavenly Writing*, Francesca Rochberg was already entrenched in discussions and disciplines well outside of Assyriology. A MacArthur fellow, a former member of the Institute for Advanced Study, a Guggenheim fellow, and an honoree of the American Philosophical Society, Chessie is considered a model humanist and intellectual among her peers.

Her achievements, intellectual goals, and her effect on diverse scholarly communities are in themselves a lesson on the limits of disciplinary thinking in the modern humanities. Her disciplinary breadth is reflected in in her scholarship, particularly in her engagement with the history and philosophy of science, which we honor in the contributions to this volume. The contributions span the diachronic and geographic sphere of the ancient Mediterranean world, from Italy to Egypt to, of course, Babylonia and Assyria. But perhaps more significantly, the contributions reflect Chessie's interests in matters of ancient knowledge. Following an initial appraisal of ancient "science" by Sir Geoffrey Lloyd, the contributions in the first half explore practices of knowledge in Assyriological sources. The second half of the volume focuses specifically on astronomical and astrological spheres of knowledge in the ancient Mediterranean.

Finally, to and for her students, Chessie is an exemplary mentor—she guides by example. She inspires her students to increase their intellectual horizons and to read widely. How else is one to broaden one's wisdom? Or "enlarge his ears" (*urappīšu uzunšu*), as scribes used to say. In all of this thinking and imitation, one ends up becoming a historiographer, writing, as she does, genealogies of ideas, and in the process, ever testing one's inherited categories and assumptions. She tends to do all of this, of course, with a light smile and the occasional "if I can do it you can do it." We remain uncertain as to whether that phrase is a devastating fib or a comically tragic challenge. But of this we are certain: she guides with a "strong hand" (and occasionally, a broken foot).

Chessie is a tireless a scholar, an *ummânu mudû*, and the most generous of mentors. Nevertheless, an attentive warning label should accompany her work—one that needn't be as mean-spirited as the Assyrian curses which caution that Šamaš will carry off the eyes of a negligent scribe (*Šamaš igīšu litbal*).

Warning labels and ancient curses, after all, can also be thought of as oaths in reverse. Seen in this light, what this volume calls for is a firm acknowledgement that we listened to and honor Chessie Rochberg, and that, in the style of an oath, *we, the editors and contributors, promise to tread with care in matters of ancient knowledge*.

Ann Arbor, Michigan
Berkeley, California
Chicago, Illinois

Abbreviations

The abbreviations employed in this volume follow *The Assyrian Dictionary of the University of Chicago* (CAD U and W 2010) and the *Oxford Classical Dictionary* 4th edition (2012). Other abbreviations are listed below. Additional abbreviations adopted by the contributors are defined at the beginning of their contributions

AAA	*Athens Annals of Archaeology*
AchHist	*Achaemenid History, Leiden*
ADFU	*Ausgrabungen der Deutschen Forschungsgemeinschaft in Uruk-Warka*
ADOG	*Abhandlungen der Deutschen Orient-Gesellschaft*
AfOB	*Archiv für Orientforschung Beiheft*
AHES	*Archive for History of the Exact Sciences*
AJSL	*American Journal of Semitic Languages and Literatures*
AMI	*Archäologische Mitteilungen aus Iran Ergänzungsband*
ASJ	*Acta Sumerologica*
AUUSUU	*Acta Universitatis Upsaliensis, Studia Semitica Upsaliensia*
BabBib	*Babel und Bibel*
BagF	*Bagdhader Forschungen*
BagMB	*Bagdader Mitteilungen Beiheft*
BAM	*Die babylonisch-assyrische Medizin in Texten und Untersuchungen*
BdE	*Bibliothèque d'Études*
BibOr	*Biblica et Orientalia*
BiMes	*Bibliotheca Mesopotamica*
BiOr	*Bibliotheca Orientalis*
BUS	*Brown University Studies*
CDAFI	*Cahiers de la Délégation Archéologique Française en Iran*
CLeO	*Classica et Orientalia*
ColAnt	*Colloquia antiqua*
CM	*Cuneiform Monographs*
CUA	Catholic University Cuneiform Texts
CUSAS	*Cornell University Studies in Assyriology and Sumerology*
FuB	*Forschungen und Berichte, Staatliche Museen zu Berlin*
GBAO	*Göttinger Beiträge zum Alten Orient*
GMTR	*Guides to the Mesopotamian Textual Record*
HACL	*History, Archaeology, and Culture of the Levant*
HANEM	*History of the Ancient Near East Monographs*
HÄB	*Hildesheimer ägyptologische Beiträge*

HdOr	*Handbuch der Orientalistik*
Hist.	*Historia: Zeitschrift für Alte Geschichte*
Hist. Sci.	History of Science
HS	*Hebrew Studies*
JAJSup	*Journal of Ancient Judaism Supplements*
JANER	*Journal of Ancient Near Eastern Religions*
JANES	*Journal of the Ancient Near Eastern Society*
J GEN PHILOS SCI	*Journal for General Philosophy of Science / Zeitschrift für allgemeine Wissenschaftstheorie*
KTU	*Die keil-alphbetischen Texte aus Ugarit*
L&C	*Language and Communication*
LAPO	*Littératures anciennes du Proche-Orient*
MC	*Mesopotamian Civilizations*
MDAI	*Mémoires de la Délégation Archéologique en Iran*
Mem. Conn. Acad. Arts Sci.	*Memoirs of the Connecticut Academy of Arts and Sciences*
MLC	Morgan Library Collection
MS	*Melammu Symposia*
MVAG	*Mitteilungen der Vorderasiatischen*
MVS	*Materiali per il Vocabolario Sumerico*
OBCh	*Orientalia Biblica et Christiana*
OrNS	*Orientalis*, NS
ResOr	*Res Orientalis*
SAAB	*State Archives of Assyria Bulletin*
SAALT	*State Archives of Assyria Literary Texts*
SANER	*Studies in Ancient Near Eastern Records*
SAOC	*Studies in Ancient Oriental Civilization*
Sci Context	*Science in Context*
SH	*Scripta Hierosolymitana*
SÖAW	*Sitzungsberichte der Österreichischen Akademie der Wissenschaften*
SpTU	*Spätbabylonische Texte aus Uruk*
StBibLit	*Studies in Biblical Literature*
STMAC	*Science, Technology, and Medicine in Ancient Cultures*
StML	*Stimmen aus Maria-Laach*
StPohl	Studia Pohl
StPohl SM	Studia Pohl Series Maior
StSem	*Studi Semitici*
STT	*The Sultantepe Tablets*, I/II
StudDem	Studia Demotica
Talanta	*Talanta: Proceedings of the Dutch Archaeological and Historical Society*

TAPS	*Transactions of the American Philosophical Society*
TCBAI	*Transactions of the Casco Bay Assyriological Institute*
TCL	Textes cunéiformes, Musées du Louvre
Topoi—BSAW	**Topoi—Berlin Studies of the Ancient World**
TPS	*Transactions of the Philological Society*
TSAJ	*Texte und Studien zum antiken Judentum*
UMM	*University Museum Monograph*
Vierteljahresschr. Naturforsch. Ges. Zürich	*Vierteljahrschrift der naturforschenden Gesellschaft in Zürich*
WG	*Die Welt als Geschichte*

Publications of Francesca Rochberg

Books

1987 (editor) *Language, Literature and History: Philological and Historical Studies Presented to Erica Reiner*. AOS 67. New Haven: American Oriental Society.

1988 *Aspects of Babylonian Celestial Divination: The Lunar Eclipse Tablets of Enūma Anu Enlil*. AfO Beih. 22. Horn: Ferdinand Berger und Söhne.

1998 *Babylonian Horoscopes*. TAPS 88. Philadelphia: American Philosophical Society.

2004 *The Heavenly Writing: Divination, Horoscopy, and Astronomy in Mesopotamian Culture*. Cambridge: Cambridge University Press.

2010 *In the Path of the Moon: Babylonian Celestial Divination and Its Legacy*. AMD 6. Leiden: Brill.

2016 *Before Nature: Cuneiform Knowledge and the History of Science*. Chicago: University of Chicago.

Articles

1979 (with Zimansky, P.) "The University of Iowa Cuneiform Texts." *JCS* 31: 127–148.

1981 "Astronomy in the Ancient Near East." Pp. 526–542 in *Encyclopedia Miqraith* 8. Jerusalem.

1982 "Fate and Divination in Mesopotamia." Pp. 363–371 in *Vorträge gehalten auf der 28. Rencontre Assyriologique Internationale in Wien 6.–10. Juli 1981*, ed. H. Hirsch and H. Hunger. AfO Beih. 19. Horn: Ferdinand Berger & Söhne.

1983a "Stellar Distances in Early Babylonian Astronomy: A New Perspective on HS 229." *JNES* 42: 209–217.

1983b Libretto for George Rochberg, Songs of Inanna and Dumuzi. Bryn Mawr, PA: Theodore Presser Company.

1984 "New Evidence for the History of Astrology." *JNES* 43: 115–140.
"Canonicity in Cuneiform Texts." *JCS* 36: 127–144.

1987 "The Assumed 29th *ahû* Tablet of Enūma Anu Enlil." Pp. 327–350 in *Language, Literature, and History: Philological and Historical Studies Presented to Erica Reiner*, ed. F. Rochberg-Halton. AOS 67. New Haven: American Oriental Society.

"Elements of the Babylonian Contribution to Hellenistic Astrology." *JAOS* 108: 51–62.

"TCL 6 13: Mixed Traditions in Late Babylonian Astrology." *ZA* 77: 207–228.

1988 "Benefic and Malefic Planets in Babylonian Astrology," Pp. 319–324 in *A Scientific Humanist: Studies in Memory of Abraham Sachs*, ed. E. Leichty, P. Gerardi, and M. de J Ellis. OPSNKF 9. Philadelphia: University of Pennsylvania.

"Naburimannu." Pp. 1439–1443 in *Great Lives from History: Ancient and Medieval Series*, ed. F.N. Magill. Pasadena: Salem.

1989 "Babylonian Seasonal Hours." *Centaurus* 32: 146–170.

"Babylonian Horoscopes and Their Sources." *OrNS* 58: 102–123; pls. I–II.

1990 "Astronomy and Astrology and the Birth of Scientific Inquiry." *Bulletin of the Society for Mesopotamian Studies* 10: 25–33.

1991 "Between Observation and Theory in Babylonian Astronomical Texts." *JNES* 50: 107–120.

"The Babylonian Astronomical Diaries." *JAOS* 110: 323–332.

1992 "Astrology." Pp. 504–507 in *Anchor Bible Dictionary*, ed. D.N. Freedman. New York: Doubleday.

"Calendars (Ancient Near East)." Pp. 810–814 in *Anchor Bible Dictionary*, ed. D.N. Freedman. New York: Doubleday.

"The Cultures of Ancient Science: Some Historical Reflections." *Isis* 83: 547–553.

1993 "The Cultural Locus of Astronomy in Late Babylonia." Pp. 31–47 in *Die Rolle der Astronomie in den Kulturen Mesopotamiens. Beiträge zum 3. Grazer Morgenländischen Symposion (23.–27. September 1991)*, ed. H. Galter. Graz: Graz Kult.

"Babylonian Cosmology." Pp. 398–408 in *Encyclopedia of Cosmology: Historical, Philosophical, and Scientific Foundations of Modern Cosmology*, ed. N. Hetherington. New York: Garland Publishing.

1995 "Babylonian Astronomy and Calendars." Pp. 1925–1940 in *Civilizations of the Ancient Near East*, ed. Jack M. Sasson. New York: Charles Scribner's Sons.

Contributions to HarperCollins Dictionary of Religion, ed. J.Z. Smith. San Francisco: Harper San Francisco.

"akitu"; "Assyro-Babylonian Religion (First Millennium)"; "Esagila"; "Esarhaddon"; "Ishtar"; "Marduk"; "Oannes"; "Rosetta Stone."

1996 (with P.-A. Beaulieu) "The Horoscope of Anu-bēlšunu." *JCS* 48: 89–94.

"Personifications and Metaphors in Babylonian Celestial Omina." *JAOS* 116: 475–485.

1997 "Babylonian Astronomy." Pp. 87–92 in History of Astronomy: An Encyclopedia, ed. John Lankford. New York: Garland Publishing.

1999 "Empiricism in Babylonian Omen Texts: Problems in the Classification of Mesopotamian Divination as Science." *JAOS* 119: 559–569.

"Continuity and Change in Omen Literature." Pp. 415–427 in *Munuscula Mesopotamica: Festschrift für Johannes Renger*, ed. B. Böck, Cancik-Kirschbaum, E. and T. Richter. AOAT 252. Münster: Ugarit-Verlag.

"Babylonian Horoscopy: The Texts and Their Relations." Pp. 39–60 in *Ancient Astronomy and Celestial Divination*, ed. N.M. Swerdlow. Cambridge: MIT Press.

"The Babylonian Origins of the Mandaean Book of the Zodiac." *ARAM* 11–12: 237–247.

2000 "Scribes and Scholars: The *ṭupšār Enūma Anu Enlil*." Pp. 359–375 in *Assyriologica et Semitica, Festschrift für Joachim Oelsner anläßich seines 65. Geburtstages am 18. Februar 1997*, ed. Joachim Marzahn and Hans Neumann. AOAT 267. Münster: Ugarit-Verlag.

2001 "Mesopotamian Divination." Pp. 249–266 in *Storia della Scienza Vol. I. La Scienza antica*, ed. Vincenzo Cappelletti. Rome: Enciclopedia Italiana.

"Late Babylonian Astrology: The Horoscopes." Pp. 431–433 in *Storia della Scienza Vol. I. La Scienza antica*, ed. Vincenzo Cappelletti. Rome: Enciclopedia Italiana.

The Transmission of Babylonian Astronomy." Pp. 426–431 in *Storia della Scienza Vol. I. La Scienza antica*, ed. Vincenzo Cappelletti. Rome: Enciclopedia Italiana.

"The Historical Significance of Astronomy in Roman Egypt." *Isis* 92: 745–748.

2002 "A Consideration of Babylonian Astronomy within the Historiography of Science." *Studies in the History and Philosophy of Science* 33: 661–684

2003 "Lunar Data in Babylonian Horoscopes." *Centaurus* 45: 32–45.

"Heaven and Earth: Divine–Human Relations in Mesopotamian Celestial Divination," Pp. 169–185 in *Prayer, Magic, and the Stars in the Ancient and Late Antique World*, ed. Scott B. Noegel, Joel Walker and Brannon M. Wheeler. State College: Pennsylvania State University Press.

2004 "A Babylonian Rising-Times Scheme in Non-Tabular Astronomical Texts." Pp. 56–94 in *Studies in the History of the Exact Sciences in Honour of David Pingree*, ed. Charles Burnett, Jan Hogendijk, K. Plofker and Michio Yano. Leiden: Brill.

2005 "Mesopotamian Cosmology." Pp. 316–329 in *Blackwell Companion to the Ancient Near East*, ed. Daniel Snell. Oxford: Blackwell Publishing.

Contributions to *Encyclopedia of Ancient Natural Scientists: The Greek Tradition and its Many Heirs*, ed. by P.T. Keyser and G. Irby–Massie. New York: Routledge.
"Babylonian Astronomy" (185–86); "Balbillos" (188); "Epigenes" (290); "Kidenas" (475); "Kritodemus" (493); "Naburianos" (567); "Petosiris/ Nechepso-Petosiris" (637–638); "Sudines" (767–768); "Teukros of Babylon" (778).

2006 "Old Babylonian Celestial Divination: Texts and Context." Pp. 336–348 in *If A Man Builds A Joyful House: Assyriological Studies in Honor of Erle Verdun Leichty*, ed. M. Waters et al. CM 31. Leiden: Brill.

2007 "'The Two Cultures' and the Historical Perspective on Science as a Culture." *Forum for Public Policy Online* 2007, no. 1. http://forumonpublicpolicy.com/archive07/rochberg.pdf.
"Marduk in Heaven." *WZKM* 97: 433–442.

2008 "A Short History of the Waters of the Firmament." Pp. 227–244 in *From the Banks of the Euphrates: Studies in Honor of Alice Louise Slotsky*, ed. Micah Ross. Winona Lake, IN: Eisenbrauns.
"Babylonian Astronomy: The Hellenistic Transmission." *Mélanges de l'Université Saint–Joseph* 61: 13–32.

2009 "The Stars Their Likenesses." Pp. 1–52 in *What Is a God? Anthropomorphic and Non–Anthropomorphic Aspects of Deity in Ancient Mesopotamia*, ed. Barbara Nevling Porter. Winona Lake, IN: Eisenbrauns.
"Inference, Conditionals, and Possibility in Ancient Mesopotamian Science." *Science in Context* 22: 4–25.

2010 "Sheep and Cattle, Cows and Calves: The Sumero–Akkadian Astral Gods as Livestock." Pp. 347–359 in *Opening the Tablet Box: Near Eastern Studies in Honor of Benjamin R. Foster*, ed. S. Melville and A. Slotsky. CHANE 42. Leiden: Brill.
"If P, then Q: Form, Reasoning and Truth in Babylonian Divination." Pp. 19–27 in *Divination and Interpretation of Signs in the Ancient World*, ed. A. Annus. OIS 6. Chicago: The Oriental Institute of the University of Chicago.
"Beyond Binarism in Babylon." *Interdisciplinary Science Reviews* 35: 253–265.
"God-talk and Star-talk in Cuneiform and its Legacy in Later Antiquity." Pp. 189–200 in *Gazing on the Deep: Ancient Near Eastern, Biblical, and Jewish Studies in Honor of Tzvi Abusch*, ed. J. Stackert, B.N. Porter, and D.P. Wright. Bethesda: CDL Press.
"Babylonian Astral Science in the Hellenistic World: Reception and Transmission." CAS LMU e-Series 2010 No. 4: 1–11. http://www.cas.uni-muenchen.de/publikationen/e_series/cas-eseries_nr4.pdf.

2011 "Scientific Observation and Knowledge of the World in Cuneiform Culture" Pp. 618–636 in *The Oxford Handbook of Cuneiform Culture*, ed. E. Robson and K. Radner. Oxford: Oxford University Press.

"The Heavens and the Gods in Ancient Mesopotamia: The View from a Polytheistic Cosmology." Pp. 117–136 in *Reconsidering Revolutionary Monotheism*, ed. B. Pongratz-Leisten. Winona Lake, IN: Eisenbrauns.

"Divine Causality and Cuneiform Divination." Pp. 189–203 in *A Common Cultural Heritage: Studies in Mesopotamia and the Biblical World in Honor of Barry L. Eichler*, ed. G. Frame, E. Leichty, J. Tigay, and S. Tinney. Bethesda: CDL Press.

"Natural Knowledge in Ancient Mesopotamia." Pp. 9–36 in *Wrestling with Nature: From Omens to Science*, ed. P. Harrison, R.L. Numbers and M.M. Shank. Chicago: University of Chicago Press.

2012 "The Expression of Terrestrial and Celestial Order in Ancient Mesopotamia." Pp. 9–46 in *Ancient Perspectives: Maps and Their Place in Mesopotamia, Egypt, Greece, and Rome*, ed. R. Talbert. Chicago: University of Chicago Press.

2013 "The History of Science and Ancient Mesopotamia." *JANEH* 1: 37–60.

2014 "Periodicities and Period Relations in Babylonian Celestial Sciences." Pp. 211–218 in *Extraction and Control: Studies in Honor of Matthew W. Stolper*, ed. W. Henkelman, C. Jones, M. Kozuh, and C. Woods. SAOC 68. Chicago: The Oriental Institute of the University of Chicago.

"'A Firm Yes': Certainty in the Assyrian Culture of Knowledge." *BCSMS* 7: 5–12.

2015 "Conceiving the History of Science Forward." Pp. 515–531 in *The Frontiers of Ancient Science: Essays in Honor of Heinrich von Staden*, ed. B. Holmes and K.-D. Fischer. Beiträge zur Altertumskunde 338. Berlin: de Gruyter.

"The Babylonians and the Rational: Contexts of Rational Reasoning in Cuneiform Scribal Scholarship." Pp. 209–248 in *In the Wake of the Compendia: How Technical Handbooks and Encyclopedia Reshape Early Mesopotamian Empiricism*, ed. J. Cale Johnson. Science, Technology and Medicine in Ancient Cultures 3. Berlin: de Gruyter.

"Canon and Power in Cuneiform Scribal Scholarship." Pp. 217–229 in Problems of Canonicity and Identity Formation in Ancient Egypt and Mesopotamia, ed. K. Ryholt and G. Barjamovic Copenhagen: Museum Tusculanums.

2016 "The Brown School of the History of Science: Historiography and the Astral Sciences." Pp. 5–17 in *The Circulation of Astronomical Knowledge in the Ancient World*, ed. J.M. Steele. Time, Astronomy, and Calendars: Texts and Studies. Leiden: Brill.

"A Critique of the Cognitive-historical Thesis of The Intellectual Adventure of Ancient Man." Pp. 16–28 in *The Adventure of the Human Intellect: Self, Society, the Divine in Ancient World Cultures*, ed. Kurt A. Raaflaub. The Ancient World: Comparative Histories. Malden: Wiley Blackwell.

(with Raaflaub, Kurt A.) "Introduction." Pp. 1–15 in *The Adventure of the Human Intellect: Self, Society, the Divine in Ancient World Cultures*, ed. Kurt A. Raaflaub. The Ancient World: Comparative Histories. Malden: Wiley Blackwell.

"Where were the Laws of Nature before there was Nature?" Pp. 21–39 in Laws of Heaven—Laws of Nature: The Legal Interpretation of Cosmic Phenomena in the Ancient World, ed. K. Schmid and C. Uehlinger. OBO 276. Fribourg: Academic Press; Goettingen: Vandenhoeck & Ruprecht.

"Foresight in Ancient Mesopotamia." Pp. 157–175 in *Foresight*, ed. L.W. Sherman and D.A. Feller. Darwin College Lecture Series. Cambridge: Cambridge University Press.

2017 "An Ancient Celestial Empire of Benevolent Knowledge." Pp. 225–252 in *Studies on the Ancient Exact Sciences in Honor of Lis Brack-Bernsen*, ed. Mathieu Ossendrijver and John M. Steele. Berlin Studies of the Ancient World 44.

"These Bones Live!" *KNOW: A Journal on the Formation of Knowledge* 1: 67–83.

Forthcoming

(with Alan C. Bowen, editor) *Ancient Astronomy in its Mediterranean and Near Eastern Contexts*. Leiden: Brill.

"Science and Ancient Mesopotamia." In *Cambridge History of Science*, ed. A. Jones and L. Taub. Cambridge: Cambridge University Press.

"The Astral Sciences of Mesopotamia." In *Oxford Handbook of Science & Medicine in the Classical World*, ed. Paul T. Keyser and John Scarborough. Oxford: Oxford University Press.

"Astrology and Astronomy." In *A Handbook of Ancient Mesopotamia*, ed. G. Rubio. Berlin: de Gruyter.

(with Alan C. Bowen) "Introduction." In *Hellenistic Astronomy: The Science and Its Contexts*, ed. A.C. Bowen and F. Rochberg. Leiden: Brill.

"Traces of Babylonian Planetary Theory in Greco-Latin Sources." In *Ancient Astronomy in its Mediterranean and Near Eastern Contexts*, ed. A.C. Bowen and F. Rochberg. Leiden: Brill.

"Babylonian Celestial Divination and Nativities." In *Ancient Astronomy in its Mediterranean and Near Eastern Contexts*, ed. A.C. Bowen and F. Rochberg. Leiden: Brill.

"Categories, Kinds, and Determinatives." In *Die Sprache des Bewusstseins und das Bewusstsein von Sprache im Alten Orient*, ed. J. Cale Johnson.

"*Ina lumun attalī Sin*: On Evil and Lunar Eclipses." In *The Sources of Evil: Complexity and Systematization, Differentiation and interdependency in Mesopotamian Exorcistic Lore*, ed. D. Schwemer. AMD. Leiden: Brill.

"The Catalogues of Enūma Anu Enlil." In *Assyrian and Babylonian Scholarly Text Catalogues: Medicine, Magic and Divination*, ed. U. Steinert. Die babylonisch-assyrische Medizin in Texten und Untersuchungen. Boston: de Gruyter.

Reviews

1982 Review of *Ephemeriden von Sonne, Mond, und hellen Planeten von -1000 bis -601* by Hermann Hunger-Rudolf Dvorak. *ZA* 72: 299–300.

1985 (with N. Swerdlow) Review of *Ephemeriden von Sonne, Mond, und hellen Planeten von -1000 bis -601* by Hermann Hunger-Rudolf Dvorak. *JNES* 44: 237–239.

1990 Review of *Astronomical Diaries and Related Texts from Babylonia* vol. I, II by Abraham Sachs/ Hermann Hunger. *Orientalia NS* 58: 551–555.

1995 Review of *Archaic Bookkeeping: Writing and Techniques of Economic Administration in the Ancient Near East* by H.J. Nissen, P. Damerow, R.K. Englund. *ISIS* 86: 209–310.

1999 Review of *The Concept of Fate in Ancient Mesopotamia of the First Millennium: Toward an Understanding of Simtu* by Jack N. Lawson. *JNES* 58: 54–58.

2002 Review of *Astronomical Papyri from Oxyrhynchus* (*P. Oxy. 4133–4300a*) by Alexander Jones. *JNES* 62: 44–47.

2008 Review of *Calendars and Years: Astronomy and Time in the Ancient Near East* by John M. Steele. *Journal of the History of Astronomy* 39: 541–534.

2012 Review of *Astrology: From Ancient Babylon to the Present* by P.G. Maxwell-Stuart. *Times Literary Supplement.*

2013 Review of *Poetic Astronomy in the Ancient Near East: The Reflexes of Celestial Science in Ancient Mesopotamia, Ugaritic, and Israelite Narrative, History, Archaeology, and Culture of the Levant* by Jeffrey Cooley. *Journal for the History of Astronomy* 45: 127–129.

2013 "The Scientific Imagination of the Other." Review of *Being, Humanity, and Understanding: Studies in Ancient and Modern Societies* by G.E.R. Lloyd. *Metascience* 23: 165–168.

2015 Review of *Pathways into the Study of Ancient Sciences: Selected Essays by David Pingree* by Pingree, David, eds. Isabelle Pingree and John M. Steele. *Journal for the History of Astronomy* 47: 102–105.

2017 "Neugebauer and Premodern Science." Review of *A Mathematician's Journey: Otto Neugebauer and Modern Transformations of Ancient Science,* eds. A. Jones, C. Proust, and J. Steele. *Journal for the History of Astronomy* 48: 354–357.

CHAPTER 1

Where Next for "Ancient Science"?

G.E.R. Lloyd

At the outset, History of Science, as is well known, was marked by pronounced positivist assumptions and motivations, which in some quarters at least it still has not completely shaken off. From Whewell's *History of the Inductive Sciences* (1837) onwards, the new academic discipline was to be the study of progress, the cumulative acquisition of greater understanding of nature and of how to control it. In my own University of Cambridge the subject belonged initially to the Department of Natural Sciences and a high proportion of the students, still in the 1960s when I began lecturing, came from other areas of science itself, rather than from history let alone from philosophy, even though both the main Cambridge practitioners, Mary Hesse and Gerd Buchdahl, insisted that history and philosophy must join forces in what was to be a philosophically sophisticated historical study.

The popular textbooks originally focussed attention on anticipations of later knowledge, and what could not be fitted into some narrative of discovery and the onward and upward march of human understanding tended to be neglected, when not totally ignored. Condescending remarks were made about individual mistakes (a flat earth and the notion that the heart is the seat of cognition) but whole programmes were all too easily dismissed as pseudo-science (the enterprise of predicting events on earth from observation of the heavens, the ambition to turn other materials into gold). And early history of ancient science was undoubtedly competitive, with advocates for different ancient societies giving them better or worse marks, depending on how far they could be represented as anticipating "us." Those trained in Greek and Latin classics were particularly slow to wake up to what Egypt or Mesopotamia or India or (perhaps especially) China had to offer.

By the time I wrote a couple of slim handbooks on Greek science in the early 1970s many of the earlier assumptions were being challenged, even though the source material that was generally available was minuscule compared with today. If you will pardon a little bit of autobiography, let me take stock of where, at that stage, I thought the subject had got to. In my opening chapter to the first book, "the background and the beginnings," I had some comments on the Near East, for example on the technological advances made in both Egypt and Mesopotamia in pottery, textiles, metallurgy and so on. I recognised the

 | DOI 10.1163/9789004363380_002

importance of both Egyptian and Mesopotamian medicine and noted that the developments in mathematics and astronomy were even more important. Of course Neugebauer's *Exact Sciences in Antiquity* was already into its second edition in 1957, though his three-volume magnum opus, the *History of Ancient Mathematical Astronomy* did not appear until 1975. I mentioned the Venus tablet from the reign of Ammisaduqa but put it that most of the eclipse observations that Ptolemy was later to draw on were carried out either in attempts to foretell the fortunes of the kingdom or the king or to establish the calendar (Lloyd 1970: 7).

At the same time I was conscious of the over-enthusiasm with which some Greek writers, Plato, Aristotle and Simplicius, for instance, referred to the amazing record of Egyptian and Babylonian celestial observations and I duly threw some cold water on the idea of the super accuracy of their work. I was also aware that many of the data recorded are not the result of direct observation: some are just the result of applying previously determined tables. Of course, I was conscious that many cuneiform texts date from the Hellenistic period, much later than the Milesians with whom I was principally concerned at the beginning of my book. But I cited Neugebauer on the question of how far back we can push the application of a systematic mathematical theory to the data that had been collected, recording that he doubted that this was achieved before about 500 B.C. But for all my hesitation on a number of issues, I homed in on two main positive conclusions. I quote (Lloyd 1970: 7): "First the Babylonians had conducted extensive observations of a limited range of celestial phenomena long before Greek science began. And secondly, with the records they accumulated they were in a position to predict certain phenomena." I went on to insist on the greater difficulty of predicting solar eclipses visible at any given point on the earth's surface, contrasting that with the more straightforward task of predicting lunar eclipses, but noting they did that "not on any geometrical model" but on purely arithmetical procedures.

I am now taken aback at how close I came, there, to identifying the important features and yet how I missed the key point. What I should have pressed was not the arithmetical versus geometrical contrast but the distinctions *within* different types of phenomena that were the aims of prediction. It was one thing to interpret celestial signs as indicating changes in the fortunes of kings or kingdoms on earth: quite another to predict accurately *that* an eclipse would occur or to say when, after a period of invisibility, a planet would become visible again. Of course the scribes thought they had good grounds for both types of prediction. But they become increasingly confident in their predictions of celestial phenomena from the 8th or at least the 7th century BCE. Good or bad harvests and the fluctuations in the fortunes of kingdoms were always going to be open to doubt. But to predict planetary movements and to GET THEM RIGHT was impressive: the wow factor, by which I mean the prestige attached to success, was obvious to them as it is to us.

But then having missed that key contrast, how did my account proceed? Most of the rest of the opening chapter in my 1970 book was devoted to identifying the elements of the Greek breakthrough, the ones that could be used to justify some claim that it was the Greeks, after all, who were the first scientists. It was not that I ever signed up to any version of the "Greek miracle": that always struck me as (at most) a restatement of the problem and so could be no part of any kind of explanation. But I was overly concerned with origins and breakthroughs, focussing on what I then called the "discovery" of nature (I would now call it an invention) and on the development of a certain style of debate where what was demanded was the justification of whatever account was offered, *logon didonai*.

That all now seems a very long time ago (and indeed it was). Two fundamental changes have taken place in the intervening near half century. The first is that the source material now available for the study of all ancient societies, Mesopotamia and China perhaps especially, is of a different order of magnitude from what one could draw on then. The *Alter Orient und Altes Testament* series took off in the 1970s. The first volume of Reiner and Pingree's *Babylonian Planetary Omens* came out in 1975. The *State Archives of Assyria* series started in 1987. The first wave of works devoted to primary materials was then supplemented by a second of books and articles offering overviews of large areas of the subject, even Mesopotamian astronomy as a whole, Swerdlow, Robson, Brown and a sequence of marvellous studies by Chessie Rochberg herself.[1]

But then the second perhaps even more important change is a matter of shifts in the conceptual or categorical framework in which the study and interpretation of ancient materials are conducted. In my handbooks, I wrote about the mathematics, the astronomy, the medicine, the science, even, in Greece and other societies as if those terms could be applied to our data more or less unproblematically. Nowadays the dangers of doing so are recognised far more widely (if still not universally). To avoid the pitfalls of anachronism we must get back to how the ancients themselves described what they were doing, their

1 See especially Swerdlow 1998, Robson 1999, Brown 2000. Rochberg's *Aspects of Babylonian Celestial Divination* dates from 1988, her *Babylonian Horoscopes* from 1998, *The Heavenly Writing* from 2004, her *In the Path of the Moon* from 2009, while her most recent book *Before Nature: Cuneiform Knowledge and the History of Science* was published in 2016. Meanwhile on China Needham's series of studies, begun in 1954, has recently published its 25th volume (Needham 1954–).

"actors" categories, in other words, rather than our "observers" ones. Even when we use words derived from Greek via Latin, mathematics, astronomy and so on, the ancient source term is far from an exact equivalent to our derived one, and that is before we stray outside the Greco-Latin heritage and ask what exactly the Chinese, for example, meant by 算 術 (*suan shu*) (or 數 術 *shu shu*) or 天 文 (*tian wen*) or 曆 法 (*li fa*).[2]

But we can and must go further. To the varieties of concepts in play in different ancient natural languages we have to add the different acceptances of the relevant terms *as between* different groups or individuals using the *same* vocabulary in the given natural language. Greek *mathēmatikē* will do very nicely to illustrate the point for sometimes it has a very general sense (think of Iamblichus),[3] sometimes one tied to the different branches of the study of numbers and shapes. *Astronomia* in turn is always the study of the stars (as also is *astrologia*), but that meant very different things to different people, crudely astrological prediction to some, astronomical observation and explanation to others, both to many more.

A first methodological principle, then, I would say, is that we have to take into account what ancient investigators themselves said they were doing, how they defined the subject-matter of their inquiries and what they were hoping to achieve. Of course some ancient authors are more explicit and self-conscious about their aims than others. But where they do not themselves tell us in their own words what they are doing and why, it is all the more important to examine their practices and to beware of applying our latter-day norms and expectations.

Yet there is an obvious converse danger from an over-zealous application of that important methodological principle. Use actors' categories, the mantra goes. But shall we not end up just speaking ancient Greek or Chinese or Akkadian or whatever? We transliterate the especially problematic terms, *qi* 氣, *dao* 道, even *logos*: that is perfectly legitimate. But we have, at some point, to gloss those transliterations with some maybe quite elaborate explanatory

2 *Suan shu* is conventionally rendered "the art of reckoning," *shu shu* as "calculations and methods." *Tian wen* literally means "the patterns in the heavens" and *li fa* stands for calendar studies. But on the range of possible meanings for *shu* ("calculations") see for example Ho 1991. The diversity in ancient mathematical traditions is explored in Chemla 2002, and I commented on the different methods and aims adopted in other intellectual inquiries across ancient societies in Lloyd 2009.

3 Iamblichus, writing at the turn of the third and fourth centuries CE, claimed that *mathēmatikē* provided the key to the understanding of nature as a whole. But for him "mathematics" included number symbolism as well as what we would count as arithmetic and geometry. See *On the Common Mathematical Science* Ch. 32.

notes and paraphrases in our own natural language. Our own readerships speak those living languages after all, not the ancient forms of them we study.

So one aspect of that problem is the translation of ancient categories into English, say, or *Putonghua*. But another is the difficulty of comparing ancient investigations with one another, where I come back to the two points, of their variety and of the difficulty of finding some common ground that will enable the comparison to get started. The more one insists that each version of what the study of the heavens is for must be treated on its own terms, the more it may look as if we are dealing with radical incommensurabilities. Did not Kuhn continue to insist, even in his last revisions of the scientific revolution (Kuhn 2000), that competing paradigms are indeed incommensurable? That makes the point of the difficulty of comparing and evaluating with maximum emphasis, but maybe not with total clarity. Incommensurability implies, to be sure, that there is no common measure (as between the side and the diagonal of a square, e.g.). But no common measure does not imply no possibility of comparison (after all we can say that the diagonal is longer than the square). We do not bother to talk of incommensurability when we are dealing with things in different categories, as between the length of a piece of wood and a night, to cite an ancient Chinese example (*Mozi*).[4]

It is time for a concrete example, and let me take "medicine" in all its ambivalence.[5] The first step is to be wary of drawing some line between what we might recognise to be efficacious and what any ancient writer may have done. "Medicine" deserves its scare quotes then. But we can still see certain commonalities in the different ambitions entertained by different individuals and groups in different ancient societies on the question of "health" and "well-being" and their opposites "disease" or "illness." The first point is that in all ancient societies I am familiar with there are terms that correspond at least approximately to "health" and "disease," and the second that the terms in question apply far beyond the immediate context of individual human beings who may be thought in need of treatment by whoever set themselves up as able to deliver it, "healers" of whatever sort. The "soundness" of the "body politic" is much talked about in ancient Greece, but in India and China too what is healthy is used as a way of talking about political and ethical situations as well as physical or psychological ones (to lapse into observers' categories to make the point). "Diagnosing" what has gone wrong and offering suggestions

4 The *Mozi* canons date from around 300 BCE. This is B 6, (cf. Graham 1989: 416) which goes on with a second example, asking what there is more of, knowledge or grain.

5 I have reviewed the extensive evidence for ambivalence in the Greek concepts of well-being and illness in Lloyd 2003.

for how to "remedy" it are common ways of talking about good and evil in general. These are what Lakoff and Johnson (1980) would call "metaphors" that are good to think with: though I prefer to talk of the semantic stretch of the relevant vocabulary, encompassing many different fields of experience besides what we consider the province of *our* biomedical practitioners.

So eschewing modern assumptions about medicine and health and disease, we can still get our comparison between ancient ideas going by looking at what indeed were the assumptions about well-being in general. Of course we should not be misled by the fact that masses of those assumptions would be deemed mistaken by our medical practitioners. Modern biomedicine has its ideas on the subject, but to apply them straightforwardly to ancient materials is to ignore the first methodological principle that we must first consider what the ancients themselves have to say about what they were doing.

It so happens that (as is well known) we can find traces of more than one ancient debate on this very subject. What has often been represented as the start of "rational medicine"—in ancient Greece—provides a test case. The Hippocratic treatise *On the Sacred Disease* wades into a polemic against those the writer labels charlatans, quacks, magicians, who claim that the disease in question is caused by the gods, and also that they, the purifiers, know how to cure it. The Hippocratic author himself describes the condition he is talking about with some care and in some detail—which allows us to be fairly confident he was talking about Grand Mal epilepsy among other conditions: more important, he insists that it has a natural cause, like every other disease. That's where the new concept of "nature" comes into play, in driving a wedge between the phenomena that do and those that do not have natural causes. But what those who enthuse about this being the start of "scientific medicine" do not always pay attention to is that when it actually comes to being able to diagnose why epileptic fits occur the Hippocratic author relies very largely on his own imagination. It is caused by the phlegm blocking the blood-vessels in the brain, and the disease affects the phlegmatic more than it does the bilious. But that is of course pure speculation. Even worse, when at the end of the treatise he claims that epilepsy like every other disease is curable, provided you seize the right moment and make adjustments to control regimen, that is, one has to say, pure wishful thinking.

Now we do not have what the purifiers themselves would have said in their own defence: as so often with ancient sources there is an unevenness in the materials available to us. But we certainly know that the well-established practices of temple medicine, in the shrines dedicated to Asclepius, Apollo and others, continued to flourish long after the attack on quacks and magicians in that Hippocratic work. They remained popular until the end of late pagan antiquity—and in a modified form on into the early Christian period indeed[6]—as we can see from the enthusiasm with which the learned orator Aelius Aristides declares his support in the 2nd century CE (demonstrating that the clientele for temple medicine included members of the literate elite). He does not explain or rationalise temple medicine, but the main point that weighs with him was his conviction that god was on his side. He was not shaken by the fact that, no sooner relieved of one complaint, he succumbed to another: the god certainly did not manage to secure his *long-term* well-being. But his occasional recoveries he put down not to any human doctors—they are regularly criticised as far inferior to Asclepius—but to the god himself. We might comment that the comfort on offer from temple medicine—IF you believed—was a matter of psychology. But then it is not just foolish to include psychology as part of what contributes to overall well-being.

Faced with the rivalry between Hippocratic or Galenic medicine on the one side and that on offer in the shrines of Asclepius or from itinerant purifiers on the other, we can do better than just say that the latter was based on rank superstition. We should here take a leaf out of the anthropologists' book, for what they have said about much magic is that it is to be judged not so much by the canon of efficacy as by that of felicity (cf. Tambiah 1973). Think of the practice, still common among some Christians in some parts of the world, of showering the bride with confetti. That act is performed, when it is performed, not so much in any belief that the bride's fertility will actually be increased. Rather the thought may be that without the confetti the wedding would not be a proper one. The act is appropriate, felicitous, rather than efficacious.

Evaluation, I would say, is inevitable, but that still leaves open the question of the criteria to use. It is not just in Greece that competing views were expressed in this area. The Greeks had those who doubted supernatural interventions in disease and health. In China, in the third century BCE Xunzi pours cold water on the belief that there are demons and spirits. When a person, having contracted a certain numbness ("rheumatism") as a result of dampness, beats a drum and boils a sucking-pig (as offering to the spirits to obtain a cure) then there will necessarily be the waste of a worn-out drum and a lost pig, but the person concerned will not have the happiness of recovering from his sickness.[7]

6 The processes by which Christ and his saints came to take over (in some sense) the roles that had previously been played by Asclepius and other pagan healing gods and heroes are outlined in Nutton 1988: Ch. 10 and cf. Lloyd 2003: Ch. 9. Some of the evidence concerning the beliefs and practices of Aelius Aristides, as recorded in his *Sacred Tales*, is rehearsed in Lloyd 2003: Ch. 8.

7 See Xunzi 21, Knoblock and Riegel 1988–1994, Vol, 3: 109. A second prominent Chinese sceptic was Wang Chong the author of the *Lun Heng* in the first century CE.

But as in Greece so also in China both sceptics and believers were generally unmoved by the arguments of their opponents. Indeed, if we adopt the efficacy/felicity contrast, they were, if anything, at cross purposes.

But might this not be taken to imply that everyone is always right on their own terms—which would be to carry Davidson's principle of charity in interpretation to absurd lengths (Davidson 2001)? Well, I would certainly insist that we have first to see what those terms are. When they are correctly assessed, there is greater room to conclude that people are not just hopelessly stupid on their own terms—as some quick diagnoses of irrationality might suggest. But that is not to commit to their being no possibility of error, diagnosed whether by the actors themselves, or by outside observers.

This takes me to a topic of particular interest to Rochberg, the references in the Mesopotamian omen literature to what we should consider impossible phenomena. These include the sun appearing at night, the shadow of the earth moving in the wrong direction (that is from west to east) in a lunar eclipse, and especially eclipses occurring on days other than at the full or the new moon. Rochberg sets out the possibilities with scrupulous care (e.g. 2004: 250f.; 2016: Ch. 6). Maybe the scribes who refer to those eclipses did not know they were impossible, though certainly at some point the "saros" cycle was well established. Maybe these were additional omens, introduced on the "basis of the requirements of formal schemata into which phenomena were arranged" (where she expresses a degree of agreement with Leichty 1970: 20). A third possibility that Rochberg had expressed earlier in her book (2004: 200) is that the idea was that the gods can do anything. Marduk can destroy order as well as create it, so the scribes had better cater for all eventualities (where we might bear in mind that a belief in God's omnipotence and in miracles survived long after the so-called "scientific revolution" in the West).

Maybe more could be made of a point about the potential ambiguity in play when these phenomena are assessed for whether they are observable or not (*amāru*, Rochberg 2004: 250). One criterion for whether something is observable is that it is so if it can be observed *when* it occurs. But in a much weaker sense something can be thought of as observable because IF it did occur, it would be something that could be seen or heard. I am, of course, no expert, as Rochberg is, to say whether that is relevant to the texts in question (which certainly do not explicitly make any such distinction). But one point that certainly emerges from her account, with which I agree, is that what is at stake here is the applicability of the term "empirical" to these Mesopotamian studies (Rochberg 2004: 296). What was considered worthy of attention and for what reasons? The heavens are full of signs. We should not be surprised if their classification and use do not tally with those we may be used to in our scientific discourse.

So to take stock of where we have got to by this stage, on incommensurability my take is that it does not preclude, rather it presupposes, comparison. On evaluation I hold that it is inevitable, given that no description can be entirely neutral, nor entirely value free. All the more reason then to be wary of anachronism, of imposing latter-day expectations and assumptions in coming to terms with what we find in other groups or societies. But where does that leave "science" itself? My question was, where is ancient science, our discipline, headed, or where should it be? With "astronomy" and "medicine" you recall I suggested we retreat from whatever expectations we might have about them, influenced by where they stand today, and consider what the ancients thought they were doing, where "astronomy" is a place-holder for whatever the ancients talked about under the rubric of the study of the heavens, and "medicine" can be unpacked into its component elements, the understanding, diagnosis and treatment of whatever militates against well-being, however that was conceived. The differences in how indeed that was conceived have of course profound repercussions both for the "diagnoses" attempted and the "treatments" then undertaken. But where "science" itself is concerned, we have to adopt a rather different tactic, since ancient categories are, as we have just seen, even further from ours than is the case with "astronomy" and "medicine."

There is of course the view that there is no such thing as ancient science, and that nothing rates as such before, say, Galileo, or the other denizens of the scientific revolution, whenever we date that. If we go along with that, there is no more to be said. But I would maintain there are good reasons NOT to go along with that. "Science" I would say should not be defined in terms of its end-results (for they are always subject to revision) but rather in terms of its aims and methods. If we accept that argument, we can certainly study what ambitions ancient writers set themselves, in terms of the understanding of the external world, explaining how it works, classifying objects and events, predicting the future (to mention aims) and how they set about observing, testing their ideas, checking and criticising them (in terms of the methods they used). We can investigate who was involved in such activities, the institutions within which they worked, their audiences and who they were trying to persuade. Rochberg herself has done marvellous pioneering studies tackling such questions, the ways in which the scribes and scholars were trained, their interactions and their dealings with the kings. Her most recent book (2016) is an expert examination of the original aims of the scribes when they observed, predicted and explained, doing so without targeting "nature," for which they had no concept.

When we engage in such inquiries across different ancient societies, we discover a very considerable variety in the possible answers that our evidence for different groups at different periods throws up.[8] Evidently there was no single royal road to "science" (understood in the way I have suggested) nothing inevitable about how "it" developed. It is not that the way in which it was practised was determined by any single factor, economic, technological, political, institutional or whatever, or by any single combination of such factors. Rather we should assign different weights to different factors in different contexts.

The political environment of the Greek city-state seems to me to be particularly important in considering how certain styles of debate came to be practised there in other areas of investigation. But with a very different political set-up, observing, checking, and yes debating were practised in ancient Mesopotamia, Egypt, India, China with often impressively durable results (though the debates took rather different forms in different societies). Others have focussed on economic factors, where it is obvious enough that a certain economic surplus is a necessary condition for many of the activities that were carried out. Whether monetary supply was a determining influence in the development of Greek inquiries (as Seaford 2004, reviving an old Marxist argument, has suggested) is another matter. Yet others have mounted strong arguments for the role of literacy, though levels of literacy in our different ancient societies are a matter largely of guesswork, and I would say that the idea of the superiority of an alphabetic to other scripts was an argument that was rather overdone by Goody in his *Domestication of the Savage Mind* (1977). Then some again believe that the development of monumental architecture was a key factor in promoting abstract analysis (Hahn 2001). There are plenty of hypotheses in the field, although none contrives to command a general consensus and most suffer from more or less overreaching themselves.

I am not going to claim that I can now suddenly reverse that situation and establish robust causal links between the explanatory factors we might invoke and what there is to explain. But, like Rochberg herself, I am confident that there is some explaining to do and that that varies across our different ancient societies, across disciplines, and across periods, and this provides me with my platform for trying to summarise, now, what I would suggest about "where next" for "ancient science."

Of course the basic tasks of collecting, sifting, scrutinising, and editing the primary source materials must and will continue. There is no reason to believe

8 Some of the main similarities and differences in the sciences in ancient Greece and China, and possible reasons for them, are discussed in Lloyd and Sivin 2002. In Lloyd 2002 and more recent studies (Lloyd 2014) I have brought Mesopotamian and Indian data into the picture, in the former instance drawing heavily on the work of Rochberg among others.

that we shall not need to revise our views as more evidence comes on stream, just as we have revised them in the past few decades.

But as far as the interpretation of those materials goes, the first essential, I would say, is to get the explananda clear. The old preoccupation with identifying a single origin for science and even a single causal factor for that origin is really not helpful and flies in the face of those divergences and varieties I have been insisting on. The key questions are not so much *that* observations began to be carried out (they always were), but *what* observations were made and how they were used, not *that* explanations were attempted (they always were) but the modes of explanation that came to be favoured, not *that* predictions were made (again they always were) but what was thought to be predictable. This is not to deny discontinuities, ruptures, and breakthroughs, but there are also continuities. Prediction especially gives us a link with what *we* still recognise as science, though looking for where we can assimilate ancient investigations to ours is subject to all the dangers I have identified. In the case of prediction, in particular, we have learnt to acknowledge that other modes of prediction were not superseded by the new.

The experience of different ancient civilisations clearly diverges, and this gives us an ideal opportunity to investigate *why*. Here so far from needing to be rather hesitant about the value of our discipline and a bit defensive about the quality of the material we are dealing with, we can be positive about some of the advantages that come with the study of *ancient* diversities. Which features of each situation in each society contributed to the development of their particular styles of inquiry, in relation to the heavens, for instance, or disease and health, or the plant and animal kingdoms? We are, as I said, unlikely ever to be able to give hard-edged explanations that resolve those problems. But at least we can see what goes with what, what styles of observation and prediction served which purposes in which societies at which periods. Thus I have argued that Greek-style axiomatic-deductive demonstration in mathematics and other fields owes much to the dissatisfaction expressed in some quarters with the merely persuasive arguments that were deployed in the law-courts, the political assemblies and elsewhere.

"Ancient science" in my view is not just essentially heterogeneous, its study is essentially comparative and interdisciplinary, and the scrutiny of the divergent patterns we find in ancient societies can be all the more valuable for bringing to light correlations and correspondences for which the study of modernity provides no exact counterpart. If we no longer insist on seeing "science" as a unitary phenomenon, with a single "method" applicable across the board, if we desist from seeing ancient investigations as having the goal of reaching where we are today, then we can begin to do justice to the complex ambitions, successes and failures, of earlier endeavours, each carried out in their own particular institutional, political, technological and intellectual framework. The pay-off is that we can see how different features of those frameworks stimulated or inhibited work in specific fields, with potential lessons for us to draw today for how aspects of the framework within which modern scientific investigations are undertaken help or hinder the ways in which they develop or may stagnate.

Of course for the interdisciplinary comparative potential of the study of ancient science to be fully realised, we need to draw on all the resources that are provided by specialists in one field or one society or one period or another. Francesca Rochberg has provided a shining example of what can be achieved, starting from Mesopotamia, but with constant reference to the bigger picture of what 'science' is. I am privileged to have been asked to provide a few thoughts for this collection of essays in her honour.

Bibliography

Brown, David

2000 *Mesopotamian Planetary Astronomy-Astrology*. Groningen: Styx.

Chemla, Karine ed.

2012 *The History of Mathematical Proof in Ancient Traditions*. Cambridge: Cambridge University Press.

Davidson, Donald

2001 *Essays on Actions and Events* 2nd ed. Oxford: Oxford University Press.

Goody, Jack

1977 *The Domestication of the Savage Mind*. Cambridge: Cambridge University Press.

Graham, Angus C.

1978 *Later Mohist Logic, Ethics and Science*. London: School of Oriental and African Studies.

Hahn, Robert

2001 *Anaximander and the Architects*. Buffalo: State University of New York.

Ho Peng-Yoke

1991 "Chinese science: the traditional Chinese view." *BSOAS* 54: 505–519.

Knoblock, John and Riegel, J.

1988–94 *Xunzi: A translation and study of the complete works*, 3 vols. Stanford: Stanford University Press.

Kuhn, Thomas S.

2000 *The Road Since Structure*, ed. James Conant and John Haugeland. Chicago: University of Chicago Press.

Lakoff, George. and Johnson, M.
1980 *The Metaphors We Live By*. Chicago: University of Chicago Press.
Leichty, Erle
1970 *The Omen Series Šumma Izbu*. TCS 4. Glückstadt: J.J. Augustin.
Lloyd, Geoffrey E.R.
1970 *Early Greek Science, Thales to Aristotle*. London: Chatto and Windus.
1973 *Greek Science after Aristotle*. London: Chatto and Windus.
2002 *The Ambitions of Curiosity*. Cambridge: Cambridge University Press.
2003 *In the Grip of Disease: Studies in the Greek Imagination*. Oxford: Oxford University Press.
2009 *Disciplines in the Making*. Oxford: Oxford University Press.
2014 *The Ideals of Inquiry*. Oxford: Oxford University Press.
Lloyd, Geoffrey E.R. and Sivin, N.
2002 *The Way and the Word*. New Haven: Yale University Press.
Needham, Joseph. et al
1954– *Science and Civilisation in China*, 25 vols. to date. Cambridge: Cambridge University Press.
Neugebauer, Otto
1957 *The Exact Sciences in Antiquity*, 2nd ed (1st ed. 1952). Providence, R.I.: Brown University Press.
1975 *A History of Ancient Mathematical Astronomy*. 3 vols. Berlin: Springer.
Nutton, Vivian
1988 *From Democedes to Harvey*. London: Variorum.
Reiner, Erica, and Pingree, D.
1975–2005 *Babylonian Planetary Omens*, 4 vols. Malibu: Undena Publications.
Robson, Eleanor
1999 *Mesopotamian Mathematics*. Oxford: Oxford University Press.
Rochberg, Francesca
1998 *Babylonian Horoscopes*. TAPS 88. Philadelphia: American Philosophical Society.
2004 *The Heavenly Writing: Divination, Horoscopy, and Astronomy in Mesopotamian Culture*. Cambridge: Cambridge University Press.
2009 *In the Path of the Moon*. Leiden: Brill.
2016 *Before Nature: Cuneiform Knowledge and the History of Science*. Chicago: University of Chicago Press.
Rochberg-Halton, Francesca
1988 *Aspects of Babylonian Celestial Divination; the Lunar Eclipse Tablets of Enūma Anu Enlil*. AfOB 22. Horn: Ferdinand Berger.
Seaford, Richard
2004 *Money and the Early Greek Mind*. Cambridge: Cambridge University Press.

Swerdlow, Noel

1998 *The Babylonian Theory of the Planets*. Princeton: Princeton University Press.

Tambiah, Stanley J.

1973 "Form and meaning of magical acts: a point of view." Pp. 199–229 in *Modes of Thought,* ed. R. Horton and R. Finnegan. London: Faber and Faber.

Whewell, William

1837 *A History of the Inductive Sciences*. London: J.W. Parker.

CHAPTER 2

Crow Omens in Mesopotamia

Ann Kessler Guinan

I would like to dedicate this short study to my friend and colleague Francesca Rochberg whose work on divination, observation of the natural world, and appreciation of crows have all been an inspiration to me. This work is also dedicated to her friend, Jim, the baby crow she rescued and raised. Together they showed me what should have been obvious—the value of getting up from my desk to pay careful attention to what was happening outside my back door and then taking what I observed back to the way I understood and translated texts.

My interest in crows dates back to my work on crow omens and began as a way of investigating how right/left duality operated as a mechanism of divinatory interpretation. In first millennium omen texts, crow is the negative counterpart of falcon. Crow functions to turn an auspicious prediction into an inauspicious one and the reverse. A comparison of two omen texts (BM 108874 reverse lines 20–21 and *sakkiku* 2, lines 3–4) which contain parallel and contrasting sets of falcon omens demonstrates a mechanism of divinatory interpretation in the process of development and provides a way of tracing the path of the crow into the Mesopotamian augural lexicon.

Raven or Crow?

The *Corvus* genus includes ravens, crows, jackdaws, and rooks and belongs to the family *Corvidae*. The Corvid species found in Mesopotamia are as follows: (1) the rook (*Corvus frugilegus*), (2) the Mesopotamian crow (*Corvus coronae capellanus*), (3) the jackdaw (*Corvus monedula*), and (4) the Raven (*Corvus*

* This paper is a revision of work presented at two workshops: "Ancient Near East and South Asia Joint Special Session on Omens" at the 222nd Meeting of the American Oriental Society, Boston 2012 and Cultural Connections between Ancient Mesopotamia and Ancient India, The University of Copenhagen, Copenhagen 2014. I am grateful to Beth Jewell for editing all the various iterations of this work. I would also like to thank Laurie Pearce for reviewing and editing the sources. The mistakes are my own.

 | DOI 10.1163/9789004363380_003

corax).[1] Crows and ravens have a large brain relative to their body. "Mentally, crows and ravens are more like flying monkeys than they are like other birds" (Marzluff and Angell 2005: 40). Ravens are solitary and larger than crows. They prefer wild uninhabited areas and can live in a wide variety of habitats—arctic tundra, taiga, forests, grasslands, and deserts. They are scavengers; when living close to human settlements they may be seen foraging around rubbish dumps, but they are not granivorous. Ravens have the most complex vocalizations of any bird; their caw is deep and throaty. Crows on the other hand are creatures of the human habitat. They are omnivores. Crows have not only adapted to urban environments, they are uncannily attuned to human beings. They can remember individual faces for up to two years, use tools to work out eight-step problems and are known to steal bright objects (Ackerman 2016: 58, 107). Fledglings commonly learn to fly not from trees, but from roofs of houses. Crows also have a complex vocabulary but their caws are high-pitched and insistent. On the ground, they have a characteristic jerky, bobbing gait (Mazluff and Angel 2005: 54).

Sumerian terminology distinguishes the terms uga for "raven" and $buru_5$ for "crow."[2] Jeremy Black makes a good case for reading $buru_4$ as "rook.".[3] The Akkadian word *āribu* is used for both.[4] The first millennium omen texts use the logogram UGA^{MUŠEN}.[5] Preserving the philological rather than the ornithological distinction, modern scholars commonly translate UGA as raven in first millennium texts. However, omens that use the logogram UGA^{MUŠEN} in the first millennium and are based on the observations made in cities, houses, or anywhere in the human habitat, are clearly references to crow. The crow is a familiar presence in everyday life and the quintessential inauspicious omen. The raven is more mysterious and is associated with a deity or is a messenger of a deity (Verderame 2017: 402–403).

Crow apparently doesn't have the poetic gravitas of raven. According to George's translation of the SB Gilgamesh flood story, the bird that is set free to seek habitable land after the flood is a raven.

---

1 See Black 1996: 39, n.58.

2 See "Rabe" (s.v. "vögel") (RIA 11 2006: 210–213); PSD B: s.v. "$buru_4$" and "$buru_5$"; Veldhuis 2004: 226–233; 299–302.

3 The flocking behavior indicates that $buru_4$ is 'most certainly the rook' (Black 1996: 39); Rooks are tolerant of falcons breeding in their rookeries and, in addition, the breeding distribution of certain falcons is linked to the presence of rookeries (Madge and Burn 2001: 150).

4 CAD s.v. *āribu*.

5 However see KAI 30: 3 which uses two terms.

I brought out a raven setting it free
Off went the raven and it saw the waters receding.
It was eating, *bobbing up and down*, it did not come back to me.
TABLET XI, 154–156, GEORGE 2003: 712–713, 888

His translation "bobbing up and down" perfectly captures the jerky gait of the crow.[6]

Two Sets of Falcon Omens

Two separate omen texts include parallel and binary sets of falcon omens. The relationship of these omen sets that contrast with one another suggests that the scribe of the second set was using the first to generate new interpretations.

The first source from this time period is BM 100874, published by Nicla De Zorzi, a collection of falcon omens which, according to its colophon, is a copy from the land of Subartu and dated to the third year of the Kassite king Melišipak (1186–1172 BCE)(De Zorzi 2009: 93, 95). The tablet is primarily about falcons but omens 3, 6 and 24 are omens from crows—which, although not paired with falcons, reverse the polarity of the outcome (De Zorzi 2009). Her work is reminiscent of the work of the famous Babylonian scholar, Esagil-kīn-apli. It is a careful disentangling of the threads she received from "times of old": namely, Gadd's 90 year old composite copy.[7]

The other source, published by Nils Heeßel comes from the second tablet of *enūma ana bīt marṣi āšipu illaku* "If an exorcist goes to the house of a sick person," the small, two tablet sub-series appended to the beginning of the large diagnostic series, *sakkiku*. The diagnostic series was created by the editorial work of Esagil-kīn-apli who untangled the twisted threads received from an earlier time and reordered the series. According to Mesopotamian tradition, he was chief scribe (*ummānu*) during the reign of Adad-apla-iddina (1068–1047).[8] The omens of *sakkiku* 2 "If a man goes to the house of a sick person" are not observations of a sick person's symptoms, but are rather signs seen outside

6 His philological discussion of KI.TA-*ma i-šá-'-i* is based on Gadd's composite copy in CT 40 48 (George 2003: 889).

7 Gadd's copy in CT 40 combines BM 108874 with a Neo-Assyrian text, K.6278+. De Zorzi has disentangled the text (see the copy, Fig. 4, 135, and list of variants in De Zorzi 2009: 117–118).

8 There are no Esagil-kīn-apli texts datable to the 11th century. Some modern scholars have questioned the historicity of Esagil-kīn-apli, but others follow the late cueiform tradition and maintain that he was an *ummānu* during the reign of Adad-apla-iddina (Finkel 1988; Heessel 2001/2002; 2010; Robson 2008; Rutz 2011; Frahm 2011; forthcoming, see below note 24).

his house and are, therefore, conceptually different from the rest of the series. In addition, there are numerous parallels between the sub-series and the first millennium omen series, *šumma ālu*, but the direction of the borrowing is not clear (Freedman 1998: 12; Heeßel 2001/2002: 25). Both De Zorzi and Heeßel pointed out that the same set of falcon omens occur in both, De Zorzi noting the reverse the direction of flight. De Zorzi says direct borrowing between the two sources is hard to imagine (De Zorzi 2009: 118), but Heeßel says Esagil-kīn-apli may have based his work on existing traditions. My understanding of the chronological relationship between these two texts and to the later first millennium omens is based on the two-value logic of divinatory interpretation.

Both texts are based on the fundamental principles of divinatory interpretation: right/left duality[9] and the rules of binary arithmetic [+ times—=—] and [—times—= +].[10] The contrasting omens of each set use the same divinatory lexemes: the sick person's chances of recovery are connected to the observation of a falcon behind his house and the direction of the falcon's flight either to the right or to the left. According to BM 108874, r. 20, the flight of the falcon from left to right is positive (the sick person will recover). It requires the second, opposing omen r. 21—the flight in the opposite direction, from right to left—is negative (the sick person's illness will be long).

> r. 20. *šumma* (DIŠ) *ina bīt* (É) *amēli* (NA) *marṣu* (GIG) *mariṣma* (GIG-*ma*) *ina še-rim ina ku-tál bīt* (É) *marṣi* (GIG) *surdû* (SÚR.DÙ[mušen]) *ištu* (TA) *za-mi-i ša šumēli* (150) *ana za- mi-i ša imitti* (15) *ītiq* (DIB) *marṣu* (GIG) *šū* (BI) *ár-hiš itebbi* (ZI-*bi*).
> r. 21. *šumma* (DIŠ) *i-na bīt* (É) *amēli* (NA) *marṣu* (GIG) *mariṣma* (GIG-*ma*) *ina še-rim ina ku-tál bīt* (É) *marṣi* (GIG) *surdû* (SÚR.DÙ[mušen]) *ištu* (TA) *za-mi-i ša imitti* (15) *ana za-mi-i ša šumēli* (150) *ītiq* (DIB) *marṣu* (GIG) *šū* (BI) *murussu* (GIG-*su*) *irrik* (GÍD.DA).

> r. 20. If in the house of a man a sick person is sick and then in the morning behind the house of the sick person a falcon crosses from the outer left corner to the outer right corner—as regards that sick person: he will quickly recover.
> r. 21. If in the house of a man a sick person is sick and then in the morning behind the house of the sick person a falcon crosses from the outer right corner to the outer left corner—as regards to that sick person: his illness will be long. (De Zorzi 2009: 93, 95).

9 See Peek 1991: 205 and the essays that deal with divination in Needham 1973.

10 See Starr 1974; Mayer 1987; Guinan 1989, 1996; Heeßel 2000, 2001/2002; 2004; De Zorzi 2006.

The omens in *sakkiku* 2, 3–4 reverse the direction of the flight (from right to left in the first omen and left to right in the second), while preserving the order of the apodoses (the sick person will recover, his illness will be long). The scribe appears to have generated the new omens by taking the two omens of BM 108874 as a unit and inverting them.

> 3. *šumma* (DIŠ) *ina še-rì ina ku-tál bīt* (É) *marṣi* (lúGIG) *surdû* (SÚR.DÙmušen) *ištu* (TA) *za-mì-i šá imitti* (15) *ana za-mì-i šá šumēli* (150) *ītiq* (DIB-*iq*) *marṣu* (GIG) *šū* (BI) *ár-hiš itebbi* (ZI-*bi*).
> 4. *šumma* (DIŠ) *ina še-rì ina ku-tál bīt* (É) *marṣi* (lúGIG) *surdû* (SÚR.DÙmušen) *ištu* (TA) *za-mì-i šá šumēli* (150) *ana za-mì-i šá imitti* (15) *ītiq* (DIB-*iq*) *murussu* (GIG.BI) *irrik* (GÍD.DA).

> 3. If in the morning behind the house of a sick person a falcon crosses from the outer right corner to the outer left corner, that sick person will quickly recover.
> 4. If in the morning behind the house of a sick person a falcon crosses from the outer left corner to the outer to the outer right corner, his illness will be long. (Heeßel 2001/2002: 28, 37).

Although the process makes sense as a divinatory operation, the results are contradictory—the protasis of *sakkiku* 2, 4 is identical to BM 1008874 r.20. The flight of the falcon from left to right cannot mean both quick recovery and long illness. Reversals of meaning are part of the apparatus of divinatory interpretation, but the logical contradiction produced by this strategy leaves no interpretative ambiguity. The problem is resolved by the introduction of a second binary operator—the crow.

In the sequences which follow, the scribe of *sakkiku* 2 opens up new interpretive possibilities by adding omens which expand the complexity of divinatory binary logic.

Lines 13–16 are a set of four omens derived from the mournful cry of a crow. All the omens have the same inauspicious prediction: "wailing." A crow crying to the right of a man predicts wailing for the sick person; a crow crying to the left of a man predicts wailing for his adversary. Similarly, a crow crying in front of a man predicts wailing for the sick person, while a crow crying behind a man predicts wailing for his adversary. The scribe appears to be keeping two interpretive scenarios open: on one hand, the cry of a crow is an inauspicious constant, but, on the other hand, an omen that is inauspicious for the adversary of a divinatory subject is auspicious for that subject which in this case is the sick person. It should be noted that the protases of all four omens refer only to a neutral "man." It isn't hard to see why—the introduction of a negatively charged component (a sick person) into the protasis would result in mind-boggling complexity.[11] The parallel omens from Uruk forgo the sick person altogether and refer only to a man and his adversary (SpTU 2 32: r. 1–4).

> *sakkiku* 2, 13–16
> *šumma* (DIŠ) *a-ri-bu na-as-siš ina imitti* (15) *amēli* (NA) *is-si* (GÙ-*si*) *marṣu* (GIG) *šū* (BI) *bi-ki-ti immar* (IGI).
> *šumma* (DIŠ) *a-ri-bu na-as-siš ina šumēl* (150) *amēli* (NA) *is-si* (GÙ-*si*) *bēl* (EN) *dabābišu* (INIM-*šú*) *bi-ki-ti immar* (IGI).
> *šumma* (DIŠ) *a-ri-bu na-as-siš ana pān* (IGI) *amēli* (NA) *is-si* (GÙ-*si*) *marṣu* (GIG) *šū* (BI) *bi-ki-ti immar* (IGI).
> *šumma* (DIŠ) *a-ri-bu na-as-siš ana arkat* (EGIR) *amēli* (NA) *is-si* (GÙ-*si*) *bēl* (EN) *dabābišu* (INIM-*šú*) *bi-ki-ti immar* (IGI). (Heeßel, 2001/2002: 30–31).

> If a crow cries out mournfully to the right of a man, that sick person will experience wailing.
> If a crow cries out mournfully to the left of a man, his adversary will experience wailing.
> If a crow cries out mournfully in front of a man, that sick person will experience wailing.
> If a crow cries out mournfully behind a man, his adversary will experience wailing.

If my proposition is correct that the scribe of *sakkiku* 2 is producing his interpretations by using the omens of BM 108874 as the point of departure—then the BM text has to be primary. The colophon that dates the text to the reign of Melišpak (1186–1172 BCE) and gives the name of the scribe Bēl-nadin-šumi also states that the text is a copy (GABA.RI) from the land of Subartu.[12] It is entirely possible that the scribe of *sakkiku* may have had access to a source that was much earlier. If this is the case, it lends support to the 11th century date for the scribe of *sakkiku* 2.

---

11 Lines 5–12 are omens from vultures. The apodoses make predictions for the sick person and reverse the positive and negative value of right and left. According to Heeßel "the man" is implied and *marṣu* GIG functions as a negative operator of the pair: man/sick person (Heeßel 2001/2002: 25–26).

12 See De Zorzi on the changing meaning and location of Subartu (De Zorzi 2009: 91 n38).

Neo-Assyrian K.6278 + Rm 2. 389

K.6278 + Rm 2. 389 is a Neo-Assyrian version of the same source; it was used by Gadd for his composite copy. The tablet duplicates ten of the BM 108874 omens. De Zorzi points out that the order of the omens differ and, more importantly, the direction of the bird's flight is reversed. K.6278+:27, 29–34 duplicates *sakkiku* 2: 2–5 and the direction of flight is the same (see line 34), but the apodoses of the omens are not preserved. The text may be a version of *šumma ālu* Tablet 67 that deals with falcons, but until the bird omen tablets are reconstructed, its placement in the series cannot be determined. The parallels to *sakkiku* 2 are significant and show the first millennium scribes to be following both the order of the omens and their meanings that were established in *sakkiku* 2 (De Zorzi 2009: 93, 95; 116–117). It appears that the first millennium Assyrian scribe of K.6278+ had access to versions of both sources.

First Millennium *šumma ālu*

Crow enters the first millennium augural lexicon as the inauspicious counterpart of falcon. Falcon and crow are paired binary operators—references to one requiring references to the other. The series *šumma ālu* preserve two falcon omen tablets that are matched by crow omen tablets. The general topic of each of the four tablets can be determined and they share parallel and contrasting sequences of omens. The preserved texts also contain sequences of omens which duplicate the two tablets previously discussed. There is clearly a system of augural logic at work but, in their present state of reconstruction, the four *šumma ālu* tablets don't allow for its analysis.[13]

1. Tablet 66: falcons observed in the human habitat (i.e., houses, cities, and roads).[14]
2. Tablet 67: The action of crows in front of an army.
3. IM 74500 (Sumer 34 1978: 40–65): crows observed in the human habitat (i.e., houses, cities, and roads).
4. Tablet 79: The action of falcons in front of an army.

13 The wait will not be long; Freedman and De Zorzi are working on the *šumma ālu* bird omens and plan to produce an edition.

14 Freedman 1998: 339–340; De Zorzi 2009: 89–91.

There are in addition two Neo-Assyrian letters that pair crow and falcon. The first (SAA X 58; http://oracc.org/saao/saa10/P334229) is a letter from Balassî to the king who had apparently written to ask about omens that pertained to the observation of a crow (Ú.NAGA).[15] The scribe answers by quoting two omens. The first, lines 7–10, is a direct quote from IM 74500 19:

> *šumma* (DIŠ) *āribu* (Ú.NAGA) *mì-im-ma*
> *ana bīt amēli ú-še-ri-ib*
> *amēlu* (LÚ) *šū* (BI) *mi-im-ma la šu-a-tú*
> *qāssu* (ŠU-su) *ikaššad* (KUR-*ád*)
>
> If a crow (Ú.NAGA) brings something into the house of a man, that man will obtain something that is not his.

The protasis of the second omen that appears in lines 11–12 refers to "either a falcon or crow" (*lu* SÚR.DÙ^{mušen} *lu a-ri-bu*). Because the bird is flying into a man's house "crow" is the preferred translation.[16]

SAA X 238 (http://oracc.org/saao/saa10/P333976), a ritual for expelling evil demons, is even more telling. Among all the complex ritual prescriptions is the provision for the exorcist to hold a crow on the right and a falcon on the left.[17] The right–left reversal may be a property of the ritual or it may reflect the position of the exorcist in relation to the patient.

Conclusion

The Esagil-kīn-apli corpora has been the subject of much scholarly attention. The discussion has focused on the multi-directional transmission of the corpora over a long period of time, across geographic borders, and through changing political hegemonies and scribal cultures.[18] My contribution is based on the logics of divinatory interpretation and is, therefore, speculative. Tablet 2 of *sakkiku* stands in relation to the 11th century text BM 108874 and to the Neo-Assyrian text, K.6278+ as a pivotal moment in the development of Mesopotamian divination. The work exhibited in *sakkiku* 2 is unlike either of the other

15 SAA X 58 lines 5, 7; Borger, AOAT 305 2003: 342.
16 Lines 13–16 (Frahm 2011: 284).
17 SAA X 189: 16–17.
18 Frahm forthcoming. I am indebted to Frahm for allowing me to read and cite his paper.

two sources and reveals a true scholar of divination working out the problems of divinatory logic.

It is the hypothesis of this paper that *sakkiku* 2 was the creation of the 11th century scholar, Esagil-kīn-apli, who along with his editorial work, or more likely because of it, became deeply engaged with the logics of divinatory hermeneutics. And I further hypothesize that the first millennium *šumma ālu* omens that parallel *sakkiku* 2 were created in the first millennium by scribes who were employing principles of divinatory interpretation. The relationship of *sakkiku* 2 to K.6278+ supports the contention that later scribal schools picked up where Esagil-kīn-apli left off, adding new omens and vastly expanding texts according to principles of divinatory logic he established.[19] And perhaps by introducing crow as the binary counterpart of falcon, Esagil-kīn-apli made the work of his successors much easier.

Bibliography

Ascher, Marcia

2002 *Mathematics of Elsewhere: An Exploration of Ideas Across Cultures*. Princeton: Princeton University Press.

Ackerman, Jennifer

2016 *The Genius of Birds*. New York: Penguin Random House.

Black, Jeremy A.

1996 "The imagery of birds in Sumerian poetry." Pp. 23–46 in *Mesopotamian Poetic Language: Sumerian and Akkadian*, ed. M.E. Vogelzang and H.L.J. Vanstiphout. CM 6. Groningen: Styx.

Borger, Rykle

2003 *Mesopotamisches Zeichenlexikon*, AOAT 305. Münster: Ugarit-Verlag.

De Zorzi, Nicla

2009 "Bird Divination in Mesopotamia: New Evidence from BM 108874." *Kaskal* 6:85–135.

Finkel, Irving J.

1988 "Adad-apla-iddina, Esagil-kīn-apli, and the Series SA.GIG." Pp. 143–159 in *A Scientific Humanist: Studies in Memory of Abraham Sachs*, ed. E. Leichty, M. deJ. Ellis, and P. Gerardi. OPSNKF 9. Philadelphia: The University of Pennsylvania Museum.

19 The expansion of falcon and crow omens across four separate tablets reflects the form divinatory systems develop over long periods of time. See, for example, Ascher 2002: 5–37 (Chap. 1 "The Logic of Divination").

Frahm, Eckart

2011 *Babylonian and Assyrian Text Commentaries: Origins of Interpretation*. GMTR 5. Münster: Ugarit-Verlag.

forth. "The 'Exorcist's Manual': Structure, Language, 'Sitz im Leben'." In *Sources of Evil*, ed. G. Van Buylaere, D. Schwemer, and M. Luuko. Leiden: Brill.

Freedman, Sally M.

1998 *If a City Is Set on a Height*. OPSNKF 17. Philadelphia: The University of Pennsylvania Museum.

Guinan, Ann K.

1989 "The Perils of High Living: Divinatory Rhetoric in *Šumma Ālu*." Pp. 227–235 in *DUMU-E$_2$-DUB-BA-A: Studies in Honor of Åke W. Sjöberg*, ed. H. Behrens, D. Loding and M.T. Roth. OPSNKF 11. Philadelphia: The University of Pennsylvania Museum.

1996 "Left/Right Symbolism in Mesopotamian Divination." *SAAB* 10:5–10.

Heeßel, Nils

2000 *Babylonisch-assyrisch Diagnostik*. AOAT 43. Münster: Ugarit-Verlag.

2001–02 "'Wenn ein Mann zum Haus des Kranken geht…': Intertextuelle Bezüge zwischen der Serie *šumma ālu* und der zweiten Tafel der Serie SA.GIG." *AfO* 48–49:24–49.

2004 "Diagnosis, Divination and Disease: Towards an Understanding of the Rationale behind the Babylonian Diagnostic Handbook." Pp. 97–116 in *Magic and Rationality in Ancient Near Eastern and Graeco-Roman Medicine*, ed. H.F.J. Horstmanshoff and M. Stol. Studies in Ancient Medicine 27. Leiden: Brill.

2007 *Divinatorische Texte I: Terrestrische, teratologische, physiognomische und oneiromantische Omina*. KAI I. Wiesbaden: Harrassowitz.

2010 "Neues von Esagil-kīn-apli. Die ältere Version der physiognomischen Omenserie *alamdimmû*." Pp. 139–187 in *Assur-Forschungen*, ed. S. Maul and N. Heeßel. Wiesbaden: Harrassowitz.

Mayer, Werner R.

1987 "Ein altbabylonischer Opferschaubericht aus Babylon." *Or*NS 56: 245–262.

Needham, Rodney, ed.

1973 *Right and Left: Essays in Dual Symbolic Classification*. Chicago: University of Chicago Press.

Madge, Steve and Burn, Hillary

2001 *Crows and Jays*. Princeton: Princeton University Press.

Mazluff, J.M. and Angel, T.

2005 *In the Company of Crows and Rravens*. New Haven: Yale University Press

Peek, Philip M.

1991 "African Divination Systems: Non-Normal Modes of Cognition." Pp. 194–212 in *African Divination Systems: Ways of Knowing*, ed. P. Peek. Bloomington: Indiana University.

Robson, Eleanor

2008 "Mesopotamian Medicine and Religion: Current Debates, New Perspectives." *Religion Compass* 2: 455–483.

Rutz, Matthew T.

2011 "Threads for Esagil-kīn-apli: The Medical Diagnostic-Prognostic Series in Middle Babylonian Nippur." *ZA* 101: 294–308.

Starr, Ivan

1974 "In Search of Principles of Prognostication in Extispicy." *HUCA* 45: 17–23.

Veldhuis, Niek

2004 *Religion, Literature, and Scholarship: The Sumerian Composition Nanše and the Birds. With a Catalogue of Sumerian Bird Names*. CM 22. Leiden: Brill.

Verderame, Lorenzo

2017 "The Seven Attendants of Hendursaĝa: A study of animal symbolism in Mesopotamian Culture." Pp. 396–415 in *The First Ninety Years: A Sumerian Celebration in Honor of Miguel Civil*, ed. L. Feliu, F. Karahashi, and G. Rubio. SANER 12. Berlin: De Gruyter.

CHAPTER 3

Old Signs in New Dress? On the Meaning of Inanna's Symbol as Sign and "Presence" in Early Babylonian Divination

Abraham Winitzer

I

Not long ago, I was privileged to share a stage with our honoree at a seminar devoted to divination and the interpretation of signs in antiquity. The paper I gave on that occasion related to matters of theology, specifically on statements concerning the "presence" of the gods in the Mesopotamian omen literature,[1] while that of our honoree was philosophically inclined, with respect to the formal logic of this same literature and this logic's significance.[2] The latter topic, and more generally the place of the Mesopotamian omen collections in a consideration of that civilization's scientific expression, has been of considerable interest to Francesca Rochberg, who has been peerless in her articulation of the basic challenges this involves. One such challenge, as Rochberg's recent *Before Nature* explains with admirable lucidness,[3] involves Mesopotamia's insistence on a divine realm integral to the natural world, the typical domain of scientific inquiry.

As I tried to suggest in my contribution to that seminar on divination, it is exactly the formal logic of the omen lists delineated by Rochberg that presents a considerable exception to the supernatural realm's ubiquity within Mesopotamia's literature—even though the omen literature itself does not exclude mention of the gods, let alone the assumption of a divine realm underlying its

* Earlier versions of this paper have benefited especially from the insightful comments of A. Huang, G. Marchesi, G. McGarry, and A. Wilent and the collations of A. Cavigneaux.
Aside from the standard abbreviations, note also the following:

AOAT 50 = Hrůša 2010;

LAOS 4 = Wasserman 2016;

CUSAS = Cornell University Studies in Assyriology and Sumerology;

OBE = Jeyes 1989.

1 Winitzer 2010.

2 Rochberg 2010.

3 Rochberg 2016b.

 | DOI 10.1163/9789004363380_004

entire enterprise. And thus one must ask how these factors should figure in the consideration of this literature within the context of Mesopotamian science. Is there a place for theology in this civilization's conception of scientific reasoning?

The essence of this question has been an occasional concern to our honoree,[4] who could no doubt offer it a far more learned answer than I, should she opt to center on it with the boldness characteristic of so much of her writing. It is in that spirit that I offer the following modest essay to her fête, on an intriguing instance of a divine-presence statement in the omen literature and its significance for a small *crux interpretum* in Mesopotamian semiotics. The statement's actual and conceptual setting within the omen collections suggests that even the presence of gods and their requests were subject to scientific analysis.

II

In a study twenty years ago P. Steinkeller presented a compelling argument in support of an equation long suspected, namely that the conspicuous if enigmatic visual motif of a "volute-like reed structure" apparent in artistic scenes from archaic Uruk the archaic emblem of Inanna.[5] This widely assumed identification was based on the image's formal resemblance to the archaic form of the sign MÙŠ, used to write the name of the goddess, but had lacked textual corroboration. Now Steinkeller drew attention to an ED literary text that challenged the traditional interpretation of the image.[6] The text establishes a poetic parallel between Inanna's great emblem (urin-gal) or spire (zà-è) and a bar-su, or, alternatively, bar-sikil:[7]

---

4 See, e.g., Rochberg 2011; and especially 2016a and 2016b: 7–11, 38–58 (though in fact the entire latter work is relevant in this respect) for her analysis of important earlier Assyriological efforts to understand the place of Mesopotamia's mythology in that civilization's broader intellectual vision.

5 Steinkeller 1998: 90. For earlier opinions about the identity of her emblem, see n. 10 below.

6 An overview of the text, a hymn to Ama'ušumgal'anna, is found in Fritz 2003: 169–172. Three MSS. are extant for the relevant portion here. My sigla follow those in Steinkeller 1998: 90 and essentially those in Fritz 2003: 169:

source A = ARET 5 20 + Archi 1992: 36–37, pls. 7–8;
source B = ARET 5 21 + Archi 1992: 38–39, pls. 9–10 + Archi 1989: 125;
source C = Biggs 1974: 278.

7 The spelling BAR-SU is attested in two textual MSS., both from Ebla (= sources A and B); the term is rendered BAR-⌜SIKIL⌝ in the text version from Abū Ṣalābīkh (= source C).

(1)

A vii 7	za-gir bar-su gú ^{d}inanna lá-lá
B viii 5	za-gir bar-su gú ^{d}inanna lá-lá
C iii 1–2	za-gìn bar-⸢sikil⸣ gú ^{d}inanna lá
	“the lapis lazuli / lustrous ... binding the neck of Inanna,”

(2)

A viii 1	urin gal-gal é ^{d}⸢inanna⸣-dar an-si-ga
B viii 6	[..............] é ^{d}⸢inanna⸣-dar an-si-ga
C iii 3	urin gal-gal kul-ab$_4$-da si-ga
	“The great urin emblems set up at the temple of Inanna / at Kulaba,”

(3)

A viii 2	za-i gal-gal gul-la-ab-da-si-ga
B viii 7	za-i gal-[ga]l gul-[l]a-[...]
C	omits line
	“The great spires set up at Kulaba.”

Steinkeller identified these terms as variant spellings of Sum. $^{(\text{túg})}$bar-si(g), Akk. *parsikkum* (*parš/sīgum*), a type of headdress,[8] known in different varieties and qualities and from varying historical periods.[9] In the passage above,

8 PSD B, 126–127: “sash, shawl”; AHw, 836: “Kopfbinde, Mütze”; Waetzoldt 1980–83:198: “eine ganz allgemeine Bezeichnung für ‘Band, Binde’”; Asher-Greve 1995–96:184 n. 22: “the covering of the substructure with a special material or ... the appearance of the crown”; Selz 1997: 190 n. 68: “turban” (for men$_x$-bar-su); Steinkeller 1998: 92: “a type of scarf or shawl”; and note George 2003: 719 and n. 49 who, translating the *parsīgum* mentioned in Gilgameš XI 257 and 266 as “kerchief,” likens the item to the *kūfiyya* still popular in the modern Arab world; also Durand 2009: 82 (and see further nn. 32 and 34 below).

9 For third-millennium Sumerian references, see Waetzoldt 1980–83: 198; PSD B, 126–127; and more recently, see the references in the essays concerning the early periods in Michel and Nosch 2010. For second- and first-millennium references, see generally CAD P, 203–205 and Durand 2009: 78–83, along with the relevant essays in the aforementioned volume by Michel and Nosch. The contribution of Reid and Wagensonner 2014 provides evidence of the use of *parsikkus* as bandages in medical contexts.

From lexical lists as well as administrative documents it is clear that there were many *parsikkum* varieties and types. Waetzoldt (ibid.) cites different qualities (gal, nì-la, sal-la, sig$_5$, sùd-a, tur-tur) made of both wool and linen. In Hh XIX 249f. (MSL 10, 135) one finds no less than eight different kinds, thirteen in the forerunner from Ras Šamra (MSL 10, 151 [nos. 162–173]). *Parsikkus* appear frequently in Mari administrative texts, in different types (e.g., *ḫamdûm, uṭuplum, naṣmadum, saqqum, šutûm*) and qualities (sag, ús, gal, etc.); see further Durand, ibid. And one reads of two types of *parsikkus* (of wool or flax) in the

this headdress is described as "binding the neck (gú … lá) of Inanna." At least in this ED tradition, then, the great emblem of Inanna was conceived as a kind of garment that ornamented the divine image.[10]

This important finding raised the question of the scope of this tradition and the conceptualization behind it, as well as its continuity in later periods of Mesopotamian civilization. Clearly a symbolic vestige of this equation was fossilized in the sign MÙŠ used throughout the cuneiform "stream of tradition" as one of the writings for the goddess Inanna (^{d}MÙŠ [= ^{d}inin/^{d}inanna]).[11] Still, in the absence of explicit testimony from later periods one would be hard-pressed to demonstrate that this early understanding of MÙŠ as a representation of Inanna's headdress survived over the centuries. Indeed, the additional meaning for the writing MÙŠ as Sum. suḫ, "diadem" (Akk. *tiqnum*)[12]—the latter an amply attested term for an ornament adorning divine, royal, or even priestly heads[13]—indicates that the understanding of Inanna's/Ištar's unique divine emblem had been diluted to the point where it could be equated with the more common, even generic, diadem that adorned gods and men alike. In the first millennium a similar situation presents itself: *parsikku*s are attested as constituent parts of the *urigallu*-standards of several gods, including Ištar's.[14] Yet it seems clear that by this time the *parsikkum* had been reduced to a mere decoration and had lost its original significance as the very headdress of Inanna herself.

In short, there can be little doubt of a tradition that maintained some relationship between *parsikkum* and Inanna's/Ištar's standard throughout much of Mesopotamian civilization. What remains uncertain, however, is whether the ornamentation of the deity by a *parsikkum* was consciously recalled as the source of her symbol as well as her sign, MÙŠ (see Figure 3.2 below).

---

remarkable administrative text describing Ištar's dress at Lagaba (Leemans 1952: 2, 12–13, ll. 16–17).

10 I thus follow Steinkeller 1998: 89–90 in dismissing early opinions, most significantly those first promoted by W. Andrae (see ibid. for extensive bibliography), advancing the interpretation of this image as a gatepost with rolled-up mat. However, it should also be recalled that this idea was not understood in purely functionalist terms. Jacobsen 1976: 36, for instance, read it in the context of the early communal storehouses of Inanna, his Ninanna(k), "the Lady of the date clusters,"—in short, as part of his ideas about the earliest, pre-anthropomorphic, phase of Mesopotamian religious thought.

11 See, e.g., Glassner 2000: 227–228.

12 Steinkeller 1998: 93 and n. 40.

13 Steinkeller 1998: 93–95 and nn. 42–43.

14 Beaulieu 2003: 15.

The following presents testimony from early Babylonian divination literature pertinent to this question, beginning with a datum that connects the *parsikkum* and Inanna (§ III). This association finds support when considered in its broader literary context, alongside additional omens that mention headdresses or headgear, whether in descriptions of ominous phenomena (§ III) or as items requested by deities (§ IV). In light of this testimony, it would seem as if the question posed above would be answerable in the affirmative: the symbolic meaning of Inanna's archaic emblem was indeed preserved in later tradition. But in fact things prove more complicated, since the question concerned the *consciousness* of the tradition, in other words the awareness or knowledge of its theological roots, not (merely) of yet another clever instance of Mesopotamian semiotics, an accidental recovery of a lost tradition. A reflection on these matters (§ V) refers us back to the issue raised at the outset of this essay, concerning the nature of the omen collections and the challenge they pose at the crossroads of theology, philosophy, and science.

III

The initial datum appears in a textual passage that connects the sign for Inanna, MÙŠ, and the *parsikkum* that it signified, at least in its early conceptualization. This comes from an OB collection of extispicy omens published almost fifty years ago by J. Nougayrol,[15] the initial lines of which are reedited below.

MAH 15994:1–8

1. [DIŠ] *i-na* ⸢egir⸣ *na-aṣ-ra-<ap>-tim ša i-mi-tim* gì[r *la*]-⸢*ri*⸣-*am ir*-⸢*ši*⸣
2. gìr ^{d}GÌR.i[ri$_{11}$$^{?}$].gal
3. DIŠ *i-na* egir *na-aṣ-ra-<ap>-tim ša i-mi-tim* 2 gìr *iš-ka-ra-at* ⸢é⸣.[gal]
4. ^{d}iškur *i*-⸢*ra-ḫi*⸣-*iṣ*
5. DIŠ ⸢*i*⸣-*na* egir *na-aṣ-ra-ap-tim ša i-mi-tim* ⸢gìr⸣ *ki*-⸢*ma*⸣ BU UR bar.⸢si *ma*⸣-*za-az* ^{d}inanna(MÙŠ)
6. DIŠ *i-na* egir *na-aṣ-ra-ap-tim ša i*-⸢*mi-tim ši-rum*⸣ *ki*-⸢*ma ḫa*⸣-*sí*-⸢*is*$^{!}$ *sà*⸣ *-am-mi-im*

   <<*be*$^{?}$-⸢*el*$^{?}$⸣[(...)]>>
7. *be-el du-um-qí-ka ù be-el lu*-⸢*um-ni*⸣-*ka ni-ta*$^{!}$-*am i*-⸢*la*⸣-*wu-ka*
8. *ša-nu šum-šu še-um* ⸢*ù*⸣ in *i-ša*-⸢*qì*⸣-*lu*

15 Nougayrol 1969.

1–2. [(If)] in the back of the Crucible of the right side a Foot has a fork ⇒ (it is) a foot-mark representing Nergal.

3–4. (If) in the back of the Crucible of the right side (there are) two Feet ⇒ Adad will devastate the *iškaru*-fields of the palace.

5. (If) in the back of the Crucible of the right side a Foot (looks) like a **pu/ūru-headdress* with a *parsikku*-sash ⇒ (it represents) the presence of Ištar.

6–8. (If) in the back of the Crucible of the right side a piece of flesh (looks) like a lyre's ear ⇒ your friend and your foe will surround you; its second interpretation: grain and straw will be weighed.

Commentary

1–2. The reading of the DN follows collation by Cavigneaux;[16] also Glassner (2009: 21 n. 8): gìr ^{*d*}gìr-⸢*ra⸣-*gal*.[17] Whatever the identity of the partly broken sign, the writing surely represents one of the variant spellings for Nergal attested from OB on,[18] probably here ^{d}GÌR.iri$_{11}$.gal.[19] The sequence of Nergal → Adad in ll. 1–4, suggesting (or elsewhere describing) an oncoming plague followed by a natural disaster, finds parallels in extispicy apodoses (see Koch-Westenholz 2000: 112 [6:66–67; *Manzāzu*; and note the full sequence, which runs: Nergal → Adad → Ištar], 194 [27:12; *Padānu*], 305 [57:59; *Pān tākalti*]). The association between a *larûm* on a foot and Nergal may be unique, though is obviously predicated on the physical resemblance of the image of a fork to that of Nergal's weapon, i.e., his iconic three-headed scimitar (on which see Wiggermann 1998–2001: 223–236 and figs. 1–3; cf. already Nougayrol 1969: 156). What appears to be an illustration of three fork-marks next to three omens describing *larûs* in a commentary on *Padānu* (CT 20 28, 10–11; Koch-Westenholz 2000: 236 [42:50–51]), underscores this resemblance; one segment of this illustration appears in Figure 3.1:

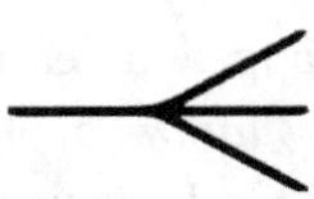

FIGURE 3.1
Illustration of a fork-mark in a SB omen commentary text

16 Generously shared with us via letter, dated 5 July, 2005.

17 *Pace* Nougayrol 1969: 156, who posited that ⸢lugal.ir$_{9}$.ra.gal⸣ must reflect a contamination of Lugal(g)irra and Irragal, representing Nergal, and rendered the DN as "Lugal(g)irragal"; similarly Richter 1994: 241 n. 87.

18 See, e.g., those mentioned in Wiggermann 1998–2001: 216.

19 In this period, perhaps to be read: ^{d}nergal$_{x}^{iri_{11}\text{-}gal}$. On the writing of Nergal in general, see Wiggerman 1998–2001: 215–218.

This image surely relates to the later tradition of illustrations appearing in SB extispicy commentaries, among which *kakku*s are especially prominent and are qualified on at least one occasion as a "weapon of Nergal" *kakki Nergal* (see Frahm 2011: 182). Still it seems likely that in the present instance it is the more specific referent that triggered the association with the deity.

3–4. The apodosis *iškarāt ekallim Adad iraḫḫiṣ* is attested elsewhere in OB extispicy, for example, YOS 10 43:6; 46 iv 34, the protases of which liken a mark to a "male gamepiece" with "knots" (*kīma passim zikarim ṭurrī īšu*). Landsberger (1960: 118, 126) suggested a basic phallic shape for this image. If correct, could the present omen be understood similarly, with the protasis's two "feet" figuring as another euphemism for "testicles" (cf. *rēšān*, lit., "heads"; *kalīt birkī*, lit., "kidney(s) of the 'lap'") and explained via paronomasia, by analogy to *iškum*, "testicle" (i.e., *passum zikarum* ↔ [*iškum*] ↔ *iškarum*)?

5. Nougayrol's (1969: 153; cf. 1976: 347) original reading of this difficult line as *b/pu-ur* BAR.S[I (?)], which was not accompanied by a translation, was followed by various attempts to make sense of it, including the uncertain suggestions by Jeyes (1989: 166: *pu-ur-pár-ḫ*[*i*! *m*]*a-za-az* ᵈinanna; the *nomen regens* a variant of *papparḫûm*, "purslane")[20] and CAD Š/2, 306 (*bu-ur-x-x*; no translation given), as well as the more recent efforts by Cavigneaux and Glassner. Cavigneaux, following his collation (GIR₃ KI? ZU? BU UR MAŠ X [(X)] X *-za-az* ᵈINANNA), posited the following reading for the line's crucial part: ...*ki-<ma> ṣú-pu-ur* MAŠ.K[AK? X? *m*]*a-za-az* ᵈINANA. But the MA sign of *kīma*, even if partly effaced, is nonetheless clear, appearing directly above the same sign in the same word in the following omen (ll. 6–7; also ll. 9, 11, 14); that it should be taken as the much-less expected ZU thus seems highly unlikely. Glassner (2009: 21 n. 48) read the difficult portion as: G[I]R₃ **ki*-[*m*]*a* **ṣú*-**pu*-**ur* *MAŠ.⸢*X⸣ [*m*]*a-za-az* ᵈINANNA, translating "un pied est comme l'ongle de... —position d'Inanna." Glassner's suggestion, which admits a partially broken MA *and* an undamaged *ṣú* before the following BU/PU, is supported neither by Nougayrol's copy nor by Cavigneaux's collation. Both of these readings appear at least partly motivated by the assumption that the simile's reference is *ṣuprum*, "nail, claw, hoof" (seemingly of a gazelle, Sum. mašda [MAŠ.KAK], in Cavigneaux's suggestion). But in fact this putative "silhouette" in an omen context is otherwise unattested.[21]

20 References to a *papparḫītum* decoration occur in similes in omen protases, and in one instance, OBE 14 ob. 19, it predicts the presence of a god, though this is Allat.

21 One does come across a *ṣupur rīmtim*, "cow hoof," in l. B 4′ of H 83, an OB text from Meturan (Cavingeaux and Al-Rawi 1993: 97–100) described by its editors as a hemerology (ibid., 95); but cf. CAD R, 359a, which takes the text as a royal ritual. A likening of part of a lung to a *ṣupur šilāḫi*, or "claw of a *šilāḫu*-bird," appears in one of the Tigunānum

Thus my reading favors Nougayrol's original proposal, with the understanding of BAR.S[I] as Sum. bar.si, Akk. *parsikkum*, as in the ED literary text cited by Steinkeller. Unlike nails (or claws or hooves), head coverings are well attested in OB omens. Various types of headdress are invoked in efforts to describe ominous shapes, and the interpretations, following a line of reasoning similar to that found in our text, usually predict royal power or fortune:

> '(If) (concerning) the gallbladder—its head (is) like a *kubšu*-turban (*kīma kubšim*) ⇒ the prince/king will find luck' (YOS 10 28:2 // Scheil B:38).[22]

> 'If the heart is surrounded by a fleshy membrane (which looks) like a *kubšu*-turban (*kīma kubšim*) ⇒ the king will defeat his enemies' (YOS 10 42 i 31–32).

Another instance appearing to liken a configuration in the exta to an element in a crown recalls a historical omen of Sargon:

> 'If the gallbladder is full of *titurru*-ornaments, like the *tigitallu* of a *crown* (*kīma tigitalli agêm*) ⇒ (it is) an omen of Sargon, who ruled totality' (YOS 10 59:9).[23]

Also noteworthy is the following, which metaphorically relates the configuration of smoke observed from incense to a tiara:

> 'If the incense-smoke has (the shape of) a crown (*agâm*) ⇒ (it forecasts) pride' (HMA 9-2433:32[24]).

---

omen collections (CUSAS 18, Appendix, No. XII iii 21) and another cited in George 2013: 120 (Tsukimoto No. 2: 24–25), though whether data from these texts can be introduced into a discussion of mainstream Mesopotamian divination is unclear.

22 Note, too, a SB version of this tradition in BRM 4 13:47, cited by our honoree in Rochberg 2010: 21.

23 Cf., however, CAD T, 398, 438, which offers *a-pí-im* as an alternative reading to the more common *a-ga-im*, possibly following a suggestion to relate the preceding *tigitalli* with *tigidlû*, a noted musical instrument, on the basis of the attestation in a Mari letter of another musical instrument, spelled *ti-in-gi-ta-li-im* (Reiner 2007: 55; cf. Shehata 2007: 524 n. 16; Mayer 2009: 434). On the latter, see Ziegler 2007: 50 and n. 199; for an explanation of its spelling, as well as another variant form, *ti-mi-it-*[*t*]*a-lu-ú* (Diri Nippur [MSL 15, 20]), see Shehata 2007: 524 n. 16.

24 Previously UCLM 9-2433. See Lutz 1929, partially re-edited with duplicates and with considerable improvements, in Finkel 1983–84.

Interest in headdresses in early Babylonian divination literature is manifest in omen apodoses as well, reflecting further a *Sitz im Leben* in which concern for the welfare of divine statues was a concern. One omen forecasts that a headdress will be requested (< *erēšum*) of the omen's client, presumably by a deity:[25]

> '(If concerning) the sheep—its nape is dark ⇒ the sheep's owner will be asked (for) a *pursāsu*-headdress' (YOS 10 47:19 [OB *Šumma immerum*]).

For additional examples of such requests (*erištum*) for headdresses in this literature, see below, § IV. For now it may be said that the appearance of head coverings in omens is sufficiently common and varied to warrant the reading of bar.si as *parsikkum* in this context.

The identity of the lexeme represented by the signs BU UR is more difficult to determine. None of the homonyms of *burum* or *būrum*, nor of *pūrum*, makes much sense in the present context.[26] That the signs should be read as a spelling for Sum. b/muru (i.e., bu-ur), equated in lexical lists with *manzāzum*,[27] seems far-fetched. My translation takes the form as a construct of **pu/ūrum* and tentatively renders it as a headdress of some sort. A precedent for this opinion is supplied by a passage from YOS 11 87, an OB love incantation describing various head and bodily coverings, that contains the the terms **pu/ūrum* (l. 22) and *parsikkum* (l. 23) in two similes in contiguous lines:

> 'Behold me (rather), as *šummannu*-rope!
> May she know me, as a ***pu/ūru***-headdress!
> Why, my love, as (with) a ***paršīgu***-sash
> did you bind your head?
> (Why), as (with) a *nēbaḫu*-sash,
> did you gird your hips' (YOS 11 87:21–26)

An analysis of ll. 22–23 in the context of the text as a whole, which includes yet additional references to bodily coverings and ornaments, appears in the Appendix below. Here it is sufficient to state that this proximity lends support

---

25 So CAD E, 285, s.v. *erēšu* mng. 3.

26 For *burum/būrum*, following the order of CAD B, 339–343: *buru*, the common surface measure term (bùr); *būru* A, "young calf"; *būru* B, "pit, hole"; *būru* C, "starvation" (NA and rare); *būru* D, "a type of song" (SB; see Groneberg 2003: 65); *būru* E (*pūru*), "sky" (Sum. loan, in lexical lists). The writing of *burû* A, "reed mat," which appears as *būru*, occurs only in NB and thus must also be excluded as a possibility here. For *pūrum*, following the order of CAD P, 526–529: *pūru* A, "bowl" (bur); *pūru* B, "lot."

27 See Antagal G 40 (MSL 17, 222); Diri IV 315 (MSL 15, 164).

to the reading proposed above for the omen's protasis because it establishes a close connection between terms represented by BU UR and bar.si.

The context of the OB love incantation suggests that **pu/ūrum* represents a type of headdress. Although this form is otherwise unattested, it is similar to *uprum*, a well-known term for a headdress, and its grammatically feminine counterpart *upurtum*, a larger and more ornate head covering.[28] From textual evidence it is clear that *upru-* and *upurtu-*headdresses adorned divine statues and human heads.[29] This item, it is submitted, stands behind the simile in our text as well as that in the passage from the incantation cited above.

If *uprum* relates to the writing of BU UR in this line, it is still necessary to explain the anomalous form. If the object of the simile in YOS 11 87:22 is accepted as identical to the bound form in l. 5 in our text, then the latter cannot be taken as a mistake, to be corrected via emendation (i.e., as *<ú>-pu-ur*).[30] Rather, one must entertain the possibility that the lexeme, which is already attested in the alternates *uprum/ḫuprum/uprûm/ḫuprûm*, gave rise to yet another, **pu/ūrum*, via apheresis.[31] While the evidence presented here is not conclusive, the mention of a *ḫuprû*-headdress—as a divine request of Ištar—in another OB omen discussed below means that this possibility cannot be dismissed lightly.

If **pu/ūrum* is indeed an alternate form of *uprum*, the construct *pu/ūr parsikkim* occurring in l. 5 may seem redundant, since most of the translations for bar.si/*parsikkum* noted earlier indicate that it is a close synonym of *uprum*. Yet not infrequently headdresses were comprised of more than one component.[32] It is thus plausible that the construct describes here a more generic headdress

28 For *uprum*, see CAD U/W, 189; Waetzoldt 1980–83: 200 ("Kopfband" or "Kopftuch"); *pace* AHw, 1425, "eine Haartracht mit Tuch," which does not rightly encompass the semantic range of the term. For *upurtum*, see already AHw, 1426: "eine Perücke?"; so Durand 1991, 2009: 45–46. And as CAD U/W, 193 makes clear, the term must also be understood as a headdress.

29 For depictions of supernatural beings wearing *uprus*, see, e.g., Lambert 1975: 116, A l. 15 (Nabû); Farber 2014: 82 (SB Lamaštu I 103), 145 (STT 145: 14'–15'), 271 (SKS, ll. 1–2).

30 *Pace* Winitzer 2010: 189.

31 For a general word on apheresis in Akkadian, see GAG §14a–b; but cf. Parpola 1984: 203–204 n. 15, which provides additional examples from different periods and contends that the phenomenon in Akkadian is more common than otherwise believed.

32 See esp. Waetzoldt 1980–83: 203; also Asher-Greve 1995–96: 184 and n. 23. The same of course holds for headdresses in the Arab world, with the modern *kāfiyya* (*kūfiyya*) along with the *ʿaqāl*, or the ringed cord or rope atop the *kāfiyya*, being a prime example; see further n. 35 below. In early Islam a common head-cloth known as a *miǵfar* or *ǵifāra* (< Sem. *√*ǵfr*, the basis of Akk. *apārum* and *uprum*) was worn under a helmet as part of one's headgear and was allegedly donned by the Prophet himself. For a similar understanding of the famous depiction of Inanna's head coverings in Innana's Descent, see already Jacobsen 1987: 207, 211, 213.

(**pu*/*ūrum* [*uprum*]) decorated or crowned with a *parsikkum* sash. I thus tentatively suggest a translation of *pu*/*ūr parsikkim* as "ornamented headdress," that is, a headdress (**pu*/*ūrum*) ornamented with a sash (*parsikkum*).

It is likely that, in the context of the omen, *pu*/*ūr parsikkim* signifies not an actual headdress but rather a stylized rendering of such apparel, in keeping with the common use of real-life imagery to describe abstract shapes in scholarly literature.[33] Given that *pu*/*ūr parsikkim* appears amidst omens whose protases or apodoses describe various doubled features—including a bifurcated Foot (ll. 1–2); two Feet (ll. 3–4); friend and foe (ll. 6–7); and grain and straw (below, l. 8)—it is tempting to speculate that the stylized rendering of the headdress stressed its two separate components, or perhaps the loose ends of the **pu*/*ūrum* descending from the upper loop or band (the *parsikkum*).[34] If so, the image may very well have approximated the shape of the MÙŠ sign in its earliest periods (Figure 3.2A):

6–8. Cf. the collations of Cavigneaux (*ši-rum ki-ma ša-zi*-S[A$^{?}$ S]A$^{?}$-*am-mi-im*) and also Glassner (2009: 21 n. 48): DIŠ *i-na* EGIR *na-aṣ-ra-ap-tim ša i*-[*m*]*i-t*[*im*]

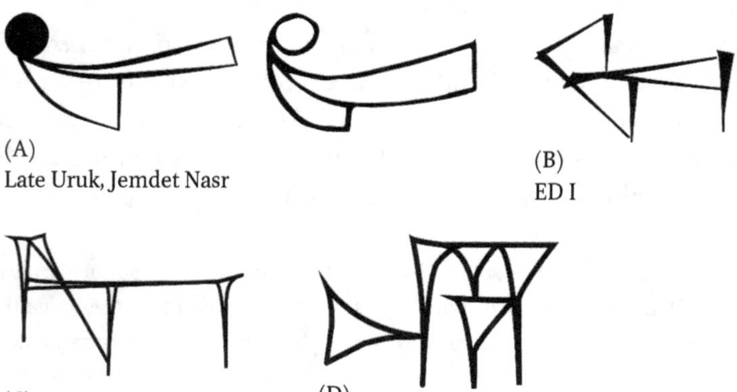

FIGURE 3.2 *Representations of the Sign* MÙŠ

33 This is almost certainly the case for the mention of the lyre's ear in the following entry (ll. 6–7), which relates to the better-known *apsamikkum*, or concave square, a geometrical figure employed in Mesopotamian mathematics from the OB period onward, including first-millennium extispicy, with little regard for its folk etymology ("window of a lyre"); see Robson 1999: 50–54; 2008.

34 See the description in Steinkeller 1998: 96, and also figs. 1–7. It is not impossible that the *parsikkum*, being the more impressive of the two items, became the general term for the entire group via metonymy, thus its mention without futher qualification in the ED hymn above. Instructive in this regard is the common usage of *kāfiyya* in the Arab world, which typically implies both the *kāfiyya* itself and the *ʿaqāl*.

š*[*i*]-rum* **ki*-**ma* ⌜**ḫa*⌝-*sí*-**ís* [**s*]*a-am-mi-im* / <<*aš-r*[*i*]>> / *be-el du-um-qí-ka ù be-el lu-um-n*[*i*]*-ka* **ni*-⌜**ta*!⌝-[**a*]*m i*-**la*-**wu-ka* / *ša-nu šum-šu še-um ù* *IN *i-šà*(sic)*-qì-lu.*

Ḫasīs sammîm. Following Glassner's collation, though what is visible of the last sign of *ḫasīs* may well be the left side of SAḪAR or GIŠ (for *ís* or *is*, respectively). For this simile in OB extispicy, see YOS 10 24 r. 31; 61 ob. 5; OBE 2:4; BM 130838 r. 17–18 (OB report);[35] but not, *pace* OBE 113, YOS 10 42 iii 54 or BM 22694 r. 6 (Aro and Nougayrol 1973: 43).

Nītam lawûm. The reading follows the collations of both Cavigneaux and Glassner.

Iššaqqilū. On the form (expected: *iššaqqal*), see CAD Š/2, 12; and cf. GAG § 90c for the possibility of an unexpected *i*-vowel in N-durative of *a-u* vowel-class verbs from later periods, perhaps indicative of a "gesprochenen Murmelvokal." Note, however, that the other examples of this variant for *šaqālum* cited in CAD Š/2, 12–13 all stem from other OB omen apodoses concerning the economic collapse theme. It is possible then that all these attestations represent reflexes of a stock expression that had become frozen and are not necessarily to be connected with similar-looking variants occurring in post-OB Akkadian.

<<***be*-⌜*el*⌝[(...)]**>>. Against the readings proposed by Cavigneaux and Glassner, I assume that this text's scribe initially attempted to render the beginning of the following apodosis indented under l. 6, as is the case in l. 2 for the first omen (ll. 1–2). It is unclear from the copy whether the scribe sought to erase this once the apodosis was reconfigured to begin in l. 7.

IV

The reading proposed above for l. 5, which posits Ištar's headdress as signifier of that deity's presence, finds support when considered alongside a small group of omens that seem to echo this relation. These omens derive from a less conventional technique of OB divination: interpreting the physiognomy of sacrificed birds.[36] They were the subject of a study by Durand, who called attention to several omens describing the requests of various deities for particular headdresses or headgear.[37] In the larger of the two extant collections

---

35 Edited as text no. 4 in Richardson 2002: 241–242.

36 The understanding of this technique has recently advanced, thanks especially to George 2013: 112–113, 114–116, 314–316 (editions of new omen collections concerning birds and references to others to be published); Maul 2013: 131–153, 348–352 (but cf. Winitzer, forthcoming).

37 Durand 1997: 273–282. Two collections of these omens are known, one in duplicate: YOS 10 51//52 and MAH 15897. For an overview of the former, see Edzard 2004: 545–548;

(YOS 10 51 // 52) occurs a series of four such omens, whose apodoses are quoted below:

(1) Protasis ⇒ *errišti Ištar ana ḫuprê* (YOS 10 51 i 10–13 // 52 i 10–13)
(2) Protasis ⇒ *erišti qarnīn ša*[38] *Bēlet ekallim* (YOS 10 51 i 14–18 // 52 i 14–17)
(3) Protasis ⇒ *erišti pursāsim* (YOS 10 51 i 19–22 // 52 i 18–21; MAH 15897:3–4)
(4) Protasis ⇒ *erišti taddî*[39] *Išḫara lubuštam irriš*[40] (YOS 10 51 i 23–27 // 52 i 22–26; MAH 15897:5–6)

Though the apodoses vary in their phraseology, a basic syntagm may nonetheless be posited for all four, which may be rendered as:

erišti X *ana* Y // *erišti* Y (*ša*/*ana* X)
(It is) a request of deity X for Y // (it is) a request of Y (of/for deity X)

where Y {*ḫuprûm*, *qarnū*, *pursāsum*, *taddûm*} represents the item requested by deity X.

The four items comprising the set Y, especially the terms *ḫuprûm* and *pursāsum*, should recall the headdresses and shawls noted above. The first, *ḫuprûm*, "headgear,"[41] here representing the request of none other than Ištar, is either related to *uprum* or, as Durand seemed to suggest,[42] is simply its biform; either way, both forms are surely derivative of Sem. *√*ǵpr*.[43] The third,

for an edition of the latter, see Nougayrol 1967: 23–25. Durand 1997: 280–282 set up two categories, differentiating between headdresses ("parures de tête") and garments ("vêtements somptueux"); these encompass three of the four items discussed below. The fourth, *qarnū*, though seen by Durand as belonging to a series of terms describing the ornamentation of the heads of cult statues, was treated earlier in a different context.

38 So YOS 10 51 i 17–18; the writing in YOS 10 52 i 17 has *e-ri-iš-ti qá-ar-*ni_6 ^{d}⌜nin⌝.é.gal and can be read as a construct without emendation (cf. Durand 1997: 276: *qá-ar-*nin_x(NIM) <*ša*> ^{d}nin-é-gal; Edzard 2004: 548: *qarnim ša* [sic] *Bēlet ekallim*).

39 On the basis of the explanation below, I follow Durand 1997: 281 and n. 33, against the reading *ša-di-im* "montagne," proposed in Nougayrol 1967: 23; also CAD E, 299; Prechel 1996: 62 and n. 199; Edzard 2004: 546; and cf. CAD Š/1, 59, which explains the term as a reference to the netherworld.

40 I follow the version in MAH 15897 l. 6 (*Išḫara lubuštam* <*i*>*-ri-iš*), and emend YOS 10 51 i 27, which reads *lu-bu-uš-tam i-*⌜*ri*⌝*-iš-ši*!(TI) for *irrissi* (cf. *ikrib ikribīšu ilšu i-ri-iš-su* [YOS 10 51 i 8–9 // 52 i 8–9]); *pace* CAD I, 284: *i-ri-iš-*<<*ti*>>. The parallel in YOS 10 52 i 26 is broken.

41 AHw, 1425: "eine Kopfbedeckung."

42 Durand 1997: 281.

43 On reflexes of the Sem. *√*ǵpr* in Akk., see Kogan 2001: 279–281 (nos. 23 and 25); 2002: 316 (no. 25), and thus add *ḫuprûm* to the examples listed there.

pursāsum, equated to *upurtum* in lexical lists[44] and here possibly referring to the nape or rear of a statue's headdress,[45] is also described as a request, presumably of a deity. Likewise, the terms *qarnū* and *taddûm* represent articles associated with headdress: the two horns (*qarnū*) in the request of Bēlet-ekallim are to be understood as those that typically mark the headgear of deities in the visual record;[46] the request of a *taddûm* is glossed by the phrase *Išḫara lubuštam irriš*,[47] "Išḫara requests (an article of) clothing," and has been identified as the *taddû*-habit known from texts dealing with the textile industry at Mari.[48]

The set Y thus represents a paradigmatic elaboration of a sort common in divination literature,[49] in this case on the idea of particular headdresses as objects of divine requests. The association of divinities with headdresses surely reflects the practice of clothing and ornamenting divine statues with garments, head coverings, and jewelry, either as part of the cult image's induction or later preservation, or its veneration in daily rituals and ceremonies.[50] Numerous terms for headgear, headdresses, and other decorative bands occur in the descriptions of divine statues in the so-called *Göttertypentext*;[51] among these are *qarnū*, *pursāsum*, and even a certain *apparrītum*, a hairstyle of some type related to *apārum* ("to provide with a headdress") and, indeed, *uprum*/*uprûm*.[52]

Also noteworthy are the deities mentioned. Aside from the actual mention of Ištar in the first entry, one finds in the second a reference to *Bēlet ekallim*, the

44 See MSL 10, 36; AOAT 50, 140–141 (Malku VIII 59).

45 AHw, 881: "etwa 'Nackenmähne' (an einer Statue)."

46 So Durand 1997: 276 n. 16.

47 Reading the DN as part of the gloss on account of the mimation on *taddûm* in YOS 10 51 i 26; alternatively: "(it is) a request of a *taddûm* of Išhara; she requests (an article of) clothing."

48 Durand 1997: 282. For attestations of the term, see Durand 1997–2000, 1: 271–278 (nos. 133–39); Durand 2009: 122–123.

49 For a general word on paradigmatic elaboration in the early Babylonian divination literature, see Winitzer 2017: 262–280. A considerable portion of what follows the quoted portion in YOS 10 51 // 52 represents a paradigmatic elaboration of requests and statements of divine presence; see further Maul 2013: 140–141.

50 For an overview of the ornamentation of divine statues during the OB period, see Maggio 2012. For one example providing a detailed accounting of the ornaments and clothing associated with the cult of Ištar, in this case from Ištar-of-Uruk in the Neo-Babylonian period, see Beaulieu 2003: 140–159.

51 Köcher 1953: 57–107.

52 See the derivation proposed in CAD A/2, 179, s.v. *apparrû*; more recently, Kogan 2001: 278–281; 2002: 316. On *apparītum*, see AHw, 59; CAD A/2, 179; and add KTT 373:5 (writ. *ḫa-ap-pa-ar-ri-tum*).

"Lady of the Palace," presumably a reflex of the deity from Mari known by this title, although the likelihood that this title represents yet another form of Ištar should not be ruled out.[53] The same may hold for Išḫara, mentioned in the final entry: in the second millennium she is assimilated to Ištar in a number of capacities, perhaps none more pervasive than as patroness of nuptials.[54]

Thus it may well be the case that the divine names enumerated in these apodoses are to be taken not as distinct deities but rather deities that had either assimilated to the same goddess, Ištar, or were understood in this way.[55] Whether this grouping was motivated by the deities mentioned (set X) or their requests (set Y) is difficult to determine. Either way, it seems that the references to headdress or headgear requests in these omens reflects an aspect of the divine image's cult, perhaps Ištar's.

Further, it would appear that these requests in the bird omens reflect a concern for the deity's presence, which is the explicit topic of certain omens associated with other divinatory techniques.[56] Indeed, *erištu*-statements elsewhere in OB divination literature sometimes follow proclamations of divine presence. A passage of OB lecanomancy omens proves instructive in this regard:

1. *šumma ištu qabliat ummatim šulmum išḫiṭ-ma*[57] *ibteli* ⇒ *manzāz Sîn erišti kaspim*
2. *šumma šamnun mê ina nadêka* ***qarnīn***[58] *irtaši* ⇒ *manzāz Sîn*
3. *šumma ana qabliat šamnim mê ina nadêka rebiat šamnim iprus* ⇒ *manzāz Šamaš* (ᵈutu) *erišti šamšim* (*ša-am-ši-im*)
4. *šumma* (*šamnum*) *ana qabliat šamnim mê ina nadêka* ***kīma kakkabim*** *išḫiṭ* ⇒ *manzāz Šamaš*

53 Note in this regard the reference to a visit by Dumuzi to Bēlet-ekallim's temple in Mari (see Durand 1987: 93). At Mari the goddess behind this appellation cannot be explained merely as the local form of Ištar, because the latter is widely attested in various guises, in one of which she herself is a frequent visitor to a temple of *Bēlet ekallim* (Durand 1987: 91–93). Nor is the appellation tantamount to the "Ištar of the Palace" (*Ištar ša ekallim*), since the appearance of both independently in the "Pantheon tablet" (Lambert 1985: 525–539; Durand 2008: 255–264) suggests that these must be construed as mutually exclusive.

54 See Lambert 1976–80: 176–177; Prechel 1996: 58–60; George 2003: 190; Durand 2008: 262–263.

55 On the oddity of these "northern" deities recalled in this Babylonian setting, see Durand 1997: 276–278; Maul 2013: 133.

56 Discussed in Winitzer 2010: 186–190.

57 One MS has: *ipṭuram-ma*, "detached."

58 One MS has: *qannīn*.

(1) "If from the middle of the mass a(n oil-)bubble rose up and has burst ⇒ (it represents) the presence of Sîn: a request of silver.
(2) If the oil, in your pouring water (upon it), has taken (the shape of) **two horns** ⇒ (it represents) the presence of Sîn.
(3) If in your pouring water into the middle of the oil one fourth of the oil separated ⇒ (it represents) the presence of Šamaš: a request of the sun disk.
(4) If in your pouring water into the middle of the oil (the oil) rose **like a star** ⇒ (it represents) the presence of Šamaš" (Pettinato, Ölwahrsagung I 57–60).

As I have contended elsewhere,[59] the request statements in omens 1 and 3 appear to provide commentary on statements of divine presence whose protases are formulated in less explicit terms than is the case in omens 2 and 4, which describe their findings according to established shapes (i.e., "horns" and "star").

The Ištar omen discussed in § III above shares features with both the oil and bird omens discussed in this section. In terms of its formulation, it displays the pattern observed in lecanomancy omens 2 and 4, whose protases appeal to metaphorical formulations similar to the *parsikkum* simile and whose apodoses do not include request statements. In terms of content, it recalls the bird omens mentioning headdresses and headgear, which figured as the subjects of requests connected either implicitly or explicitly to divinities. The common denominator uniting all of these omens is their concern with divinities: they include statements that can be characterized as unusual in the broader context of the omen collections[60] and that appear to be triggered or qualified by references to items related to the divinities' cultic representation. My reading of the Ištar omen above thus finds corroboration elsewhere in OB divination literature in terms of form, its content, and even its concern with the goddess herself.

V

The question remains whether the association of the *parsikkum* with Inanna/Ištar in the context of OB divination literature actually recalled the original theological and graphic relations between sign and signified, well over 1,000 years after this equation was forged. That no evidence can be mustered to

59 Winitzer 2010: 187–189.

60 As discussed ibid.

suggest any such memory for the interim period militates against this possibility. A more likely explanation is that the diviner made a rediscovery of sorts: a real or imagined visual sign interpreted as a headdress prompted him to fashion an interpretation recalling Ištar, the celebrated goddess famed for such garbs, in a manner that actually echoed the *parsikku*-headdress' original signification and graphic representation—though the these denotations were unknown to him.

Still, in this instance the diviner buttresses his semiotics on a *verbal* representational level. The Ištar omen thus differs from the omen that equates a *larûm* with Nergal (ll. 1–2), in which the signification operates at the *visual* level alone: the diviner associates a fork-like feature with Nergal's fork-like weapon solely on the basis of its shape; indeed, Nergal's weapon is otherwise never described as a *larûm*. The medium of divination thus provided the grounds for what might be described as an interpretive re-creation, with a new verbal semiotic system in Mesopotamia reinventing what had been established initially with the birth of writing. Small wonder, then, that the development of this semiotic system followed a course impressively similar to that witnessed in the case of Mesopotamian writing and abandoned the visual sign in favor of the verbal, which offered far greater possibilities for the divination literature's interpretive framework.[61] The association between a *parsikkum* and Ištar's presence reflects one example of this point, as do the additional omens quoted above relating specific headgear to particular deities.

Indeed, here it is appropriate to step back and consider the larger body of evidence examined, which includes all the omens concerning divine headdress from several branches of divination. What was the origin of these omens? Do they hint at a real interest on the part of divination literature with divine statues and their welfare? This possibility may seem dubious, at least initially, owing to the overall rarity of concern with such matters in divination literature.[62] Then again rarity does not mean absence, and in any case there is no denying

61 On this point, and on the place of visual images in the early divination literature more generally, see Winitzer 2017: 370–393, esp. 392–393.

62 An especially telling indication of this point comes from the *Šumma ālu* omens concerning Marduk's statue during its procession in the course of the Akītu festivities (studied in Sallaberger 2000). These omens, whose protases ostensibly consider the divine statue in various guises, in fact prove wholly detached from any genuine theological expression for the cult statue in their interpretations. But cf. Maul (2013: *passim*), who not only insists on situating the divinatory act in ancient Mesopotamia squarely within a religious-cultic framework, but also seems to assign cultic and religious queries in divination a standing no less significant than those concerning military or political matters (e.g., Maul 2013: 25).

the fact that divinatory queries roughly contemporary with the OB omen collections did consider cultic matters on occasion, including the welfare of divine statues. References in various year names to the determination of cultic details—often the installation of priests or priestesses—via extispicy furnish a quick reminder of this overall point. To offer but one example:

> 'Year (when) Išbi-Erra selected the *engaba*-priest of Inanna by means of extispicy' (máš-e ì-pàd) (Išbi-Erra 13).

Of course interest in cultically oriented extispicy extends well beyond its attestation in year names. Literary sources describe the turn to this form of divination to ascertain the will of the gods at pivotal moments. Such occasions include Gudea's receipt of the plan for the building of Ningirsu's temple in Lagash, as well as the notorious depiction of Narām-Sîn's attempt to procure a favorable omen for temple restoration in the hope of staving off Agade's destruction:

> He performed extispicy in a kid and his omen was favorable (maš-a šu ì-gíd maš-a-ni ì-sa_6). He cast grain on to…, its appearance was right. Lying down Gudea rested, and a "word" came up to him: The building of his master's House, the separating of the Eninnu from heaven and earth—was it (not) before his eyes? It made him extremely happy.[63]
>
> GUDEA CYLINDER A XX 5–12

> Performing extispicy (máš-àm šu-gíd-dè) with regard to the temple, the (omen for) building of the temple was not present in the extispicy (máš-a nu-mu-un-dè-gál); performing extispicy (máš-àm šu-gíd-dè) for a second time with regard to the temple, the (omen for) building of the temple was not present in the extispicy (máš-a nu-mu-un-dè-gál).[64]
>
> CURSE OF AGADE, ll. 94–97

The care of divine statues, though perhaps not on a par with such historic matters, was still the subject of divinatory investigation. Several Mari letters that report on extispicies conducted concerning the (re-)plating of a divine statue record repeated examinations and opposing results and make it clear that this was a matter of obvious significance:

63 See Edzard 1997: 81.

64 Following Cooper 1983: 54–55.

'Zikri-Ḫanat [and] I made an extispicy (*têrtam nīpuš*) [concerning] the plating of the face of Bēlet-bīrī, and I sent a report of those extispicies (*ṭēm têrētim šināti*) to [my] lord. Now I returned to the results of my initial round (*ana qātīya atūr*)[65] and concerning the plating of her face with silver and gold made an(other) extispicy, but the(se) extispicies were not sound (either) (*têrtam ēpuš-ma* [*têr*]*ētum ul šalmā*)' (ARM 26/1 132:5–18).

'Further: the work [of the face/statue[66]] of Bēlet-bīrī concerning [its] manufacturing is finished; and concerning the plating [of her] face they made an extispicy and the extispicy was sound (*têrtam īp*[*uš*]*ū-ma têrtam šalmat*), and so Erēssum-mātum spoke to me [thus]: [...that] I will plate [...]...[...that with] silver I will plate [...]' (ARM 26/1 134:12′–22′).[67]

One is left to wonder about what specifically was deemed unsound and sound in these reports, although the logical assumption is that these judgments were based on the findings in the extispicy exams. Such seems to be the case for the oil omens cited above, which connect findings typical and atypical in lecanomancy to statements about the presence of deities and their requests for items undoubtedly associated with their cult.

In any case, against this background the possibility seems real that the references to headdresses or headgear in the omens discussed above reflect genuine concern for divine statues and their cults. Whether this means that these omens are grounded in actual examinations of the sort referred to in the Mari missives is, of course, a different issue altogether, and not one that can be settled easily. It seems more likely that, in the context of the paradigmatic elaborations in the bird omens discussed above, cultically focused omens represent the product of deductive reasoning and not empirical observation, or what our honoree describes as matters of conception, not perception.

But even if these omens are the product of reflection and not inspection, it is important to recognize that this explanation, if correct, would not invalidate these omens as witnesses to a certain religious praxis in Mesopotamia; it would simply shift the setting of this activity to the scholastic domain. In that

65 See Durand 2008: 513.

66 On the restoration possibilities here, see Heimpel 2003: 226 n. 154.

67 For this reading, see Durand, 1988: 299 n. d; *pace* Heimpel 2003: 226. In connection to these letters, see also ARM 26/1, 133 and ARM 26/2, 294, in which the statue's appearance, now reportedly "perfected" (*šuk*[*lulat*]), appears ready to undergo a mouth-opening ritual.

context the diviner, employing the language of divination and the format of the omen collections, strove to make sense of the world around him, including the realm of the gods and their cultic representation. Why this interest in theology among diviners did not develop into a program for using the hermeneutics of the divination literature to articulate more nuanced statements about the divine realm remains an open question.[68] Whether, especially in light of this modest output, such thinking about divinities in the context of divination should ultimately be attributed to semiotics and not theology is also debatable.[69] So, ultimately, is the matter of how all this figures into the question of the scientific nature of the chief output of Mesopotamian divination—the omen literature. But on this I defer to the insight and reflections of our honoree.

Appendix: YOS 11 87[70]

Text and Translation

1.	*erēmu erēmu*	Love-charm, Love-charm —
2.	*qarnāšu ḫurāṣum*	Its two horns are gold
3.	*zibassu*! *uqnum ellum*	Its tail is pure lapis lazuli —
4.	*šakin ina libbim ša Ištar*	It is placed on Ištar's bosom.
5.	*assīšim-ma ul itūram*	I called to her, but she did not turn back to me;
6.	*aḫzu šī-ma ul ippalsam*	I hissed, but she did not look at me.
7.	*š[um]ma qašdat limqu[t]*	If she is consecrated,
8.	*[d]ādūša*	May her darling fall;
9.	*[š]umma nadiat muppirša*	If she is set for the cloister,
10.	*limqut*	May her veiler fall.
11.	*batultum māratu awīlim*	O, nubile one, nobleman's daughter —
12.	*ana rigmīya*	At my clamor,
13.	*ana rig<mī> šagīmīya*	the clamors of my call —
14.	*<l>imqut līšum*	May the dough fall
15.	*[ina] qātīša*	[from] her hand.
16.	*limqut [ṣ]uḫārum*	May the youth
17.	*ša aḫīša*	at her side fall (as well).
18.	*ē [tar]qusīm bītki*	May you not tie shut your house from me!
19.	*ana [r]iksim [ša] qātīki*	May you not look

68 For one reflection on it, see Winitzer 2010: 192.

69 On the problems inherent in this type of dichotomy, see Winitzer 2017: 1–9.

70 An edition of this text appears now as no. 26 in Wasserman 2016, which contains many insights about the early love-literature corpus generally.

20.	*⸢ē⸣ t[ap]pa[l]sī*	at the sash [of] your hand!
21.	*kīma šu[mm]annim itaplasīnni*	Behold me (rather), as a *šummannu*-rope!
22.	*kīma pu/ūrim lū idînni*	May she know me, like a *pu/ūru-headdress*!
23.	*am-mīni râmī kīma par[šī]gim*	Why, my love, as (with) a *paršīgu*-sash
24.	*tarkusī rēški*	did you bind your head?
25.	*kīma [n]ēba[ḫ]im*	(Why), as (with) a *nēbaḫu*-sash,
26.	*tēz[iḫ]ī-ma qablīki*	did you gird your hips?
27.	*...[...]ya k[īm]a šamnim*	My ...[...], as oil,
28.	*[...]...šī [i]n[a x]⸢x⸣[x]īki*	...[...*with*] your [...]...?
29.	*[alk]ī-ma[an]a ṣērīya*	[*Come then*], toward me!
30.	*[ ]x-im-m[a š]umim*	[...] ... *name.*

Notes

l. 1. ***erēmu (irēmu/ir'imu).*** On this enigmatic item, concrete on some occasions and more abstract on others, see George 2009: 53, *ad* CUSAS 10 8 (= LAOS 4 2), l. 22. Here it seems to refer to a physical entity, favoring the view that it was conceived as a piece of jewelry (Groneberg 1997: 300; 2001: 106–111), at least on some occasions. Whether the material *irēmum* adorned the neck, head, or both remains unclear. Quite possibly this apparent imprecision owes to the fact that the *irēmum* was defined by its size and function, not by where it was placed on the body; see Beaulieu 2003: 10 for a similar view concerning first-millennium cult-statue ornaments.

l. 3. ***zibassu.*** The text has *zibassi,* though this is surely an error, probably in anticipation of ll. 5, 6 and the many other upcoming references to the female, and should be emended as in Wasserman 2016: 252. The suggestion in Cavigneaux 1996: 36 n. 10 that the form is terminated by the rare—*ši* 3fs suffix (in genitive forms; see GAG § 42g) is grammatically untenable owing to the logical masculine antecedent.

l. 4. Cf. Foster 2005: 199 ("placed in Ištar's heart"); Wasserman 2016: 253 ("placed in the heart of Ištar"). Groneberg (2001: 111) reads "in der Mitte des Körpers" and speculates that perhaps this even meant the lap or genitalia region, as is seemingly the case in the same image from MAD 5, 8 (= LAOS 4 22), ll. 3–4: *in sà-qì-[sa u-ša-a]b*, "it rests on [her] lap." The depiction here seems to echo the complex practices involving the adornment of cult statues with breast ornaments, which were frequently made of gold and further embellished with jewels, beads, stones, and engravings of real and mythological figures.

l. 6. ***aḫzu.*** The copy supports the reading of *a-aḫ-zu-ši* in Foster 2005: 200; *pace* Wasserman 2003: 163; 2016: 252 (*a-mur-*{MA?!}*-ši-i-ma*). Note especially the parallel between *ḫazûm* in this line and *šasûm* in the previous one (l. 5; cf. CT 38 26:25–28; 39:23 [*Šumma ālu*], where *ḫazûm // šasûm*). Still one expects the verb to take the dative suffix. In theory *aḫzūšim-ma* is possible, but then the choice of I over IM seems odd.

ll. 7–10. These lines find an important parallel in the recently published OB love incantation CUSAS 10 11 (= LAOS 4 20), 12–14:

12. *šumma qašdat imdāša l[i-x-x]*
13. *šumma nadiat bibilša l[i-x-x]*
14. *šumma kezret liqabbir aštammaša elīya limqut*

12. If she is consecrated as a holy woman may her *support* [...]!
13. If she is set for the cloister may her gift [...]!
14. If she is *set* for harlotry may she *close down* her tavern and fall upon me!

The tradition shared by the two passages reveals differences as well. If one follows George 2009: 68, ll. 11–12 of CUSAS 10 11 appear to describe the provisions given to females designated for some special status, with the respective women the subject of the broken precative forms (which are thus understandably—if unexpectedly—not restored as intransitive forms of *maqātum*); by contrast, in place of such allotments our text presents different males as the subject of *limqut* in ll. 7 and 10 (see further below). Yet notwithstanding these differences, a common theme runs through all these lines concerning the suitor's wish that the girl abdicate her commitment by letting the male "fall." From this angle it is significant to recall that her consecration into such roles is expressed via *našûm*/*nišītum*, lit., "to rise"/"rising" (Wilcke 1982: 440–450; Barberon 2012: 183–192; Stol 2016: 1054, 1087). Is it possible that the choice of *maqātum* in these lines was intended to resonate with this idea, albeit in the opposite direction? But cf. ll. 13–17.

ll. 7–8. *qašdat*. The rendering of the form here as "consecrated" follows convention. For recent discussions of the *qadištum* in the OB period and differing conceptions of the religious nature of this figure, see esp. Durand 2008: 406–410; Barberon 2012: 179–183, 207–208; Stol 2016: 1084–1090.

***limqut dādūša*.** Assuming its last sign is read correctly, the 3ms precative (// l. 10) calls for a singular subject and thus *dādum* must have the sense of "darling, favorite," even though it exhibits here the form known from the PN *Dādūša*. The identity of this figure in this context is unclear, especially given the uncertainty about the nature of *qadištu*s. Given the parallelism between ll. 7–8 and 9–10 in this text as well as the focus on the woman's provisions in the parallel tradition in CUSAS 10 11, however, it would seem that the term here has more to do with aspects of caretaking than sexuality.

ll. 9–10. Previous treatments (e.g., Wasserman 2003: 159) understood the penultimate word, spelled *mu*-BI-*ir*-*ša*, as *mubbirša*, "her accuser," a reference to one making charges concerning the girl's "fallen (state)," or virginity (see especially Lambert (?) *apud* Foster 2005: 199). With the benefit of the CUSAS 10 11 (LAOS 4 20) passage, however, it is certain that *nadiat* in this passage means "to become consecrated as a *nadītu*" (so already Cavigneaux 1996: 36: "une nonne"; now Wasserman 2016: 253). As such, George 2009: 68 n. 2 and Wasserman 2016: 254–255 opt to take the word as the D-stem participle of *epērum*, "to provision," suggesting that this refers to the one responsible for the *nadītu*'s needs.

As both note, however, the verb is otherwise unattested in the D stem. This last point, as well as the expectation that this image should relate to the one in the previous line (so George 2009: 68 n. 2), leads me to propose a third option: the participle *muppirša* refers to someone who places on (*uppurum*) the woman a headdress as the marker of her status in marriage or consecration.

The place of head-coverings as markers of the married or consecrated status of different classes of women in Mesopotamia is well established. This is true with respect to both veils (for which see especially MA laws §§ 40–41 and the dossier concerning the marriage of Šibtu and Zimri-Lim, discussed by Durand 1988: 99–112, esp. 103–104; Michel 1997; see also Fleming 1992: 188, *ad* Emar 369: 61) and headdresses. When the latter are mentioned, the accompanying verb is typically *apārum* in the G or D. Noteworthy examples include the mention of headdresses in descriptions of dowries:

> One servant girl, PN$_1$, [...] garments apart from the one she is wearing, [...] headdresses apart from the one with which she is covered ($^{\ulcorner\text{túg}\urcorner}$bar-si$^{\text{ḫi.a}}$ *ezub ša aprat*) ... all this is the dowry of PN$_2$, the *šugītum* ... which PN$_3$, her father, and PN$_4$, her mother, gave to her ... (BE 6/1 101, ll. 1–2, 13–17).[71]

> Two garments which she is wearing, two headdresses <with which> she is covered (2 bar-si$^{\text{ḫi.a}}$ <*ša*> *aprat*) ... All this is what her father, PN$_1$ ... gave to PN$_2$, the *šugītum*, his daughter ... (PBS 8/2 252, ll. 1–2, 7–10).[72]

Also instructive is the famous birth scene in ATRAḪASIS, describing the midwife's preparations of the woman in labor, who is described as a *qadištum*:

> (With) her face beaming, joyful
> She covered her head (*i'pur*[73] *qaqqassa*), performed the midwifery,
> She girded her loins (*qablīša īteziḫ*) (while) pronouncing the blessing,
> ...
> May the midwife rejoice in the house of the *sacred* (*qadišti*) woman (I 283–287).[74]

Against this background in mind, it seems preferable to read here a participial form of *uppurum* "to cover."

---

71 Edited in Dalley 1980: 62–63.

72 See Westbrook 1988: 129.

73 On this reading of the verb (writ. V'-*pu-ur*), see Shehata 2001: 78–79; also CAD Š/1, 16). The original reading of *uppur* (D), "and covered head," is possible (as seemingly reflected in Stol 2000: 115; Foster 2005: 237), but does not accord with the rest of the passage, which seems to describe the scene from the perspective of the midwife's actions.

74 On the meaning of *qadištum* in this instance, see Stol 2000: 116 n. 33; also Foster 2005: 238.

l. 11. *māratu*. On preserved nom. case vowels in construct forms, see GAG § 64a.
l. 13. *rig<mī> šagīmīya*. So Van Dijk 1985: 50; cf. *rigim šá-gi-meš* (BM 34982:3). Wasserman 2016: 252 emends: {RI-IG} *ša-gi-mi-ia*.
ll. 14–17. *limqut*. The emendation follows Foster 2005: 199; Wasserman 2016: 252–253; cf. Cavigneaux 1996, who reads a preterite; see below. The restoration *līšum* [*ša*] *qātīša* is also possible.
l. 16. *ṣuḫārum*. Following Cavigneaux (1996: 36): [*ṣ*]*ú*!-*ḫa-ru-um*; Wasserman 2016: 252.
ll. 16–17. Cf. the translation of Cavigneaux (1996: 36): "La jeune fille, la fille du bourgeois, à mon cri, à mon hurlement?, la pâte, lui est-elle tombée de la main? Que tombe ... qui est à son côté!"
ll. 18–20. *ē tarqusīm bītki*. Following Cavigneaux 1998, who compares YOS 11 29:20: *at-ti a-na ra-mi-ia e tar-ku-sí-im bi-it-*[*ki*]; Wasserman 2016: 252. For *rakāsum* as "to tie shut," see esp. *bāba arakkas*, "tie shut the door," Lambert Love Lyrics 122:11–12; so Foster 2005: 200: "Don't lock your house against me"; cf. Wasserman ibid.: 253, 255 ("do not *attach* your house to me"), who understands the verb in the extended sense of "to attach one to a place, duty, assignment"; see below.
***riksum*.** Wasserman 2016: 253 tentatively renders the word as "*tasks*" in keeping with his sense of what precedes. But this does not conform to the word's semantic range. No better is the established sense of the word as "rules, laws," which would refer to the family's vision for the future nuptials (or ministry) of their daughter (cf. Cavigneaux 1996: "lien"), but is unlikely in construct with "hand(s)." The image seems to anticipate the references to other terms of apparel in the following lines, which justifies the rendering "sash"; cf. Foster 2005: 200: "latschstring."
***ē tappalsī*.** So Cavigneaux 1998; Wasserman 2016: 252, both of whom offer alternatively: ⸢*la*⸣ *t*[*a-ap*]*-pa-*[*la*]*-sí-i*. The distinction is minor, though a vetitive here is perhaps slightly preferable following the vetitive in l. 18 and the four precatives in ll. 7, 10, 14, 16; but cf. l. 21.
l. 21. *kīma šummannim*. So Groneberg 2001: 108; Wasserman 2003: 164; cf. Foster 2005: 200. The understanding of *šummannum* as a (tethering) rope is provisionally accepted here on the basis of two apparent occurrences of this simile in OB incantations: *kīma šu-mu-nim lunē'e kibiska*, "as a *tether rope* may I turn back your way" (Whiting 1985: 181, ll. 15–18; CAD Š/3, 280); *kīma šu-mu-un-ni-im ḫu-BU-ú-ša-ki ē tatbukīm* "May your ... not *cast from*/*to* me as a *tether rope*" (Wilcke 1985: 200–201, l. 35 [= LAOS 4 27, l. 35]). Note, however, that CAD Š/3, 280 gathers these three attestations under *šummannu* (B), of unknown meaning but distinct from the homonym *šummannu* (A), "(tethering) rope" (ibid., 279). Wasserman's (2016: 253) qualification of the term here as a cow's halter seems motivated by his reading of the next line; but cf. below.
l. 22. *kīma pu/ūrim lū īdînni*. Groneberg 2001: 108 and n. 63 and Wasserman 2003: 164 and n. 35; 2016: 252–253, 255, following Cavigneaux 1996: 36, translate respectively: "wie ein Kalb belecke mich!"; "lick me like a calf!" This solution, with the verb a form of *lêkum*, "to lick," is theoretically possible, though by no means certain, with CAD listing

only one attestation of it in the D-stem (*ulâk*, concerning an ox [SB]). But a more serious obstacle involves common sense: the simile is wholly inconsistent with the rest of the text.

The simile's immediate context concerns matters of bodily girding or covering and situates the word in a sequence that runs: *riksum ša qātim* (l. 19) → *šummannum* (l. 21) → … (l. 22) → *paršīgum* (l. 23) → *nēbaḫum* (l. 25). Moreover, the alleged bovine image is otherwise unknown in love poetry, which contains only one potential but broken reference to a calf (LAOS 4 1 iii 22). Wasserman 2016: 255 recalls in this context the mention of *rīmtum* in CUSAS 10 11 (= LAOS 4 20), l. 9, though in that text the referent is the desired female, not her suitor, and the word is meant in a semi-mythological or at least undomesticated sense, parallel to the *ašnugallum* of the previous line. Nor does it seem likely that this is a variation on the mother-child theme seen in the love incantation MAD V 8 (LAOS 4 22), ll. 22–24 (*pace* Groneberg 2001: 108). In those instances the bucolic imagery, which better suits the scene's outdoor setting, in the "garden of Sîn," describes both mother and child (goat-kid; ewe-lamb; jenny-foal), with the former encircling the latter in the manner fancied by the young male. Nothing of the sort is at issue here.

More likely this line forms a couplet with the preceding (in agreement with the overall parallelism of the text) and thus expresses a sentiment similar to it, which has the suitor begging for the female's attention (*itaplasīnni*). What, then, is the verb in l. 22? If one assumes the form to be an imperative, then the options are next to nil. However, one could take the initial LU separately (so van Dijk 1985: 50), as an asserverative, in which case what follows could be either imperative or imperfect.

Unlikely possibilities in this case include forms of *nâkum* ("to have illicit sexual relations") and *eqûm* ("to daub [salves, etc.]"). But a simple and sensible option does exist, one that may not demand (what in any case would be the slightest) emendation: *idûm*. As is well known, injunctive forms of this verb frequently take a preceding *lū* to convey direct and indirect commands (e.g., *lū īde* "(he) should know (this)!"; *lū tīde* "take notice!"; *lū īdûšu* "(they) should know it!").[75] It is doubtful that "to know" was meant euphemistically here, though that is within the verb's semantic range (see, e.g., CH § 130). A more intriguing option would allude to the earlier statements concerning consecration, an act that included the announcement (*wuddûm*) of the females' dowries (Wilcke 1982: 441). Read in this way, the suitor implores to be "discovered" in place of the alternatives marking consecration or domesticity.

With *lêkum* rejected, no justification remains for reading what precedes as "calf." As noted, the immediate context (*riksum*, *šummannum*, …, *paršīgum*, *nēbaḫum*) here strongly favors another type of bodily covering. The evidence from the Ištar omen

75 For additional examples, see CAD I, s.v. *idû*.

studied above supports this further (though the perils of circular reasoning are noted). The writing, it is proposed, reflects **pu/ūrum*, another attestation of an alternate to *uprum/ḫuprum/uprûm/ḫuprûm*,[76] thus a headdress of some sort.

l. 26. *tēzihī-ma qablīki*. The same image employing *ezēḫum* appears in ATRAḪASĪS I 286 (cited above), taken in Stol 2000: 116 as possibly referring to stone amulets placed on the *qadištum*, or woman in labor.

ll. 27–28. On the first word, see the collation reported in Foster 2005: 200: *z*[*i-mi*]*-ja*?; cf. Wasserman 2016: 253: ⸢*ṣí*⸣-[*ḫa-ti*]*-ia* "*my love making*." The initial word of l. 28, ending with *ši*, is likely a 2fs verbal form (cf. ll. 24, 26). Wasserman 2016: 253 has [*n*]*a-ši* [*i*?*-n*]*i-ki*, but does not translate this (cf. van Dijk 1985: 50), though the idea of sight and the idiom of lifting one's eye (*īnam našûm*) would accord with much of the earlier part of the poem and the interpretation offered above. But cf. Foster 2005: 200: "Stroke your [body] with [the happy glow] of my fa[ce] as if it were oil?"

Bibliography

Archi, A.

1989 "Integrazioni a testi eblaiti." *Or*NS 58 (1989): 124–127.

1992 "Transmission of the Mesopotamian Lexical and Literary Texts." Pp. 1–40 in *Literature and Literary Language at Ebla*, ed. P. Fronzaroli. *QdS* 18. Florence: Department of Linguistics, University of Florence.

Aro, J. and Nougayrol, J.

1973 "Trois nouveaux recueils d'haruspicine ancienne." *RA* 67: 41–56.

Asher-Greve, J.

1995–96 Reading the Horned Crown: A Review Article. *AfO* 42/43: 181–189.

Barberon, L.

2012 *Les religieuses et le culte de Marduk dans le royaumme de Babylone*. Archibab 1. Mémoires de N.A.B.U 14. Paris: Sepoa.

Beaulieu, P.-A.

2003 *The Pantheon of Uruk During the Neo-Babylonian Period*. CM 23. Leiden/Boston: Brill/Styx.

Biggs, R.D.

1974 *Inscriptions from Tell Abū Salābīkh*. OIP 99. Chicago: University of Chicago Press.

Cavigneaux, A.

1996 "Notes Sumerologiques." *ASJ* 18: 31–46.

1998 "YOS 11 n° 29: 19–21 // n° 87: 18–20: seduction et thérapie!" *NABU* 1998/74.

76 See above and n. 31.

Cavingeaux, A., and Al-Rawi, F.
1993 "New Sumerian Literary Texts from Tell Haddad (Ancient Meturan): A First Survey." *Iraq* 55: 97–105.
Cooper, J.
1983 *The Curse of Agade*. Baltimore: The Johns Hopkins University Press.
Dalley, S.
1980 "Old Babylonian Dowries." *Iraq* 42: 53–74.
Durand, J.-M.
1987 "L'organisation de l'espace dans le palais de Mari: le témoignage des textes." Pp. 39–110 in *Le système palatial en Orient, en Grèce, et à Rome*. Actes du Colloque de Strasbourg 19–22 juin 1985, ed. E. Lévy. Leiden: Brill.
1988 *Archives épistolaires de Mari* I/1. ARM 26. Paris: Éditions Recherche sur les Civilisations.
1989 "Minima emariotica." *NABU* 1989/55.
1991 "Perruques." *NABU* 1991/52.
1997a "La divination par les oiseaux." *MARI* 8: 273–282.
1997b–2000 *Documents épistolaires du palais de Mari*. 3 vols. Paris: Éditions du Cerf.
2008 "La religion amorrite en Syrie à l'époque des archives de Mari." Pp. 161–716 in *Mythologie et religion des Sémites occidentaux*, vol. 1: *Ébla, Mari*, ed. G. del Olmo Lete. OLA 162. Louvain: Peeters.
2009 *La nomenclature des habits et des textiles dans les textes de Mari*. MDBP 1; ARM 30. Paris: CNRS.
Edzard, D.O.
2004 "Altbabylonische Literatur und Religion." Pp. 481–640 in D. Charpin, D.O. Edzard, and M. Stol, *Mesopotamien: Die altbabylonische Zeit*. OBO 160/4. Fribourg: Academic Press; Göttingen: Vandenhoeck & Ruprecht.
Farber, W.
2014 *Lamaštu: An Edition of the Canonical Series of Lamaštu Incantations and Rituals and Related Texts from the Second and First Millennia B.C.E.* MC 17. Winona Lake, IN: Eisenbrauns.
Finkel, I.L.
1983–84 "A New Piece of Libanomancy." *AfO* 29/30: 50–55.
Foster, B.
2005 *Before the Muses: An Anthology of Akkadian Literature*. 3rd ed. Bethesda, MD: CDL Press.
Frahm, E.
2011 *Babylonian and Assyrian Text Commentaries: Origins of Interpretation*. GMTR 5. Münster: Ugarit-Verlag.

Fritz, M.

2003 *"...und weinten um Tammuz": Die Götter Dumuzi-Ama'ušumgal'anna und Damu*. AOAT 307. Münster: Ugarit-Verlag.

George, A.R.

2003 *The Babylonian Gilgamesh Epic: Introduction, Critical Edition and Cuneiform Texts*. 2 vols. Oxford: Oxford University Press.

2009 *Babylonian Literary Texts in the Schøyen Collection*. CUSAS 10. Bethesda, MD: CDL.

2013 *Babylonian Divinatory Texts Chiefly in the Schøyen Collection*. CUSAS 18. Bethesda, MD: CDL Press.

Glassner, J.-J.

2000 *Écrire à Sumer: L'invention du cunéiforme* Paris: Seuil.

Groneberg, B.

1997 "Ein Ritual an Ištar." *MARI* 8: 291–303.

2001 "Die Liebesbeschwörung MAD V 8 und ihr literarischer Kontext." *RA* 95: 97–113.

2002 "'The Faithful Lover' Reconsidered: Towards Establishing a New Genre." Pp. 165–183 in *Sex and Gender in the Ancient Near East: Proceedings of the 47th Rencontre Assyriologique Internationale, Helsinki, July 2–6, 2001*, ed. S. Parpola and R.M. Whiting. Helsinki: Neo-Assyrian Text Corpus Project.

Heimpel, W.

2003 *Letters to the King of Mari: A New Translation, with Historical Introduction, Notes, and Commentary*. MC 12. Winona Lake, IN: Eisenbrauns.

Hrůša, I.

2010 *Die akkadische Synonymenliste* malku = šarru: *Eine Textedition mit Übersetzung und Kommentar*. AOAT 50. Münster: Ugarit-Verlag.

Jacobsen, T.

1976 *The Treasures of Darkness: A History of Mesopotamian Religion*. New Haven: Yale University Press.

1987 "Pictures and Pictorial Language (the Burney Relief)." Pp. 1–11 in *Figurative Language in the Ancient Near East*, ed. M. Mindlin, M.J. Geller, and J.E. Wansbrough. London: SOAS.

Jeyes, U.

1989 *Old Babylonian Extispicy: Omen Texts in the British Museum*. PIHANS 64. Istanbul: NINO.

Koch-Westenholz, U.

2000 *Babylonian Liver Omens: The Chapters* Manzāzu, Padānu *and* Pān tākalti *of the Babylonian Extispicy Series mainly from Aššurbanipal's* Library. CNI 25. Copenhagen: The Carsten Niebuhr Institute of Near Eastern Studies.

Köcher, F.

1953 "Der babylonische Göttertypentext." *MIO* 1: 57–107.

Kogan, L.

2001 "*ġ in Akkadian." *UF* 33: 263–298.

2002 "Additions and Corrections to '*ġ in Akkadian.'" *UF* 33: 315–317.

Lambert, W.G.

1975 "The Problem of the Love Lyrics." Pp. 98–135 in *Unity and Diversity: Essays in the History, Literature, and Religion of the Ancient Near East*, ed. H. Goedicke and J.J. Roberts. Baltimore and London: The Johns Hopkins Univ. Press.

1976–80 "Išḫara." Pp. 176–177 in *RlA* 5. Berlin: de Gruyter.

1985 "The Pantheon of Mari." *MARI* 4: 525–539.

Landsberger, B.

1960 "Einige unerkannt gebliebene oder verkannte Nomina des Akkadischen." *WZKM* 56: 109–129.

Leemans, W.

1952 *Ishtar of Lagaba and Her Dress*. SLB I/1. Leiden: Brill.

Lutz, H.F.

1929 "An Old Babylonian Divination Text." *UCP* 9/5: 367–377 + pls. 2–3.

Maggio M.

2012 *L'ornementation des dieux à l'époque paléo-babylonienne: Étude du matériel ayant appartenu aux dieux d'après des documents de la pratique: Réflexions sur le don, l'ornementation des statues divines et la conservation des objets précieux.* AOAT 393. Münster: Ugarit-Verlag.

Maul, S.M.

2013 *Die Wahrsagekunst im Alten Orient: Zeichen des Himmels und der Erde*. Munich: Beck.

Mayer, W.R.

2009 "Akkadische Lexikographie: *CAD* T and Ṭ" *Or*NS 78: 423–439.

Michel, C.

1997 "Un témoignage paléo-assyrien en faveur du port du voile par la femme mariée." *NABU* 1997/40.

Michel C., and Nosch, M.-L. (eds.)

2010 *Textile Terminologies in the Ancient Near East and Mediterranean from the Third to the First Millennia BC*. Ancient Textile Series 8. Oxford/Oakville: Oxbow.

Nougayrol, J.

1967 "'Oiseau' ou Oiseau?" *RA* 61: 23–38.

1969 "Nouveaux textes sur le *ziḫḫu* (I)." *RA* 63: 149–157.

1976 "Les 'silhouettes de référence' de l'haruspicine." Pp. 343–350 in *Kramer Anniversary Volume*, ed. B. Eichler. AOAT 25. Kevelaer: Butzon & Bercker; Neukirchen-Vluyn: Neukirchener Verlag.

Parpola, S.

1984 "*Likalka ittatakku*: Two Notes on the Morphology of the Verb *alāku* in Neo-Assyrian." *StOr* 55: 185–209.

Prechel, D.

1996 *Die Göttin Išḫara: Ein Beitrag zur altorientalischen Religionsgeschichte*. ALASPM 11. Münster: Ugarit-Verlag.

Reid, N. and Wagensonner, K.

2014 "My Tooth Aches So Much." *CDLB* 2014/3.

Reiner, E.

2007 "Supplements to *Chicago Assyrian Dictionary* T (Volume 18)." *JNES* 66: 47–62.

Richardson, S.

2002 "Ewe Should Be So Lucky: Extispicy Reports and Everyday Life." Pp. 229–244 in *Mining the Archives: Festschrift for Christopher Walker on the Occasion of His 60th Birthday, 4 October 2002*, ed. C. Wunsch. Dresden: ISLET.

Richter, T.

1994 "Zu einigen speziellen Keulenmarkierungen." *AoF* 21: 212–246.

Robson, E.

1999 *Mesopotamian Mathematics 2100-1600 BC: Technical Constants in Bureaucracy and Education*. OECT 14. Oxford: Clarendon.

2008 "The Long Career of a Favorite Figure: The *apsamikku* in Neo-Babylonian Mathematics." Pp. 211–226 in *From the Banks of the Euphrates: Studies in Honor of Alice Louise Slotsky*, ed. M. Ross. Winona Lake, IN: Eisenbrauns.

Rochberg, F.

2010 "'If P, then Q': Form and Reasoning in Babylonian Divination." Pp. 19–27 in *Divination and the Interpretation of Signs in the Ancient World*, ed. A. Annus. OIS 6. Chicago: The Oriental Institute of the University of Chicago.

2011 "Divine Causality and Babylonian Divination." Pp. 189–203 in *A Common Cultural Heritage: Studies on Mesopotamia and the Biblical World in Honor of Barry L. Eichler*, eds. G. Frame et al. Bethesda, MD: CDL, 2011.

2016a "A Critique of the Cognitive-historical Thesis of *The Intellectual Adventure*." Pp. 1–28 in *The Adventure of the Human Intellect: Self, Society, and the Divine in Ancient World Cultures*, ed. K. Raaflaub. Chichester, UK: Wiley Blackwell.

2016b *Before Nature: Cuneiform Knowledge and the History of Science*. Chicago: University of Chicago Press.

Sallaberger, W.

2000 "Das Erscheinen Marduks als Vorzeichen: Kultstatue und Neujahrfest in der Omenserie Šumma ālu." *ZA* 90: 227–262.

Selz, G.

1997 "'The Holy Drum, the Spear, and the Harp': Towards an Understanding of the Problems of Deification in Third Millennium Mesopotamia." Pp. 167–209 in *Sumerian Gods and Their Representations*, ed. I.L. Finkel and M.J. Geller. CM 7. Groningen: Styx.

Shehata, D.

2007 "Privates Musizieren in Mesopotamien?" *WZKM* 97: 521–529.

Steinkeller, P.

1987 "The Name of Nergal." *ZA* 77: 161–168.

1990 "More on the Name of Nergal and Related Matters." *ZA* 80: 53–59.

1998 "Inanna's Archaic Symbol." Pp. 87–100 in *Written on Clay and Stone: Ancient Near Eastern Studies Presented to Krystyna Szarzyńska on the Occasion of Her 80th Birthday*, ed. J. Braun et al. Warsaw: AGADE.

2004 "Studies in Third Millennium Paleography, 4: Sign KIŠ." *ZA* 94: 175–185.

Stol, M.

2000 *Birth in Babylonia and the Bible: Its Mediterranean Setting*. CM 14. Groningen: Styx.

2016 *Women in the Ancient Near East*. Boston: de Gruyter.

Van Dijk, J.

1985 "Comments on Texts." Pp. 17–52 in J. van Dijk, A. Goetze, and M.I. Hussey, *Early Mesopotamian Incantations and Rituals*. YOS 11. New Haven: Yale University Press.

Wasserman, N.

2003 *Style and Form in Old-Babylonian Literary Texts*. CM 27. Leiden: Brill/Styx.

2016 *Akkadian Love Literature of the Third and Second Millennium BCE*. LAOS 4. Wiesbaden: Harassowitz Verlag.

Waetzoldt, H.

1980–83 "Kopfbedeckung." Pp. 197–203 in *RlA* 6. Berlin: de Gruyter.

Whiting, R.

1985 "An Old Babylonian Incantation from Tell Asmar." *ZA* 75: 179–187.

Wiggermann, F.A.M.

1998–2001 "Nergal A. Philologisch." Pp. 215–223 in *RlA* 9. Berlin: de Gruyter.

2000 "Lamaštu, Daughter of Anu: A Profile." Pp. 217–252 in M. Stol, *Birth in Babylonia and the Bible: Its Mediterranean Setting*. CM 14. Groningen: Styx.

Wilcke, C.

1985 "Liebesbeschwörungen aus Isin." *ZA* 75: 188–209 + 2 pls.

Winitzer, A.

2010 "The Divine Presence and Its Interpretation in Early Mesopotamian Divination." Pp. 177–197 in *Divination and the Interpretation of Signs in the Ancient World*, ed. A. Annus. OIS 6. Chicago: The Oriental Institute of the University of Chicago.

2017 *Early Babylonian Divination Literature: Its Organizational Framework and Generative and Paradigmatic Characteristics*. AMD 12. Leiden: Brill.

Forthcoming "Conceptions of Mesopotamian Divination." *AfO*.

Ziegler, N.

2007 *Les musiciens et la musique d'après les archives de Mari*. FM 9. Paris: SEPOA.

CHAPTER 4

Material, Constellation, Image, God: The Fate of the Chosen Bull According to KAR 50 and Duplicates

Alan Lenzi

Although best known for her work on Mesopotamian celestial divination, astrology, and astronomy, Francesca Rochberg has also made significant contributions to conceptualizing the nature of the relationship between the gods and celestial bodies.[1] Just as she championed the idea that the Mesopotamians were capable of doing science,[2] she has argued convincingly that they were capable of sophisticated thinking about their gods. With Chessie's work as inspiration, the following study examines the ritual use of a bull in the making of a divine image as presented in the incantation in KAR 50 and its duplicates, a text related to the *kalû*'s ritual covering of the *lilissu*-drum. The exposition pays particular attention to the various interconnections between deity, celestial body, manufactured image, and the raw materials used to manufacture the image—all present in the incantation—in order to understand how the ritual language effects its goal. I offer these thoughts to Chessie with the greatest admiration and as a tribute to her stellar career, marked by both intellectual depth and breadth.

Covering the *Lilissu*-Drum Ritual: Textual Discovery and Present Focus

Lamentation-experts or *kalû*s were ancient Mesopotamian cultic personnel/scholar-scribes who sang or recited ritual laments—*eršemma*, *eršaḫunga*, *balaĝ*, and *šu'ila* were the main genres, all of which were written in the Emesal sociolect of Sumerian—to soothe and appease divine wrath during various calendrical and occasional rituals.[3] One of the musical instruments the *kalû*

1 See Rochberg 1996, 2007, 2009, 2010, 2010a, 2011.

2 Rochberg 2004.

3 See Löhnert 2011 for a brief overview of ritual lamentation, including the work of the *kalû*. Gabbay 2014: 63–79 offers a full discussion of the *kalû*. Gabbay 2014a focuses on the *kalû* in Assyria.

 | DOI 10.1163/9789004363380_005

used to accompany his laments was a *lilissu*-drum, a copper or bronze kettle drum that had a bull's hide for its head. When a new drum was made, when a drum head wore out, or when a temple was dedicated[4] the *kalû*[5] enacted a ritual process to prepare and install a new head for the drum using a bull's hide as the covering and its tendon as a means of securing it over the opening of the *lilissu*-drum.[6]

The text upon which the present article focuses, KAR 50 and its duplicates, is only one witness to this covering process among several other related texts, whose discovery spans the history of Assyriology. These witnesses come from some of the most significant first millennium scribal repositories, sometimes separated in their times of flourishing by hundreds of years. Although all of the texts bear witness to a common ritual goal, they attest to a diversity of ritual details involved in covering the *lilissu*-drum. A brief survey of this material provides some context and perspective on KAR 50 and its duplicates.

In 1875 Henry Rawlinson published the first known witness related to the covering ritual in the fourth volume of his *Cuneiform Inscriptions of Western Asia*.[7] The tablet, K.4806 (now + K.9421), came from the royal library and archive at Nineveh and bears a concatenation of (mostly bilingual) incantations with some ritual instructions inserted in between. Based on information in its text and colophon, this tablet is the first in a series of tablets—not yet identified—containing a version of the entire covering ritual.[8]

In 1901 Zimmern published another fragment from Nineveh related to the covering ritual, K.6060 (now + K.10820),[9] which is a partial duplicate to KAR 50 (= VAT 8247), discovered during German excavations at Ashur (1903–1914).

4 See Gabbay 2014: 121–123.

5 In some cases, his colleague the *āšipu*, "conjurer," was also involved. See Gabbay 2014: 126, n. 428 on KAR 60 and Linssen 2004: 94, 99, who cites two astronomical diaries that indicate the involvement of the conjurer in the ritual.

6 See Gabbay 2014: 118–139 for a full discussion of the *lilissu*-drum, including the rituals surrounding its covering and the related explanatory text, in the context of a broader discussion of musical instruments the *kalû* used in temple liturgies. Gabbay 2015: 192–196 treats the ritual briefly in a study of whispering incantations into an ear.

7 Rawlinson 1875, pl. 23, no. 1. Early treatments of the tablet are listed in Bezold 1891: 664 and Borger, HKL I, 403. Borger joined this piece to K.9421 some time later (see Borger, HKL III, 344, s.v. K 4806). A copy of the fragment (*sans* K.4806) is in Linssen 2004, plate III. A photograph of the joined pieces is available at http://cdli.ucla.edu/P395721.

8 See Linssen 2004: 275–282 for the most recent edition.

9 Zimmern 1901, no. 56 (edition: 170–171; copy: plate LV). Borger joined the tablet to K.10820 some time later (see HKL III, 349, s.v. K 6060). A copy of the joined pieces is in Linssen 2004, plate II and a photograph is available at http://cdli.ucla.edu/P396341.

Both KAR 50 and K.6060+ bear the text of a bilingual incantation spoken over the bull and then the brief instructions for performing the ritual in which the incantation was used. Aside from another duplicate to these tablets noted below (BM 33343), this incantation is not found in other witnesses related to the covering process. The final line of KAR 50 is a catchline to another (lost) tablet that continued its version of the ritual.

KAR 60 (= VAT 8022), another piece related to the covering ritual, was also discovered at Ashur, probably in the same house as KAR 50.[10] (Both KAR 50 and 60 were published in autograph by Ebeling in 1919.[11]) Despite its similarities in purpose to KAR 50, KAR 60 gives only ritual instructions (which are lengthier than KAR 50's) and cites prescribed incantations by incipit. Although this witness is an abbreviated version of the much longer version of the ritual preserved on TCL IV 44, to be described presently, it also bears a unique element among witnesses in that it requires the *kalû* to deny his responsibility for his actions ritually and to attribute them performatively to the gods (rev. 2–4a).[12]

Besides the texts from Nineveh and Ashur, several tablets related to the ritual covering have surfaced among tablets from Uruk. The first ones discovered among these were two tablets from the Anu temple, TCL VI 44 (= AO 6479) and TCL VI 47 (= O.175). The former appeared in Thureau-Dangin's *Rituels Accadiens* in 1921 (pp. 3–6, 10–21; reprinted 1975) and then in 1922 in copy only as TCL VI 44. This tablet preserves the fullest version of instructions for the ritual performance currently known and includes an appendix that lists the materials needed for its execution.[13] The latter, TCL VI 47, is not a ritual text *per se* but an explanatory text that interprets the covering ritual in terms of a mythological scene, which is illustrated with a drawing on the reverse of the tablet. This text, which probably also came from the Anu temple in Uruk (though there is some uncertainty),[14] was first published in 1919 (Thureau-Dangin 1919: 144–156) and then again in 1922 as TCL VI 47. The late Babylonian tablets AO 17626, acquired by the Louvre in 1935,[15] and BM 54119, acquired

10 Pedersén has determined that both pieces come from the well-known house belonging to a family of exorcists (1986: 61).

11 See Borger, HKL I, 98, s.v. 50) and 60) for early literature.

12 Cited below. BaMB 2, no. 5, rev. 8 may have contained something similar, but the lines containing the instructions to this effect seem to have fallen out of the present copy. See Linssen 2004: 274, the note on lines 24–25. For more on this action in the context of a discussion of ritual denials in ancient Mesopotamia, see Lenzi 2010: 161–162.

13 The most recent edition is Linssen 2004: 252–262.

14 See Linssen 2004: 325, n.27 for references and Gabbay 2014: 264.

15 See Nougayrol 1947: 30–32 for its initial publication. The tablet may have come from Nippur (see Gabbay 2014: 130, n. 470, citing Œlsner 1982: 94–95).

by the British Museum in 1882,[16] provide partial duplicates to TCL VI 47 as does another quite fragmentary tablet, BaMB 2, no. 8.[17] This last tablet was one of five Hellenistic tablets found in Anu's Rēš-temple at Uruk in the winter of 1959–60.[18] Van Dijk and Mayer published all of these in 1980 as BaMB 2, nos. 5–9 (= W.20030/4, W.20030/1, W.20030/89, W.20030/122, and W.20030/83, respectively). Most important among these are nos. 5 and 6.[19] No. 5 shows various parallels with both TCL VI 44 and KAR 60, though in terms of length and detail it falls somewhere between the two.[20] And no. 6, first edited by Mayer in 1978,[21] contains the text of a bilingual incantation-prayer—in fact, a *šuʾila*, according to its rubric, which seems to be identical with the one prescribed (cited by incipit) in TCL VI 44, obv. ii 15–16. Like the incantation in KAR 50 and duplicates, it was spoken over the bull to prepare it for ritual slaughter; unlike our incantation, it was addressed to the great gods rather than the bull itself.

BM 33343, the most recent addition to the material related to the covering ritual, is a Late Babylonian tablet listed in Irving Finkel's unpublished catalog of Babylonian tablets in the British Museum. It duplicates KAR 50 and K.6060+ and is published here for the first time.[22]

In all, therefore, fourteen tablets comprise the textual materials related to the ritual covering of the *lilissu*-drum at present, but many of these parallel one another and are often quite fragmentary. The material is thus less extensive than the number of witnesses might suggest.

---

16 It is part of the 82-5-22 collection, "which includes texts from Babylon, Šahrinu and Cutha in addition to Sippar," according to MacGinnis (1999), who edits the tablet.

17 See Linssen 2004: 261–262 for a transliteration of the fragment. Livingstone 1986 (2007): 187–204 offers an edition of TCL VI 47 and AO 17626 with some discussion of the explanatory material. Gabbay 2014: 130–137 provides the fullest exposition.

18 van Dijk and Mayer 1980: 13.

19 No. 7, a small fragment, parallels the ending of KAR 60 (see Linssen 2004: 265 for a transliteration of the obverse). Note especially both the prohibition of the *galamāḫu* from eating the meat of the bull (KAR 60, rev. 14 // BaMB 2, no. 7: 6′) and the prohibition on showing anyone the tablet (KAR 60, rev. 15 // BaMB 2, no. 7: 7′). No.9's connection to the ritual is only tentative due to the fragmentary nature of the piece. See Linssen 2004: 280 for a transliteration.

20 See Linssen 2004: 270–274 for the most recent edition and Farber 1987: 234–235 for a translation.

21 Mayer 1978: 232–237.

22 Note, however, that Gabbay utilizes the tablet in his citations of KAR 50 and parallels (see 2014: 124, n.409 and 125, nn.416 and 418 and Gabbay 2015: 194–195). I thank the Trustees of the British Museum for granting me access to examine and permission to publish the tablet.

Both the general continuity of the tradition over the centuries—from Neo-Assyrian Ashur to Hellenistic Uruk—as well as the fluidity in details (e.g., length, wording, ritual actions, incantations recited, etc.) are features in the evidence worth noting here briefly, reminding us that even as the scribes honored the past through copying—sometimes producing nearly exact copies—they were also liable to introduce expansions, abbreviations, variations, or innovations—none of which is easy to classify as such with certainty. Another striking feature in the evidence is the fact that different versions of the ritual could co-exist in the same library, as seems to have been the case with KAR 50 and 60 as well as with materials from Uruk. Contextualizing KAR 50 and parallels within this material requires us to consider in a careful, self-conscious manner both our text's distinctive features as well as its shared characteristics in achieving the ritual goal common to all of the sources. In other words, we should not simply treat all of the related ritual materials as if they are one large though fragmented source, each equally relevant to our understanding of the details of the one particular text that holds our present attention.[23]

Thureau-Dangin's treatment in *Rituels Accadiens*[24] was the standard edition of the texts related to the covering ritual until the appearance of Linssen's study in 2004, which incorporated most of the new pieces from Uruk.[25] Shibata's forthcoming volume (Heidelberger Emesal-Studien 3) on Sumerian *šu'ila*s will treat BaMB 2, no. 6 anew. And the present article incorporates BM 33343 into the other textual witnesses for the unique incantation that is the focus of the present article. With this philological foundation in place, we are in a position to ask how these ritual materials did what they were supposed to do.

Of course, understanding what a ritual does and how it does it is a rather complex, multi-perspectival task,[26] which might include, for example, a discussion of what the ritual does to/for the people/objects acted upon during the ritual (psychologically, socially, economically, symbolically, etc.), to/for the

23 Gabbay's compact analysis offers an insightful, bird's eye perspective on the materials and explanatory text (2014: 124–139), drawing on the various versions of the ritual to piece together an understanding of the common ritual goal. The present contribution suggests we ought also to consider how each major source *distinctively* accomplishes (or contributes to) this common ritual goal, with a focus here on KAR 50 and duplicates.

24 Thureau-Dangin 1975 (1921): 1–33, which Sachs 1969 translates into English.

25 Linssen 2004: 92–100 discusses the kettle drum, its use, and the various materials for covering it, which is now superseded by Gabbay 2014: 118–139. The editions of the texts are on pp. 252–282.

26 Bell 1997 remains an excellent introduction to ritual studies. Grimes 2014 offers an introduction to the discipline with more emphasis on fieldwork.

people performing/officiating over the ritual;[27] to/for the people watching the participants (both officiants and non-expert participants); to/for the broader society (or social order or dominant fraction of society or political elites), affecting those not even present; to/for the (perception of the) natural order of things (e.g., cosmic maintenance, insurance of order, divine placation, etc.); and so forth. If our evidence for the covering ritual processes were sufficient for a full anthropological investigation (and I do not at all think that it is),[28] the results of our investigation would go far beyond the confines of a short article. We must therefore narrow our goals in light of the limitation of evidence (and space) and sharpen our focus in terms of our present interpretive interest. Moreover, given the diversity of textual materials associated with the covering, a narrower focus in terms of the data is also necessary.

As stated from the start, the matter of interest here is the ritual use of a bull in the making of a divine image and the various interconnections between deity, celestial representation, manufactured image, and materials used to produce the image. Although we could understand these things via the actions described in the various versions of the ritual instructions (e.g., KAR 60, BaMB 2, no. 5, or TCL VI 44), the focus here will be on the incantation recorded as a part of the abbreviated version of the ritual procedure in KAR 50 and its duplicates in K.6060 + K.10820 and BM 33343.[29] In the following thoughts, I argue that this incantation ritually incorporates (the hide and tendon of) a live bull into a divine entity—a deified drum[30]—via performative assertions and persuasive

27 For example, one could interpret the participation of the lamenters (*kalû*) in the present ritual (or in all of their ritual activity) in line with Bell's argument in *Ritual Theory, Ritual Practice* (1992) as the active construction and perpetuation of the ritual official's expertise and authority in society even while they are effecting a ritual transformation or placating divine anger.

28 The most basic hurdle to such an investigation is the reasonable assumption that what the texts record—in their variety of length, detail, and inclusion of particular elements—does not convey all of the information about the ritual procedures (both what was done and precisely how) that would be necessary to answer the kinds of questions posed in a full investigation (see Grimes 1995: 24–39 for the spectrum of questions a full investigation might consider).

29 Other materials relating to the ritual covering of the *lilissu*-drum will come into our purview only as deemed necessary to support an interpretation of the present text via illustration or analogy.

30 The drum is referred to as ^{d}*lilissu* in TCL VI 44, rev. i 17, where the drum is listed with deities who are to receive an offering assemblage (*riksu*) and just before the drum is incorporated into the temple liturgy. This is the only time the word takes the divine determinative in the various versions of the ritual instructions (for the broader significance, see Porter 2009, 178–179). However, note that both instances of *lilissu* on the reverse

analogies that give the relevant bovine materials and the resulting manufactured object celestial and mythological credentials worthy of divinity. In doing so, the incantation shows structural and thematic similarities to the incantations prescribed in the *Mīs pî* ritual[31] and evokes complex, symbolic interconnections between deity, celestial representation, manufactured image, and materials used to make the image.

The Text

As briefly mentioned above, the text is known in three exemplars from three different sites. KAR 50 (= VAT 8247) comes from Neo-Assyrian Ashur; K.6060 + K.10820 from Neo-Assyrian Nineveh; and BM 33343 from late period Babylonia. In a textbook example of scribal conservationism, all three exemplars preserve essentially the same text, despite the several intervening centuries. The text, as witnessed by each of these tablets, runs as follows:[32]

A = KAR 50 (= VAT 8247)
B = K.6060 + K.10820
C = BM 33343 (approx. 5.9 × 8.8 cm in landscape format; the copy follows the article)[33]

A 01 én gu$_4$-maḫ za dím-ma dingir-gal-[gal-e-ne]
C 01 [é]n gu$_4$-maḫ za dím-[ma] dingir-gal-gal[34]-e-ne

of the ritual explanatory text TCL VI 47 include the divine determinative. See Livingstone 1986 (2007): 194–195. On the deification of musical instruments, see Gabbay 2014: 90–91, in a fuller discussion (84–153) of musical instruments, especially drums, used by the *kalû*. His discussion includes significant details about the *lilissu*-drum and the present ritual (see especially 118–139).

31 Gabbay has explored the many similarities of the materials related to the *lilissu* covering ritual with the *Mīs pî* ritual in 2014: 120–139, though he has not treated KAR 50 in specific detail. For his insights into the role of the incantations within the ritual actions, see especially 2014: 125–126 with footnotes.

32 MS B and C have been collated with the original tablets in the British Museum. The transliteration of MS A is based on Ebeling's copy in consultation with Linssen 2004: 267.

33 The tablet's reverse is quite convex so that its center area is at least twice as thick as its edges. The lines on the reverse are inscribed straight across the tablet, but the copy shows the lines of text apparently following an arc across the tablet, starting around line 3. By line 7, the arc is quite pronounced. This distortion in the copy is due to the fact that I traced a digital photograph of the tablet in order to make the copy, and the convex shape of the tablet distorted the image.

34 There seems to be an erasure here.

A 02 *at-ta gu-um-m*[*a-ḫ*]*u bi-nu-ut* [DINGIR.MEŠ GAL.MEŠ]
C 02 *at-ta gu-um-ma-ḫ*[*u*] *bi-nu-ut* DINGIR.MEŠ GAL.MEŠ

A 03 dím-me-bi kin-ma dingir-ga[l]-gal-e-n[e …]
C 03 dím-me-bi kin-ma dingir-gal-gal-e-ne tùm-ma

A 04 *tab-ba-ni-ma ana š*[*i-pir* DING]IR.MEŠ GAL.[MEŠ …]
C 04 *tab-ba-ni-ma a-na ši-pir* DINGIR.MEŠ GAL.MEŠ *tab-ba-bal*[35]

A 05 an alam garza ^{d}a-nu-um$^{!}$-ke$_{4}$ g[ar$^{?}$-…]
C 05 an alam garza ^{d}a-nu-um-ke$_{4}$ gar-bi an-ta-bi

A 06 *ina* AN-*e ṣa-lam-ka a-na pa-ra-aṣ* ^{d}*A-nu-*[*ti* …]
C 06 *i-na šá-me-e ṣa-lam-ka a-na pa-ra-aṣ* ^{d}*A-nu-tum šá-kin e-la-tu*

A 07 u$_{4}$ an ^{d}en-líl-lá ^{d}en-ki ^{d}nin-maḫ nam dingir-gal-[gal-e-ne …]
C 07 u$_{4}$ ^{d}a-num $^{d+}$en-líl $^{d+}$en-ki ^{d}nin-maḫ nam dingir-gal-gal-⸢e-ne-X-X-X⸣

A 08 *i-nu* ^{d}MIN ^{d}MIN ^{d}MIN *u* ^{d}MIN *ši-mit* DINGIR.MEŠ GAL.MEŠ [*i-ši-mu*]
C 08 *i-nu* ^{d}*A-num* $^{d+}$EN.LÍL *u* ^{d}*É-a u* ^{d}*Be-let-ì-lí ši-mat* DINGIR.MEŠ GAL.M[EŠ *i-ši-mu*]

A 09 kuš-zu sa ḫal-ga dingir-gal-gal-e-ne nam tar-tar-[…]
C 09 kuš-zu sa ḫal-ga dingir-gal-gal-e-ne nam tar-t[ar-e]-⸢ne$^{?}$⸣

A 010 *ma-šak-ka ši-ir-an-ka ana pi-riš-ti* DINGIR.MEŠ GAL.MEŠ *iš-*⸢*šá*⸣[36]*-a-mu*
C 010 *ma-šak-ka ši-ir-an-ka ana pi-riš-ti* DINGIR.MEŠ GAL.MEŠ *iš-*[*šá-a-m*]*u*

A 011 u$_{4}$-ḫal dingir-bi u$_{4}$ da-ir-e[š] za gi[37]-na
B 03′[38] … ding]ir-bi u$_{4}$ da-r[í-éš …]
C 011 u$_{4}$-ḫal dingir-bi u$_{4}$ da-rí-éš za gi-na

35 This is an N durative from *babālu*. See GAG §103x.
36 This sign is not on Ebeling's copy. I am following Linssen 2004: 267.
37 Ebeling's copy seems to show an erasure between the two signs.
38 The Ninevite text preserves a distinctive order here vis-à-vis the witnesses from Ashur and Babylonia.

A o12 *ina pi-riš-ti* DINGIR.BI *ana* U$_4$.MEŠ *da-ru-ú-tú ku-ú-ni*
B o4′ [*in*]*a pi-riš-ti* DINGIR *šá-a-šú* [...]
C o12 *ina pi-riš-ti* DINGIR *šá-a-šu* <*ana*$^?$> U$_4$.MEŠ *da-ru-ti ku-ú-ni*

A o13 alam dingir ki šeš-bi nam tar-tar-e-ne
B o1′ [ala]m ⸢dingir ki⸣[39] [...]
C o13 alam dingir ki šeš-bi nam tar-tar-e-ne

A o14 *ṣa-lam šu-ú it-ti* DINGIR.MEŠ ŠEŠ.MEŠ-*šú ši-mit ši-mu*
B o2′ *ṣa-lam šu-ú it-*⸢*ti*⸣ DI[NGIR ...]
C o14 *ṣa-lam šu-ú it-ti* DINGIR.MEŠ ŠEŠ.MEŠ-*šú ši-mit liš*[40]*-mu*

A o15 dingir-bi ⸢zag-gu⸣-la-bi mu-un-sikil mu-un-dadag
B o5′ [dingir-b]i ul-la-bi m[u-...]
C o15 dingir-bi zag-gul-la-bi mu-un-sikil mu-un-dadag-ga

A o16 DINGIR *šu-ú e-šèr-ti-šú li-lil li-bi-ib*
B o6′ DINGIR *šu-ú ina a*[41]*-šér-ti-šú li-*[...]
C o16 DINGIR *šu-ú ina a-šèr-ti-šú li-lil* ⸢*li*⸣*-bi-ib*

A o17 eme-ḫul-gál bar-šè ḫé-en-ta-gub
B o7′ [eme-ḫ]ul-gál bar-šè ḫ[é-...]
C o17 eme-ḫul-gál bar-šè ḫé-em-ta-gub

A ø[42]
B o8′ *li-šá-nu le-mut-tú ina a-ḫa-t*[*i* ...]
C o18 *li-šá-nu le-mut-tum ina a-ḫa-a-ti li-iz-ziz*

A r1 DÙ.DÙ.BI *ina* ITI SI.SÁ *ina u*$_4$*-mi* DÙG.GA *ina šèr-ti*
B o9′ [DÙ.D]Ù.BI *ina* ITI SI.SÁ *ina u*$_4$-[*mi* DÙG.GA] | o10′ [*ina*] *šèr-ti*
C r1 DÙ.DÙ.BI *ina* IT[I SI.S]Á *ina u*$_4$*-mi* DÙG.GA *ina šèr-ti*

39 Only the smallest hint of the bottom part of these signs is visible on the tablet (not entirely included on Linssen's copy, 2004, plate II). See the collation provided below.

40 This seems to me to be a graphic error on the part of the copyist as ŠI and LIŠ are relatively similar in NB script.

41 Linssen's copy shows an E sign here (2004, plate II). But the tablet shows *ina a-*.

42 Niek Veldhuis in a personal communication makes the reasonable suggestion that the Sumerian phrase was so formulaic in incantations that the copyist of MS A did not see the need to include its translation.

A r2 *la-am* ^{d}UTU KUR-*ḫa* 3 GI.DU$_{8}$ *ana* ^{d}*É-a* ^{d}UTU
B o10′ *e-ma* ^{d}UTU KUR-*ḫa* 3 ⸢GI⸣.D[U$_{8}$] | o11′ [*ana*] ^{d}*É-a* ^{d}UTU
C r1 ⸢*e*⸣-[*m*]*a* ^{d}UTU KU[R]-⸢*ḫa*⸣ | r2 3 GI.⸢DU$_{8}$ *ana*⸣ [^{d}]⸢*É-a*⸣ ^{d}UTU

A r3 *u* ^{d}ASAL.LÚ.ḪI *tar-kás* ZÚ.LUM.MA ^{zì}EŠA DUB-*aq*
B o11′ *u* ^{d}ASAL.LÚ.ḪI *tar-kás* | o12′ ⸢ZÚ⸣.LUM.MA ^{zì}EŠA DUB-*aq*[43]
C r2 *u* ⸢^{d}⸣A[SAL.LÚ.ḪI] *tar-*[*ká*]*s* | r3 ZÚ.LUM.MA [^{z}]^{ì}E[ŠA DU]B-[*aq*]

A r4 NINDA.Ì.DÉ.A LÀL Ì.NUN.NA GAR-*an* 3 UDU.NÍTA.MEŠ BABBAR.MEŠ
B o12′ NINDA.Ì.DÉ.A | o13′ LÀL Ì.NUN.NA GAR-*an* 3 UDU.NÍTA.MEŠ BABBAR.MEŠ
C r3 [NINDA.Ì.D]É.A L[ÀL] ⸢Ì⸣.NUN.[N]A | r4 GAR-*an* 3 ⸢UDU⸣.NÍTA.[MEŠ BABBAR.MEŠ]

A r5 BAL-*qí* ^{uzu}ZAG.LU *ḫi-in-ṣa u šu-me-e* TE-*ḫi*
B o14′ BAL-*qí* ^{uzu}ZAG.LU *ḫi-in-ṣa* | r1 *ù šu-me-e* TE-*ḫi*
C r4 ⸢BAL⸣-*qí* [^{u}]^{zu}[ZA]G.LU [*ḫi*]-⸢*in-ṣa ù*⸣ *šu-me-*⸢*e*⸣ | r5 TE-*ḫi*

A r6 NÍG.NA ^{šim}LI GAR-*an* ^{zì}MAD.GÁ *ta-sár-raq*
B r1 NÍG.NA ^{šim}LI GAR-⸢*an*⸣ | r2 ^{zì}MAD.GÁ *ta-sár-raq*
C r5 N[ÍG.NA ^{š}]^{im}[LI] ⸢GAR⸣-*an* ^{zì}M[AD.G]Á [*ta*]-⸢*sar*⸣-[*raq šid-*][44]

A r7 *šid-du* GÍD.DA-*ad* ZÌ.DUB.DUB.BU ŠUB.ŠUB-*di*
B r2 *šid-di* GÍD.DA-*ad* ZÌ.DUB.DUB.BA | r3 ŠUB.ŠUB-*di*
C r6 -*du* G[ÍD].D[A-*a*]*d* ZÌ.[DU]B.D[UB.BA] [ŠUB].Š[UB-*di*]

A r8 GU$_{4}$.MAḪ *ana* KI *par-si tuš-za-az-ma ana* ŠÀ GEŠTUG^{II} 15-*šú*
B r3 GU$_{4}$.MAḪ *ina* KI *par-si tuš-za-az-ma* | r4 ⸢*a-na*⸣ ŠÀ GEŠTUG^{II} 15-*šú u* 2,30-*šú*
C r6 [GU$_{4}$.MA]Ḫ *ina* KI *par-*[*si*] | r7 *tuš-*⸢*za-az*⸣*-ma* [*a-na*] ŠÀ [GEŠTUG^{II} 15]-⸢*šú u*⸣ [2,30]-⸢*šú*⸣

43 Not a fully formed AK sign.

44 This mid-word line break is quite unexpected but seems the only way to account for the sign at the beginning of the next line.

A r9 3.TA.ÀM *tu-làḫ-ḫàš gu-ma-ḫu* KUD-*is-ma*
B r4 3.TA.ÀM | r5 [*tu-là*]*ḫ-ḫàš gu-ma-ḫa* [K]UD-*is-ma*
C r7 ⸢3⸣.T[A.ÀM] | r8 ⸢*tu-làḫ*⸣-[*ḫàš g*]*u-ma-*[*ḫ*]*a* [KU]D-*is-*⸢*ma*⸣

A r10 ⸢*ma-áš-ku*⸣ *u šér-a-nu ana šip-ri šá-a-šú* [T]I-*qí*
B r6 [*ma-áš*]*-ka* ⸢*u šér*⸣-*a-*⸢*na*⸣ […
C r8 ⸢KUŠ⸣ [*u* SA? *ana šip-ri šá-a-šú*] T[I?-*qí*]

A r11 ÉN KAR.ZA.GÌN.NA ⸢KAR x⸣ […][45]
C A ruling follows line 8 and then remains of a colophon too fragmentary to transliterate.[46]

Translation:[47]

1/2. You, choice bull, *are* the creation/product of the great gods.
3/4. You were created and are being brought forward for the work[48] of the great gods.

---

45 The catchline and reference to another (unknown) incantation indicates that this version of the ritual continued. For references to "quay of lapis lazuli," see Linssen 2004: 269, note on line 11. Is this incantation the same one cited by incipit (also broken) in the Ninevite version of the *Mīs pî* ritual (see Walker and Dick 2001: 54, line 21)?

46 Very little can be made out. See the copy.

47 The translation follows the Akkadian version, which will be the point of reference in the remainder of the article.

48 The precise meaning of *šipru* here is unclear. Does the word have a more restricted sense with only the present ritual task of preparing a new drum head in view or does it convey a more extensive idea, e.g., that the bull is to be used in future service to the gods in the form of a drum head on the divine *lilissu*-drum? CAD Š/3, 81 seems to take something like the latter view, rendering *šipir ilī rabûti* as "to be used as an artifact for the great gods," though this translation has the finished product that the bull will provide in view rather than the bull's on-going service in the form of that product. It may be that *šipru* was chosen for this context precisely because of its broad meaning, which encompasses the present ritual task as well as the resulting service/product for the divine drum and liturgy. "Work" seems to be a similarly broad term in English and thus an appropriate rendering. Similar questions about the meaning and rendering of *šipru* could be raised with regard to how the term is used in the ritual text for the repair of a divine image, edited in Walker and Dick 2001: 228–245 (see the full exposition in Boden 1998: 127–169), and the Akkadian version of the incantation spoken over the bull in BaMB 2, no. 6, obv. 12 and rev. 4 (*udduš šiprīkunu* and *ana šipri epšētīku*), for which see Mayer 1978: 433 and Shibata, f.c., no. 22.

5/6. In the heights of the heavens[49] your *ṣalmu*[50] is treated as[51] a supreme divine office (lit. rite of Anu-ship).
7/8. When Anu, Enlil (var. and), Ea, and Belet-ili [assigned] the fate (var. fates) of the great gods,
9/10. Your hide *and* your tendon were assigned to the secret *council* of the great gods.
11/12. Be firmly established (*kūnī*), *O fate*,[52] in the secret *council* of that god for days evermore.
13/14. *As for* that *ṣalmu*, its assigned fate[53] *is* with the gods, his brothers.
15/16. *As for* that god, may his shrine be clean *and* pure.
17/18. May the evil tongue stand aside.
Rev.
Its ritual:[54] In an auspicious month on a good day in the morning before (var. whenever) the sun rises, you set up three offering tables for Ea,

---

49 Thanks to the gap that BM 33343 fills at the end of this line, we can now reconstruct the entire line, which raises the question of how to understand an ... an-ta-bi / *ina šamê ... elâtu*. Literally, these phrases are rendered: "(of / in) the heavens ... its upper part" (Sum.) / "in the heavens ... (its) upper part" (or simply "above," see Gabbay 2014: 125, n.418 and Gabbay 2015: 195) (Akk.), though a freer rendering, treating the phrase as a kind of broken poetic equivalent to *ina elât šamê*, "in the heights of the heavens," seems reasonable. In this case, the line could be describing the *ṣalmu*'s zenith as a constellation (Sum. an-pa) and/or simply its position high in the sky (see, e.g., Horowitz 1998: 144–145, 227, and 236–238).

50 I leave this term untranslated since I believe its multivalence in the incantation is essential to the incantation's understanding. Generally, *ṣalmu* means "image," which can apply to, e.g., a statue/figurine or a constellation. For its full semantic range, see CAD Ṣ, 78–85; note also *ṣalmu*, "black," in CAD Ṣ, 77–78.

51 For the various semantic nuances of *ana* X *šakānu*, see CAD Š/1, 148–149.

52 We might have ignored the final vowel in *ku-ú-ni* if it were only attested in the LB manuscript. But it is also attested in the Ashur manuscript. Thus, the imperative is *kūnī*, a feminine singular, and cannot refer to the bull or the two body parts, as other translations seem to imply. The only other feminine noun in the vicinity to which the imperative could be addressing is *šīmtu* in line 7/8, two lines back. As the fate of the bull is the theme of lines 7/8–9/10 and the topic comes up again in line 13/14, it is reasonable to think that the fate assigned to the hide and tendon could be addressed in the present line rather than the hide and tendon or the bull itself.

53 I render *šīmit šīmu* as a head noun in construct with the following 3ms predicative, which shows the subjunctive. Literally, the phrase means "the fate which it (i.e., the *ṣalmu*) is assigned." Compare this rendering with Linssen 2004: 268, "the fate ... has been decreed" and Sachs 1969: 336, who takes it as a plural imperative apparently, despite the final vowel: "(O great gods,) fix the destiny of this image."

54 The ritual instructions are exclusively in Akkadian. They start on the first line of the reverse of both KAR 50 and BM 33343, though this is not the case in K.6060 + K.10820.

Shamash, and Asalluhi. You scatter dates *and sasku*-meal. You set out a *mersu*-cake *made with* honey *and* ghee. You sacrifice three white sheep.[55] You present the shoulder, fatty tissue, and roasted meat as an offering. You set up a censer of juniper. You strew *maṣḫatu*-flour. You draw the curtains.[56] You repeatedly pour out small heaps of flour. You stand the choice bull in a secluded place and into its right (var. and its left) ear you whisper *the above incantation* three times. You slaughter the choice bull *and* take the hide and tendon for that work.

A catchline follows in MS A: "Incantation: The shining quay (or, quay of lapis lazuli) …"

The Ritual Context of the Incantation

The ritual instructions in KAR 50 and duplicates are short compared to other versions of the instructions and only include the initial ritual setting for the recitation of the incantation, that is, a meal for Ea, Shamash, and Asalluḫi and the seclusion of the bull before the bull's ritual slaughter. Nothing about the preparation of the bull's hide and tendon, the covering of the drum, or the drum's ritual initiation is included in the text. Presumably, as MS A's catchline suggests, another tablet (or more) continued the ritual proceedings. Besides their brevity, the instructions are also unique in that they prescribe several distinctive actions or configurations of actions in comparison to the other witnesses. For example, the prescribed time ("in the morning before or whenever the sun rises") for the ritual performance is not explicitly stated in the other instructions; the prescription to set up altars (only) to Ea, Shamash, and Asalluḫi and to offer three sheep to them is not found in the other witnesses;[57] and none of the other sources requires the scattering of dates and *sasku*-meal and the

Because lineation is not shared in the sources, I have not indicated it in the translation.

55 One sheep for each of the three gods mentioned earlier.

56 Or, as Maul has argued, the phrase means something more general, "you draw a line or boundary," which could have been done with flour (Maul 1994: 55–56). In any case, the bull's position is marked off in a special way from its broader surroundings, as is clear in a subsequent instruction: *gumāḫu ana ašri parsi tušzaz-ma*, "you stand the choice bull in a secluded place."

57 Compare TCL VI 44, rev. i 16–19 for the closest prescription. All of the sources agree about which parts of meat are to be offered when an offering is made (see TCL VI 44, obv. i 31–32, ii 3–4, 5–6, 33–35; rev. i 18–19; BaMB 2, no. 5, obv. 13–14; K.4806+, rev. i 27–28).

setting out of a *mersu*-cake made with honey and ghee.[58] When viewed in a more general manner, however, the main ritual actions prescribed in KAR 50 and duplicates are found in the other ritual instructions: the setting up of various offerings,[59] the seclusion of the bull,[60] whispering into its ear(s) the prescribed incantation(s),[61] and the bull's ritual slaughter.[62] To understand the more substantial distinctiveness of KAR 50 and duplicates we must turn to its uniquely prescribed incantation.

The Origin and Use of the Bull

The incantation in KAR 50 and duplicates establishes the origin of the bull in its very first line via a simple predication (*attā ... binûti ilī rabûti*) with a vocative (*gumāḫu*) in apposition to the leading pronoun. A parallel to this syntax occurs in, for example, a line from *The Tamarisk and the Date Palm*: *attā bīnu iṣṣē lā ḫišeḫte*, "you, Tamarisk, *are one of* the useless/undesirable trees."[63] This understanding of the opening line stands in contrast to its typical translation, which makes the bull the predicate and puts the following phrase in

58 Instead, other instructions prescribe the setting out or scattering of a *mersu*-cake made of honey, ghee, dates, and *susku*-meal. See TCL VI 44, obv. i 18; BaMB 2, no. 5, obv. 12–13; and K.4806+, rev. i 26–27.

59 These occur extensively in all of the sources.

60 The "secluded place" in KAR 50 and duplicates may very well be a circumlocution for the *Bīt mummi* prescribed in the other sources (so Linssen 2004: 96–97). See TCL VI 44, obv. i 7; BaMB 2, no. 5, obv. 7; BaMB 2, no. 7, rev. 7′; K.4806+, rev. ii 25; and the rubric of the incantation preserved in BaMB 2, no. 6, rev. 13. Note also that two other sources mention the drawing of curtains (TCL VI 44, obv. ii 8 and K.4806+, rev. ii 1–2) while three mention the drawing back(?) (*tunāḫ*) of curtains (KAR 60, rev. 4; BaMB 2, no. 5, rev. 9; and K.4806+, rev. ii 24). The fact that some sources mention the undoing of an act without prescribing its initial performance ought to indicate clearly that our ritual instructions do not always prescribe everything that the officiants were supposed to do.

61 See TCL VI 44, obv. ii 9–12; KAR 60, obv. 10′–13′; BaMB 2, no. 5, rev. 2–4; and K.4806+, obv. i 17–18, 26. See Gabbay 2015 for whispering incantations. Gabbay argues that whispering an incantation marks a significant transition in a ritual—often moving something from mundane to divine status—and highlights the secrecy and intimacy attached to the incantation as well as the mediatory role of the one hearing it.

62 See TCL VI 44, obv. ii 16; KAR 60, obv. 15′; and BaMB 2, no. 5, rev. 5–6. K.4806+ does not mention the slaughtering of the bull. Presumably the instructions on a subsequent tablet in its series would have.

63 See Lambert 1960 (1996): 162–163: 22.

apposition: "you *are* a choice bull, the creation/product of the great gods."[64] This is a possible translation, of course, but this translation undermines the purpose of the line, in my view, by backgrounding *binûti ilī rabûti* in the second half of the line when, in fact, this phrase completes the line's main point and establishes an important theme of the incantation, i.e., associating the bull with the divine realm. A similar opening line in an incantation addressing a clod of dirt is instructive in this regard: *kirbān binût Apsî attā-ma*, "O clod, you *are* the creation/product of the Apsu."[65] Here the vocative and predication are unambiguous. And the association of the clod with the Apsu, Ea's divine realm, is clearly the most important point of the line. Given the syntactic parallel in the line from *The Tamarisk and the Date Palm* cited above and this conceptual parallel, *gumāḫu* in the first line of the present incantation is best taken as a vocative that specifies the incantation's subject (the referent of *attā*): it is a kind of first-rate bull that was often used in sacrifices.[66] *Binût ilī rabûti* completes the predication.

On this view, the opening line makes an extraordinary claim: it asserts the divine origin of the bull. Although *binûtu* is most often used to describe suprahuman beings and kings, the term is also used of animals, plants, and inanimate objects.[67] Labelling the bull a *binûtu*, therefore, does not necessarily give it an inherent distinction. That work, rather, is done by the phrase that follows. When *binûtu* is in construct with another term, the *nomens rectum* identifies the source whence the creation/product came. Sometimes the source is a deity (e.g., *binût Anim*, "creation/product of Anu"),[68] a region of the world broadly conceived (e.g., *binût šamê, binût tâmti, binût Apsî, binût Ḫamānim*, "creation/product of the heavens," "creation/product of the deep," "creation/product of the Apsu," and "creation/product of the Amanus"), or a social institution (e.g., *binût bīt redûti*, "creation/product of the house of succession"). When a god (or gods) is named in the second term, a special relationship to the named deity—a kind of divine election—is claimed for the one being described. When King Ashurbanipal, for example, claims about himself, *anāku binût Aššur u Ninlil*, "I *am* the creation of Ashur and Ninlil," he asserts his special relationship to

64 See, e.g., Linssen 2004: 267; Sachs 1969: 336; and Thureau-Dangin 1975 (1921): 23.

65 This incantation is actually embedded within another incantation to Shamash, which is part of a ritual for the removal of the evil of a bad dream. See Butler 1998: 275 and 298.

66 See CAD G, 132–133.

67 See CAD B, 243–244, where all of the following examples are cited.

68 A temple sometimes stands metonymically for the deity (e.g., *binût Ešarra*).

these deities.[69] The same understanding may be applied to the more general assertion *binût ilī rabûti* in the present incantation. In this case, the phrase asserts a special relationship between the bull brought forward for the ritual and the divine realm, that is, all the gods of the pantheon. In fact, given its position as the opening line of the incantation, it seems most fitting to understand the simple predication *attā ... binût ilī rabûti* as performatively instantiating this relationship (i.e., establishing it in the act of asserting it), which is then elaborated upon in the following lines. If this is the case, then the incantation is creating a cosmological connection between the bull present during the ritual and the gods, a point that the next line affirms and the following ones develop. The line effects this performative instantiation through the successful use of what Stanley Tambiah calls a persuasive analogy in magic and ritual.[70]

As I have explained elsewhere, "[f]or Tambiah, the key to the effectiveness of magic is its ability to create persuasive analogies. That is, magic establishes an analogical relationship between paired items that takes into account both similarities and differences in order to transfer, for example, a desired attribute from one item having such an attribute to the other lacking it or to obtain an anticipated result for one item that is achieved routinely by the other."[71] The transfer is effected via simile, metaphor, metonymy, and other culturally conventional associations. Tambiah emphasizes that such transfers may be accomplished by means of both word (i.e., incantation) and deed (i.e., ritual acts), though we are only interested here in the former.[72] Tambiah also asserts that "to test whether [the analogies] are true or false in a referential or assertive sense or whether the act has effected a result in terms of the logic of 'causation' as this is understood in science" is inappropriate.[73] Rather, "[t]he ... objectives in (magical) ritual are 'persuasion,' 'conceptualization,' 'expansion of meaning,' and the like, and the criteria of adequacy are better conveyed by notions such as

69 See similarly the phrase *annanna apil annanna rēʾû binût qātīkunu*, "so-and-so, son of so-and-so [the king's name and filiation], the shepherd, the creation/product of your hands," in the new Akkadian *šuʾila*-prayer to the Three Paths of Heaven, which was recited in the royal *Bīt salāʾ mê* ritual (Lenzi 2013: 6, rev. ii′ 8′). The gods' hands stand metonymically for the gods themselves and highlight the divine creative act.

70 See Tambiah 1985: 60–86, a chapter entitled "Form and Meaning of Magical Acts" and Lenzi 2010: 148–150, where I condense the relevant ideas for the purposes of understanding an Akkadian ritual denial in various incantations in terms of a persuasive analogy rooted in Mesopotamian legal pronouncements.

71 Lenzi 2010: 148–149 (see also n.43) with reference to Tambiah 1985: 71–72.

72 Tambiah 1985: 80.

73 Tambiah 1985: 81.

'validity,' 'correctness,' 'legitimacy,' and 'felicity' of the ceremony performed."[74] In other words, persuasive analogies are successfully deployed in rituals because they are accepted as effective by all those involved.

The opening line's persuasive analogy is quite simple, comprising a metaphorical predication intended to provide the bull with a pedigree that is not derivable (or verifiable, for that matter) from the individual bull's personal history. To state it more abstractly, the vertical relation in the first column of the table below is perceived as a problem to be overcome ritually by means of the persuasive analogy in the second column, whose vertical relation offers the solution.

present case	**desired outcome**
bull	creation of the gods
~~ordinary birth~~ →	divine origin

Such persuasive analogies are the stock in trade of incantations. The present example is the first of several in KAR 50 and duplicates.

A striking feature about the incantation after its opening line is that it no longer refers to its addressee (the bull) *as* a bull. Note how the next line of the text uses second person verbs in which the "you" is unspecified. The effect is to background the bull as the addressee. Moreover, since the recipients of human speech are typically humans or anthropomorphized deities, this backgrounding may be tacit encouragement to take an anthropomorphic view of the bull (that is, it hears human speech) in preparation for its incorporation into a god, the divine drum.

The bull, the incantation asserts in line 3/4, has been (s)elected for the present ritual from the beginning of its existence:

> 3/4. You were created and are being brought forward for the work of the great gods.

74 Tambiah 1985: 77. For the role of analogy in Mesopotamian understandings of *ṣalmu*, a key term in the discussion that follows, see Glassner 2017, where he states: "Le mode de pensée dominant en Mésopotamie est l'analogisme. La resemblance n'y est pas le lien primordial qui unit le reel à sa representation, elle n'est que l'une des figures de la constellation à laquelle l'analogie donne naissance … Partant, la relation entre *ṣalmu* et son referent ne saurait se réduire à un rapport univoque ou unidimensionnel" (218).

It is not unusual for *gumāḫū* used in rituals to be described with adjectives denoting their desirability (e.g., *marû*, "fat," *paglu*, "massive," *pūṣam lā īšu*, "spotless," *rēštû*, "first-rate," *šuklulu*, "perfect i.e., ungelded").[75] But the bull used in this ritual, if other versions of the ritual may be used illustratively, had to have quite special physical requirements. For example, the fullest version of the ritual, TCL VI 44, states:[76]

1. *e-nu-ma* LI.LI.ÌZ ZABAR *a-na a-*[*ra-mi* [77] IGI-*k*]*a*
2. GU$_4$ *šuk-lu-lu* GE$_6$ *šá* SI.⸢MEŠ⸣ *ù* gišU[MBIN].⸢MEŠ⸣ *šal-mu* TA SAG.DU-*šú*
3. EN *ap-pi* KUN-*šu* lúUM.ME.A *m*[*u-d*]*u-ú it-ta-nap-la-as-su-ma*
4. *šum-ma zu-mur-šú* GIM ESIR GE$_6$ *a-na par-ṣi u* KI.DU.DU-*de-e*
5. *il-leq-qu*[78] *šum-ma* 7 SÍK BABBAR-*tum* GIM MUL *ta-kip ina* gišNÍG.GIDRU
6. *ma-ḫi-iṣ ina qin-na-zi la-pit ana par-ṣi u ki-du-de-e ul il-leq-qé*

> When you intend to cover a bronze *lilissu*-drum, a knowledgeable scholar will thoroughly inspect an ungelded black bull—whose horns and hooves are whole[79]—from its head to the tip of its tail. If its body *is* black like pitch, it may be taken for the rites and ordinances. If it is speckled with *even*[80] seven white hairs like a star, or has been beaten with a staff, or lashed with a whip, it may not be taken for the rites and ordinances.

If the ritual practitioners using KAR 50 and duplicates abided by this or some similar demand, the bull would have needed to come from a closely guarded herd. Since temples kept herds, this would have not been a difficulty.[81] But this reading may be too literalistic. This one bull as a recently born contemporary

---

75 See CAD G, 132.

76 The transliteration is based on Linssen 2004: 252 in consultation with the published copy in TCL VI 44. The translation is my own, though it differs only insignificantly from Linssen's (2004: 255). A nearly identical prescription occurs in the opening six lines of the shorter version of the ritual, BaMB 2, no 5. See Linssen 2004: 270–271.

77 CAD A/2, 228 adds *ana* here.

78 Mistake for *illeqqe*.

79 Compare the sacrificial animals for the daily offerings in Uruk, "whose horns and hooves are perfect" (*šá* SI *u* UMBIN *šuk-lu-lu*; Linssen 2004: 175, rev. 29, 32 for the text and 179 for the translation).

80 With CAD P, 332.

81 See, e.g., Bongenaar 1997: 416–422 for herdsmen of the Ebabbar temple in Sippar and Rochberg 2010: 351 for the broader point of temple cattle pens (*tarbaṣu*). Note also Roch-

member of its species may not be (exclusively) in view so much as the fact that this bull represents the *kind* of bull the gods created for this very work. In other words, in keeping with the idea of the first line, we are dealing with the bull's place within the cosmological order, which, as lines 7/8–9/10 indicate, was ordained for this purpose at the foundation of the world. This second line, therefore, forms a kind of chronological transition between the contemporary setting and primordial time introduced clearly with line 7/8.

Line 3/4 also clearly indicates that the opening several lines of the incantation form what scholars of ancient magic call a *historiola*, a short mythological narration in an incantation or spell that in the act of speaking it (performatively) evokes the power of the mythical past and applies it to the contemporary moment through the creation of an analogy;[82] in this case, the analogy (a kind of chronological persuasive analogy, in Tambiah's terms) is "just as the gods created and fated that mythologically paradigmatic bull for this task, so this present bull is likewise created and fated."[83] As described below, this *historiola*

---

berg's comments on the term's appropriateness as a metaphor applied to the celestial description of the gods.

82 For the theory and practice of *historiola*, see Frankfurter 2001.

83 Other incantations used in various ritual texts related to covering the drum also present a *historiola* for the bull. Most notably is the bilingual BaMB 2, no. 6, obv. 21/22–27/28 (prescribed by incipit in TCL VI 44, obv. ii 15–16), which states (following Mayer 1978: 433 provisionally; see Shibata, f.c., no. 22):

21. gu$_4$-gal gu$_4$-maḫ ⸢á⸣-úr-kur$_4$-kur$_4$-ra s[u]-bi mu-un-šu-du$_7$
22. MIN MIN *šá meš-re-⸢e⸣-ti pu-ug-gu-lu zu-mur-šú šuk-lu-lu*
23. ^{d}nanna amaš-kù-⸢ga-ta⸣ mí-zi-dè-eš du$_{11}$-ga
24. *<ina su>-pu-ri* ^{d}30 *ki-niš i'-al-du*
25. ga-zi in-na$_8$-na$_8$ me-te-nam-dingir-[r]a tùm-ma
26. *ši-iz-bi ki-nim i-⸢ni$_7$-qu⸣ šu-lu-uk ana si-mat* DINGIR-*ti*
27. ^{d}gašan-amaš-kù-ga bùluĝ-ĝá ki-kù-ga-ta mu-un-dím-ma
28. *<tar>$^{!}$-bit* ^{d}MIN *šá ina áš-ri el-lim ib-ba-nu-ú*

Following the Akkadian, this may be rendered:
21/22. A great bull, a choice bull, whose limbs are massive *and* whose body is perfect,
23/24. Who was born legitimately in the fold of Sin,
25/26. Who sucked true milk, suitable for divinity,
27/28. Rearling of Gashan-Amash-Kuga, who was created in a pure place …

Note also the Akkadian incantation prescribed by incipit in TCL VI 44, obv. ii 10–12; KAR 60, obv. 12′–13′; and BaMB 2, no. 5, rev. 4, whose full text is given in K.4806+, obv. i 19–25 (see Linssen 2004: 275 for the edition). The first few lines form the *historiola* and run as follows:

extends five lines (1/2–9/10) and thus accounts for just over half of the incantation's content.

Before reaching back to the original time of creation the incantation's *historiola* looks first to the heavens in line 5/6 to define further the mythological and cosmological place of the bull. The text states:

> 5/6. In the heights of the heavens your *ṣalmu* is treated as a rite of Anu-ship.

As the bull is in the process of becoming part of a divine image, the *ṣalmu* in view in this line may be the final product of the entire ritual, the manufactured *lilissu*-drum, which as a divine image, this line asserts, has its origins in the heavens. "Image," it should be noted, is the typical meaning of *ṣalmu* throughout the *Mīs pî* ritual texts and incantations, referring to the divine image that that ritual prepares. In light of this (and the parallels with the *Mīs pî* ritual noted below), "image" seems to be the most obvious meaning here in our incantation. Without discounting or contradicting this idea, we might also entertain the idea that the bull is linked here by assertion to a celestial body, the constellation (*ṣalmu*) Taurus, as TCL VI 47 explains.[84] Like the bull, the constellations were a product of special creation at the origins of the world and had cosmological significance (see, e.g., *Enūma eliš* V 1–2 and *Enūma Anu Enlil* 22, the conclusion of MS E).[85] The connection between the two would have been made on the basis of the fact that both the bull and the constellation are bovine, obviously; but we might also consider the possibility that the connection was made via the word *ṣalmu*, which alongside the meanings "image" and "constellation" also means "black," the prescribed color for the bull in the ritual instructions elsewhere.[86] Note also that these ritual instructions go on to state after the color prescription that the bull is not to have even "seven white hairs

---

19. GU$_4$ *i-lit-ti An-zi-i at-ta-ma*
20. *a-na par-ṣi ki-du-de-e na-šu-ka*
21. *a-na da-riš* ᵈNIN.GIŠ.ZI.DA *ib-ri-ka*

19. O Bull, you *are* the offspring of Anzu,
20. For rites *and* rituals they have lifted you up,
21. Ningizzida *is* your companion for evermore.

84 The reverse of explanatory text TCL VI 47 presents a diagram of the ritual set-up, in which the bull is labeled "the constellation Taurus" (^{mul}gu$_4$-an-na). See Livingstone 1986 (2007): 194 and 195, line 4.

85 For the latter, see Rochberg-Halton 1988: 270–271.

86 See TCL VI 44, obv. i 4 and BaMB 2, no. 5, obv. 4.

like a star" (7 *šārātu peṣâtu kīma kakkabi*),[87] which may also hint at the bull's celestial connection.[88] It would seem then that the bull's color is standing in metonymically for the entire animal: "(your) black one" = the bull.[89] And thus there are three possible referents for *ṣalmu* in our line: the manufactured image, the constellation Taurus, and the black bull itself, all of which are placed ritually in the heavens by line 5/6. The implicit persuasive analogy may be represented as follows:

present case	**present case**	**desired outcome**
black (*ṣalmu*) "star-less" bull	*ṣalmu* (manufactured image)	*ṣalmu* (constellation Taurus)
material	product	
~~terrestrial, ∴~~	~~terrestrial, ∴~~ →	celestial, ∴
~~mundane~~	~~mundane~~ →	divine
~~star-less~~	→	Pleiades
~~undefined *šipru* (line 3/4)~~	~~no position~~ →	supreme position (rite of Anu-ship)

The bull is to the drum as material is to product. Both are terrestrial and therefore mundane in nature and need the qualities that the association with celestial Taurus can provide, namely, an association with the divine realm. Notably this celestial connection helps us understand the prescriptions related to the bull's color in some of the ritual instructions mentioned above. The bull's lack of seven white hairs and the simile "like a star" (singular) might be a reference to the Pleiades, which consists of a cluster of seven stars and, not coincidentally, appears in the constellation Taurus. How do we understand a pitch black bull and its celestial counterpart with seven stars? I suggest that this is part of the symbolic transfer of the "star-less" bull on earth to the heavens, where it obtains both its celestial and divine bona fides (its star and place among the gods).

87 See TCL VI 44, obv. i 5 and BaMB 2, no. 5, obv. 5.

88 I thank Eduardo Escobar for pointing this out.

89 To object to this metonymic connection on the basis of the presence of the pronominal suffix is to take the hermeneutical process at play in such linguistic maneuvers too literally. Commentaries and explanatory texts often simply ignore what does not fit the interpretative goal.

The *ṣalmu*, the incantation asserts here in line 5/6, occupies not only a heavenly position but also a superlative rank in cosmic governance; that is, it is treated (note the timelessness of *šakin*) as a rite of Anu-ship (*paraṣ Anūti*), in which "Anu-ship" designates the highest order or rank.[90] In making this celestial connection and assertion of rank for the *ṣalmu*, the incantation makes the *ṣalmu* (bull and image) worthy of divinity.[91] Moreover, as a result of this line's assertions, the incantation defines the bull's relationship to the gods in a closer though also more ambiguous manner than previous lines had suggested.

Before moving on to line 7/8, it is worth reflecting on this claim about the celestial / divine origin of the *ṣalmu*, since the movement from mundane to divine lies at the heart of the problem that the covering ritual must resolve. The bull that was taken for the ritual does not remain a bull; rather, it (or parts of it: hide and tendon) will be incorporated into an object that will be in some sense a divine entity, the *lilissu*-drum. How does one make something divine from mundane materials? When the covering ritual is framed in this way, one sees that it confronts the same issue as does the *Mīs pî* ritual, in which a divine image, made from mundane materials, was ritually prepared to embody in a manufactured object a deity. The *Mīs pî* ritual series was performed on the manufactured image to prepare it to enter the temple as the materially present divinity. Interestingly, in some versions of the covering ritual the *mīs pî* ritual action (not the entire ritual complex) is also performed on the bull just before it is slaughtered and again to the divine drum before its use in the temple.[92] A couple of other features in the covering ritual show clear parallels with the manufacture and preparation of divine images. For example, the covering ritual was performed, according to some versions of the ritual, in the *Bīt mummi*,[93] the "temple workshop," which was the same place the divine images

90 Possession of the *paraṣ Anūti* implies significant power; see, e.g., the description of Ishtar in two inscriptions of Esarhaddon, cited in Beaulieu 2003: 118, and the description of Qingu in *Enūma eliš* IV 82 (compare I 159).

91 Note that BaMB 2, no. 6 actually mentions me-te-nam-dingir-ra tùm/túm-ma / *simat ilūti*, "worthy or/suitable for divinity," with regard to the bull twice (obv. 25/26 and rev. 6/7).

92 For the performance of *mīs pî* on the bull, see BaMB, no. 5, rev. 1, KAR 60, obv. 7′, and TCL VI 44, obv. ii 8; the latter text also prescribes the *mīs pî* ritual action on the divine drum in rev. i 23. K.4806+, rev. i 23 only mentions the performance of the *mīs pî* on the drum.

93 Our ritual version does not include this detail, though as mentioned above "the secluded place" may imply it. The other text from Ashur (KAR 60) may have originally included *Bīt mummi*. In its current condition, however, the text begins after such a detail would have occurred in the instructions. See, therefore, the ritual instructions in the manuscripts from Uruk (TCL VI 44, obv. 7; BaMB 2, no. 5, obv. 7; and BaMB 2, no. 7, rev. 7′) and Nineveh (K.4806+, rev. ii 25). Note also the rubric of the incantation preserved in BaMB 2,

were manufactured and repaired.[94] And shortly after the slaughter of the bull, the ritual instructions in one version of the covering ritual (KAR 60, rev. 2–4) prescribe the lamenter to deny his part in the ritual actions. The text states:

> 2. *ù* EGIR-*šú ki-a-am* 3-*šú i-qab-bi*
> 3. [*e*]*p-še-e-ti an-na-a-ti* DÌM.ME.ER KÌLIB.BA *i-tep-pu-uš*
> 4. *a-na-ku ul e-pu-uš*
>
> After that, he says three times the following: 'The totality of the gods have done these rituals; I did not do (them).'[95]

The same kind of denial is prescribed for the craftsmen working on the divine image in the Ninevite and Babylonian Ritual tablets for the *Mīs pî* ritual.[96] In all of these cases, participants disavow their role in the material creation of objects that had become representatives or symbols or receptacles[97] for the divine presence. Recognizing these parallels between the materials related to the covering ritual and the *Mīs pî* ritual justifies considering how the latter might inform our understanding of KAR 50 and duplicates.[98] In this regard, it is significant to note that many of the incantations used in the *Mīs pî* ritual, like our incantation, contain a *historiola* in their opening lines. Although these are most prevalent among incantations addressed to preparatory/purificatory items or items to be used for the divinity (e.g., throne and crown), some occur in incantations treating the image itself.[99]

---

no. 6 (rev. 13): ŠU.ÍL.LÁ *šá* GU$_4$.MAḪ *šá* É *mu-um-mu*, "*This is* the great bull's hand-raising prayer for the ritual workshop."

94 See Walker and Dick 2001: 52, n.34.

95 BaMB 2, no. 5 may also have preserved this instruction. See footnote12 above.

96 Walker and Dick 2001: 66: 179–186 and 80: 49–52. Note also the *šu'ila* in Incantation Tablet 3, lines 84–86.

97 How to capture the way(s) the Mesopotamians conceived the relationship between the image (statue) and the deity is an open question to my mind. See the conclusion for more on this.

98 See also Gabbay 2014: 120–139, who has emphasized the parallel with reference to the broader textual materials related to the covering ritual.

99 A survey of the material edited in Walker and Dick 2001 finds *historiolae* in the following lines: Incantation Tablet 1/2, §A, 1–2 ("Asarre/Marduk saw your pure clay in the Apsu"), 9–19, 21–29, 34–46, 49–58 (treating reed); §B, 1–9 (tamarisk), 15–21 (soapwort), 27–33 (reed), 39–45 (alkali), 78–82 (syrup), 90–96 (ghee), 99 (sesame oil); §C, 1–9 (incense), 15–34 (Girra/fire), 38–47 (water basin), 51–93 (water basin); KAR 229: 3–8 (water?); STT 199 obv. 13–31 (tree), rev. 22′–39′, 43′–45′ (water); Incantation Tablet 3, 50–91, 100–113; Incantation Tablet 4, §A, 1–8, 23–33 (divine image); §C, 1–6 (divine dwelling), 11–14 (throne);

In lines 7/8–9/10 the incantation's *historiola* explicitly moves back to primordial times to assert that the bull (or rather, parts of it) has (have) a place among the gods in the divine assembly. It is worth noting again that although the bull is the addressee of these lines, it is not named. The line simply uses nouns with pronominal suffixes to refer to the relevant parts of the bull, *mašakka*, "your hide," and *šer'ānka*, "your tendon," just as line 5/6 used *ṣalamka*, "your image/constellation." Unlike line 5/6, however, the bull-as-raw-material is solely and unambiguously in view. The lines read:

> 7.-8. When Anu, Enlil (var. and), Ea, and Belet-ili [assigned] the fate of the great gods,
> 9.-10. Your hide *and* your tendon were assigned to the secret *council* of the great gods.

Casting a retrospective glance back to the moment that Anu, Enlil, Ea, and Belet-ili[100] established the various roles of the other gods—a time of primordial decisions that determined the nature of things, the incantation asserts that the bull's hide and tendon, essential to the production of the *lilissu*-drum, were assigned a place in the divine assembly, designated here simply as *pirištu*, "secret council."[101] The parts named in this line are also those focused on in the ritual instructions,[102] which instruct the bull's tanned hide to be used to cover the drum and its left tendon to be used to secure it. The hide and the tendon were therefore the only parts of the bull that entered ritual service with the drum and the only parts needing to be singled out in the incantation as having access to the secret council of the gods.[103] This access places the bull, or rather, its essential parts, in the divine domain, just as line 5/6 did for its *ṣalmu*. It is

---

Incantation Tablet 5, §A, 6–13 (crown); §B, 8–21(?) (throne), 1–61 (water basin/incantation priest).

100 Note that these are not the same deities invited to the ritual meal in the instructions: Ea, Shamash, and Asalluḫi.

101 On translating *pirištu* as "secret council," see Lenzi 2008: 27–45 for the OB evidence, mostly from Mari, and 50–52, with reference to the divine assembly in mythological texts. The discussion of Anzu III 29–31 in light of I 98–99 on page 50 is especially relevant for understanding the juxtaposition of *pirišti ilī rabûti* in line 9/10 with *itti ilī aḫḫēšu* in line 13/14.

102 See the ritual instructions in KAR 50, rev. 10 (with broken parallels in K.6060+, rev. 6 and BM 33343, rev. 8); also note the ritual instructions in the different versions of this ritual: KAR 60, rev. 8–9; TCL VI 44, ii obv. 25–30; BaMB 2, no. 5, rev. 11–12.

103 As Gabbay 2014: 127–128 observes, some versions of the ritual instructions (TCL VI 44 obv. ii 19, 32 and BaMB 2, no. 5, rev. 15–16) prescribe that the rest of the bull be buried (and thus sent to the netherworld) and thereby made unavailable for further human use. The

noteworthy that the celestial aspect of the divine realm, though probably implied in the terminology used, is not explicit.

Line 11/12 returns the time frame to the contemporary moment to apply the mythology that was recounted in the *historiola* to the actual bull involved in the ritual performance. In keeping with the bull's fading presence in the incantation, the line seems to be addressed not to the bull itself but implicitly to the primordial fate decreed earlier for its hide and tendon.

> 11.-12. Be firmly established, *O fate*, in the secret *council* of that god for days evermore.

As the chart shows below, lines 11/12 shares the same basic elements as 7/8–9/10, but the elements are now modulated (and presented in a slightly different order) so that what was decreed in general, in the past, and among the great gods as a whole is now invoked for the particular tendon and hide used in the ritual, in the present (with implications that extend into the future), and for the specific god of concern in the incantation.

Elements in Lines 7/8–9/10 (as they appear):	**Elements in Line 11/12 (as they appear):**
A. Time Frame (past): *īnu Anu Enlil Ea u Belet-ilī šimit ilī rabûti išīmū*, "when Anu, Enlil, Ea, and Belet-ili assigned the fates of the great gods"	B.′ Subject of the Verb (implied): the fate of the hide and tendon as described in line 9/10[a]
B. Subject of the Verb: *mašakka šerʾānka*, "your hide *and* your tendon"	C.′ Location of the (Specific) Fate: *ina pirišti* ili šâšu, "in the secret *council* of that god"
C. Location of the (General) Fate: *ana pirišti* ilī rabûti, "to the secret *council* of the great gods"	A.′ Time Frame (present and future): *ana ūmī darûti*, "for evermore" (i.e., from now on)
D. Verb (past, completed action): *iššamū*, "were assigned" (as a fate)	D.′ Verb (present, commanded action): *kūnī*, "be firmly established"

a As the imperative is feminine singular, *šīmtu* must be on the receiving end of the command. There are no other options.

---

explanatory text TCL VI 47 seems to support the idea that the bull enters the netherworld, if Gabbay's interpretation is accepted (2014: 137–138).

The incantation commands the permanent establishment of the general fate that was decreed for this kind of bull's hide and tendon in line 9/10 to this particular bull standing in the present execution of the ritual via an implied persuasive analogy:

present case	**desired outcome**
present hide and tendon	primordial hide and tendon
no assigned fate →	assigned fate
in the secret council of "that god" (divine drum)	in the secret council of the great gods

As Frankfurter has pointed out, when a *historiola* is utilized in ritual speech that includes directive utterances (imperatives and jussives), "the *historiola* … works in these cases as a guarantee or rationale, an explicit precedent, for the directive utterance, the command, which is the central speech-act."[104] Line 11/12 is therefore the turning point in the incantation.

The bull's hide and tendon become in a ritual sense fated to become (part of) "that god," which must refer to the final outcome of the ritual process, the divine *lilissu*-drum, into whose image the bull's hide and tendon are being incorporated.[105] The connection of "that god" here to the divine *lilissu*-drum is supported by the fact that this is the only sensible referent for the phrase as it is used later in line 15/16 (see below). It is worth noting that dingir(-bi) / *ilu* (*šū*), "(that) god," is also used in the *Mīs pî* ritual tablets and incantations to refer to the divine image during its production, even *before* it has been introduced into the shrine.[106]

Having applied the *historiola* in line 11/12 to the raw material from the bull, the incantation now pronounces in line 13/14 the successful result of its application with respect to the manufactured product, the image.

> 13.-14. *As for* that *ṣalmu*, its assigned fate *is* with the gods, his brothers.

104 Frankfurter 2001: 469.

105 See likewise Gabbay 2015: 213.

106 See, e.g., lines 55–60 in the Ninevite Ritual Tablet (Walker and Dick 2001: 57, with n.67). Also, note Incantation Tablet 3, §B, where "this statue" (line 35) is also called a god (line 38; Walker and Dick 2001: 134, 149) and the opening line of a bilingual *šu'ila* in §B, line 49, én u$_4$ dingir dím-ma alam sikil-la šu du$_7$-[a] / *īnu ilu ibbanû ṣalmu ellu uštaklilu*, "when the god was fashioned, the pure statue completed" (Walker and Dick 2001: 136, citing their translation here from p. 149).

Conspicuously, the *ṣalmu* mentioned in this line is distanced from the bull—which is hearing the incantation whispered in its ear, it should be remembered—through the use of the far demonstrative adjective *šū*, "that." This contrasts sharply with the earlier use of the term in line 5/6, where *ṣalmu* is closely associated with the bull: *ṣalamka*, "your *ṣalmu*." It seems then that just when the final outcome of the ritual, the god, comes into view explicitly (line 11/12), grammatical reference to the bull disappears completely from the incantation, replaced by "that image" (line 13/14) and eventually "that god" (line 15/16). If the incantation is about incorporating the bull into an image that will host or be in some sense a god, this is precisely what we should expect.

Further defining the incantation's unfolding concept of fate, the incantation claims that "that *ṣalmu*" is fated to stand among the gods, its brothers. As this same idea is asserted for the divine image in the *Mīs pî* ritual,[107] we are warranted to understand the *ṣalmu* here to be the manufactured drum. The essential parts of the bull (its hide and tendon) have successfully been incorporated into it (the raw material has become product) and the drum has successfully been deified, ready for temple service among the other gods. Alongside this, the use of *ṣalmu* here may have another purpose since the word harks back to line 5/6. Just as line 11/12 claims for the present time and to the specific bull the general, primordial fate decreed for its hide and tendon described in lines 7/8–9/10 of the *historiola*, so line 13/14, I suggest, specifies the general, celestial claim of the (ambiguous) *ṣalmu* described in line 5/6 of the *historiola* (namely, its role as an authoritative *parṣu*, "rite or office") for the specific *ṣalmu*, unambiguously called now "that" (manufactured) image, in the present iteration of the ritual. If this is the case, then it is significant that it does so in language that elides explicit celestial reference, though the idea of the fraternity of the gods might imply it. This elision of explicit celestial reference, I suggest, continues what was started in line 7/8 and is due to the fact that the incantation is moving the *ṣalmu* to its final resting place, its shrine, which is notably another locale for the fraternity of the gods but quite terrestrial. This movement is brought to completion with line 15/16, where "that god" heads the line in parallel with "that *ṣalmu*" in line 13/14, suggesting a very close connection between drum and deity:

107 See the Ninevite Ritual Tablet, lines 89, 165, and 168 [reconstructed] and Incantation Tablet 3, §C, lines 6 and 10 (see Walker and Dick 2001: 44, 59 and 146–147, 152–153). Note especially the context of Tablet 3, §C, line 10: 8. *ultu ūme annî š*[*īmatīka*] 9. *ana ilûti* [*limmanū-ma*] 10. [*i*]*tti ilī aḫḫēka* [*ta*]*ttamnu*, "From this day let your fate be counted as divinity; among your brother gods may you be counted" (see Walker and Dick 2001: 146, 153 for text and translation, respectively).

15.-16. *As for* that god, may his shrine be clean *and* pure.

Something like this line (i.e., a wish for purity) as well as the one that follows in 17/18, "may the evil tongue stand aside,"[108] are commonly found near the end of incantations used in the *Mīs pî* ritual (among others).[109] Thus we have another structural and thematic parallel in KAR 50 and duplicates with the *Mīs pî* incantations. Interestingly, in the attestations of these phrases in Tablet 1/2 of the *Mīs pî* ritual—the tablets that provide the wording of *Weihungstyp* incantations addressing preparatory/purificatory materials used in the ritual such as reed, tamarisk, soapwort, and incense—the Sumerian refers to the deity/image with dingir-ra, "for the god":

> dingir-ra an-gin$_7$ ḫé-en-kù-ga

> May it/he [i.e., the addressee of the incantation] be pure for the god like the heavens.[110]

When the *Mīs pî* ritual incantations begin to treat the divine image itself in Tablet 3, the same phrase occurs with a slight alteration so that it resonates strongly with line 15/16 in KAR 50 and duplicates:

---

108 An evil tongue may be associated with gossip, slander, a curse, or witchcraft (see CAD L, 209–215, s.v. *lišānu*). Although the line's precise meaning is unclear, it may express something along the lines of "the divine image is ritually purified and no one shall gainsay it."

109 See Walker and Dick 2001 at Incantation Tablet I/II, §A, 7, 20, 31–32, 47; §B, 10–13, 22–25, 34–37, 73–76, 86–88, 96–97, 101–102; §C, 10–13, 35–36, 48–49, x+2–x+6; KAR 229: 9; STT 199: 9–11; Tablet III, §B 1–4, 38–41, 94–96; Tablet IV, §B, 28ab–29ab; §C, 19ab–20ab; Tablet V, §B, 4–6, 38–40; §C, 7–9, 18; and Tablet VI/VIII, 58–66.

110 See specifically Walker and Dick 2001 at Tablet 1/2, §B, 10, 22, 34, 73 and §C, 10, 35, 48, who translate the phrase differently, namely, "may the god become pure like the heaven" (e.g., p. 101, line 10). My translation is based on two facts: first, a parallel phrase in Tablet I/II, §C, x+2 (Walker and Dick 2001: 109) explicitly makes the incantation's addressee the subject of the verb in an Akkadian translation of the phrase: [a-gúb-ba : *e-gub*]*-bu-u ki-ma* AN-*ú* [*li-lil* : ḫé-en-kù-ga], "[May] the holy water basin [become pure] like heaven" (Walker and Dick 2001: 109, 112; the restorations are justified given the formulaic character of the line and the ones following); and second, in one version of the wish for purity (Tablet I/II, §B, 86–87), in which house deities are the agent of the Sumerian verbs used to express purity and the deity is the one addressed, the verbs show a second person dative in the verbal prefix chain: ḫé-ri-ib-kù-ga ḫé-ri-íb-sikil-⌜la⌝ ḫé-ri-íb-dadag-ga, "may he purify it for you, may he cleanse it for you, may he make it bright for you" (Walker and Dick 2001: 99, 102). These do not provide perfect parallels to the phrase cited above. But they suggest that the proposed translation is reasonable.

ilu šū kīma šamê līlil (line 38, Akk. only)
dingir-bi an-gin$_7$ ḫé-en-kù-ga (line 94, Sum. only)

May that god be clean like the heavens.

This close parallel with the *Mīs pî* incantations treating the divine image itself suggests that line 15/16 in KAR 50 and duplicates has the covering ritual's final product in view, too, the divine drum.[111]

The close connection and even interchange of *ilu* and *ṣalmu* occurs rather promiscuously in the *Mīs pî* ritual. As KAR 50 and duplicates share so many other features with this ritual series, the same interchange should not be very surprising in our text or related materials.[112] What *is* remarkable is the way the term *ṣalmu* in line 5/6 of KAR 50 and duplicates activates references to the bull (the raw material), the image (the manufactured representation), and an associated constellation (Taurus). This pregnant use of *ṣalmu* early in the incantation is not abandoned in line 13/14, where the term seems to be limited to the drum; rather, the multivalent use lies in the background so that the image now bears[113] the divinely-charged (5/6; and fated, 7/8–9/10) raw materials from the bull and is worthy, like Taurus in the heavens, of a divine office. That drum is ready to serve that god (13/14–15/16), or, as some might prefer to express it: that drum is ready *to be* that god.

Concluding Remarks

Although KAR 50 and duplicates offer useful perspectives and data for our continued attempts to understand the ontology of divinity in ancient Mesopotamia, including how gods related to their manufactured image and/or

111 Note the parallel in a *namburbi*, which also foregrounds the final ritual outcome: the person for whom the ritual was performed. The incantation is parallel to the *Mīs pî* incantation, except for the initial word, which is restored as it is because the text treats a victim of snake bite: [*amēlu*] *šū k*[*î šamê l*]*īlil kî erṣeti lī*[*bib k*]*î qereb šam*[*ê l*]*immir* [*l*]*išānu* [*le*]*muttum ana aḫâti lizzizu* (for *lizziz*), "*As for* that man, may he be clean like the heavens, may he be pure like the earth, may he shine like the midst of the heavens, may the evil tongue stand aside." See Maul 1994: 302–303 for the text and Walker and Dick 2001: 135, n. 21 on their MS N.

112 BaMB 2, no. 6 also uses *ilu šū* / dingir-bi near its conclusion (rev. 7/8). See Mayer 1976: 433 and Shibata, f.c., no. 22.

113 The incantation is treating the process as completed even though the ritual itself and thus the divine drum, as the instructions make clear, will not be completed at the recitation's conclusion.

associated celestial body, it would be presumptuous to think this one incantation can resolve the matter definitively.[114] Moreover, I am not yet prepared to stake out a position in this fascinating and confounding intellectual puzzle. However, the network of interconnections that the ritual language forges in KAR 50 and duplicates between deity, celestial body, manufactured image, and raw materials used to manufacture the image, summarized in paradigmatic fashion in the chart below, does inform us about a couple of related concerns. I offer these as modest concluding remarks.[115]

First, the bull is not a god or divine.[116] The bull's backgrounding throughout the incantation and the distancing "that image" in line 13/14 indicate this rather clearly to my mind. The bull itself is a special creation of the gods (*binût ilī rabûti*), which provides raw material and the potential for metaphorical associations through which it effects the constitution and nature of the god, the divine drum (e.g., the drum acquires a connection to the divine realm via its *ṣalmu*'s metaphorical association with Taurus and the sound that the divine drum makes is metaphorically understood as the bull's voice).[117] As raw material, it is comparable in some respects to the *mēsu* tree, whose wood, known as the *šīr ilī*, "flesh of the gods" (*Erra and Ishum* I 150),[118] was used to make anthropomorphic images of deities.

Second, symbolism and metaphor are useful ways to think about the constitution of deity in ancient Mesopotamian ritual texts. Our honoree has treated this issue in several publications in relation to various text genres, especially celestial omens,[119] and I think the findings here support her position that the relationships are usefully considered via symbol and metaphor. The

114 Admittedly, I had my hopes when I began to study this text carefully. For a survey of various options of the relationship between a deity and a divine image as well as the related issue of the relationship between a deity and celestial body, see Rochberg 2009: 84–89 for her fullest statement; see also Walker and Dick's brief discussion with literature (2001: 6–8) and Pongratz-Leisten 2011: 138–152.

115 I would like to thank at this point our honoree who through a friendly correspondence has challenged my thinking on the issue of the ontology of divinity in ancient Mesopotamia and unbeknownst to her has helped me formulate my thoughts that I offer to her in this article.

116 Contra Gabbay, who, if I have understood him correctly, equates the bull with the divine drum, which he argues on the basis of the ritual activities as prescribed in the instructions (e.g., TCL VI 44); see Gabbay 2014: 127 and 2015: 192.

117 For the latter point, see Gabbay 2014: 124, 127, though I do not agree that the bull must itself be divine to make such a contribution.

118 See Cagni 1969: 75 for the text.

119 See, e.g., Rochberg 1996, 2007, 2009, 2010, 2010a; and 2011.

manufactured image, the celestial body, and the god are all mentioned and related in various ways via performative assertions and conventional persuasive analogies in KAR 50 and duplicates. Based on what we can see in this text, our understanding of the three entities are enriched through explication of the symbolic and metaphorical associations. How this translates into or contributes to a Mesopotamian ontology of divinity remains an open question worthy of future exploration. But for now, in my present thinking, we do not surpass

<table>
<tr><th>relation to bull</th><th>substantive</th><th colspan="2">locale</th><th>company</th><th>object</th></tr>
<tr><td rowspan="4">Familiarizing</td><td>attā, gumāḫu,
“you, choice bull” (line 1/2)</td><td colspan="2">site of ritual, presumably Bīt mummi</td><td>great gods</td><td>binûtu,
“creation/ product”</td></tr>
<tr><td>“you” (line 3/4)</td><td colspan="2">terrestrial locale implied (created) / site of ritual (being brought)</td><td>great gods</td><td>šipru, “work”</td></tr>
<tr><td>ṣalamka,
“your image, constellation, black one” (line 5/6), multivalent</td><td colspan="3">ina šamê, “in the heavens”
clearly celestial
divine company implied</td><td>paraṣ Anūti,
“supreme divine office”</td></tr>
<tr><td>mašakka šer’ānka, “your hide (and) your tendon (lines 7/8–9/10, 11/12)</td><td rowspan="2">Ambiguous
(possibly celestial or terrestrial)</td><td>pirištu,
“secret council”</td><td>great gods, that god</td><td>šīmtu, general “fate(s)”

described, specific “fate” applied</td></tr>
<tr><td rowspan="2">Distancing</td><td>ṣalmu šū, “that image” (line 13/14), definite</td><td colspan="2">itti ilī aḫḫēšu,
“with the gods, his brothers”</td><td>šīmtu,
resulting “fate”</td></tr>
<tr><td>ilu šū, “that god” (line 15/16)</td><td colspan="3">ina ašertīšu, “in his shrine”
clearly terrestrial
divine company implied
locale = object</td><td></td></tr>
</table>

the evidence to see these relationships in the context of rituals like KAR 50 and duplicates as *at least* symbolic and metaphorical.

Excursus

The time explicitly given in the instructions of KAR 50, rev. 1–2 to perform the ritual is unique among the three textual witnesses for this incantation and ritual in that KAR 50 specifies the ritual ought to take place "in the morning before the sun rises" (*ina šērti lām Šamaš inappaḫu*).[120] The instructions also specify that the ritual ought to be done on a favorable day and month (KAR 50, rev. 1; likewise, K.6060+, obv. 9′ and BM 33343, rev. 1).[121] In first millennium Mesopotamia, Taurus would have been visible in the sky before sunrise from late May (rising just prior to sunrise) to late November (setting a couple of hours before sunrise), moving from the eastern to the western horizon as time elapsed. It may be then that as the present incantation was recited over the bull, Taurus was supposed to be visible in the sky.[122] This, however, would not have been required in the instructions for the parallel texts, K.6060+, obv. 10′ and BM 33343, rev. 1, which only prescribe the ritual to take place "when the sun rises" (*ēma Šamaš inappaḫu*).

Interestingly, astronomical diaries record the performance of the covering ritual on two dates, the 24th day of Araḫsamna in year 41 of the Seleucid era, that is, November 26, 271 BCE, and the [Xth] day of Ayyaru in year 85 of the Seleucid era, that is, May X, 226 BCE. If the latter date was late in the month, when Taurus would have begun to appear in the eastern sky just before sunrise, then Taurus would have been visible in the sky before sunrise on both dates of the ritual performance.[123]

120 The other two witnesses prescribe *ina šērti ēma Šamaš inappaḫu*, "in the morning whenever the sun rises." Linssen believes morning as the implied time of the ritual performance is "clear from context" in the other versions of the instructions (2004: 97).

121 See also TCL VI 44 i 7 (day only); KAR 60, rev. 12; BaMB 2, no. 5, obv. 7 and rev. 14; and BaMB 2, no. 7, obv. 7′.

122 For reflections on the role of stars, planets, and constellations in ritual texts from ancient Mesopotamia generally, see Reiner 1995.

123 Both dates are mentioned by Linssen 2004: 94–95 with references to the astronomical diaries in which the dates occur and secondary literature. The second date's association with the covering ritual is not certain, though mention of the *Bīt mummi*, the sacrifice of a lamb, the kettle drum, and the recitation of incantations are quite suggestive. Exact date conversions were made with the help of the Babylonian calendar converter created by R.H. van Gent of University of Utrecht (see http://www.staff.science.uu.nl/~gent0113/

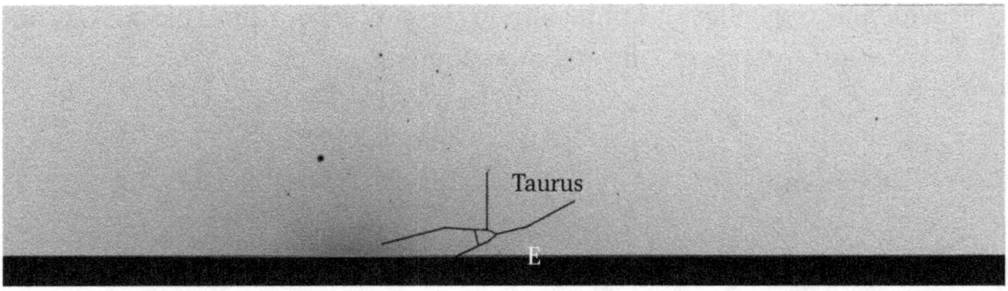

FIGURE 4.1 *Taurus rising shortly before sunrise in late Ayyaru, 226 BCE (May 29th).*[124]

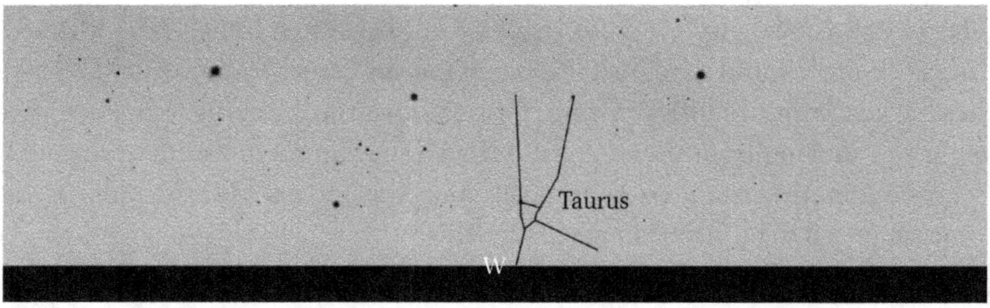

FIGURE 4.2 *Taurus beginning to set a couple of hours before sunrise in late Araḫsamna, 271 BCE (November 26th).*

Is it significant that in both cases the constellation would have been near the (opposite ends of the) horizon?[125] Of course, we do not know what version of the ritual was used in both instances and thus we cannot be certain that performing it before sunrise, an element expressed explicitly only in KAR 50, was considered important.[126]

babylon/babycal.htm). I used Stellarium (www.stellarium.org) to estimate the position of Taurus as seen from Babylon in the mid- to late first millennium and confirmed these findings with my colleague in the Physics department, Prof. Jim Hetrick, using Digitalis's portable planetarium running Nightscape software. The screenshots are from Stellarium, both depicting the sky at Babylon at 4:10am local time.

124 MUL.APIN records the visibility of the "Jaw of the Bull" on the 20th of Ayyaru (Hunger and Pingree 1989: 40 at ii 39).

125 If so, then this would require a non-literal understanding of line 5/6, at least with reference to the constellation.

126 I would like to thank Uri Gabbay and Evelyne Koubkova for their comments and suggestions on a penultimate version of this paper.

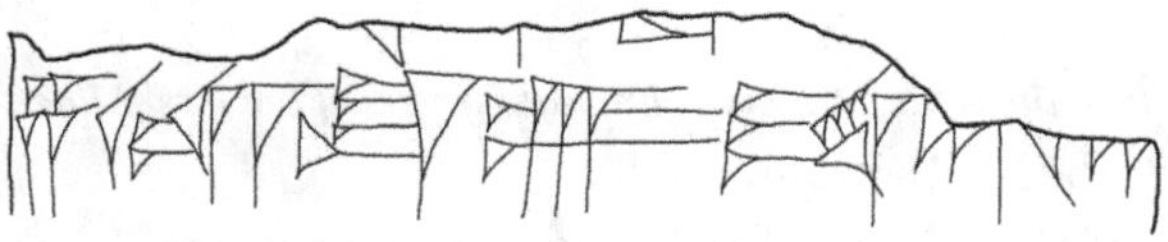

FIGURE 4.3 *Collation of K.6060 + K.10820, line 1 (line 2 is given for perspective) 3.7 cm*

Bibliography

Beaulieu, Paul-Alain

2003 *The Pantheon of Uruk during the Neo-Babylonian Period.* CM 23. Leiden: Brill.

Bell, Catherine

1992 *Ritual Theory, Ritual Practice.* Oxford: Oxford University Press.

1997 *Ritual: Perspectives and Dimensions.* Oxford: Oxford University Press.

Bezold, Carl

1891 *Catalogue of the Cuneiform Tablets in the Kouyunjik Collection of the British Museum*, Volume 2. London: Harrison and Son.

Boden, Peggy Jean

1998 "The Mesopotamian Washing of the Mouth (*mīs pî*) Ritual: An Examination of Some of the Social and Communication Strategies which Guided the Development and Performance of the Ritual which Transferred the Essence of the Deity into Its Temple Statue." Ph.D. Dissertation, Johns Hopkins University.

Bongenaar, A. C.V.M

1997 *The Neo-Babylonian Ebabbar Temple at Sippar: Its Administration and Its Prosopography.* Istanbul: Nederlands Historisch-Archaeologisch Instituut.

Cagni, Luigi

1969 *L'Epopea di Erra.* StSem 34. Rome: Istituto di Vicino Oriente.

Ebeling, Erich

1919 *Keilschrifttexte aus Assur religiösen Inhalts.* Volume 1. WVDOG 28. Leipzig: J.C. Hinrichs.

Farber, Walter

1987 "Rituale und Beschwörungen in akkadischer Sprache." Pp. 212–281 in *Religiöse Texte: Rituale und Beschwörungen I*, ed. Walter Farber et al. Gütersloh: Gütersloher Verlagshaus Gerd Mohn.

Frankfurter, David

2001 "Narrating Power: The Theory and Practice of the Magical *Historiola* in Ritual Spells." Pp. 457–476 in *Ancient Magic and Ritual Power*, ed. Marvin Meyer and Paul Mirecki. Leiden: Brill.

Gabbay, Uri

2014a *Pacifying the Hearts of the Gods: Sumerian Emesal Prayers of the First Millennium BC*. HES 1. Wiesbaden: Harrassowitz.

2014b. "The *kalû* Priest and *kalûtu* Literature in Assyria." *Orient* 49: 115–144.

2015 "Ancient Mesopotamian Cultic Whispering into the Ears." Pp. 185–220 in Marbeh Ḥokmah*: Studies in the Bible and the Ancient Near East in Loving Memory of Victor Avigdor Hurowitz*, ed. S. Yona et al. Winona Lake, IN: Eisenbrauns.

Glassner, Jean-Jacques

2017 "Comment Presentifier l'invisible? Reflexions autour des Termes *Ṣalmu*, *Tamšīlu* et *Uṣurtu*." *JCS* 69: 213–220.

Grimes, Ronald L.

1995 *Beginnings in Ritual Studies*, Second Edition. Columbia: University of South Carolina Press.

2014 *The Craft of Ritual Studies*. Oxford: Oxford University Press.

Hunger, Hermann and David Pingree

1989 *MUL.APIN: An Astronomical Compendium in Cuneiform*. AfOB 24. Horn: Ferdinand Berger & Söhne Geselleschaft.

Lambert, W.G.

1960 *Babylonian Wisdom Literature*. Oxford: Oxford University Press. Reprinted, Winona Lake, IN: Eisenbrauns, 1996.

Lenzi, Alan

2008 *Secrecy and the Gods: Secret Knowledge in Ancient Mesopotamia and Biblical Israel*. SAAS 19. Helsinki: The Neo-Assyrian Text Corpus Project.

2010 "*Šiptu ul Yuttun*: Some Reflections on a Closing Formula in Akkadian Incantations." Pp. 131–166 in *Gazing on the Deep: Ancient Near Eastern and Other Studies in Honor of Tzvi Abusch*, ed. Jeffrey Stackert, Barbara Nevling Porter, and David P. Wright. Bethesda: CDL Press.

2013 "A New Akkadian Shuila-Prayer to the Three Paths of Heaven and the Third Tablet of *Bīt salā' mê*." *Or*NS 82, 1–10.

Linssen, Marc J.H.

2004 *The Cults of Uruk and Babylon: The Temple Ritual Texts as Evidence for Hellenistic Cult Practice*. CM 25. Leiden: Brill/Styx.

Livingstone, Alasdair

1986 *Mystical and Mythological Explanatory Works of Assyrian and Babylonian Scholars*. Oxford: Clarendon. Reprinted, Winona Lake, IN: Eisenbrauns, 2007.

Löhnert, Anne

2011 "Manipulating the Gods: Lamenting in Context." Pp. 402–417 in *The Oxford Handbook of Cuneiform Culture*, ed. Karen Radner and Eleanor Robson. Oxford: Oxford University Press.

MacGinnis, John

1999 "A Cultic Handlist?" *NABU*, no. 2

Maul, Stefan

1994 *Zukunftsbewältigung: Eine Untersuchung altorientalischen Denkens anhand der babylonish-assyrisches Löserituale (Namburbi)*. BagF 18. Mainz: Verlag Philipp von Zabern.

Mayer, W.R.

1978 "Seleukidische Rituale aus Warka mit Emesal-Gebete." *Or*NS47, 431–458.

Nougayrol, Jean

1947 "Textes et documents figures." *RA* 41, 23–53.

Œlsner, Joachim

1982 "Spätachämenidische Texte aus Nippur." *RA* 76: 94–95.

Pedersén, Olaf

1986 *Archives and Libraries in the City of Assur, A Survey of the Material from the German Excavations, Part II*. AUUSUU 8. Uppsala: Almqvist and Wiksell.

Pongratz-Leisten, Beate

2011 "Divine Agency and Astralization of the Gods in Ancient Mesopotamia." Pp. 137–187 in *Reconsidering the Concept of Revolutionary Monotheism*, ed. Beate Pongratz-Leisten. Winona Lake, IN: Eisenbrauns.

Porter, Barbara

2009 "Blessings from a Crown, Offerings to a Drum: Were There Non-Anthropomorphic Deities in Ancient Mesopotamia?" Pp. 153–194 in *What Is a God? Anthropomorphic and Non-Anthropomorphic Aspects of Deity in Ancient Mesopotamia*, ed. Barbara Nevling Porter. TCBAI 2. Winona Lake, IN: Eisenbrauns.

Rawlinson, H.C.

1875 *The Cuneiform Inscriptions of Western Asia; Vol. IV: A Selection from the Miscellaneous Inscriptions of Assyria*. London: R.E. Bowler.

Reiner, Erica

1995 *Astral Magic in Babylonia*. TAPS 85.4. Philadelphia: American Philosophical Society.

Rochberg, Francesca

1988 *Aspects of Babylonian Celestial Divination: The Lunar Eclipse Tablets of Enuma Anu Enlil*. AfOB 22. Horn: F. Berger.

1996 "Personifications and Metaphors in Babylonian Celestial Omina." *JAOS* 116: 475–485.

2004 *The Heavenly Writing: Divination, Horoscopy, and Astronomy in Mesopotamian Culture*. Cambridge: Cambridge University Press.

2007 "Marduk in Heaven." *WZKM* 97: 433–442.

2009 "'The Stars Their Likenesses': Perspectives on the Relation between Celestial Bodies and Gods in Ancient Mesopotamia." Pp. 41–91 in *What Is a God? Anthropomorphic and Non-Anthropomorphic Aspects of Deity in Ancient Mesopotamia*, ed. Barbara Nevling Porter. TCBAI 2. Winona Lake, IN: Eisenbrauns.

2010a "Sheep and Cattle, Cows and Calves: The Sumero-Akkadian Astral Gods as Livestock." Pp. 347–359 in *Opening the Tablet Box: Near Eastern Studies in Honor of Benjamin R. Foster*, ed. Sarah C. Melville and Alice L. Slotsky. Leiden: Brill.

2010b "God-Talk and Star-Talk in Cuneiform and its Legacy in Later Antiquity." Pp. 189–200 in *Gazing on the Deep: Ancient Near Eastern and Other Studies in Honor of Tzvi Abusch*, ed. Jeffrey Stackert, Barbara Nevling Porter, and David P. Wright. Bethesda: CDL Press.

2011 "The Heavens and the Gods in Ancient Mesopotamia: The View from a Polytheistic Cosmology." Pp. 117–136 in *Reconsidering the Concept of Revolutionary Monotheism*, ed. Beate Pongratz-Leisten. Winona Lake, IN: Eisenbrauns.

Sachs, Abraham

1969 "Ritual to be Followed by the *Kalû*-Priest when Covering the Temple Kettle-Drum." Pp. 334–338 in *Ancient Near Eastern Texts Relating to the Old Testament*, 3rd Edition with Supplement, ed. James B. Pritchard. Princeton: Princeton University Press.

Shibata, Daisuke

f.c. *Die Šu'ila-Gebete im Emesal*. HES 3. Wiesbaden: Harrassowitz.

Thureau-Dangin, François

1919 "Un Acte de Donation de Marudk-Zâkir-Šumi." *RA*, 117–156.

1922 *Tablettes d'Uruk à l'usage des Prêtres du Temple d'Anu au Temps des Séleucides*. TCL VI. Paris: Librairie Orientaliste Paul Geuthner.

1975 *Rituels Accadiens*. Osnabrück: Otto Zeller; originally 1921.

van Dijk, Jan and Mayer, Werner R.

1980 *Texte aus dem Rēš-Heiligtum in Uruk-Warka*. BagMB 2. Berlin: Gebr. Mann.

von Soden, Wolfram

1995 *Grundriss der akkadischen Grammatik*, 3rd Edition. Rome: Editrice Pontificio Istituto Biblico.

Walker, C.B.F. and Dick, Michael B.

2001 *The Induction of the Cult Image in Ancient Mesopotamia: The Mesopotamian* Mis Pî *Ritual*. SAALT 1. Helsinki: Neo-Assyrian Text Corpus Project.

Zimmern, Heinrich

1901 *Beiträge zur Erkenntnis der Babylonischen Religion*. AB 12. Leipzig: J.C. Hinrichs.

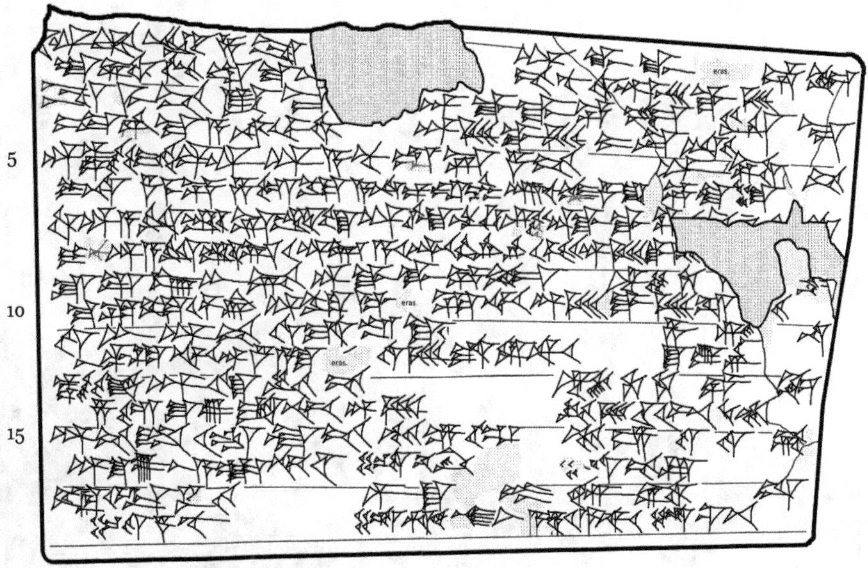

FIGURE 4.4 *BM 33343 Obverse*

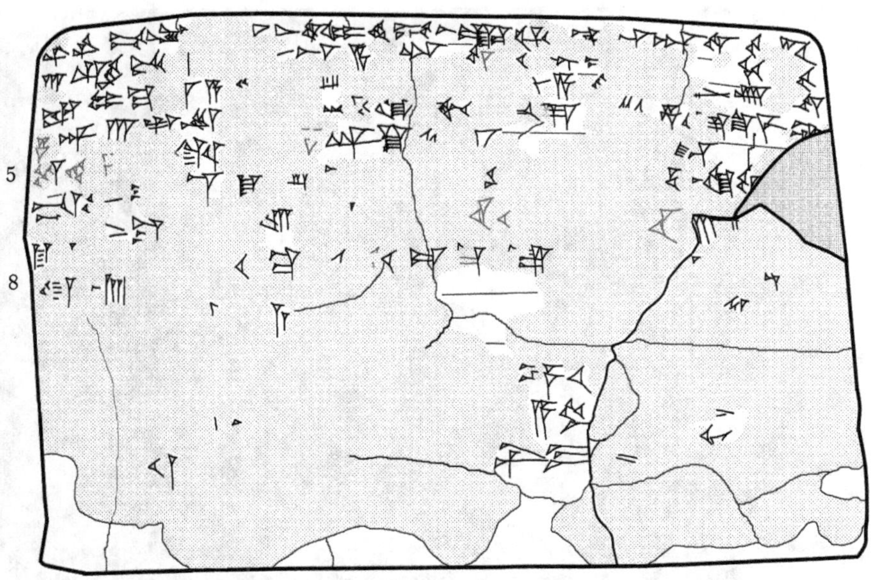

FIGURE 4.5 *BM 33343 Reverse*

FIGURE 4.6 *BM 33343 obverse, courtesy the Trustees of the British Museum*

FIGURE 4.7 *BM 33343 reverse, courtesy the Trustees of the British Museum*

CHAPTER 5

A Late Babylonian Compilation Concerning Ritual Timing and *Materia Medica*

Matthew Rutz

One of the signature features of Late Babylonian scribal culture is the tension it preserved between the pull of the received cuneiform scholarly traditions and the push of the innovative repurposing of those traditions in changing political, economic, social, and cultural contexts. The purpose of this contribution is to publish and discuss one example of the curious textual compilations that were produced in the dynamic milieu of late-first-millennium Babylonia. The fragment in question is particularly important for understanding the textual circulation of the cuneiform hemerological literature and the production of "hemerological compilations" (Jiménez 2016).

CBS 562[1] (approximately 6 × 5 × 2 cm) comes from the so-called Khabaza 1 collection held in the Babylonian Section of the University of Pennsylvania Museum of Archaeology and Anthropology, Philadelphia. That lot was acquired on the antiquities market in late-nineteenth century London and accessioned into the museum's holdings in 1888. Although certain Old Babylonian tablets in the Khabaza 1 group were most probably looted from Sippar (Kalla 1999: 206–207), there is no reason to believe that the entire lot came from that site. Thus, the original provenience of CBS 562 will likely never be known.

Early on H. Hilprecht provisionally catalogued CBS 562 as a Neo-Babylonian fragment with "Astrological(?)" contents. A Late Babylonian date can safely be inferred on palaeographic grounds, but it is now clear that the extant part of the tablet consists of three distinct compositions, two on the obverse, one on the reverse. The obverse contains examples of two of the lesser-known hemerologies: first, there is an excerpt from the end of the text commonly known as the Prostration Hemerology (col. i′), and it is followed by the beginning of the Exorcist's Almanac (cols. i′–ii′). The reverse contains the names of various *materia medica* with glosses. It is not entirely clear what these compositions

* Thanks are due to Steve Tinney and Grant Frame for open access to the University of Pennsylvania Museum's Babylonian Section.

1 Full photographic documentation is available via the Cuneiform Digital Library Initiative (P258012).

 | DOI 10.1163/9789004363380_006

are doing in close proximity, but a didactic context is possible (cf. Gesche 2000: 172–178, 214–215).

Attested in a handful of first-millennium Assyrian and Babylonian copies, the composition known as the Prostration Hemerology takes its modern name from the occurrence of precative formulations employing the verb *šukênu* that are found in certain daily prescriptions throughout the text (Livingstone 1993: 100–101).[2] Unlike the Babylonian Almanac, which relates laconic hemerological evaluations and advice for each day of the year, the Prostration Hemerology focuses only on a puzzling selection of specific days for all months, with the notable exception of Tašrītu (VII) and the intercalary months (Livingstone 2013: 161–175). Moreover, Tašrītu is not simply missing from the Prostration Hemerology, rather the text actually notes its omission: DIŠ *ina* ^{iti}DU$_6$ *ka-la ep-še-tu-šu ia-a-nu* GARZA *ana* ^{d}*en-lil*$_2$ *gu-um-mu-*⸢*ru*?/*ri*?⸣, "In the month Tašrītu there are no actions at all for him (= the client); the rites of Enlil are to be carried out in full" (K 3769 obv. 17′ and dupls., Livingstone 2013: 164). Interestingly, an appendix with more extensive cultic prescriptions for the beginning of Tašrītu appears in two manuscripts after the Prostration Hemerology (CT 51, 161 // K 2607+, see Livingstone 2013: 173). However, this appendix concerning Tašrītu should not be understood as an integral part of the Prostration Hemerology itself (so Livingstone 2013: 161) but rather as a separate composition. Removing Tašrītu from the end of the reconstructed text necessitates the removal of one of Livingstone's manuscripts, BM 34602 (Labat 1961), a fragment that contains only the material for the first ten days of Tašrītu followed by an unplaced series of entries pertaining to fire breaking out in the palace or in a house, content that is reminiscent of *Iqqur īpuš* §§65–66 (Labat 1965: 134–137). CBS 562 provides additional evidence against integrating the Tašrītu appendix into the Prostration Hemerology because an altogether different composition follows it, albeit one that is thematically related.

After the end of the Prostration Hemerology in col. i′, CBS 562 has a double ruling followed by the incipit of the Exorcist's Almanac, the text of which continues in the second column. The Exorcist's Almanac is otherwise known from three manuscripts that manifest two versions of the text as well as a commentary: a Neo-Assyrian "amulet"-shaped manuscript from Ḫuzirīna that employs the Standard Babylonian calendar (STT 2, 300; Casaburi 2002–2005;

2 An important new study of the Prostration Hemerology appeared after I had already completed and submitted the present article (Jiménez and Adalı 2015), and even though the authors independently identified CBS 562 in their revised edition, it has not been possible to systematically incorporate their work in what follows.

Glassner 2009; Geller 2014: 47–57); and two tablets that adapt the text to zodiacal astrology, one of which (BRM 4, 20) includes a *ṣâtu*-commentary as an appendix (BRM 4, 19 and 20; Ungnad 1941/44; Reiner 1993; Geller 2014: 27–46; cf. Scurlock 2005–2006).[3] STT 2, 300 is the work of a late-seventh century apprentice scribe Ninurta-[...] of the Qurdi-Nergal family (Geller 2014: 50, 57). BRM 4, 20 belonged[4] to a well-known Hellenistic ritual expert who was active in fourth-century Uruk, namely, Iqīšāya, son of Ištar-šumu-īreš, descendant of Ekur-zākir (Frahm 2011: 292–296). The related tablet BRM 4, 19 does not have a preserved colophon, but it is certainly Late Babylonian in date, and, because of its contents, it probably comes from Uruk as well. The Exorcist's Almanac makes temporal associations between ritual behaviors and various afflictions, but the precise nature of those associations has been debated. While some have interpreted the Exorcist's Almanac as a guide for timing apotropaic intervention (Reiner 1995: 109–111), a more recent consensus has emerged that many of the incantations and rituals contained in the Exorcist's Almanac are not apotropaic or prophylactic but rather aggressive in nature (Schwemer 2007: 158–163; Geller 2014: 27).[5]

CBS 562 provides two notable contributions to the study of the Exorcist's Almanac. First, although it is quite fragmentary, CBS 562 seems to rearrange the order of the some of the entries found in the complete Neo-Assyrian manuscript. STT 2, 300 takes all the material for a given month and concatenates the various days and their associated incantations/rituals into one ruled entry, with several instances in which days of the month are listed out of sequential order, for example, in Ayyaru (II), Simānu (III), Du'ūzu (IV), Tašrītu (VII), Araḫsamnu (VIII), Kislīmu (IX), and Addaru (XII) (Table 5.1). In contrast CBS 562 seems to reorder entries so that the calendrical dates are more sequential,

3 Technically the scheme employed in BRM 4, 19–20 is the so-called micro-zodiac (see, selectively, Rochberg-Halton 1988: 57–60; Monroe 2016; Wee 2016). The text of SpTU 2, 23: 1–9 provides direct evidence for the adaptation of the calendrical scheme to that of the micro-zodiac.

4 In the latest of several contributions on the topic Ossendrijver (2011: 214–215) has suggested that the "owner" of a tablet (IM PN) was someone who had advanced to a formal position within a temple's administrative apparatus, while the subordinate copyist (ŠU PN) was someone who had attained some high level of specialized scribal competency but not a temple post.

5 SpTU 2, 24: 18–19 speaks to the existence of still other versions of the Exorcist's Almanac (see below the comment on CBS 562 i' 12'); compare SpTU 1, 12: r. 17–18, a text that lists months during which certain rituals against enemies were not efficacious (NU SIG$_5$). Texts with protective measures for specific days/months are known in the Late Babylonian period as well (Finkel 2000: 212–217).

TABLE 5.1 *Days in the Neo-Assyrian (STT 2, 300) and Late Babylonian (CBS 562) Exorcist's Almanac*

	Month	STT 2, 300	CBS 562
XI	Šabāṭu	1–30	[...]
XII	Addaru	1–15, 1–30	[...]
I	Nisannu	10, 21, 21	[...]
II	Ayyaru	10, 21, 13	[10(?)], 13?, 21
III	Simānu	10, 21, 4, 30, 30	4, (4, 21), 30
II / III	Ayyaru / Simānu	1–30	[...]
IV	Du'ūzu	10, 21, 12, 21	[...]
V	Abu	10, 10, 28	[...]
VI	Ulūlu	1, 21, 24	[...]
IV / V	Du'ūzu / Abu	—	[...]
VII	Tašrītu	10, 11, 16–20, 16, 17, 7, 21	[...]
VI / VII	Ulūlu / Tašrītu	—	[...]
VIII	Araḫsamnu	10, 21, 12, 28, 28; 28; —	[...]
IX	Kislīmu	10, 21, 12; [x?]; —	[...]
X	Ṭebētu	10, 11, 11, 12	[...]
XI	Šabāṭu	10, 21	[...]
XII	Addaru	26, 28, 27	[...]
X / XI	Ṭebētu / Šabāṭu	21	[...]
XI / XII	Šabāṭu / Addaru	27; —	[...]
I–XII	Nisannu–Addaru	27, 28, 29	[...]

and CBS 562 is also not concerned with lumping all the material for a given month under a single ruled heading. Second, Frahm (2011: 31, 128) notes the disconnect between the zodiacal version and the commentary in the manuscript BRM 4, 20, and he plausibly suggests that the commentary in the appendix was on "a [textual] predecessor ... in the vein of STT 300." There is now direct evidence for such a text: CBS 562 is a Late Babylonian version of the Exorcist's Almanac that uses the hemerological associations found in the Neo-Assyrian version, not the zodiacal scheme of BRM 4, 19 and 20.

While there is some thematic unity to the two texts juxtaposed on the obverse, the reverse is not obviously related to the Babylonian hemerological traditions. What remains of the reverse of CBS 562 appears to be arranged like a Late Babylonian "cola type" commentary in one column, not two as on the obverse (Frahm 2011: 36–37). The text on the reverse contains the names of therapeutic ingredients with various glosses, and it is thus reminiscent of material found in the cuneiform pharmacopeia, such as the composition Uru-an-na = *maštakal*, a scientific edition of which has yet to appear (Kinnier Wilson 2005; Rumor 2017; cf. Böck 2011). Other plant lists produced in the course of specialized Late Babylonian scribal training typically mimicked the column-based format associated with Uru-an-na = *maštakal*, e.g., BM 42629+, BM 42625+, and BM 44204 (Finkel 2000: 185–187), and very few commentaries on the therapeutic plant lists are known, e.g., CT 41, 45 (Frahm 2011: 253–254). To further complicate the picture, laconic commentaries such as CTMMA 2, 69–70 may be commenting on lists of *materia medica* or on a therapeutic text that employs the named substances (Frahm 2011: 239–240). Thus, the fragmentary text on the reverse of CBS 562 remains difficult to place with any real precision, though that state of affairs may change once more is known about both the cuneiform pharmacopeia and the corpus of commentaries on therapeutic texts.

CBS 562 (Figures 5.1–5.2)

FIGURE 5.1 *CBS 562 obverse*
PHOTOGRAPH: M. RUTZ

FIGURE 5.2 *CBS 562 reverse*
PHOTOGRAPH: M. RUTZ

Obverse

col. i′
(lacuna)

i′ 1′ [x (x)]⸢x⸣ BAL ⸢x⸣[...]
i′ 2′ [DIŠ UD.]⸢25⸣.KAM *ina* GU$_2$ ID$_2$ *ana* ⸢^{d}⸣[*e$_2$-a liš-ken*]
i′ 3′ [K]AŠ.SAG NU NAG *mu-un-g*[*u* NU DAB-*su*]
i′ 4′ [DIŠ U]D.30.KAM *ana* ^{d}U.GUR *liš-ken* GEŠTIN NU NA[G]
i′ 5′ [*ina* I]D$_2$ ⸢*ṣi*⸣-*da-nu* NU DAB-*su*
i′ 6′ [DIŠ *ina* ^{iti}Š]E UD.⸢13⸣.KAM KU$_6$ *u* MUŠEN NU GU$_7$
i′ 7′ [SIG$_5$-*ta-šu ina*] KA UN.MEŠ GAR-*an*
i′ 8′ [DIŠ UD.20.KAM U]ZU GA *u* MUD$_2$ NU GU$_7$
i′ 9′ [*ina* TUG$_2$.ḪI.A ŠU.II-*š*]*u$_2$ la i-kap-par*
i′ 10′ [*i-rib* ^{d}G]IR$_3$ *sa-dir-šu$_2$*
i′ 11′ [*li-pit* ^{d}UTTU] *ina* E$_2$-*šu$_2$* SI.SA$_2$

---

i′ 12′ [DIŠ *ina* ^{iti}BARA$_2$ UD.10.KA]M UD.DA.KAM ŠA$_3$.BAL.BAL
i′ 13′ [x x x (x)]⸢x x x x⸣
(lacuna)

col. ii′
(lacuna)
ii′ 1′ ⸢x⸣[...]

---

ii′ 2′ DIŠ *ina* ^{iti}G[U$_4$ ...]
ii′ 3′ : ⸢12⸣[(+1?).KAM ...]
ii′ 4′ KI.SIKIL.E K[A$^?$ ...]

---

ii′ 5′ DIŠ *ina* ^{iti}GU$_4$ UD$^!$.21.KAM NA[M$^?$.ERIM$_2$ (...)]
ii′ 6′ BUR$_2$.RU.DA E[RIM$_2$$^?$ ...]

---

ii′ 7′ DIŠ *ina* ^{iti}SIG$_4$ UD.4.KAM$^{⸢:⸣ \text{UD}.⸢4⸣ : \text{UD}.⸢20⸣[(+1?) ...]}$ UD.⸢30⸣.[KAM (...)]
ii′ 8′ ḪUL.GIG.G[A ...]

---

ii′ 9′ DIŠ *ina* ⸢^{iti}SIG$_4$⸣ U[D.10(?).KAM ...]
(lacuna)

Reverse

(lacuna)
r. 1′ [...]⸢x⸣ IGI ZA ⸢x⸣[...]
r. 2′ [...]⸢x⸣ NU ⸢KA$^?$ x⸣[...]
r. 3′ [...]⸢*x*⸣-*tu*$_4$ *ša*$_2$ *uz*-⸢*nu* x⸣ [...]
r. 4′ [...]⸢x⸣ RI ⸢: *x*⸣-GAR-*ri* : *kal-ba*-⸢*nu*⸣[...]
r. 5′ [...]⸢x⸣ : AD ⸢x (x) x⸣ *ša*$_2$ ^{giš}APIN ^{u2}⸢x⸣[...]
r. 6′ [...] GAR : *a-a-ar*$_2$ ⸢x (x)⸣ : ^{u2}*el-li-p*[*i*$^?$ ...]
r. 7′ [...]⸢x⸣ *u* ŠE.NUMUN-*šu*$_2$ ⸢*sa-a*⸣-*mu* ^{u2}*tu-mam* ⸢x⸣ [...]
r. 8′ [... -*š*]*u*$_2$$^?$: ŠIM.ḪAD$_2$.A : *u*$_2$-*ru-u*$_2$: *l*[*a*$^?$-*ba-na-tu*$_4$ ...]
r. 9′ [... $^{u2(?)}$*ḫ*]*a-as-sa-ar-tu*$_4$: ⸢^{u2}x⸣[...]
r. 10′ [... ^{u2}*pu-qut-t*]*u*$_2$$^?$: ^{na4}*a-sak-ku* : ⸢x⸣[...]
r. 11′ [...]⸢x x x x⸣[...]
(lacuna)

Translation
Obverse
col. i′
(lacuna)
i′ 1′ [...] ... [...]
i′ 2′ [If the] 25th: [let him prostrate] to [Ea] on the river bank;
i′ 3′ he shall not drink first-rate [b]eer; stiff[ness will not seize him.]

i′ 4′ [If the] 30th day: let him prostrate himself to Nergal; he shall not dri[nk] wine;
i′ 5′ vertigo will not seize him [in/on the r]iver.
i′ 6′ [If in the month of Ad]daru (XII) it is the 13th day: he shall not eat fish or fowl;
i′ 7′ [his good fortune] will be established [in] the mouth of the people.
i′ 8′ [If the 20th day:] he shall not eat [m]eat, milk, or blood.
i′ 9′ he shall not rub hi[s hands on (his) clothes;]
i′ 10′ [income from Ša]kkan will be steady for him;
i′ 11′ [the work of Uttu] will prosper in his house.

---

i′ 12′ [In the month of Nisannu (I) the 10t]h [day] is the appointed time (for the ritual) to change (someone's) mood/mind;
i′ 13′ [...] ...
(lacuna)

col. ii′
(lacuna)
ii′ 1′ ...[...]

---

ii′ 2′ In the month Ay[yaru (II) ...]
ii′ 3′ : 1[3th(?) day ...]
ii′ 4′ *ardat lilî*-demon ... [...]

---

ii′ 5′–6′ In the month Ayyaru (II), the 21st day, (for the ritual) releasing (the ill effects of) an o[ath (...)]; [...] e[nemy(?) (...)].

---

ii′ 7′ In the month Simānu (III), the 4th day the 4th day : the 2[1st(?)] day [...] (and) the 30th day [(...)]
ii′ 8′ (for) hate(-magic) [...]

---

ii′ 9′ If in the month Simānu (III), [the 10th(?)] d[ay (for) ...]
(lacuna)

Reverse
(lacuna)
r. 1′–3′ (too fragmentary to translate)
r. 4′ [...] ... : ... : *kalbānu*-plant(?) [...]
r. 5′ [...] ... : ... of the plow, ...-plant [...]

r. 6′ [...] ... : bloom of ... : *ellip*[*u*(?)]-plant [...]
r. 7′ [...]... and its red seed; *tūmamu*-plant(?) ... [...]
r. 8′ [...] ... : aromatics : (means) *urû*-aromatics : (means) i[ncense(?) ...]
r. 9′ [... *ḫ*]*asarratu*-grass : ...-plant [...]
r. 10′ [... thor]n-[plant(?)] : *asakku*-stone : ... [...]
r. 11′ (too fragmentary to translate)
(lacuna)

Commentary

i′ 1′: The Prostration Hemerology contains instructions for the 20th day of the month Šabāṭu (XI) (STT 2, 302: r. 8′ // CT 51, 161: r. 3 // K 2607+ r. 1 // K 3769 r. 19), but the traces in this line are difficult to square with the other manuscripts (Livingstone 2013: 167). Perhaps CBS 562 preserves a different tradition in which the text prescribes the libation of water (e.g., A.MEŠ BAL-*qi*$_2$ or the like), which appears in the Prostration Hemerology's entry for the 20th day in the month of Ṭebētu (X) (Livingstone 2013: 166).

i′ 2′: STT 2, 302: r. 9′ likewise has the 25th day, as opposed to the other sources (CT 51, 161: r. 4 // K 2607+ r. 2), which have the 24th (Livingstone 2013: 167). Thus, the reading found in CBS 562 suggests that Livingstone's emendation of STT 2, 302 is unwarranted: together these NA and LB manuscripts appear to represent a variant tradition.

i′ 3′: Drinking first-rate beer appears in a variety of ritual texts (CAD Š/2 426 sub *šikaru*). Akkadian *mungu*, "stiffness" or "paralysis," is often associated with the effects of a perceived attack of witchcraft (CAD M/2 202–203 *mungu* A; Schwemer 2007: 106; cf. Scurlock and Anderson 2005: 249–250).

i′ 4′: STT 2, 302: r. 10′ // CT 51, 161: r. 6 // K 2607+ r. 3 all prescribe prostration to Adad, not Nergal (Livingstone 2013: 167); however, Nergal is named in the same text for the 16th of Kislīmu (IX) (Livingstone 2013: 165).

i′ 5′: The disease *ṣīdānu* is not terribly common (CAD Ṣ 171–172 sub *ṣīdānu*). There had been some question as to whether it refers to a sign such as involuntary eye movements or a symptom such as vertigo (Scurlock and Andersen 2005: 734–735 n. 58), but a Late Babylonian bilingual anti-witchcraft incantation contains a phrase that points to vertigo: [sag/igi(?)]-nigin nig$_2$-nam tuk$_4$-tuk$_4$ // *ṣīdānu murīb mimma šumšu*, "vertigo (that) makes everything shake" (BM 47451 obv. 3, Schwemer 2009: 49, 51, 54).

i′ 6′: Labat (1972: 321) already noted the connections between the Prostration Hemerology and CT 4, 5–6 (Bu 88-5-12, 11 = BM 78825), which is an apotropaic lunar eclipse hemerology written in Neo-Assyrian script but reportedly copied from a Nippur original (Koch-Westenholz 2001: 75–77; cf. Livingstone

2013: 195–198). That text consists of ritual prescriptions in the event of a lunar eclipse on the 12th, 13th, or 14th day of each of the twelve months of the Standard Babylonian calendar. Various food prohibitions are commonplace in the hemerological tradition (Livingstone 2013: 263–264; Ermidoro 2014), and in CT 4, 5–6 the same prohibition on eating fish or fowl, KU$_6$ MUŠEN$^!$(RI) NU GU$_7$ (CT 4, 6: r. 10), is prescribed in the event of a lunar eclipse on the 12th, 13th (as here), or 14th of Addaru (Koch-Westenholz 2001: 77; cf. Livingstone 2013: 197).

i′ 8′: Food prohibitions occur frequently in the hemerologies (see the comment on i′ 6′). The particular combination of meat, milk, and blood appears to be unique to the present context, but note that the eating of meat and milk is prohibited in the lunar eclipse menology: GA UZU NU GU$_7$ (CT 4, 6: r. 3; Koch-Westenholz 2001: 77; cf. "He should not eat meat," Livingstone 2013: 197). The expression *šizba akālu*, "to eat milk," is so far not found outside the Prostration Hemerology and CT 4, 5–6 (CAD Š/3 151a sub *šizbu*; cf. CAD Š/2 212a sub *šatû*); thus, Livingstone (2013: 171) proffers the translation "yoghurt" in the Prostration Hemerology.

i′ 9′: The best parallel for this line is found in CT 4, 6 for the month Šabāṭu (XI), where the prohibition is probably to be reconstructed: [*ina*(?) T]UG$_2^!$ ŠU.II-*šu*$_2$ *la i-ka-par* , "he shall not wipe his hands [on] (his) clothing" (CT 4, 6: r. 4; cf. Koch-Westenholz 2001: 77; Livingstone 2013: 197).

i′ 10′–11′: These expressions are not very common in isolation or, as here, in tandem. Regardless, the references to the deities Šakkan and Uttu imply an entirely plausible association between acquiring wool and weaving textiles (Livingstone 2013: 171).

i′ 12′: Following a double ruling is the incipit of the Late Babylonian version of the Exorcist's Almanac, which is known from the colophon of Iqīšāya's tablet: UL *u šu-ut* KA *ša*$_2$ ^{iti}BARA$_2$ UD.10.KAM UD.DA.KAM ŠA$_3$.BAL.BAL, "Lemmata and oral explanations of '(In) the month of Nisannu the 10th day is the appointed time for changing (someone's) mood/mind'" (BRM 4, 20: r. 41). The expression UD.DA.KAM, glossed *u*$_4$*-mu a-da-nu* (BRM 4, 20: r. 15), refers to the fixed day in the calendar on which a ritual was to be performed for maximum efficacy (cf. Geller 2014: 35, 47), and it should probably be related to the difficult expression *ud-da-ki-šu*$_2$ found twice in a NA compilation of anti-*zikurudû* prescriptions (BAM 5, 461 ii 37′; see Rutz 2016: 57–58; cf. Abusch and Schwemer 2016: 365, 367, 10.6.2: 42″). In Iqīšāya's tablet the expression ŠA$_3$.BAL.BAL is given the following Late Babylonian glosses (BRM 4, 20: r. 16–17; cf. Geller 2014: 32, 35): ŠA$_3$.BAL.BAL *lib*$_3$*-bi a-na na-bal-ku-tu*$_2$, "ŠA$_3$.BAL.BAL (means) 'to change (someone's) mood'" (CAD N/1 17a) and ŠA$_3$.BAL.BAL *lib$_3$-bi a-na šu-un-nu-u*, "ŠA$_3$.BAL.BAL (means) 'to confuse (someone's) mind'" (cf. CAD Š/1 408a). However, orthographies like ŠA$_3$.⸢BAL⸣.BAL-*e* (STT 1, 78 iii 1) and ŠA$_3$.BAL.BAL-*a'* (SpTU 1, 7: 19′) seem to point to a rare loanword *šabalbalû*, "(mental/emotional) confusion/disturbance" or "mood swings" (Abusch 2016: 289, 317) which so far is attested only in the anti-witchcraft series *Maqlû* (1:91; 4:12) in two litanies of malevolent ritual behaviors and their results (cf. CAD Š/1 3–4).

It is also worth noting that another innovative (or conflate) tradition is in evidence in LB Uruk as well: on a small tablet with various incantations and rituals for social success is the entry: *ina* ^{iti}APIN UD.20.KAM UD.DA.KAM ŠA$_3$.BAL.BAL / NUN *ina* E$_2$.GAL-*šu$_2$* MU-*šu$_2$* *ana* SIG$_5$-*ti$_3$* *ḫa-sa-sa*, "In Araḫsamnu (VIII) the 20th is the appointed time (for rituals) to change (someone's) mood/mind (as well as for rituals for) the prince to remember his (= the client's) name for good in his (= the prince's) palace" (SpTU 2, 24: 18–19). The related section of the NA version of the Exorcist's Almanac refers to the 10th of Araḫsamnu (as opposed to the 20th) and makes no mention of ŠA$_3$.BAL.BAL (STT 2, 300: 25–26; Geller 2014: 49).

i′ 13′: The traces are difficult but perhaps read [x x x (x)]⸢MUD$_2$$^?$ MUNUS$^?$ TAR$^?$-*si*$^?$⸣ (cf. STT 2, 300: 6; Geller 2014: 47).

ii′ 1′: Perhaps the sign is to be read N[U$^?$ ...] (cf. STT 2, 300: 6; Geller 2014: 47). Regardless, based on the subsequent entry, this line is probably the end of the section that treats the month of Nisannu (I).

ii′ 2′–6′: The order of entries appears to be switched in the Neo-Assyrian version, where the text reads (STT 2, 300: 7–8; Geller 2014: 48):

7 DIŠ *ina* ^{iti}GU$_4$ UD.10.KAM$_2$ UD.DA.KAM$_2$ ŠU.DU$_8$.A UD.[2]1.KAM$_2$ ⸢NAM.ERIM$_2$.BURU$_2$.DA⸣ ERIM$_2$ *ana* ⸢LU$_2$⸣ TAR-*si*

8 UD.13.KAM$_2$ LIL$_2$.LA$_2$.EN.NA KI.SIKIL.LIL$_2$.LA$_2$.EN.NA ⸢ZI⸣-*ḫi* DIM$_2$-*ma* AL.SILIM

In the month Ayyāru (II) the 10th day is the appointed time (for the ritual) "removing the hand"; the [2]1st day, (for the ritual) releasing (the ill effects of) an oath (and) blocking an enemy from a man;
the 13th day, (for the ritual) removing the *lilû*-demon (and) *ardat lilî*-demon—you perform (them thus), and it (i.e., the ritual) will succeed.

The *ardat lilî* was imagined as a malevolent female being created as the result of a virginal death (Wiggermann 2000: 227–228). For the 21st day the number "21" appears to be written over the UD sign on CBS 562.

ii′ 7′–8′: Directly above the main text the scribe includes a gloss that mentions the 4th and the 21st. The Neo-Assyrian exemplar appears to read (STT 2, 300: 9–10; agreeing with Glassner 2009: 76; cf. Geller 2014: 48):

9 DIŠ *ina* ^{iti}SI[G$_4$] UD.10.KAM$_2$ UD.DA.KAM$_2$ KI.⌜AG$_2$⌝ NITA *ana* MUNUS UD.21.KAM$_2$ KI.⌜AG$_2$.GA$_2$⌝ MUNUS *ana* [N]ITA UD.4.KAM$_2$ ⌜ḪUL.GIG⌝

10 UD.30.KAM$_2$ ḪUL.GIG UD.30.KAM$_2$ KI.⌜AG$_2$⌝ NITA *ana* MUNUS DIM$_2$-[*m*]*a* A[L].SILIM

In the month of Simānu (III) the 10th day is the appointed time (for the ritual to effect) love of a man for a woman; the 21st day, (for the ritual to effect) love of a woman for a man; the 4th day, (for) hate(-magic);
the 30th day, (for) hate(-magic); the 30th day, (for the ritual) love of a man for a woman—you perform (them thus), and it (i.e., the ritual) will succeed.

The witchcraft designated *zīru* (written HUL.GIG) is never clarified any further in the texts, but it must refer to ritual means thought to bring about animus and hostility in the social world of the bewitched (Schwemer 2007: 14, 121, 159–160). It is unclear whether the scribe of CBS 562 was noting the existence of an entry for the 21st day of the month, but the gloss appears to be in keeping with his concern for sequential and thematic groupings of entries in the Exorcist's Almanac.

ii′ 9′: This broken entry is difficult to square with the sources known from elsewhere. It appears to contain another entry for Simānu (III), while the Neo-Assyrian version contains a recommendation for Ayyāru and Simānu for the entirety of each month (STT 2, 300: 11; Geller 2014: 54).

r. 1′–3′: The first three lines of the reverse are too badly damaged to hazard any translation, but it is worth noting that a rare usage of *uznu* is "(part of a plant)" (AHw. 1448b; CAD U/W 371a).

r. 4′: A medicinal plant name *kalbānu* exists, but the attestations are difficult to connect with the present text (CAD K 67b; AHw. 424a).

r. 5′: The expression *ša epinni*, "of the plow," is to my knowledge not commonly found in lists of medicinal plants.

r. 6′: The *nomen regens* in the phrase is clear, i.e., *ay*(*y*)*aru*, "bloom, blossom," but unfortunately the *nomen rectum* here is damaged and many possible candidates exist (AHw. 25a; CAD A/1 229b).

r. 7′: Having red seeds is hardly diagnostic of a single specific plant, but perhaps this phrase refers to the plant name that follows. The *tūmamu*-plant is so far not attested in a syllabic writing, but a plant name is attested that has the appropriate logographic writing u2MAŠ(.TAB.BA), entered in the dictionaries under the lemma *māšu* (AHw. 631b; CAD M/1 402b).

r. 8′: The word *urû*, "(an aromatic plant)" (CAD U/W 260b sub *urû* E) is found in the lexical tradition, giš-šim-ḫad$_2$-a = *u$_2$-ru-u$_2$* (SB Ur$_5$-ra = *ḫubullu* 3, MSL 5, 102; Van Der Meer 1939: Kish 64), as well as in an entry from a Late Babylonian commentary on a medical therapeutic text, ŠIM.ḪI.A : *u$_2$-ru-u$_2$* : *la-ba-na-tu$_4$* (BRM 4, 32: 15, see Geller 2010: 169, 172).

r. 9′: The entry [*ḫ*]*a-as-sa-ar-tu$_4$* is probably to be equated with *ḫasarratu* (var. *ḫasirratu*), "(a kind of grass)" (AHw. 329b sub *ḫasa/irratu*; CAD Ḫ 122 sub *ḫasarratu*; cf. AHw 64b *arantu* I, 89a sub *aṭartu*), which is found in the lexical commentary Mur-gud (Frahm 2011: 249–253; Veldhuis 2014: 363–366) as well as in the first tablet of the series Uru-an-na = *maštakal*.

r. 10′: Note U$_2$ *pu-qut-tu$_2$* : U$_2$ NA$_4$ *a-sak-ku* in Uru-an-na = *maštakal* (CAD A/2 326b sub *asakku* A); cf. U$_2$ *pu-qut-tu$_2$* [...], U$_2$ *a-tar-tu$_2$* [...] (BM 44204 r. 6′–r. 7′; Finkel 2000: 187), a tablet marked as 2-*u$_2$* *ni*-[*is-ḫu*], "second ex[tract]," presumably of Uru-an-na = *maštakal*.

Bibliography

Abusch, Tzvi

2016 *The Magical Ceremony* Maqlû*: A Critical Edition*. AMD 10. Leiden: Brill.

Abusch, Tzvi, and Schwemer, Daniel

2016 *Corpus of Mesopotamian Anti-Witchcraft Rituals*. Vol. 2. AMD 8/2. Leiden: Brill.

Böck, Barbara

2011 "Sourcing, Organizing, and Administering Medicinal Ingredients." Pp. 690–705 in *The Oxford Handbook of Cuneiform Culture*, ed. K. Radner and E. Robson. Oxford: Oxford University Press.

Casaburi, Maria Cristina

2002–2005 "Early Evidences of Astrological Aspects in a Neo-Assyrian Medical Hemerology." *SAAB* 14: 63–88.

Ermidoro, Stefania

2014 "Food Prohibition and Dietary Regulations in Ancient Mesopotamia." *AuOr* 32: 79–91.

Finkel, Irving L.

2000 "On Late Babylonian Medical Training." Pp. 137–223 in *Wisdom, Gods and Literature: Studies in Assyriology in Honour of W.G. Lambert*, ed. A.R. George and I.L. Finkel. Winona Lake, IN: Eisenbrauns.

Frahm, Eckart

2011 *Babylonian and Assyrian Text Commentaries: Origins of Interpretation*. GMTR 5. Münster: Ugarit-Verlag.

Geller, Markham J.

2010 *Ancient Babylonian Medicine: Theory and Practice*. Chichester: Wiley-Blackwell.

2014 *Melothesia in Babylonia: Medicine, Magic, and Astrology in the Ancient Near East*. STMAC 2. Berlin: De Gruyter.

Gesche, Petra D.

2000 *Schulunterricht in Babylonien im ersten Jahrtausend v. Chr*. AOAT 275. Münster: Ugarit-Verlag.

Glassner, Jean-Jacques

2009 "Exorcisme et chronomancie selon STT 2, 300." Pp. 75–81 in *Et il y eut un esprit dans l'Homme: Jean Bottéro et la Mésopotamie*, ed. X. Faivre, B. Lion, and C. Michel. Paris: De Boccard.

Jiménez, Enrique

2016 "Loose Threads of Tradition: Two Late Hemerological Compilations." *JCS* 68: 197–227.

Jiménez, Enrique, and Adalı, Selim F.

2015 "The 'Prostration Hemerology' Revisited: An Everyman's Manual at the King's Court." *ZA* 105: 154–191.

Kalla, Gábor

1999 "Die Geschichte der Entdeckung der altbabylonisch Sippar-Archive." *ZA* 89: 201–226.

Kinnier Wilson, J.A.

2005 "Notes on the Assyrian Pharmaceutical Series Uru.an.na : *Maštakal*." *JNES* 64: 45–51.

Koch-Westenholz, Ulla

2001 "Babylonian Views of Eclipses." Pp. 71–84 in *Démons et merveilles d'Orient*, ed. R. Gyselen. ResOr 13. Leuven: Peeters.

Labat, René

1961 "Tabous de Tešrit et autres prescriptions." *Iraq* 23: 88–93.

1965 *Un Calendrier babylonien des travaux des signes et des mois (séries iqqur îpuš)*. Paris: Librairie Honoré Champion.

1972 "Hemerologien." *RlA* 4: 319–323.

Livingstone, Alasdair

1993 "The Case of the Hemerologies: Official Cult, Learned Formulation and Popular Practice." Pp. 97–113 in *Official Cult and Popular Religion in the Ancient Near East*, ed. E. Matsushima. Heidelberg: Universitätsverlag C. Winter.

2013 *Hemerologies of Assyrian and Babylonian Scholars*. CUSAS 25. Bethesda: CDL Press.

Monroe, M. Willis

2016 Advice from the Stars: The Micro-zodiac in Seleucid Babylonia. Ph.D. Dissertation, Department of Egyptology and Assyriology, Brown University, Providence, Rhode Island.

Ossendrijver, Mathieu

2011 "Science in Action: Networks in Babylonian Astronomy." Pp. 213–221 in *Babylon: Wissenskultur in Orient und Okzident*, ed. E. Cancik-Kirschbaum, M. van Ess, and J. Marzahn. Topoi—BSAW 1. Berlin: De Gruyter.

Reiner, Erica

1993 "Two Babylonian Precursors of Astrology." *NABU*: no. 26: 21–22.

1995 *Astral Magic in Babylonia*. TAPS 85/4. Philadelphia: The American Philosophical Society.

Rochberg-Halton, Francesca

1988 "Elements of the Babylonian Contribution to Hellenistic Astrology." *JAOS* 108: 51–62.

Rumor, Maddalena

2017 "The 'AŠ section' of Uruanna III in *Partitur*." *JMC* 29: 1–34.

Rutz, Matthew

2016 "Anti-witchcraft Prescriptions from Ḫattuša and Nippur." *ZA* 106: 42–61.

Schwemer, Daniel

2007 *Abwehrzauber und Behexung: Studien zum Schadenzauberglauben im alten Mesopotamien*. Wiesbaden: Harrassowitz.

2009 "Washing, Defiling, and Burning: Two Bilingual Anti-witchcraft Incantations." *Or*NS 78: 44–68.

Scurlock, JoAnn

2005–2006 "Sorcery in the Stars: STT 300, BRM 4.19-20 and the Mandaic Book of the Zodiac." *AfO* 51: 125–146.

Scurlock, JoAnn, and Andersen, Burton R.

2005 *Diagnoses in Assyrian and Babylonian Medicine: Ancient Sources, Translations, and Modern Medical Analyses*. Urbana: University of Illinois Press.

Ungnad, Arthur

1941/44 "Besprechungskunst und Astrologie in Babylonie." *AfO* 14: 251–284.

Van Der Meer, P.E.

1939 "Tablets of the ḪAR-ra = *ḫubullu* series in the Ashmolean Museum." *Iraq* 6: 144–179.

Veldhuis, Niek

2014 *History of the Cuneiform Lexical Tradition*. GMTR 6. Münster: Ugarit-Verlag.

Wee, John Z.

2016 "Virtual Moons over Babylonia: The Calendar Text System, its Micro-zodiac of 13, and the Making of Medical Zodiology." Pp. 139–229 in *The Circulation of Astronomical Knowledge in the Ancient World*, ed. J.M. Steele. Time, Astronomy, and Calendars 6. Leiden: Brill.

Wiggermann, F.A.M.

2000 "Lamaštu, Daughter of Anu: A Profile." Pp. 217–252 in *Birth in Babylonia and the Bible: Its Mediterranean Setting*, M. Stol. CM 14. Groningen: Styx.

CHAPTER 6

BM 40187: A Birthnote for Two Named Individuals

M. Willis Monroe

Introduction

Rochberg's edition of the horoscopes and birthnotes, dating from the Achaemenid to Parthian periods, provides a comprehensive basis from which to study the small corpus of horoscopic texts written for a client.[1] A general description of birthnotes is found in the introduction to Rochberg's study.[2] They are small texts which record the date of birth and in some cases the name of the individual. They are not astrological in their content, serving instead as a method to calculate an eventual horoscope. Horoscopes on the other hand record, the location of the planets (including the sun and moon), in addition to the time of birth. It is my intent to add an Achaemenid period birthnote to the corpus.[3] The text presented below is now the fifth example of a birthnote. It is not the earliest preserved birthnote; however it does record the names of two children and their dates of birth. As far as I am aware this is the earliest link between personal names and exact time of birth in the cuneiform record.

The tablet (Fig. 6.1) is small, square, and written in a clear and steady hand. The text covers the obverse and continues onto the bottom edge, but the reverse of the tablet is completely uninscribed and quite pristine. Stylistically it is very similar to Text 29 in Rochberg's catalog. However, this is the only birthnote that includes an introductory prayer. Additionally, this text uses a unique term for a period within the day, and a previously unattested female personal name. The accession information suggests that the tablet itself comes from Babylon.[4]

1 Rochberg 1998.

2 Rochberg 1998: 19.

3 The text was made known to me by Christopher Walker in the British Museum, and I am grateful to him for his assistance. All mistakes and errors remain my own. Publication of photographs of the tablet is due to the courtesy of the Trustees of the British Museum.

4 Reade 1986: xxx.

 | DOI 10.1163/9789004363380_007

Edition:
BM 40187 (1881-3-24.53)
dimensions: 3.4 × 3.4 × 1.4cm

Birthdates recorded on the tablet:[5]
Nov 18th 364 BCE (-363)
Dec 7th 358 BCE (-357)

Transliteration:
top edge.
1. *ina a-mat* ^{d}EN *u*$_3$ ^{d}AG *liš-lim* (beginning written over erasure)
obverse.
2. U$_4$.9.KAM *ša*$_2$ ITU.APIN
3. MU.41.KAM ^{m}*ar*$_2$*-tak-šat-su* LUGAL KUR.KUR
4. ^{f}*a-pa-at-tu*$_4$ DUMU.MUNUS *ša*$_2$
5. ⸢^{m}⸣^{d}AG-MU-DU$_3$![6] KI *sa-ra-qu*
6. [*ša*$_2$] *še-ḫe-e-ri a-lid*
7. [U$_4$].3.KAM *ša*$_2$ ITU.GAN *ina* MU.1.KAM
8. [^{md}*ar*$_2$]-⸢*tak*⸣-*šat-su* LUGAL KUR.KUR
bottom edge.
9. [^{f}]⸢GEME⸣-^{d}GAŠAN-*ia*$_2$ DUMU.MUNUS *ša*$_2$
10. ⸢^{md}⸣AG-MU-DU$_3$! KI *sa-ra-qu*
11. *ša*$_2$ *še-ḫe-e-ri a-lid*
reverse uninscribed

Translation:

1. By the command of Marduk and Nabu may it go well.
2. 9th of Month VIII
3. 41st year of Artaxerxes king of the world,
4. Apattu daughter of
5. Nabû-šumu-ibni (at) the sprinkling
6. of the morning was born.
7. 3rd of Month IX, 1st year
8. [of Ar]taxerxes king of the world,

---

5 These are reckoned according to Parker-Dubberstein 1961.

6 The sign is clearly NI on the tablet, but I can find no prosopographical reading for AG-MU-NI. Nabû-šumu-ibni is a relatively standard Babylonian name.

9. Amat-belitya daughter of
10. Nabû-šumu-ibni (at) the sprinkling
11. of the morning was born.

Commentary:

Obv. Top Edge: The introductory formula *ina amat Bēl u Nabu lišlim* occurs on scholarly tablets from Babylon and perhaps Borsippa.[7] Interestingly, it does not appear on any of the previously published birthnotes.

Line 4: The name ^{f}*a-pa-at-tu*$_4$ is not otherwise attested. Possible etymologies might derive from *appu* "nose" (CAD A/II pg. 184), *apû* "to be visible" (CAD A/II pg. 201), or *apû* "cloudy" (CAD A/II pg. 201). Neither of the first two etymologies are convincing, certainly "nose" would be an unfortunate choice for a young girl. The last option strikes me as more possible, perhaps the adjective describes a medical condition of the eyes of the child at birth. For a discussion of *apû*, in the form *īnū apâtu*, in the medical corpus see Fincke's summary of the evidence.[8]

Line 5: Both of the individuals mentioned on this tablet were born at the same time, *sarāqu ša šeḫeri*. This designation is unique in the corpus of horoscopic texts and in the wider astronomical and astrological literature. For the most part, the previously published horoscopes and birthnotes conform to a relatively standard group of time designations outlined by Rochberg in her study.[9] She notes that three systems are in use: a system of three watches of the night, a division of the 24 hour day into four parts using sunset and sunrise as reference points, and finally a division of the day-time into twelve hours which changed according to the season. This reference does not fit into any of those previously mentioned systems. The use of *šeḫeri* is also relatively rare: it might represent a West Semitic variant of *šēru* (CAD Š/II pg. 263). Its translation as "morning" is not in question, but the first element of the phrase is more problematic. The verb *sarāqu* (CAD S pg. 172) has the sense of sprinkling or scattering, which might suggest a reference to morning rainfall, i.e. "the sprinkling of the morning." However, the various valences of *sarāqu* only refer to this action from a human-centered perspective, i.e. it seems to have no connotation with *zunnu* "rain." Another option is to read *sarāqu* as a variant of *šarāku* present in the Neo-Babylonian period (CAD S *sarāqu* B pg. 174). *šarāku* "to offer" would also be used in the context of ritual performance like *sarāqu* above

7 Roth 1998: 2, with note 4.

8 Fincke 2000: 78–80.

9 Rochberg, 1998: 35–39 and Rochberg 1989.

(CAD Š/II 40). In either case, it seems as if the action taking place in the morning is performed by people and probably involves some sort of votive offering, perhaps a libation or scattering of flour. Some evidence of this practice comes from the ritual texts from Hellenistic Babylonia edited by Linssen. There was clearly some element of sprinkling occurring in the morning in certain ritual texts, TCL 6 48, lines 10–11 have: *ina še-ri$_2$ ma-al-ṭi-ṭi sa-ra-qa a-na* ^{d}*A-nu* "in the morning, the sprinkling of *malṭīṭu* for Anu."[10]

Discussion

Adding this fifth birthnote to the corpus shows that there is very little similarity between the five known birthnotes. The purpose these texts seems clear enough: to record the time of birth so that a horoscope could be calculated for the individual later in life. However, the means by which the existing tablets record this data differ on each exemplar. Their physical characteristics can generally be described as small and tightly written. Some are clearly rectangular, and others have a more oval shape. This range in shape is reflected in the horoscope texts as well. BM 40187 perhaps most resembles Text 29 from Rochberg's study. Interestingly, Text 29 (BM 64148), which was written in -364,[11] is the only birthnote in the current corpus written earlier than BM 40187. The physical similarity between these two texts rests in the roughly square shape, and lack of an inscribed reverse.

However, beyond general physical similarities in a few cases the corpus as a whole does not seem very homogenous. The most obvious form of diversity comes in the recording of the time of birth (see Table 6.1). One will immediately notice that all of these are different both in what they include as well as the order in which they are recorded. Only two of the texts, 31 and 32, record the year and month in the same order at the beginning of the sequence. While, admittedly, the sample size here is miniscule compared to the other genres of this period, it is remarkable that there is no similarity at all between how these texts record their most basic function. Rochberg already made mention of this in regard to the horoscopes themselves, "The horoscope texts are only

10 TCL 6 48, in Linssen 2004: 30–32.

11 The date given in Rochberg's edition is -368, which is the date of birth for the child. The text goes on to record that he was "young" (*se-ḫi-ir*) in year 40, presumably of Artaxerxes II, giving the text a *terminus post quem* of -364. This later date is quite close to the first birth noted on our text in November of -363.

TABLE 6.1 *Birthnotes*

Text[a]	(Latest) Date	Individuals	Time span	Names	Date format[b]
BM 64148 (Text 29)	-368	1	4.5 years[c]		M, D, Y, T
BM 40187	-357	2	6 years	Y	D, M, Y, T
BM 33563 (Text 30)	-292	1		Y	D, T, M, Y
BM 34693 (Text 31)	-213	1			Y, M
BM 34567 (Text 32)	-114	3	36 and 2.5 years		Y, M, D, T

a The identification in paratheses refers to the numbering in Rochberg's catalog.
b t = time of day, d = day, m = month, y = year.
c As mentioned above, despite this text having only one individual there are two dates mentioned.

in the weakest possible sense 'standardized' …"[12] This must force us to consider the idea that the birthnotes do not share a commonality of genre. Rather, they only represent an intermediary step in the production of horoscopes and thus have no formal textual categorization.

Birthnotes offer one potential route for understanding the method by which horoscopes could be computed later in life based on the initial recording of the date of birth. We know from the study of Anu-bēlšunu's horoscope[13] that birthdates facilitated the creation of horoscopes well after the client's childhood. Steele has speculated that a unique astronomical tablet from Uruk (A 3405) represents Anu-bēlšunu calculating backwards to the range of time around his birth in order to cast his own horoscope.[14] Similarly, some birthnotes are more complex than just a single recording of a date. The examples that contain multiple individuals must be written, necessarily, after the birth of the youngest individual. The range of time between the multiple births then must represent a period for which some form of record (written or otherwise) existed for the

12 Rochberg 1998: 42.
13 Beaulieu and Rochberg 1996.
14 Steele 2000: 132–135.

earlier birth(s).[15] Three of the five known birthnotes included multiple dates ranging in time from a couple years to thirty-six at a maximum. In both Text 29 and BM 40187, the dates are in chronological order. Text 32, however, presents the dates for the three individuals in reverse, from youngest to oldest.

The addition of a fifth birthnote to the admittedly small corpus reinforces the idea that the stylistic and physical characteristics of the texts differ substantially with each other. The nonuniform nature of the birthnotes suggests that these were ephemeral texts, discarded after the computation of a horoscope, or prognostication by an astrologer. This raises the issue of whether or not these texts had a genre or type. The best example of a known genre for an astronomical text is the astronomical diaries which were known as *naṣāru ša ginê*, "regular watching." Birthnotes have no standard form or style, and even the data contained within the texts can differ considerably in its specificity or terminology. The one unifying characteristic is the function of the text, which in all cases is to preserve the date of a client's birth.

FIGURE 6.1
BM 40187, courtesy the Trustees of the British Museum

15 In some cases the birthdate could be remembered, and no birthnote was required. Text 31 only records the month and year of the client's birth, a relatively simple piece of information to remember or transmit.

Bibliography

Beaulieu, P. and Rochberg, F.

1996 "The Horoscope of Anu-Bēlšunu." *JCS* 48: 89–94.

Fincke, J.

2000 *Augenleiden nach keilschriftlichen Quellen: Untersuchungen zur altorientalischen Medizin*. Würzburg: Königshausen & Neumann.

Linssen, M.

2004 *The Cults of Uruk and Babylon: The Temple Ritual Texts as Evidence for Hellenistic Cult Practises*. Leiden: Brill.

Parker, R. and Dubberstein, W.

1961 *Babylonian Chronology 626 B.C.–A.D. 75*. Brown University Studies 19[a]. Providence: Brown University Press.

Reade, J.

1986 "Rassam's Babylonian Collection: The Excavations and the Archives." Pp. xiii–xxxvi in *Catalogue of the Babylonian Tablets in the British Museum*, eds. E. Leichty and J. Reade. London: British Museum publications for the trustees of the British Museum.

Rochberg, F.

1989 "Babylonian Seasonal Hours." *Centaurus* 32: 146–170.

1998 *Babylonian Horoscopes*. Philadelphia: American Philosophical Society.

Roth, M.

1988 "Ina Amat DN_1 U DN_2 Lišlim." *JSS* 33: 1–9.

Steele, J.

2000 "A 3405: An Unusual Astronomical Text from Uruk." *Archive for History of Exact Sciences* 55: 103–135.

CHAPTER 7

Converging Fortunes—Links between Celestial and Intestinal Divination

Ulla Koch

Modern forms of astrology bear testament to the enduring power of Mesopotamian science and cultural inventiveness. Our appreciation of the Mesopotamian contributions to astrology and astronomy has grown considerably in the past decades, not least due to the work of Francesca Rochberg. In antiquity, Mesopotamian celestial divination was rivalled only by intestinal divination, extispicy. In the first, the diviner scoured the open skies; in the other he scrutinized the hidden innards of animals. Celestial signs were sent by the gods as warnings and reassurance primarily for the king; intestinal signs were accessible to everybody able to afford it, and were most often sought when a particular course of action needed to be confirmed or rejected. They were thus very different, but complementary and equally valid sources of knowledge. From the very beginning the odds were not stacked against extispicy. From the point of view of practical use and versatility, extispicy should have been the most successful—and from the outset, it definitely had the largest client base. However, while astrology grew, adapted to new religions and worldviews, and prospered to such an extent that it is still with us today, extispicy withered and is today an oddity, practiced by head hunters and their ilk, and its modern forms are little researched.[1] It is safe to say that contrary to the Babylonian celestial pursuits, Babylonian extispicy has no living relatives.[2] The influence

1 J. Abbink 1993 is one of the very few treatments of modern extispicy (reference courtesy of Selena Wisnom). Blecher in his study *De extispicio capita tria* on ancient extispicy actually included references to the contemporary anthropological record (1905: 242–245). However, in modern as well as in older literature, the practice is mostly only mentioned in passing, e.g. by William Ellis (1832: 150), who gives, a brief description of the performance of extispicy and oneiromancy before going to war.

2 Divination by entrails is in fact one of the most widespread forms of divination, probably due to its natural link with sacrifice. Extispicy as practised today, for instance by the Turkana nomads in Subsaharan Africa, is not related to Mesopotamian divination. The Turkana "reader of intestines," inspects especially the blood and arteries, but also the movements of the animal (often a goat) when slaughtered, see Barrett 1998: 112–113. The Me'en form of extispicy appears related to other kinds of African divination where the answer emerges as a joint

DOI 10.1163/9789004363380_008

of Babylonian extispicy on Greek, Etruscan and Roman practices has long been assumed and is by now beyond doubt. Recent research into papyri from the first and fourth centuries CE on Greek technical extispicy shows both unquestionable loans from Babylonian omen traditions and that the Greeks made changes and adaptions as was their wont.[3] Extispicy was practised into late antiquity, and was even allowed under the first Christian rulers, but it did not survive into the Middle Ages.

Despite their obvious differences, the two divination disciplines always touched upon each other, and by the end of the first millennium, they even began to converge, their symbols and elements becoming associated, and the practice of astrology approached that of extispicy. New forms of celestial divination were oriented not towards the king and state but towards the fate of the individual and his everyday pursuits; but perhaps the most significant shift was that some of these new practices were oracular or induced and not ominous or spontaneous. Before then, oracular forms of celestial divination existed, e.g. the observation of shooting stars to get an answer to a question, but these forms were probably folksy rather than scholarly[4] and did not draw on the astrological lore embedded in the astrological omen series *Enūma Anu Enlil*. Heaven and Earth were perceived as closely interlinked also when it came to divination. This sentiment is most famously expressed in the "Diviner's Manual,"[5] the signs of Heaven and Earth reflected each other. In the Hellenistic period the two divination genres became further associated, and in the first century CE, the configurations of stars and the appearance of the entrails could be viewed as two manifestations of the very same message.[6] The present paper endeavours to outline some of the concrete points of interdisciplinary contact between the two divination genres, and the earliest indication for convergence between them in cuneiform sources. The overarching cultural commonalities between the two divinatory disciplines and their setting is not the topic here.[7] The themes I touch upon are:

effort of interpretation between the diviner and his audience (J. Abbink 1993). This is very different from Mesopotamian extispicy, which did not have this social component.

3 See Furley and Gysemberg 2015, especially pp. 77–95.

4 This distinction between divinatory disciplines goes back to Reiner 1960.

5 Oppenheim 1974.

6 See most recently William Furley and Victor Gysembergh 2015: 13–15 and 109–114. The Greek philosopher Onesander wrote a theoretical treatise on military command in which he described the relationship between astrology and extispicy thus "…the movements of the stars in heaven…are revealed by sacrifice in an alternative manifestation through the entrails," quoted *ibid.* p. 109–110.

7 The most recent description of this vast topic is Stefan Maul 2013: 277–296.

1. Validation of celestial signs by extispicy.
2. Prediction of celestial signs by extispicy.
3. *Ikribu*-prayers to celestial bodies in extispicy rituals.
4. Interdisciplinary entries in omen series and commentaries.
5. Converging semantics: correlation of celestial and intestinal elements.
6. Converging purposes: from celestial omens to celestial oracles.

Validation of Celestial Signs by Extispicy

The earliest and most concrete connection between the two divination disciplines is undoubtedly the use of extispicy to investigate or confirm the significance of a celestial phenomenon as a sign.[8] In the Old Babylonian period there is no evidence that celestial phenomena were actually interpreted according to astrological principles. Celestial omen compendia were compiled from the Old Babylonian period on, but unfortunately only very few are preserved, and the majority of these pertain to the interpretation of lunar eclipses.[9] So far there are even fewer examples of reports of actually observed celestial phenomena from the second millennium BCE, and none were interpreted with reference to astrological omens. The extispicy expert (*bāru*) Asqudum from Mari reported a lunar eclipse "on the 14th day," which he deemed worrying (*maruṣ*).[10] Evidently this interpretation could not stand alone. In order to ascertain whether there was any real cause for concern, he performed extispicy and advised the king to have the same done. An Old Babylonian manuscript from Mari, perhaps a fragment of a letter, quotes an excerpt of a menological text concerning eclipses, but without a report of an actual observation.[11] Presumably, Asqudum had similar texts available, but he still had to rely on extispicy in order to make a certain interpretation of the eclipse. The *Mari Eponym Chronicle* mentions a solar eclipse in the year after the birth of Shamshi-Adad I. The eclipse was probably total and the chronicle is fragmentary, but it would appear that the solar eclipse was simply considered a spectacular event.[12] Solar eclipses are more frequent than lunar, but they are much more rarely visible from a particular point on earth and only played a small part in

8 Pongratz-Leisten 1999: 12–14.
9 Rochberg 2006.
10 *ARM* 26 81.
11 Charpin 1989, Heimpel 2003: 271.
12 Translated by Glassner 2004 no. 8.

omen astrology. Half a millennium later, a solar eclipse observed and reported in Ugarit still could not be interpreted solely relying on the astrological omen tradition; again its significance was investigated and confirmed by extispicy.[13] Eclipses must have been perceived as portents, and perhaps some rules of interpretation were applied, but whether the signs pertained to a chosen course of action or particular situation was confirmed by extispicy. This is what Sargon did on the famous occurrence of a lunar eclipse during his 8th campaign a thousand years later than the Mari letters.[14] Other than this, there is little evidence that extispicy was used to check the significance of a celestial sign at the Neo Assyrian court. However, one extispicy report (SAA 4 316) concerns the interpretation of a particular dream, and it seems likely that extispicy would still be used to clarify any kind of worrisome, unclassifiable, signs. A number of Middle Babylonian extispicy reports are concerned exactly with the interpretation of worrisome dreams.[15] The latest evidence for this interdisciplinary approach is Nabu-na'id, who records that he investigated a lunar sign by extispicy in connection with the investiture of his daughter as *entu*-priestess of Sin,[16] paying homage to millennia old traditions combining astrology and extispicy and of verifying the selection of an *entu* by extispicy.

Conversely, extispicy could be used to investigate whether an eclipse would take place. A collection of oracle prayers (*tāmitus*) concerns the occurrence of lunar and solar eclipses.[17] The dates stipulated for either kind of eclipse are quite within the possible range (12th–16th), so presumably the writers had knowledge of astrology/astronomy—or an astrologer had at some point posed the questions. The *tamītus* were exemplary oracle questions and do not necessarily present queries that were actually performed. They were a well-established part of the technical literature of the Seer (*bārû*) and they

13 KTU 1.78, see Dietrich and Loretz 1990: 39–86. The eclipse is most likely to be dated March 5, 1223 BCE. See the discussion of the dating by Dietrich and Loretz 1990: 62.

14 TCL 3, Oppenheim 1960: 137–138.

15 Kraus 1985 no. 17 (2), 18 (1), (3), (5).

16 The whole passage (*YOS* 1 45 I 8 ff.) is quoted and discussed by Reiner 1995: 76–77, with references to previous literature. See also Rochberg 2004: 225–226. The omen Nabû-na'id quotes is "on the 13th day of Ulūlu, the month of the work of goddesses, the Fruit became eclipsed and set while eclipsed (this means) Sin requests an *entu*-priestess." Full moon and lunar eclipses were possible on the 13th day, but the 13th is not among the canonical days for lunar eclipses in *Enūma Anu Enlil*, Rochberg-Halton 1988: 38–40. Perhaps Nabû-na'id cited an *aḫû* tradition. *Enūma Anu Enlil* tablet 7 § VI.4 contains the omen "If an eclipse occurs in Ulūlu in the morning watch (this means) Sin will request an *entu*-priestess" (Rochberg-Halton 1988: 133 and n. 32). It is the wrong watch, cites no date, but it is the right month and apodosis.

17 Lambert 2007: 42–51.

are preserved from the first millennium BCE in library editions complete with colophons referring to originals, both tablets and writing boards. There are however, no reports preserved which could attest to the actual use of extispicy to inquire into the imminent occurrence of eclipses. By the first millennium other much more reliable means of predicting an eclipse were available.

The earliest reference to a celestial sign that could have been interpreted entirely according to astrological principles stems from a Hittite text known as *Mursili's Accusations against Tawananna.*[18] The Hittite king Mursilis II (1318–1290 BCE) refers to a solar sign which his stepmother, Tawananna, interpreted to his disadvantage, apparently without recourse to extispicy: "As I (Mursilis) set off towards the land of Azzi, the Sun god sent a sign. The Queen (Tawananna) in the land of Hatti declared: 'the sign, which the Sun god sent, what did it signify? Did it not signify the death of the king?'" The last wife of Suppiluliumas I was probably a daughter of the Kassite king of Babylon, Burnaburias II. Presumably, Tawananna had brought with her the practical use of celestial divination from her father's court.

Prediction of Celestial Signs by Extispicy

In the omen compendia, eclipses were a topic of extispicy apodoses as a kind of generic misfortune.[19] Eclipses could be predicted by all kinds of signs, including other celestial signs. One extispicy commentary specifies: "In a favourable extispicy: an eclipse will not occur, in an unfavourable one: it will occur."[20] This makes sense from the point of divination hermeneutics, but does not necessarily imply any form of convergence. However, a degree of intertextuality is suggested by omens with apodoses that refer to more specific eclipses, for instance:

> If the top of the Presence is surrounded by Holes: Eclipse on the 16th day.[21]

The groove called the Presence was the crucial element in extispicy; without it the extispicy was invalid. Most of the Presence was favourable including its

---

18 Haas 2008: 84–85, 134.

19 E.g. Koch-Westenholz 2000 no. 20:71 "If the Presence is obliterated: Eclipse," no. 59:105 "If there is a Cross on the Strength: Pestilence or eclipse," and no. 64:43 "If the Well-being is like the sign "kur": Eclipse."

20 Koch-Westenholz 2000 no. 19:31.

21 From tablet 6 of *Šumma manzāzu,* Koch-Westenholz 2000 no. 6:68.

top. Holes were negative markings and associated with death and blindness throughout *Bārûtu*. It is tempting to interpret the protasis as an allusion to the eclipse myth, according to which the moon becomes eclipsed because it is surrounded by the evil Sebettu-demons, although this may be taking the interpretation too far.[22] An eclipse on the 16th day pertained to Subartu, according to one tradition.[23] It would have been an unusual occurrence, and accordingly an eclipse on the 16th is reported in SAA 8 no. 279 as a great observation (EN.NUN GAL-*ti*).

Another example, also concerning the Presence is:

> If there is a Hole in the centre top of the Presence: A high priest will die, fall of the chief temple administrator, the sacrificer will die at the beginning of the year, or his son will die, or, one of his eyes will go blind. For warfare: Defeat of the leader of the army. Eclipse of the evening watch. For the king: Revolt. Or, a weir will collapse during the damming of its canal and take me away.[24]

Again an eclipse is associated with Holes. This time the eclipse is specified as taking place during the evening watch, i.e., the first watch, which matches the position of the Hole in the first part of the Presence (*rēš manzāzi*). According to the astrological commentary *Šumma Sîn ina tāmartīšu* an eclipse of the evening watch meant plague,[25] so its significance was sufficiently sinister to fit with the rest of this apodosis.

Ikribu-prayers to Celestial Bodies in Extispicy Rituals

Every step of the extispicy ritual was accompanied by *ikribu*-prayers, mostly quite brief blessings to ensure the favourable outcome and successful performance.[26] Šamaš and Adad were the masters of sacrificial divination, and most of the *ikribus* are directed at both of them or Šamaš alone, but they were not the

---

22 *Utukkū Lemnūtu* tablet 16 (CT 16 19). For a recent translation and composite text, see Geller 2007: 178–187 and 251–257. The concluding paragraph of the Sin omens of *Enūma Anu Enlil* also refers to this myth, see Rochberg-Halton 1988: 269–270 and 216.

23 Tablet 20 of *Enūma Anu Enlil*, see Rochberg-Halton 1988: 39. The tradition is also included in the commentary *Šumma Sîn ina tāmartīšū* tablet 4 (ACH 2. *Suppl. Sin* 19:9′–20′).

24 Koch-Westenholz 2000 no. 6:11.

25 ACH 2. *Suppl. Sin* 19:9′–20′.

26 For an introduction to the *ikribu* genre, see Lenzi 2011: 46–49, 85–101 and Cooley 2011. A new edition of the first millennium *ikribus* are in preparation by the author.

only gods to be invoked during extispicy rituals. Indeed, in the Old Babylonian period, it appears quite the norm that any god of the implorer's choice could be asked for an oracle answer. The Old Babylonian collection of *ikribus* edited by Starr (1983)[27] are all addressed to Šamaš and Adad as are those from the first millennium published by Zimmern (1901). However, two Old Babylonian *ikribus* address Ishtar as Annunītum and Ninsianna.[28] Among the gods called upon in Old Babylonian and Kassite reports on acts of extispicy are Šamaš,[29] Marduk,[30] Enlil,[31] Nanāja,[32] Ištar,[33] Annunītum,[34] Sin,[35] Nergal,[36] Nusku,[37] Sirius,[38] an unnamed ancestral god,[39] and an unnamed "angry god."[40] One Old Babylonian report mentions extispicy performed for the Triad Bēlet-ilī, Šulpae and Pap-nigin-gara.[41]

The extispicy itself took place in the morning when Šamaš was visible, but the preparations for the performance took place at night,[42] so one could reasonably expect some of the *ikribu*-prayers of the seer to have been addressed to the stars as the visible celestial manifestations of gods.[43] The most famous is of course the Old Babylonian prayers to the Gods of Night, which are directed to the stars of the sky and poetically invoke the peaceful night landscape where the diviner pleads his case.[44]

27 HSM 7494, AO 7032 and AO 7031.
28 Meyer 1982.
29 Goetze 1957 no. 9 (MLC 2255).
30 Goetze 1957 no. 1 (Meissner text; new edition Koch-Westenholz 2002: 133–134), no. 3 (CBS 1462b) and no. 16 (CUA 101), Kraus 1985 no. 16 (CBS 10495).
31 Goetze 1957 no. 19 (UMM G 33).
32 Goetze 1957 no. 21 (AO 7615), Kraus 1985 no. 21 (1) (YBC 4363).
33 Goetze 1957 no. 21 (AO 7615), Kraus 1985 no. 21 (2) (YBC 4363).
34 Goetze 1957 no. 6 (CBS 1462a), *VS* 24 no. 116 (BM 17542, edited by Mayer 1987).
35 Richardson 2002 no. 2 (BM 97433).
36 Starr 1977 (YBC 16148). The name is written ^{d}GÌR.RA.GAL.
37 Kraus 1985: 150 (CBS 13517).
38 Kraus 1985: 150 (CBS 13517).
39 Goetze 1957 no. 5 (VAT 6678, new edition Koch-Westenholz 2002: 134–135). The family god called upon is referred to as the "god of her father."
40 *CT* 4 34b (BM 78680, Goetze 1957 no. 15, Koch-Westenholz 2002: 135).
41 Goetze 1957 no. 12 (YBC 5018).
42 Fincke 2009.
43 The many functions of the stars as intermediaries and incarnations of the gods are described by Reiner 1995. See also Steinkeller 2005 for a discussion of the cosmological setting of extispicy.
44 Known in two slightly different versions, for a translation and references, see Foster 1996: 148–149.

However, none of the preserved first millennium prayers to the Gods of Night, was part of the extispicy ritual.[45] The only text I know of, from the first millennium where the Gods of Night are invoked in connection with divination, is *STT* 73:110–117, which is a ritual for producing an oracle from pouring water on an ox and presenting offerings to the gods of night (Reiner 1960). Otherwise, the Gods of Night are mentioned only in apotropaic rituals, *namburbûs*,[46] including one specifically directed against the evil following flawed flesh in an extispicy performed in the house of a private person in the month of Nisan[47] and other protective (e.g. *SpTU* I 11) and apotropaic rituals such as šu-ilas and *lipšur* litanies with no reference to extispicy.[48] The "New Prayer to the Gods of Night,"[49] which has many literary similarities to the Old Babylonian prayers, was *not* part of the extispicy ritual.[50] It is yet another apotropaic ritual to be performed "so that no [evil] from unfavourable signs [will approach a person]" (K 2315+: 109). This included signs from dreams, flawed flesh (*ḫi-in-qi* udu.níta) "which keeps occurring in my house during the ritual of the extispicy" (*ni-pi-iš-ti* máš.gíd.gíd), and numerous other signs belonging to the terratological and terrestrial omen genres. As I have argued elsewhere, both *namburbûs* and the "New Prayer to the Gods of Night" probably refer to the unfavourable omen of a botched performance of the ritual of extispicy or that the flesh simply was too flawed to be read, rather than to an unfavourable extispicy answer.[51] The first incantation in the series Maqlû

45 The fragment K 7270 (Fig. 7.1) is related, it mentions the great [gods] of Anu, Enlil and Ea, that is the stars of the 3 celestial paths. However, due to its fragmentary state it is impossible to tell what the context is.
1 [DINGIR.MEŠ] GAL.MEŠ *šu-ut* [d]*[a-nim* ……]
2 [DINGIR.MEŠ] GAL.MEŠ *šu-ut* [d]E[N-LIL ……]
3 [DINGIR.MEŠ] GAL.MEŠ *šu-ut* [d]*[é-a* …..]
4 [……….] *al-si-k[u-nu-ši* ……]
5 [………….] *na-as-ḫi-i[r-an-ni* …….]
The fragment was identified by Reiner and mentioned by Oppenheim (1959: 293 n. 2) as K 7260!, who reconstructed it as referring to [MUL].MEŠ. There is however no doubt that the remains are of the sign GAL, cf. fig. 7.1 and the photo published on CDLI (cdli.ucla.edu/P397150).

46 E.g. *KAR* 38 with duplicates, Reiner 1995: 16.

47 K 3853+, Maul 1994: 439–444. Similarly, *STT* 231 obv. 31–37, see Reiner 1995 n. 69.

48 For references see Reiner 1995: 17–19.

49 K 2315+, Oppenheim 1959.

50 Reiner (1995: 67) suggests that it was, but there is no indication that it was anything but apotropaic.

51 Koch 2010.

(I 1–36)[52] is addressed to the Gods of Night who are exhorted to pass judgement on the witch or warlock who troubles the sufferer.

There are, however, a couple of first millennium *ikribu* prayers which are directed at celestial bodies.[53] They were left out in the edition of the extispicy ritual by Zimmern (1901) even though he recognised their relationship to the other extispicy *ikribus*.

The *ikribu* to Ninurta as Sirius (K 128)[54] is almost completely preserved on an oblong tablet (*u'iltu*) written in Babylonian ductus. The tablet was edited, collated and translated recently by W.R. Mayer (2005) and by Foster (1996).

Ninurta is associated not only with Sirius but also with Saturn and Mercury.[55] In astrological contexts, Saturn could represent the Sun and omens and reports that mention "the Sun" (20 or ᵈUTU) frequently actually refer to Saturn. Saturn's proper name was "the steady" (*kaiamānu*), which in the commentaries could be explained as "the star of truth and justice" (*kittu u mīšaru*).[56] It is possible that the association between the two was due to wordplay on the common root of *kittu* and *kajamānu*. Both the Sun and Saturn were thus linked to justice. Another reason for the association with the Sun could be that Sirius is the brightest star on the firmament and easily visible. The close association between Saturn/Ninurta and Šamaš may be the reason why this unique *ikribu* is directed at him.

Sirius rises heliacally and is visible in the east after a period of about 70 days of invisibility. When Sirius reappears after conjunction with the sun in the morning sky in summer it inaugurates the period of about five weeks, known in antiquity as "the dog days," when Sirius is visible for a short period in the morning together with the Sun. Due to the precession of the equinoxes, the heliacal rising is today in early August; when this *ikribu* was written, it would have occurred in early July. If the *ikribu* were to be performed at the heliacal rising of Sirius and the days following, it would only have been pertinent for a very limited time. This is not unreasonable, since the *ikribu* was not part of the

52 For a discussion see Schwemer 2010: 313–317; the most recent edition of the text is Abusch 2016.

53 Mentioned *en passant* by Nougayrol 1963 n. 1.

54 Published by Burrows 1924. It has been translated by Falkenstein and von Soden 1953: 275–277 and 389, Seux 1976: 480–482, Foster 1996: 621–622 and edited by Mayer 2005. I have referred Lambert's unpublished copy, Lambert Notebooks 1364–1366 (available at oracc.org/contrib/lambert).

55 For the celestial aspects of Ninurta as a star and arrow, see Annus 2002, 133–138.

56 Koch-Westenholz 1995: 123 with references to previous literature. Cf. also Brown 2000: 69, 74–79.

standard extispicy ritual. It was intended for a Seer who had gotten a difficult and unintelligible oracle answer and needed clarity; he therefore waits for the Sun and Sirius to rise in conjunction—their combined presence presumably reinforces the validity of the oracle.

1	^{d}MAŠ *šur-bu-ú* DINGIR *qar-du a-šá-ri*[*d*] ^{d}*A-nun-na-ki mu-ma-'i-ir* ^{d}*Í-gì-gì*	Oh exalted Ninurta, valiant god, foremost of the Anunnaki, commander of the Igigi,
2	DI.KUD *kul-la-ti sa-niq mit-ḫur-t*[*i mu-š*]*aḫ-li ek-le-ti mu-nam-mir e-ṭu-ti*	judge of the whole world, who probes disputes, a light in darkness and illuminator of shadows,
3	*pa-ri-is pu-ru-us-se-e* [Ø] *a-na* UN.MEŠ *a-pa-a-ti*	who renders the verdicts for teeming mankind!
4	*be-lí šu-pu-ú mu-ṭib* UZU KUR *šá ina ta-m*[*ar*]*-ti-šú* GIG *di-'i-i lem-nu*	My splendid lord, who cares for the land, at whose appearance illness and evil headache
5	*iḫ-ḫa-zu tub-qa-a-ti* GIG *naq-du i-tur-ru áš-ru-uš-šú*	go into hiding, the critically ill recovers!
6	*re-mé-nu-ú ga-mil nap-šá-ti mu-bal-liṭ mi-*[*i-t*]*u*	Merciful one, who spares life, reviver of the dead,
7	*ta-mi-*[*i*]*ḫ kit-ti u mi-šá-ri mu-ḫal-liq* [*ra-ag-g*]*i?*[a]	who holds truth and justice firmly and makes [.....] disappear,
8	*šu-ku-du la a-ni-ḫu šá kul-lat a-a-bi i-*[*na*]*-ru*	Tireless Arrow, who [vanqui]shes all foes,
9	*u*$_{4}$*-mu ra-*[*bu*]*-ú ta-mi-i*[*ḫ*] *ṣer-re-*[*et* AN *u* KI]	great storm, who grasps the *lead rope* [of Heaven and Netherworld]
10	DI.KUD *pu-ru-us-se-e ba-ru-ú te-re-*[*ti* x x]	judge of verdicts, seer of orac[le answers ...]
11	^{d}GIŠ.BAR *mun-na-aḫ-zu qa-m*[*u*]*-ú lem-*[*nu-ú-ti*]	Flaring-up fire who burns up the evil-doers
12	*šá* MUL.GAG.SI.SÁ *ina* AN*-e zi-kir-šu ina* [*g*]*i-mir* ^{d}*Í-gì-gì šur-bu-ú* MU*-ka*	whose name in the sky is Gagsisa, your name is the most exalted among all the Igigi.
13	*ina kul-lat ka-la* DINGIR.MEŠ *šu-*[*un*]*-na-tu* [Ø] DINGIR*-ut-ka*	Among all the gods your divinity is double,[b]
14	*ina ni-pi-iḫ* MUL.MEŠ *nu-um-m*[*u*]*-ru zi-mu-*[*ka ki-ma*] ^{d}UTU*-ši*	when the stars appear, your lustre is brilliant [like] the Sun.

15	*ta-bar-ri si-ḫi-ip da-ád-me a-*[*n*]*a nu-ri-k*[*a nam-ri ṣu-um-m*]*u-ru*	You survey the entire inhabited world, towards your [bright] light [thirs]ts
16	*te-ni-še-e-ti áš-šú na-as-s*[*u*] *dal-pu šu-n*[*u-ḫu šu-te-šu-r*]*i*[c]	mankind, in order to save the wretched, the tired, the ex[hausted]
17	*šá la i-šu!-ú mam-ma-na tu-kul-ta-šú at-ta šá?* [x x x x x] x	who have no one you are his support, he who … [………..]
18	*ta-gam-mil šá ina šur-qu pu-un-zu-ru-ma ez-bu a-?*[- x x x x x x x x]	you save, the one who was kidnapped[d] and abandoned .. [you ………..]
19	*ul-tu a-ral-le-e ta-qab-bi né-eš-su šá ina a?-*[*šar? ṣa-al-t*]*i*	You call him back to life even from the Underworld, he who in a pl[ace of strife]
r1	*la-mu-u qab-lu a-na mu-ú-tu šu-lu-ku-ma iz-ku-ru zi-kir-ka*	In the thick of battle, marked for death, calls your name
r2	*re-mi-na-ta be-lum ina dab-de-e ta-gam-mil-šú u ana-ku* DUMU LÚ.ḪAL *pa-li-ḫu* ÌR*-ka*	You save him from the fight, merciful lord. Now I, a seer, your humble servant,
r3	*ur-ri dal-pak-ma a-še-ʻi ma-ru-uš-ti di-i-nu šup-šuq-ma a-na la-ma-da áš-ṭu*	in the morning exhausted I seek (the cause for) my problem, my case remains difficult and hard to understand,
r4	*pa-ra-as ár-ka-tú né-sa-an-ni* ^{d}UTU*-ši a-ku-šam-ma ina* GI$_6$ *ú-qa-a-a* SAG*-ka*	I am far from coming to the bottom of it. During the day I come to you, at night I wait upon you,
r5	*a-na la-ma-da ar-ka-ti at-ta-ziz ma-ḫar-ka a-na šu-te-šú-ru di-i-nu*	I stand before you to understand the cause, for a right verdict
r6	*ni-iš qa-ti ra-šá-ku i-ziz-za-am-m*[*a i*]*na qé-reb* AN*-e* KÙ.MEŠ *ši-mi zi-kir* KA*-ia*	I have raised my hands in prayer. Stand by me, listen to me from the holy Heavens,
r7	*pu-ṭur ár-nu pu-su-us ḫi-ṭ*[*i*]*-ti tu-bu-uk-ku-ú-a lik-ru-bu-ka*	Absolve any guilt, remove any error, may my grain offerings give you praise
r8	*sér-qu-ú-a lìb-ba-ka li-ni-*[*iḫ-ḫ*]*u e ta-šeṭ i-ziz-za-am-ma*	May my aromatic offerings set your heart at peace, do not disdain to stand by me and
r9	[*d*]*i-na di-i-nu un-nin-ni-ia li-qé-e-ma ši-mi tés-li-ti*	Decide my case, accept my prayer and hear my invocation.

r10	[*a*]-*na mim-mu-ú ak-pu-du pu-ru-us-su-ú šu-kun-ma ina an-ni-ka ki-nu-um*	For all I have planned give me a decision and with your firm consent
r11	[*lu*]-*ma-'i-ir ur-tu u ana-ku ina* IGI-*ka* ZI-*tim lu-ṭer* ^{d}MAŠ *šur-bu-ú*	[let me] give directions. As for me, let me stay alive before you, exalted Ninurta,
r12	[DINGIR] *el-la ina* SÍSKUR GUB-*za-am-ma ina* INIM-ia_5 MU ŠU-ia_5 *u mim-ma* DIRI DÙ-*šú*	holy [god], stand by me in the sacrifice and in my prayer, when I raise my hands in prayer, and in everything I do,
r13	[*ina t*]*a-mit a-kar-ra-bu kit-ti lib-ši*	In the blessing I pronounce let there be truth!
r14	[*ik*]-*rib* MUL.GAG.SI.SÁ *ki-ma ina* ^{d}UTU.È *iz-za-zu*	*Ikribu*-prayer for Gagsisa when it stands at sun rise.

a The traces do not really fit GI.

b Marduk is granted double divinity in *Enūma eliš* I 91.

c Room enough for a combination of the readings suggested by Mayer and Burrows.

d Difficult passage. The translations by Foster and Mayer are not entirely convincing. But since *šurqu* does not otherwise mean *Raubüberfall*, I follow Foster's translation based on the literal meaning "he who was hidden away by theft," that is, "kidnapped."

Two brief *ikribu*-prayers directed at the two Assyrian Ištars are probably related to extispicy;[57] at least they contain the formula found at the end of every extispicy-*ikribu* albeit in its abbreviated form: "*ina niqê šuāti kīam.*" If the Ištars represent Venus it is not explicitly stated. Venus was associated with the female and male aspects of Ištar as a morning and evening star but not, to my knowledge, specifically with the Assyrian Ištars.[58] The two *ikribus* are preserved on an oblong tablet (*u'iltu*) written in Babylonian ductus but copied from Assyrian originals (K 20).[59] The similarity in format also suggests a relationship with the Sirius *ikribu.*

57 They are mentioned by Mayer 1976: 32 n. 63, as related to the extispicy-*ikribus*.

58 See e.g., Koch-Westenholz 1995: 125–127.

59 Copy by Pinches published by Gurney 1936–37: 368–369 pl. 6, with corrections to Pinches' original edition (1909: 67–69). I have referred to Lambert's unpublished copy, Lambert Notebooks 1364–1366.

1	^{d}*iš-tar* GAŠAN GAL-*ti a-ši-bat* É-MA[Š.MAŠ *šá qé-r*]*eb* NINA.[KI]	Ištar, great queen, who dwells in E-mašmaš in Niniveh
2	^{d}*a-nu-ni-tum šá* MÈ.MEŠ-*tu*[a] SUM-*at* [*a-gu-uḫ-ḫ*]*i*[b] ri [x]	Anunītu of battles, who gives the [……]
3	ZI-*bu*[c] *gab-šú* TAR-*at* EŠ.BAR *mu-šim-m*[*at*] NAM.M[EŠ]	An overwhelming attack, who makes decisions and decides the destinies,
4	*ḫa-i-rat* KUR.MEŠ *šá la-lu-šá ku-uz-bu* ^{d}*iš-tar muš-ta-lat* ik- x[d]	Wooed by all lands, whose charm is irresistible, Ištar who ponders ….
5	*mul-te-šir-at* DÙ URU.URU *qa-mat gi-ri-šá*	Who guides all cities, who consumes her enemies,
6	*šá si-pu-šá* DÙG.GA ^{d}*iš-tar* GAŠAN GAL-*ti*	Praising her is sweet. Ištar great queen,
7	[*a-ši*]-*bat* É-MAŠ.MAŠ *ina* UDU.SÍSKUR.NE *ki-a-am*	who dwells in E-mašmaš, in this sacrifice *and so on* (i.e., in this sacrifice stand by me, and in my prayer, when I raise my hands in prayer, and in everything I do, in the blessing I pronounce let there be truth).
8	[*ik-ri*]*b* ^{d}*iš-tar šá* NINA.KI GABA.RI BAL.TIL.KI	Extispicy-prayer to Ištar of Niniveh. Original from Aššur
9	[^{d}15 GAŠAN GA]L-*ti a-ši-bat* É-GAŠAN.KALAM.MA	[Ištar gre]at [queen] who dwells in E-gašankalamma
10	[GAŠAN GAL-*t*]*i*[e] URU-4-DINGIR *e-tel-lat*[f] ^{d}15.MEŠ	[grea]t [queen] of Arbil, Lady of the goddesses
11	[AN]-*e u* KI-*tim qa-i-šat* TI.LA	[…. lady of Hea]ven and Netherworld, who bestows life,
12	[ ] *la-ab-bi te-let* ^{d}15	[.….] ….. able Ištar
13	[*š*]*á la* BAL-*ú qí-bit-sa*	[……….] whose command does not bend
r1	[*la*] *ut-tak-ka-ru qí-bit* ŠÀ-*šá*	[………] whose decrees are irrevocable
r2	[*š*]*á i-lam-ma-du* [*milikša*] *a-a-um-ma*	no one can fathom [her plans].
r3	[^{d}]15 GAŠAN GAL-*ti a-*[*ši-bat* É]-GAŠAN.KALAM.MA	Ištar, great queen who dwells in E-gašankalamma

r4	*ina* UDU.SÍSKUR NE-*i* [*k*]*i-a-am*	in this sacrifice *and so on*
r5	*ik-rib* ^{d}15 GABA.RI URU-4-DINGIR	*Ikribu*-prayer to Ištar. Original from Arbil.

a *Taḫāzu* is almost always singular, but used in plural in connection with Ištar of Arbil in a treaty *ABL* 1105 (*SAA* 2 9) r25 and with Annunitu in the commentated bilingual hymn to Nanāja (*JNES* 33: 227 line 35; Reiner 1974).
b *Nadīnat aguḫḫi* is an epitheton of Ištar in *KAR* 306:24. The meaning of *aguḫḫu* is unclear and I have no suggestions for the last word.
c Pinches and Gurney read *zi-bu* and translated "jackal" and "vulture" respectively. This word is rarely attested and almost always spelt with "zi-i-," probably to distinguish it from the logogram ZI. *Tību* is attested in transferred meaning as an epitheton of Hammurabi and Tukulti-Ninurta (*CAD* s.v. *tību* mng. 3').
d Sign begins with two horizontal wedges.
e The traces could fit either TI and U.
f Assyrianism.

Two further manuscripts contain *ikribu*-prayers to celestial bodies (A K 3794 + BM 99127 (KI 1904-10-09, 159)[60] and B K 2751 + K 2792 + K 7973 + K 9247 + K 10011 + K 13789).[61] MS A contained a series of *ikribus*; the first (obv. 1–6) is addressed to Adad. The rubric reads (A obv. 6): *ik-rib sa-raq* NÍG.[NA-*ki*] "blessing for sprinkling the censer," a common ritual procedure also in the extispicy ritual. The second (obv. 7–31) is addressed to Sin, the third to Šulpaea (obv. 32–38), perhaps Jupiter. The preserved parts of MS B contain only *ikribus* to Sin. According to the remains of one rubric (B rev. ii 24) one of them was to be performed on the 15th day:[*ik-rib* ....] UD-15-KAM. The 15th day is mentioned as one of the favourable days for performing extispicy in the Seer's vade-me-cum *KAR* 151, perhaps the other *ikribus* of the text were related to some of the other favourable days. Extispicy was generally forbidden on the days most closely associated with the phases of the lunar cycle (7th, 14th, 21st and 28th).[62]

The sparse evidence from the Sirius and Sin *ikribus* suggests that celestial bodies were called upon in the extispicy ritual ad hoc and in accordance with their actual appearance in the sky.

60 Published by Perry 1907 no. 5b Taf. I pp. 23–24 (K 3794) and Langdon 1915 (Ki 1904-10-09, 159). I have referred to Lambert's unpublished copy of the joined fragments, Lambert Notebooks 1387–1389. Foster has recently translated lines 7–30, 1996: 665–666.

61 Published by Perry 1907 no. 5c Taf. II (K 2792+7973). I have referred to Lambert's unpublished copy of the joined text, Lambert Notebooks 1389–1391.

62 See e.g. Koch 2013: 132–134.

Interdisciplinary Entries in Omen Series and Commentaries

The formal similarities between the divination manuals of the two disciplines extend to the omen-format and divination arithmetic, the use of stock apodoses, a high degree of standardization, serialization, and diffuse technical terminology that necessitated the development of an extended ancillary literature. For most of their history, the two disciplines were mastered by different professionals. While it was not common, in the first millennium the same man could master, if not the practice, then at least the literature of both disciplines. The manuals were copied and commented upon by the same group of learned men for their own collections and for the libraries of kings and temples. The most prominent example of this is of course the polyhistor Nabû-zuqup-kēnu, who copied texts of all genres including *Bārûtu* and *Enūma Anu Enlil*. Despite the formal similarities and the overlap in specialists and tradition, the two genres were hardly ever combined in the omen series, and cross-references are very rare indeed.

The calendar is a superb tool for divinatory purposes with its mathematical structure and inherent links to religious symbolism and the cult. It is most naturally linked to celestial phenomena, but the calendar worked as a unifying vehicle for all kinds of omina. The Babylonian calendar was constructed according to the movements of the Moon and Sun, and it was regulated by the observation of certain other celestial bodies. This makes it very obvious that astrological omens already from the Old Babylonian period should be structured according to which month they were observed.[63] Eventually, both celestial omens and terrestrial omens were incorporated into the calendrical omen series *Iqqur īpuš*, especially the series *Enūma Anu Enlil* and *Šumma ālu*. Indeed, the hemerologies and menologies cut across all divination genres. There are even examples of extispicy omens interpreted according to which month the observation was made; of all the thousands of extispicy protases, however, only two examples were selected for inclusion in the *Iqqur īpuš* sections of the royal hemerology *Enbu bēl arḫi*.[64] In *Enbu bēl arḫi* each month was described first with a section of rules, injunctions, rituals and taboos of each day of the month, then a section of *Iqqur īpuš* omens pertaining to that month (*série mensuelle*). The two extispicy omens selected pertained to damage to the Gall Bladder and the Finger (*processus pyramidalis*) respectively:

63 Charpin 1989, Heimpel 2003: 271.

64 The most recent edition is Livingstone 2013: 199–248. Livingstone did not recognize the passages citing extispicy omens.

> [If] the Gall Bladder is destroyed [in the month of Simānu]: His heart will not be content.
> [If] the Finger is destroyed [in the month of Simānu]: His heart will not be content.[65]
> If the Gall [Bladder is destroyed] in the month of Araḫsamnu: His heart will not be content.
> [If the Finger is destroyed in the mo]nth of Araḫsamnu: His heart will not be content.[66]
> [If] the Gall Bladder is destroyed [in the month of Ṭebētu]: The man's wife will die.
> [If] the Finger is destroyed [in the month of Ṭebētu]: Adad will flood the land.[67]

At the time Labat edited the texts,[68] the two extispicy omens were only attested in the *Iqqur īpuš* section of *Enbu bēl arḫi* and not in the standard versions of *Iqqur īpuš*. However, the omens are also attested in *Iqqur īpuš* versions from Babylonia and Sippar.[69] The *série génerale* is represented only in excerpts. Anyway, it is certain that the omens also existed as part of the extispicy series itself. Of course, it is impossible to know whether they originated in the extispicy series or were constructed for *Iqqur īpuš* and transferred into extispicy compendia from the calendrical texts. However, whereas it can be difficult to tell whether a given fragment is part of *Enūma Anu Enlil, Šumma ālu* or *Iqqur īpuš,* there can be no doubt that these fragments were not part of a calendrical text, since the omens begin with the indicator BE rather than DIŠ. The reason these two parts of the Liver were selected may be that the Gall Bladder was associated with the king: one of its names was the Shepherd—a title any good king aspired to. Extispicy omens pertaining to the Gall Bladder often concern the royal family, the king and throne.[70] Part of the Finger was called the Palace,

65 K 4068 iv 21–22. The passage is included in Livingstone's edition of *Enbu bēl arḫi,* but misunderstood, 2013: 212.

66 K 3269 iv 22–23. Labat 1965 pl. xliv. The passage is included in Livingstone's edition of *Enbu bēl arḫi,* but misunderstood, 2013: 223. Livingstone reads the remains of the sign ZÉ as I.

67 K 2809 iv 12–13. Labat 1965 pl. xlv. The passage is included in Livingstone's edition of *Enbu bēl arḫi,* but misunderstood, 2013: 227.

68 Labat 1965: 139 n. 4.

69 The unpublished manuscripts BM 65137 and BM 65570 are mentioned by Reiner 1995 n. 284.

70 Jeyes 1989: 62–63.

and the Finger seems generally to have been associated with life at court, especially its more sinister aspects.[71]

K 11142 (Fig. 7.2) contains the omens pertaining to a missing Finger each month. The apodoses are missing and the small fragment simply lists:

1′ [BE *ina*] ITI.BÁ[R ŠU.SI *ḫal-qat* ....]	[If in] Nisan [the Finger is missing ....]
2′ [BE] *ina* ITI.GU$_4$ ŠU.SI *ḫal-*[*qat* ....]	[If] in Ajjaru the Finger is mis[sing ....]
3′ [BE] *ina* ITI.SIG$_4$ ŠU.SI *ḫal-qat* [.........]	[If] in Simanu the Finger is missing [..........]
4′ [BE] *ina* ITI.ŠU ŠU.SI *ḫal-qat* [...........]	[If] in Du'ūzu the Finger is missing [..........]
5′ [BE] *ina* ITI.NE ŠU.SI *ḫal-qat* [...........]	If in Abu the Finger is missing [..........]
6′ BE *ina* ITI.KIN ŠU.SI *ḫal-qat* [...........]	If in Ulūlu the Finger is missing [..........]
7′ BE *ina* ITI.DUL ŠU.SI *ḫal-qat* [...........]	If in Tašrītu the Finger is missing [..........]
8′ BE *ina* ITI.APIN ŠU.SI *ḫal-qat* [...........]	If in Araḫsamna the Finger is missing [..........]
9′ BE *ina* ITI.GAN ŠU.SI *ḫal-qat* [...........]	If in Kislīmu the Finger is missing [..........]
10′ BE *ina* ITI.AB ŠU.SI *ḫal-qat* [...........]	If in Ṭebētu the Finger is missing [..........]
11′ BE *ina* ITI.ZÍZ ŠU.SI *ḫal-qat* [...........]	If in Šabaṭu the Finger is missing [..........]
12′ BE *ina* ITI.ŠE ŠU.SI *ḫal-qat* [...........]	If in Addāru the Finger is missing [..........]

The omens pertaining to a missing Gall Bladder are better preserved and attested on two manuscripts. MS A is part of a larger compendium of Gall Bladder omens, though its position in the extispicy series is uncertain. MS B is an excerpt which only cites the calendrical omens, so they were evidently viewed as an entity. It runs:[72]

71 Jeyes 1989: 65–71 and Koch 2000: 69–70.

72 MS A K 1813+: 13′–25′ (*CT* 30 12) and MS B K 717 (*CT* 28 44).

> If in Nisan the Gall Bladder is destroyed: [.......] .. either his son or his daughter.
> If in Ajjaru the Gall Bladder is destroyed: The sacrificer will die.
> If in Simānu the Gall Bladder is destroyed: The enemy will reach my meadowland and there will be hunger.
> If in Du'ūzu the Gall Bladder is destroyed: The man's god will desire a gift on the command of Dumuzi.
> If in Abu the Gall Bladder is destroyed: Destruction of cattle, the patient will recover.
> If in Ulūlu the Gall Bladder is destroyed: There will be .... in the land and the man's household will diminish.
> If in Tašrītu the Gall Bladder is destroyed: Either the man's brother or the man's wife will die.
> If in Araḫsamna the Gall Bladder is destroyed: The man will experience a lot of worry and .....
> If in Kislīmu the Gall Bladder is destroyed: There will be enmity in the land on the command of Nergal.
> If in Ṭebētu the Gall Bladder is destroyed: The man's wife will die.
> If in Šabaṭu the Gall Bladder is destroyed: There will be death [in the la]nd.
> If in Addāru the Gall Bladder is destroyed: The man will die and [hi]s h[ousehold] will be scattered.
> If the Gall Bladder is destroyed in a favourable month: The sacrificer will die and the enemy will devour the go[ods of the land].

The omens concerning Simānu and Araḫsamna are different from the passage from *Enbu bēl arḫi*, only the Ṭebētu omen is the same. The passage in *Iqqur īpuš* was evidently not derived directly from these.

I know of only one interdisciplinary entry in the commentaries. A commentary to an extispicy text explains the interpretation of two particular parts of the liver as favourable when they face to the right with the rising of Arcturus/Boötes:[73]

> "Why is it favourable when the Presence and the Well-being point to the right?" "Because Arcturus is visible at its rising. Arcturus rises."

This seems a *non sequitur* and I have no good explanation why it made sense once. Perhaps it was simply a parallel. Arcturus is the brightest star North of

73 Koch-Westenholz 2000 19:27.

the celestial equator and it was one of the Enlil and Akkad stars, which means it pertained to the Babylonians, i.e., "us."[74] If Arcturus was bright at its rising, which was positive, it would be a good sign for "us." In the same way, the Presence and the Well-being were basically favourable parts of the liver and pointing to the right was a favourable sign.

Interdisciplinary entries like these are very rare, so it is impossible to establish any kind of pattern, but perhaps there was some sort of association between Arcturus and the Presence and Well-being. Could this passage be evidence for a beginning convergence of elements? In view of their common background and the overlap in scholars mastering both text groups, it seems strange that there are in fact so few extant examples of scholarly cross-references between astrology and extispicy in the commentaries.

Converging Semantics, Correlation of Celestial and Intestinal Elements

The *Gestirn-darstellung* texts are probably the most famous example of the systematic correlation of disparate cultural phenomena in late Babylonian material. Days of the months, medical remedies, celestial signs, stones and rituals are listed without any real explanation as to the purpose or reason for the associations. One late Babylonian text from Uruk (*SpTU* 4 159)[75] copied by the prolific Iqīšāja correlates the main parts of the liver with gods, months and stars:

> The Presence (corresponds to) Enlil, month Nisan, [the Hired Man (Aries)], Dumuzi, the sign NA is to be read *manzāzu* [.......].
> The Path (corresponds to) Šamaš, the month Ajjāru, the Bull of Heaven (Taurus); SI.SÁ means "to be straight" concerning walking. Bull ... [......]
> The Pleasing Word (corresponds to) Nusku, the month Simānu, True Shep[herd of Anu] (Orion); KA is to be read *pû ṭābu* .....[.....]
> The Strength (corresponds to) Ninurta, the month Du'ūzu, the Crab (Cancer), the Plow-star, the Heroic, Šulpaea.
> The Palace Gate (corresponds to) Bēlet-ekalli, the month Abu, the constellation King of the Palace, the Bow with the King (Regulus, part of Leo).
> The Well-being (corresponds to) Adad, the month Ulūlu, the Raven (Corvus).

74 Gössmann no. 385. Not many omens pertaining to Arcturus/Bootees are preserved.

75 Translated Koch-Westenholz 2000: 24–25. A collation of the original photo would be useful.

> The Gall Bladder (corresponds to) Anu, the month Tašrītu, the Scales (Libra), *mu-kin-na-at* zi ad [.......].
> The Finger [.............] the month Araḫsamna, mul .... [Scorpio?] mul .....
> The Increment [.............] the month Ṭebētu [..........] Bēlet-balāti (Vega).
> The Yoke [............] the month Šabāṭu, the Giant (Aquarius), [.....] the constellation Yoke of the land (Arcturus).
> The River of the Liver [..........] the month Addaru, the Ta[ils? (Pisces)], Venus, [....]

Sometimes the text adds a brief explanatory passage but generally we are hard put to find any reasons behind the associations. The parts of the liver are listed in the order they were recorded in extispicy reports. The month Kislīmu is omitted, presumably by mistake. The list of months and stars is probably derived from a list of stars rising heliacally in each month found in the astronomical compendium MUL.APIN I ii 36 iii 12.[76] The stars selected from the list are close to, but not identical with the Zodiac. Interestingly, another small fragment from Uruk belonging to the Enmeduranki texts[77] also contains similar identifications of parts of the liver but only with gods. The fragment MLC 2204[78] stems from a tablet written by Anu-bel-šunu and belonging to Nidinti-Anu of the Sin-leqe-unnini clan. Only three lines with possible correspondences are preserved (l. 3–5):

> [................. the Ga]ll Bladder (corresponds to) Anu (^{d}60), the Path to the right of the Gall Bladder is Sin (^{d}30)
> [................................] (corresponds to) Ea? (^{d}IDIM)
> [................................] (corresponds to) Ningirsu.

The correspondence between the Gall Bladder and Anu is the same for the two texts, but the Path to the right of the Gall Bladder equals the Well-being which in *SpTU* 4 159 is correlated with Adad and the constellation Corvus. This appears related to *melothesia* in Hellenistic astrology whereby parts of the body came to be associated with planets and stars, the liver itself with Jupiter according to one tradition.[79] The correlation of celestial bodies with parts of the liver which began in the Hellenistic period extended into a concept

76 Hunger and Pingree 1989.

77 Lambert 1998.

78 Identified by Enrique Jiminez who kindly allowed me to cite the manuscript prior to publication.

79 Reiner 1995: 59–60 quotes some examples, illnesses of particular intestines associated with planets in cuneiform medical texts. Furley and Gysembergh 2015: 9, 13–15.

whereby the celestial bodies directly influenced and determined the appearance of the liver and its parts.[80]

Converging Purposes—From Celestial Omens to Celestial Oracles

In its application, celestial divination changed in primary purpose from being concerned primarily with the public domain to include the life of the individual. A number of texts deal with specific areas of life, including the fate and character of a newborn, and some texts briefly instruct the diviner how to look at the sky for information about, for instance, war or trade. The clear distinction between omen and oracle became blurred in the late period. Since the diviner in effect observed the sky seeking knowledge pertaining to a particular question rather than scanning it for any kind of divine message, astrology in effect became oracular like extispicy. Some texts even use the formula found in extispicy "you can make a pronouncement" (*qība tašakkan*).

For instance, a late Babylonian text from Uruk provides the following instructions for the astrologer:

> "If you want to make a pronouncement (*qība tašakkan*) for the domain of the market for corn (gloss: *break*) you look for the course of the planets and appearance and disappearance, turning points and conjunction" (*SpTU* 1 94:1).

Instructions to the Seer for when and how it was possible to "make a pronouncement," either for a specific topic or in general, are found throughout Chapter 10 of the extispicy omen series, *Bārûtu*, and the esoteric *niṣirti bārûti* texts.[81]

Texts such as this and the various forms of personal astrology that developed in the Hellenistic period illustrate a fundamental shift in celestial divination, whereby it approached the versatility of oracular forms of divination like extispicy.[82] It culminated with horary astrology that became an independent branch of astrology designed to determine the suitability of a particular moment for a particular undertaking.

---

80 Detailed by Hephaistion of Thebes in "On Auspicious and Inauspicious Days" from the 5th century BCE, Furley and Gysembergh 2015:14–15, see pp. 111–112 for an excerpt of the relevant text passages.

81 Koch 2005 *passim*.

82 For forms of personal astrology especially early horoscopes, see Rochberg 1998.

Closing in

The influence of Babylonian astrology/astronomy on Hellenistic and Greek practices has long been recognised, and the nature of the relationship between Mediterranean practices of extispicy with Babylonian has been discussed since the first cuneiform texts pertaining to extispicy were deciphered.[83] However, the relationship between astrology and extispicy has hardly ever been touched. This may be due to the nature of the sources—Greek astrological treatises are many and long, while extispicy treatises apparently were non-existent. The sources for extispicy from the Classical and Imperial world which are utilized and which are familiar to most scholars have mostly been scattered references in works by historians and litterati.[84] But extispicy treatises did exist. The existence of Greek technical extispicy texts had been largely forgotten, but they have recently been brought back to our attention by Furley and Gysembergh (2015) who have collected and edited most of the sources for technical extispicy attested on Greek papyri from Roman Egypt. The manuscripts date to the Imperial and early Christian periods and treat both the liver and the lungs.[85] Even though all forms of divination including astrology were more or less banned off and on in the Imperial period, extispicy seems to have enjoyed a privilege as "harmless versions of divination." It was considered part of the traditional and official Roman religion that was afforded some venerance even after the introduction of Christianity. In 371 Valentinian I sanctified the tradition of "benevolent haruspicina." In the code of Valentinian the main distinction was between *haruspicina* and *maleficium*, between white and black magic, not between true religion and paganism.[86] The extispicy papyri do not mention any gods at all and seem to consider the markings on the entrails not as the result of a sacrifice but simple observation of correlated things. They are arranged according to topics of life; mentioned are auspiciousness of a given moment, business and trade, social relations, military and agricultural matters, and any given endeavour (*praxis*).[87] Clearly extispicy was at some point

83 Lenormant 1873 and 1875.

84 The sources are conveniently collected by Blecher 1905: 3–22.

85 Furley and Gysembergh 2015 contains an edition of three Greek extispicy papyri from Roman Egypt dating to the 1st, 2nd and 4th centuries CE respectively. These all pertain to the liver, a related papyrus dealing with the lungs is still unpublished. As the authors state, these texts are the very last and only Greek technical extispicy texts.

86 Hunt 2010: 145–146.

87 Furley and Gysembergh 2015: 2–4.

utilised both by private persons and in connection with public and military affairs much as horary astrology.[88]

Even though they remained distinct practices, astrology and extispicy always touched, and over the centuries grew to become ever more intertwined. From the very superficial points of contact in the early 2nd millennium BCE, they ended as two parts of a whole. As expressed by the first century writer Onosander:

> And it seems to me that the movements of the stars in heaven, their risings and settings, the inclinations of their triangular or quadrangular formations and diameters, are revealed by sacrifice in an alternative manifestation through the entrails.[89]

Like Mesopotamian extispicy, classical and imperial extispicy provided "no" or "yes" answers to questions pertaining to the advisability of a specific course of action, a useful tool for any kind of decision making; as in Mesopotamian extispicy, the answer was derived from the scrutiny of multiple changes in the organ, each of which had a particular significance. Celestial divination came to approach this oracular form in the practices of personal astrology that began to emerge in the late Babylonian and Hellenistic periods, the understanding of which to a large extent we owe to the work Francesca Rochberg.

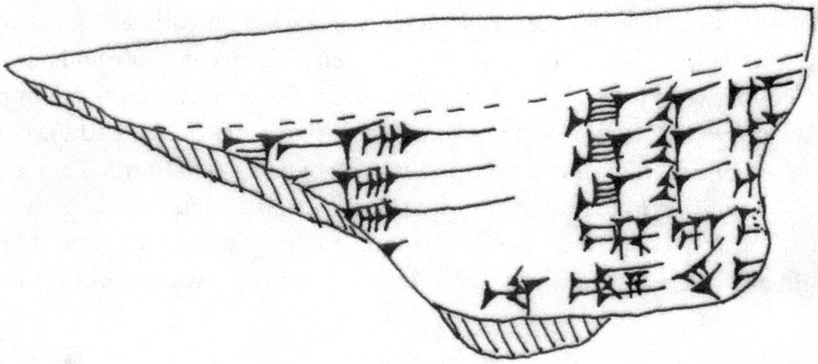

FIGURE 7.1 *K 7270.*
COURTESY THE TRUSTEES OF THE BRITISH MUSEUM

88 This versatility calls to mind the traditional list of purposes for extispicy found for instance in *Multābiltu*: "for the well-being of the king, for warfare, for the campaign, for taking a city, for healing the sick, for rain, for undertaking an enterprise and whatever else." Koch 2005: 11. The list is found throughout the extispicy texts.

89 Cited from Furley and Gysembergh 2015: 110.

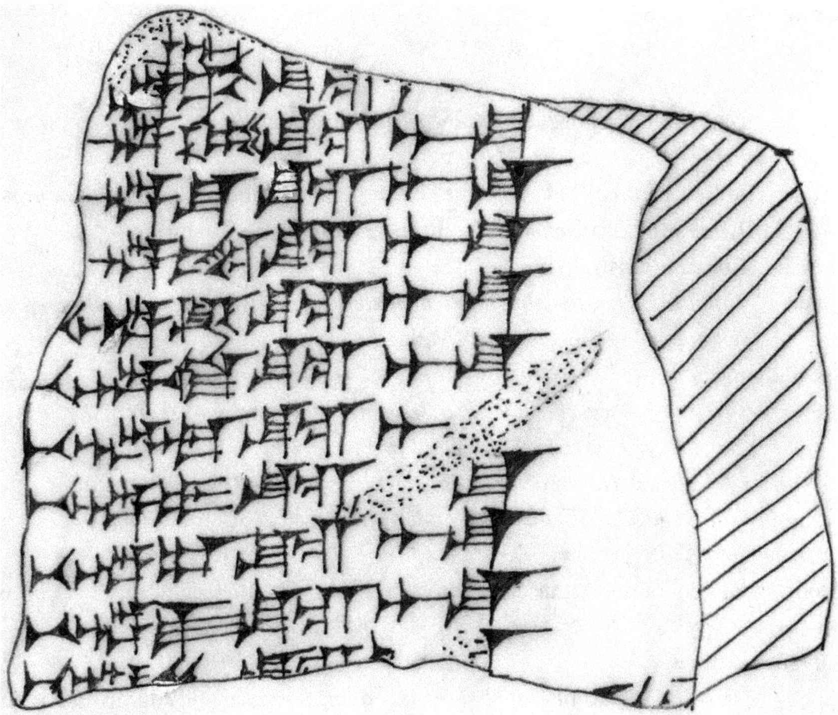

FIGURE 7.2 *K 1142.*
COURTESY THE TRUSTEES OF THE BRITISH MUSEUM

Bibliography

Abbink, J.
1993 "Reading the Entrails: Analysis of an African Divination Discourse." *Man.* NS 28: 705–726.

Abusch, Tzvi
2016 *The Magical Ceremony Maqlû: A Critical Edition.* AMD 10. Leiden: Brill.

Annus, Amar
2002 *The God Ninurta in the Mythology and Royal Ideology of Ancient Mesopotamia.* SAAS 14. Helsinki: Neo-Assyrian Text Corpus Project.

Barrett, Anthony Joseph
1998 *Sacrifice and Prophecy in Turkana Cosmology*. Limuru: Kolbe.

Blecher, Georg
1905 *De Extispicio Capita Tria.* Giessen.

Brown, D.
2000 *Mesopotamian Astronomy-Astrology*. CM18. Groningen: Styx.

Burrows, Eric

1924 "Hymn to Ninurta as Sirius." *JRAS* Cent. Suppl.: 33–40 and plt. II–III.

Charpin, Dominique

1989 "Les mois Uwarum et Sebûtum." *NABU*, no. 93.

Cooley, Jeffrey

2011 "A Prayer to the Gods of Night." Pp. 71–82 in *Reading Akkadian Prayers and Hymns: An Introduction.* Atlanta: Society of Biblical Literature.

Dietrich Manfred and Oswald Loretz

1990 *Mantik in Ugarit. Keilalphabetische Texte der Opferschau–Omensammlungen, Nekromantie.* Münster: Ugarit-Verlag.

Ellis, William

1832 *Polynesian Researches, during a Residence of Nearly Eight Years in the Society and Sandwich Islands*, vol. 4. London: Fisher, Son and Jackson.

Falkenstein, Adam and Wolfram von Soden

1953 *Sumerische und akkadische Hymnen und Gebete*. Zürich: Artemis.

Fincke, Jeanette C.

2009 "Ist die mesopotamische Opferschau ein nächtliches Ritual?" *BiOr* 66: 519–558.

Foster, Benjamin R.

1996 *Before the Muses. An Anthology of Akkadian Literature.* 2nd edition. Bethesda: CDL Press.

Furley, William and Victor Gysembergh

2015 *Reading the Liver*. Tübingen: Moehr Siebeck.

Geller, Markham

2007 *Evil Demons. Canonical Utukkū Lemnūtu Incantations.* SAACT 5. Helsinki: The Neo-Assyrian Text Project.

Glassner, Jean-Jacques

2004 *Mesopotamian Chronicles,* ed. Benjamin Foster. Atlanta: Society of Biblical Literature.

Goetze, Alfred

1957 "Reports on Acts of Extispicy from the Old Babylonian and Kassite Times." *JCS* 11: 89–105.

1968 "An Old Babylonian Prayer of the Divination Priest." *JCS* 22: 25–29.

Haas, Volkert

2008 *Hethitische Orakel, Vorzeichen und Abwehrstrategien.* Berlin: Walter de Gruyter.

Heimpel, W.

2003 *Letters to the King of Mari. A New Translation, with Historical Introduction, Notes, and Commentary.* Winona Lake, IN: Eisenbrauns.

Hunger, Hermann and David Pingree

1989 *MUL.APIN. An Astronomical Compendium in Cuneiform.* AfOB 24. Horn: Verlag Ferdinand Berger und Söhne.

Hunt, David

2010 "Christianising the Roman Empire: the evidence of the Code." Pp. 143–158 in *The Theodosian Code. Studies in the Imperial law of late antiquity*, 2nd ed., ed. Jill Harries & Ian Wood. London: Duckworth 1993.

Jeyes, Ulla

1989 *Old Babylonian Extispicy. Omen Texts in the British Museum.* Leiden: NINO.

Koch-(Westenholz), Ulla Susanne

1995 *Mesopotamian Astrology. An Introduction to Babylonian and Assyrian Celestial Divination.* Copenhagen: Museum Tusculanum.

2000 *Babylonian Liver Omens. The Chapters Manzāzu, Padānu and Pān tākalti of the Babylonian Extispicy Series mainly from Aššurbanipal's Library. CNI Publications* 25. Copenhagen: Museum Tusculanum.

2005 *Secrets of Extispicy. The Chapter Multābiltu of the Babylonian Extispicy Series and Niṣirti bārûti Texts mainly from Aššurbanipal's Libraries. AOAT* 326. Münster: Ugarit-Verlag.

2010 "Three Strikes and You're Out! A View on Cognitive Theory and the First-Millennium Extispicy Ritual." Pp 43–60 in: *Divination and Interpretation of Signs in the Ancient World*, ed. Amar Annus. OIS 6. Chicago: The Oriental Institute of the University of Chicago.

2013 "Concepts and Perception of Time in Mesopotamian Divination." Pp. 127–142 in *Time and History in the Ancient Near East. Proceedings from the 56th Rencontre*, ed. L. Felieu et al. Winona Lake, IN: Eisenbrauns.

Kraus, F.R.

1985 "Mittelbabylonische Opferschauprotokolle." *JCS* 37: 127–218.

Labat, René

1965 *Un calendrier babylonien des travaux des signes et des mois (séries iqqur īpuš).* Paris: Librairie Honoré Champion.

Lambert, W.G.

2007 *Babylonian Oracle Questions.* MC 13. Winona Lake, IN: Eisenbrauns.

Langdon, Stephen

1915 "A Fragment of a Series of Ritualistic Prayers to Astral Deities in the Ceremonies of Divination." *RA* 12: 189–192.

Lenormant, Francois

1873–75 *Choix de textes cunéiformes inédites ou incomplètement publiés jusqu'à ce jour.* Paris: Maisonneuve.

1875 *La divination et la science des présages chez les Chaldéens.* Paris: Maisonneuve.

Lenzi, Alan

2011 *Reading Akkadian Prayers and Hymns: An Introduction.* Atlanta: Society of Biblical Literature.

Livingstone, Alasdair

2013 *Hemerologies of Assyrian and Babylonian Scholars.* CUSAS 25. Bethesda: CDL Press.

Maul, Stefan

1994 *Zukunftsbewältigung.* BagF 18. Mainz: Verlag Philipp von Zabern.

2013 *Die Wahrsagekunst im Alten Orient.* München: C.H. Beck.

Mayer, W.R.

1976 *Untersuchungen zur Formensprache der babylonischen 'Gebetsbeschwörungen.'* StPohl SM 5. Rome: Biblical Institute.

1987 "Ein altbabylonischer Opferschau aus Babylon." *Or*NS 56: 245–262.

2005 "Das Gebet des Eingeweideschauers an Ninurta." *Or*NS 74: 51–56.

Meyer, Leon de

1982 "Deux priers *ikribu* du Temps d'Ammī-ṣaduqa." Pp. 271–278 in *Zikir Šumim: Assyriological Studies Presented to F.R. Kraus on the Occasion of his Seventieth Birthday,* ed. G. vanDriel, Th. J.H. Krispijn, M. Stol, and K.R. Veenhof. Leiden: Brill.

Nougayrol, Jean

1963 "Aleuromancie babylonienne." *Or*NS 32: 381–386.

Oppenheim, A. Leo

1959 "A New Prayer to the Gods of Night." *AnBib* 12: 282–301.

1960 "The City of Aššur in 714 B.C." *JNES* 19: 133–147.

1974 "A Babylonian Diviner's Manual." *JNES* 33: 197–220.

1977 *Ancient Mesopotamian: Portrait of a Dead Civilization.* Revised edition by Erica Reiner. Chicago: University of Chicago Press.

Perry, Edmund Guthrie

1907 *Hymnen und Gebete an Sin.* Leipzig: J.C. Hinrichs'sche Buchhandlung.

Pinches, Theophilus G.

1909 "The Goddess Ištar in Assyro-Babylonian Literature." *PSBA* 31: 57–69.

Pongratz-Leisten, Beate

1999 *Herrschaftswissen in Mesopotamien.* SAAS 10. Helsinki: Neo Assyrian Text Corpus Project.

Reiner, Erica

1960 "Fortunetelling in Mesopotamia." *JNES* 19: 23–35.

1974 "A Sumero-Akkadian Hymn of Nanā." *JNES* 33: 221–236.

1995 *Astral Magic in Babylonia.* TAPS 85, Part 4. Philadelphia: American Philosophical Society.

Richardson, Seth

2002 "Ewe Should Be So Lucky..." Pp. 229–244 in *Mining the Archives, Festschrift for Christoffer Walker on the Occasion of his 60th Birthday,* ed. Cornelia Wunsch. Dresden: Islet.

Rochberg (-Halton), Francesca

1988 *Aspects of Celestial Divination: The Lunar Eclipse Tablets of Enūma Anu Enlil.* AfOB 22. Münster: Ugarit-Verlag.

1998 *Babylonian Horoscopes.* TAPS 88/1. Philadelphia: American Philosophical Society.

2004 *The Heavenly Writing: Divination, Horoscopy and Astronomy in Mesopotamian Culture.* Cambridge: Cambridge University Press.

2006 "Old Babylonian Celestial Divination." Pp. 336–348 in *If a Man Builds a Joyful House: Assyriological Studies in Honor of Erle Verdun Leichty,* ed. Ann K. Guinan et al. Leiden: Brill.

Schwemer, Daniel

2010 "Empowering the Patient: The Opening Section of the Ritual Maqlû." Pp. 310–339 in *Pax Hethitica. Studies on the Hittites and their Neighbours in Honour of Itamar Singer,* ed. Yoram Cohen et al. Wiesbaden: Harrassowitz.

Seux, Marie-Joseph

1976 *Hymnes et prieres aux dieux de Babylonie et d'Assyrie.* LAPO 8. Paris: Éditions du Cerf.

Starr, Ivan

1977 "Extispicy Reports from the Old Babylonian and Sargonid Periods." Pp. 201–208 in *Essays on the Ancient Near East in Memory of Jacob Joel Finkelstein,* ed. Maria de Jong Ellis. Hamden: The Connecticut Academy of Arts and Sciences.

1983 *The Rituals of the Diviner.* Malibu: Undena Publications.

Steinkeller, Piotr

2005 "Of Stars and Men: The Conceptual and Mythological Setup of Babylonian Extispicy." Pp 11–47 in *Biblical and Oriental Essays in Memory of William L. Moran,* ed. Agustinus Gianto. BiOr 48. Rome: Pontificio Istituto Biblico.

Zimmern, Heinrich

1901 *Beiträge zur Kenntnis der babylonischen Religion.* AB 12. Leipzig: J.C. Hinrichs'sche Buchhandlung.

CHAPTER 8

Sumerian Divination

C. Jay Crisostomo

Sumerian and Divination

Adam Falkenstein began his seminal article on divination: "Wohl in keinem Gebiet der babylonischen Überlieferung ist der Unterschied zwischen der sumerischen und der akkadischen Tradition so ausgeprägt wie in dem Sektor ... dem der 'divination', umschrieben wird" (Falkenstein 1966: 45). Now, more than fifty years since this article was published, the sentiment behind Falkenstein's statement remains true. Whereas Falkenstein explored the traces of divinatory practice in Sumerian texts, I am more interested in the issue of language use. As far as the language of divination in the extant cuneiform texts from the ancient Near East, Akkadian is exclusive—or at least nearly so. This contribution discusses the exceptions, the rare examples of unilingual Sumerian or bilingual Sumerian-Akkadian divination. Although we share many common philosophical and linguistic interests, it is rare that my work on multilingualism perfectly intersects with Chessie Rochberg's interests in the history of astronomical sciences and cuneiform divination. It is fortunate that I have the opportunity to offer this singularity to her, a wonderful teacher, champion, and intellectual—almost as fortunate as I am to have been her student.

Divination undoubtedly had a long history in the region of Mesopotamia before it was ever textualized (Falkenstein 1966; Richardson 2010). Less clear, however, are the languages that tradition was conducted in or associated with—maybe Sumerian, possibly some Semitic languages including forms of Akkadian, perhaps others. Divination is referenced in Sumerian texts such as the Gudea cylinders and Šulgi B or in Ur III year names (see Michalowski 2006: 247–250; Koch 2015: 59–63). Presumably, divination began as some sort

* I had the pleasure of reading divinatory and astronomical texts with Chessie Rochberg while I was a student at Berkeley. I hope that this attempt reflects well upon her training and insight. Many thanks to Wayne Horowitz for insights on certain points and to Ulla Koch for reading and commenting upon early drafts. I also thank the Trustees of the British Museum for permission to publish K.2241+ here and the staff of the Department of Middle Eastern Studies for allowing me to collate the tablets mentioned here. All errors in fact or interpretation are mine alone.

 | DOI 10.1163/9789004363380_009

of oral tradition, although, as S. Richardson argues, the relationship between these traditions and the scholarly divination that emerges in the early second millennium is suspect (Richardson 2010: 226–239). But we need not dive into divinatory traditions or origins here. My focus is on language use, specifically, language use that we can observe and analyze in texts.

When divination became textualized in the early second millennium, when omen reports were written down, and when serialized lists of omens became compendia of sorts, these were written in Akkadian rather than Sumerian, the traditional language of listing scholarship, such as the lexical lists. Divinatory texts are markedly Akkadian. It seems as though Sumerian is deliberately excluded from divinatory scholarship. Although the reasons for this possibly linguistic ideologically driven output are unclear, Richardson's recent proposal connecting the creation of divinatory literature directly to political concerns at the end of the Old Babylonian period provides major clues for consideration.[1] Whatever the reasons, textualized divination is nearly always written in Akkadian.

The Beginnings of a Written Sumerianized Divination Tradition

In the Festschrift honoring Erle Leichty, Piotr Michalowski discussed the possibility of an omen report written in Sumerian. Michalowski argues convincingly that this represented some scribal attempt to render technical Akkadian extispicy terminology into Sumerian. That report is attested on two different tablets and treated slightly differently in each.

37 X1 igi niĝ$_2$-sa$_6$-⸢ga⸣-ni ĝa$_2$-ra mu-un-ši-in-ni-bar
X2 *omits*
38 X1 ša$_3$-ne-ša$_4$-ĝu$_{10}$ šag$_4$ kug$_3$-bi-še$_3$ mu-un-ĝar
X2 *omits*
39 X1 kiĝ$_2$-gi$_4$-a-ĝu$_{10}$ uzu silim-ma-ke$_4$ ma-an-ĝar
X2 [kiĝ$_2$-gi$_4$-a]-⸢ĝu$_{10}$⸣ uzu silim-ma im-ma-an-ĝar

1 Richardson goes so far as to argue that "the divinatory craft was appropriated by competing Amorite courts, hungry for legitimizing devices" and further that "The project to deliberately encode and control this common culture form enabled Old Babylonian kings to define alternative access to divine knowledge" (Richardson 2010: 239). This argument strongly suggests that the use of Akkadian as opposed to Sumerian would underscore the otherness of this literature in contrast to the more prevalent Sumerian scholarly/literary form. Richardson's argument remains quite provocative.

40 X1 uzu zi-da-na uzu gub$_3$-⸢ba$^?$-ĝa$_2$$^?$⸣ a$_2$-diri u$_3$-ni-ak
X2 [… z]i-da a$_2$-gub$_3$-bu-ba a$_2$-diri u$_3$-mu-ni-ak
41 X1 ^{giš}tukul a$_2$ zi-da-ĝa$_2$ gu$_3$-bi zi-da ul gur$_3$-ru mi-ni-šum$_2$
X2 *omits*
42 X1 ^{giš}tukul gub$_3$-bu-na gu-da la$_2$-la$_2$ gu$_2$-ri-bi ĝar-ra
X2 *omits*
43 X1 lu$_2$ hul-ĝal$_2$-ĝu$_{10}$ šu-ĝa$_2$ i$_3$-ĝa$_2$-ĝa$_2$ saĝ ĝiš bi$_2$-ra
X2 [.......]-⸢ĝal⸣-e šu-ĝa$_2$ ba-ni-in-ĝal$_2$ saĝ ĝiš bi$_2$-ra-ra
44 X1 un kukku$_2$-ga-ke$_4$ zalag-še$_3$ e$_3$-de$_3$ dur$_2$ silim-⸢ma⸣ nu$_2$-de$_3$
X2 [..............]-⸢ga⸣-ke$_4$ zalag-še$_3$ e$_3$-⸢de$_3$ dur$_2$ silim⸣-ma nu$_2$-de$_3$
45 X1 ^{d}utu en ka-aš-bar an-ki-ke$_4$ a$_2$-aĝ$_2$-ĝa$_2$ im-ma-an-ĝar
X2 *traces*

X1:

(Enlil) has cast his favorable glance upon me, set his holy heart on mercy, and established for me my favorable omen. And after I had the follow-up reading made concerning his *pars familiaris* and my *pars hostilis*, the weapon-mark on my right side, its trunk is straight, and that (good news) gave me joy; on the weapon-mark on his (i.e., Išbi-Erra's) left side a filament is suspended, it is placed (against) the other side. (The message was:) "My enemy shall fall into my hands, he shall be killed, the people, having come out of darkness into the light, will lie in peaceful habitations." Utu, lord who makes the decisions of the heavens and the earth, has provided (this) omen.

X2:

(Enlil) has established for me my favorable omen. And after I had the follow-up reading made concerning right side and the left side, (the message was:) "My enemy shall fall into my hands, he shall be killed, the people having come out of darkness into the light, will lie in peaceful habitations." …[2]

This omen is embedded within a literary letter from the Ur III king Ibbi-Sîn to Puzur-Numušda, governor of Kazallu (CKU 24, Michalowski 2011: 463–482). As Michalowski emphasizes, this omen was "clearly invented by some northern scribe, who cleverly invented a Sumerian technical vocabulary for extispicy" (Michalowski 2011: 479).[3] This Sumerian omen is thus merely a construct, a

2 Edition and translation based on Michalowski 2006: 251–252 and Michalowski 2011: 471–474.

3 Although Michalowski had earlier argued, "Within the narrow constraints of the present discussion one can only argue that all of these strands point in the same direction: the Larsa

bold attempt at replication and translation. Moreover, this clever fabrication remains the only Old Babylonian attempt at Sumerian divination.

With the exception of a single Kassite school text (Veldhuis 2000: 74) and—if they count as evidence—the Sumerian and bilingual menologies that eventually become absorbed into *Astrolabe B* (Horowitz 2010, 2013, 2007),[4] we have nothing further resembling Sumerian divination for about 1,000 years. CKU 24 and the singular Kassite omen—both products of scribal schools—hint at an attempt to graft Sumerian into the divinatory tradition. At the least, such few examples are highly marked. A handful of Sumerian-only and bilingual Sumerian-Akkadian omen compendia provide us an opportunity to examine writing practices and norms associated with scholarship, specifically divinatory scholarship, in cuneiform cultures.

A Corpus of Sumerian Divinatory Texts

I know of eleven texts attesting Sumerian and bilingual divination, beginning with the Kassite school text and continuing into Hellenistic Uruk (see table 1.0). These texts represent extispicy, behavioral omens, diagnostic omens, and especially celestial omens, the latter primarily attested at Nineveh.[5] Of these eleven, three appear to be monolingual Sumerian; the rest seem to be bilingual. Unfortunately, most of these are highly fragmentary. I first present a few remarks and observations. In the appendix of this votive to Chessie, I offer an edition of the longest of these omens, K.2241+ from Nineveh. Editions (some rudimentary) of the other ten are available online on the Bilinguals of Late Mesopotamian Scholarship project (oracc.org/blms/), using the six-digit P-number provided in the table below inserted following the forward slash in the url (example: oracc.org/blms/P394288). Photographs of each are available on CDLI, again by inserting the appropriate P-number after the url cdli.ucla.edu/.[6]

origin of the convoluted Sumerian omen embedded in the literary letter" (Michalowski 2006: 255).

4 For full editions of all astrolabe texts, see Horowitz 2014.

5 These celestial omens were initially indicated in SpTU 1 (Hunger 1976: 90) and listed in Reiner 1998 and again in Koch 2015: 149 n.404. Some are cited in BPO 2 and used in the later volumes of the CAD.

6 Further metadata are available on the British Museum website (britishmuseum.org), including measurements.

Number	Publ.	Type	Period	Place	Lang.	P-number
UM 29-13-542	Veldhuis 2000, fig. 8	extispicy	MB	Nippur	Sum.	P230847
K.2241+	present article	celestial	NA	Nineveh	bil.	P394288
K.2286	unpub.	celestial	NA	Nineveh	bil.	P394321
K.3094	unpub.	celestial	NA	Nineveh	bil.	P394802
K.8634	unpub.	celestial	NA	Nineveh	bil.	P394802
82-3-23, 120	unpub.	celestial	NA	Nineveh	Sum.	P452506
ND 4536+	CTN 4 89	behavioral	NA	Nimrud	Sum.	P363503
W 22246,c	SpTU 1, 85	behavioral	LB	Uruk	bil.	P348506
W 22307,13(+) W 22307,66	SpTU 1, 86 (+) SpTU 1, 145[a]	diagnostic	LB	Uruk	bil.	P348507 (+) P348564
W 22660,5	SpTU 3, 86	diagnostic	LB	Uruk	bil.	P348690
W 17360,ad	AUWE 23, 324	celestial	LB	Uruk	bil.	P349317

a On the possible indirect join, see Cavigneaux 1988. See also corrections and additional reconstructions in Geller 1991–1992.

Some Observations on Sumerian Divination

The first and unsurprising observation of this corpus is that the Sumerian rarely reflects Old Babylonian literary standards. In all cases, the Sumerian is obviously constructed off the accompanying Akkadian or, similar to the Old Babylonian Sumerian omen of CKU 24, some Akkadian omen or terminology in the background.

The syntax follows patterns established for Akkadian omens, the so-called "If 'P' then 'Q'" formulation also known from both Sumerian and Akkadian lists of laws (Rochberg 2009, 2010b). The subject of the protasis is topicalized. Verbal forms in the protasis are given as perfectives, imperfectives, statives (with al-), nominalized (with -a) or as participles (with -ed), which generally correspond to Akkadian preterites or statives. Verbs in the apodosis are typically imperfective corresponding to Akkadian duratives, again conforming to recognizable patterns.

Most of the translations—and I use this term in a broad sense—are simply one-to-one correspondences, a type of translation method we observe regularly in late bilinguals. The following is representative:

tukum-bi gu_2-gal kug-sig_{17} $šam_2$-$šam_2$-da lugal ma-da-bi en-⸢na⸣-[x] ti-la ma-da-bi gi-in-gur_2-ru
šum-ma ⸢LUGAL⸣ *hur-ra-ṣu i-ša₂-am* LUGAL KUR *ša₂-a-šu₂ a-di* ⸢*bal*⸣*-ṭu₃ mat-su i-kan-nu-us-su*
If a king purchased gold, the king of that land, as long as he lives, his land will submit to him
SpTU 1, 85:11′–12′[7]

This example presents the classic method of word A in Akkadian is equivalent to word B in Sumerian. So Akkadian *šumma* "if" equals Sumerian tukum-bi "if," a straightforward and unremarkable correspondence. The second word, however, provides Akkadian *šarru* "king," represented by the logogram LUGAL, equated to Sumerian gu_2-gal, a word which, although not really equivalent to "king" nevertheless refers to high status. Indeed, a Neo-Assyrian version of the lexical list *Lu* = *ša* actually provides the correspondence gu_2-gal = *šarru* (*MSL* 12, 91 A = VAT 10216 obv. i 40; oracc.org/dcclt/P373780).[8] It therefore seems as though one method of translation employed in creating these bilingual divinatory texts involves drawing on known equivalents from lexical texts or other similar strategies.

Secondly, Sumerian divination appears to be particularly connected to celestial divination and other aspects of astronomy and astrology.[9] The celestial bodies all have names written in Sumerian, several with no known Akkadian equivalents.[10] The menologies also utilize Sumerian. A Sumerian-only menology from Middle Babylonian Babylon seems to be adapted as a bilingual into *Astrolabe B* as well as into *Enūma Anu Enlil* 51 (Horowitz 2010). Moreover, *Enūma*

7 The Sumerian of the protasis of this particular omen (with DIŠ preceding) is cited in the *Diviner's Manual,* (line 3 Oppenheim 1974; noted in Hunger 1976: 90).

8 The only other copy that I am aware of which preserves this section for this version of *Lu* = *ša* I (long) is broken at this line (STT 2 373 obv i 40: [...] = ⸢*šar*⸣*-ru*; see oracc.org/dcclt/P338688). To my knowledge, the only other times the equation is made is in the citation of this omen in the *Diviner's Manual* (see preceding note) and implicitly in a colophon of *Diri* 6 from Nippur, on a tablet belonging to Taqiš-Gula written by Enlil-rema-šukun from the fifth year of the reign of one of the Artaxerxes, BAK 119: ᵐ*ar-tah-šá-as-su* GU_2.GAL KI^{me} (PTS 1, published Goetze 1945; online at oracc.org/dcclt/X459237).

9 For overviews of celestial divination, see especially Koch-Westenholz 1995, Reiner 1995, Reiner 1999, Rochberg 2004, Rochberg 2011, and Koch 2015: 147–196.

10 Jessica Baldwin (Tübingen) is currently writing a dissertation on the star signs and names.

Anu Enlil, the major celestial omina series, includes a bilingual introduction in which the Sumerian is clearly fabricated and requires an explanatory commentary to account for the translation (BM 59595, see Verderame 2002: 36–37; Frahm 2011: 136–137; Lambert 2013: 175–176).[11] Finally, as Niek Veldhuis points out, a Sumerian technical term for an eclipse, abununnu, is attested in the Old Babylonian version of the list *Diri* from Nippur, but not in any other contexts except for a late OB or MB esoteric bilingual text which likely draws on *Diri* as a scholarly source (Veldhuis 2014: 187, 263–265).[12] Our corpus includes six Sumerian or bilingual omen texts dealing with celestial divination, more than half of our corpus. Unfortunately, only one of these is substantially preserved (see the edition below).

All of the omens in this group deal with fixed stars, but very few are unbroken. Very few have parallels in published texts. The preserved omens provide protases for the following stars: Raven (^{mul}UGA), Chariot ($^{mul\ giš}$GIGIR), Scorpion (^{mul}GIR$_2$.TAB), Plow (^{mul}APIN), *pāšittu* (^{mul}KA.MUŠ.I$_3$.GU$_7$.E), Anzu bird ($^{mul\ d}$ANZUD^{mušen}), Jumping star (^{d}GU$_4$.U$_3$),[13] Mousy One (^{mul}EN.TE.NA.BAR.HUM), Flashing star (^{mul}AN.TA.SUR.RA), Doe goat (^{mul}UZ$_3$), Field (^{mul}AŠ.GAN$_2$), and Liar (^{mul}LUL.LA).[14] The order of the constellations in this group of texts differs from those on the Astrolabes or Mulapin or on any of the constellation texts published in BPO 2.

Notably, a few of these astronomical omens seem to occur in multiple copies. One particular omen is known from a constellation text edited as BPO 2 17.

> * UGA MUL.BI *ma-diš* S[A$_5$] BURU$_{14}$ ŠE.GIŠ.I$_3$ SI S[A$_2$]
> If the Raven star is very red, the flax harvest will prosper

Although fragmentary, it appears as though this omen occurs in two of our bilingual texts and one Sumerian text.

---

11 An online edition with photographs is available at the Cuneiform Commentaries Project (ccp.yale.edu/P461268).

12 See Westenholz 2005 for an edition of this bilingual text.

13 See commentary to rev. 1–5 below.

14 Lists of stars and proposed identifications may be found in Gössmann 1950, Reiner and Pingree 1981: 10–16, Koch 1989, Hunger and Pingree 1989: 137–146, Hunger and Pingree 1999: 58–73, and Horowitz 2014: 245–249.

K.2241+ o 4′–6′:	^{mul}UGA [...] *ma-ʾ-diš sa-a-*[*mu* ...][a]
AUWE 23, 324 4–5:	[...] al-⸢su$_4$-su$_4$⸣ bur$_{14}$ še-ĝiš-i$_3$ si [sa$_2$] [...] *e-bur* ŠE.GIŠ.I$_3$ ⸢*iš*⸣-[*šir*]
82-3-23, 120 1:	[... ^{mul}]⸢uga⸣ mul-bi lul-aš ⸢al⸣-[su$_4$-su$_4$ ...][b]

a Parallel noted in Reiner and Pingree 1981: 77.

b The protasis of this omen occurs several times for other constellations in this group including ^{mul}APIN (K.2241+ 24′–27′), $^{mul\ d}$ANZU^{mušen} (K.2241+ 30′–33′), ^{mul}AN.TA.SUR.RA (K.2241+ rev 10–11), ^{mul}UZ$_3$ (K.2241+ 12–13), ^{mul}AŠ.GAN$_2$ (K.2241+ 14–15). The negative version of this omen also occurs: ^{mul}KA$_5$.A (K.8634 5′).

So was there a particular tradition of bilingual divination, perhaps even a "canonical" series such as the known, but not extant series of Sumerian and bilingual proverbs and wisdom texts known as Sidu (Finkel 1986; see also Frahm 2010)? Indeed, celestial divination appears to have been regularly translated into other languages including Hittite, Hurrian, Ugaritic, and Elamite (Koch 2015: 160–162) and textual manuscripts were imported from Susa into Emar during the Middle Babylonian period (Rutz 2006). Moreover, K.2241+ was, according to the colophon, copied from an original indicating some kind of ongoing tradition for at least that particular text. I hesitate, however, to reconstruct an entire tradition or series on such scant evidence as one or possibly two omens that happen to co-occur on a few texts. From what little remains, it is clear that this particular omen is surrounded by different omens in each text.

Where these texts fit into the history of textualized celestial divination is unclear.[15] Since the section in *Enūma Anu Enlil* dealing with fixed stars (tablets 50–58)[16] is so fragmentarily preserved, it is impossible to say whether any of these texts attest a standard recension of *Enūma Anu Enlil* or whether they represent an alternative celestial omen tradition or whether these were among the many astronomical and astrological excerpts and one-offs. The corroboration of individual omens within these texts nevertheless further establishes the link between Sumerian divination and celestial divination.

Finally, each of these texts uses syllabic writing to represent Akkadian. As Mark Geller has previously observed, syllabic Akkadian is highly unusual for

15 On which, see Rochberg 1999a.

16 According to the Nineveh catalog; for literature on the difficulties in reconstructing the latter part of the series, see Koch 2015: 174 n.456.

late omen compendia (Geller 1991–1992: 181–182).[17] He further considers a bilingual diagnostic omen compared to a parallel omen from TDP, written in typical logographic fashion.

> TDP 82:24
> * $^{\text{na4}}$KIŠIB GU$_2$-*šú* DU$_8$ SA.MEŠ-*šú* GAR^{nu} *u* PA.AN.BI *ina* KIR$_4$-*šú* DAB.DAB GIG.BI NU-TIN
>
> SpTU 3, 86:3–4
> [tukum-bi $^{\text{na4}}$kišib gu$_2$]-ni al-du$_8$-a sa-sa-a-ni al-ĝar-re-eš u$_3$ pa-an-bi kir$_4$-bi ba-da-dab-dab lu$_2$-ru-ra-bi nu-ti-e
> *[šum-ma ku]-⸢nu-uk⸣-ku ⸢ki-šá⸣-di-šú pa-⸢ṭi⸣-[ir] šér-a-nu-šú šak-nu u na-pa-as-su ina ap-pi-šú ú-ṣab-bat mar-ṣa šu-ú ul i-bal-luṭ*
> If vertebrae on his neck are dislocated, his tendons are present, and his breath is caught in his nose, that sick man will not get well.

As Geller notes, the Sumerian version of the omen from SpTU 3, 86 is almost exactly the same as the logographic Akkadian version from TDP, albeit with fully expressed verbal forms and other minor elaborations. The Akkadian version of SpTU 3, 86, however, is fully syllabic. Geller emphasizes this as an example of "how late Sumerian bilinguals could have been constructed, perhaps for pedagogic purposes." I would prefer to underscore just how Sumerianized the typical omen is.

"Sumerian" Divination?

It is this notion, then, that I wish to ponder more abstractly. Given the paucity of Sumerian language divination, why do typical first millennium omens bother to create or present the appearance of Sumerian omens through the use of extensive logograms? The obvious possibility is that these divinatory scholars wanted to tap into the scholarly indexicality of Sumerian. But I think such an explanation is incomplete: it fails to account for the important implication that Sumerian—linguistically—never establishes itself as a scholarly language in the field of textualized divination.

---

17 The extensive use of logograms in the late divinatory tradition stands in contrast to the highly syllabic nature of the Old Babylonian compendia, with the relative exception of Old Babylonian celestial divinatory compendia (Rochberg 2006: 342–343; Koch 2015: 159).

The scholar-scribes who are crafting and copying these divinatory texts, whether highly logographic or bilingual or monolingual Sumerian, are no ignoramuses. They know what "good, Classical" Sumerian looks like. They know that whatever they are fabricating is not up to the standards of a literary text such as *Lugal-e* or *An-gim*, not even the late versions of these compositions.

The Sumerian omens I have remarked upon here are the exceptions that prove the general rule. I consider them a discursive semiotic, an attempt to explore the boundaries of traditional indexicality, a metadiscourse on the relationship between writing and language in cuneiform scribal culture. In these texts, the written norms of the community of late divination scholars are effectively violated. In contrast to this "Sumerian" divination, for typical scholarly divination, it is irrelevant whether omens are linguistically Akkadian or Sumerian or written primarily in logograms or syllabic signs. Textualized omens exist within the interlingual space that the cuneiform writing system allows, even encourages.

Rochberg has often reminded us of the complexity of scholarly knowledge evident in these divinatory texts from the first millennium.[18] Indeed, Rochberg has continually emphasized the interconnection between all the scholarly disciplines in cuneiform scribal culture and how cuneiform scholars created and transmitted knowledge. In addition to the various disciplines associated with the *tupšarrūtu*, these scholars were masters of cuneiform and their languages. They doubtless recognized and took full advantage of the possibilities afforded by their scholarship. In other words, scholars' ability to manipulate the script and the languages was yet another means by which they created scholarly knowledge. Or, as Rochberg astutely puts it, "[T]he crux for a cuneiform epistemology is in what was considered the object of knowledge, especially when words and signs were of central interest" (Rochberg 2014: 50).

The writers of these divinatory texts—whether they write monolingual Akkadian, monolingual Sumerian, or bilingual Sumerian-Akkadian—are not necessarily conscientious of the language ideologies we typically ascribe them. They act according to their enculturated habits, practices formulated in the act of copying, in their identification with the scribal arts. These habits and practices are in part grounded in a scholarly ideal of crafting, manipulating, and engaging with the cuneiform writing system and the interlingual space generated by that writing system in cuneiform scribal culture. This is the reason why "Sumerian" divination never catches on. The sharp distinction between Akkadian and Sumerian is perhaps more of a modern methodological delineation than necessarily an ancient one.

18 See especially Rochberg 1999b, 2000, 2004, 2010a, 2011a, 2014, 2015, 2016.

Appendix: Edition of K.2241+

K.2241 + K.2704 (approx. 12.5cm × 11cm; see Figs. 8.1 and 8.2) with the registration number A.186 was found on the Kuyunjik mound and accessioned into the British Museum at an unknown date. It represents almost the entire left half of the tablet, with only a few lines missing at the beginning and from the end of the colophon. It was written by and/or belongs to the ubiquitous scholar Nabû-zuqup-kēnu, based on a previous copy.[19] Nabû-zuqup-kēnu is associated with several copies of *Enūma Anu Enlil* and other tablets dealing with astronomical and astrological concerns, including the commentary *Sîn ina Tāmartišu*.[20] Since, as noted above, this text was apparently copied off an unknown original, it could belong to a series proper. It was certainly not a one-off, but belonged to some sort of textual tradition.

Most of the omens in this text begin the Sumerian omen with ud in the left margin.[21] The entire omen is given in Sumerian and then translated into Akkadian in the lines following. The beginning of the Akkadian translation is not so marked.

Obverse

1′. [... U]G[A...]
2′. ⸢^{mul}⸣UG[A ...]
3′. ^{mul}UGA [...]
4′. ^{mul}UGA ⸢x⸣ [...]
5′. *ma-ʾ-diš sa-a-*[*mu* ...]
6′. ZAG *šá* URU ^{mul}UGA *ina*$^{?}$ [...]
7′. *šum-ma a-ri-*⸢*im*$^{?}$⸣ MUL.MEŠ [...]
8′. *ik-ki-ra*$^{?}$ [...]
9′. ud ^{mul} ^{giš}gigir ^{d}en-me-šar$_2$-ra ⸢x⸣ [...]
10′. ^{mul} ^{d}EN.ME.ŠAR$_2$.RA *ul* ⸢x⸣ [...]
11′. ud ^{mul} ^{giš}gigir mul-bi igi-bi [...]
12′. kum$_2$ kalag-ga [...]

19 On whom, see Hunger 1972, Lieberman 1987: 204–217, Frahm 1999, PNA 2 (Baker and Pearce 2001), and Frahm 2011: 265–267.

20 See Frahm 2011: 155–160, 265–267; Koch 2015: 183 n.473, 185 n.481. Note also a tablet presenting *Iqqur īpuš* material, K.98 + MS 2226 (CUSAS 32 II.M.2, George 2016: 169–172).

21 Since the ud that begins the Sumerian line is almost never translated, it is clear that it is simply the entry marker. Other instances of UD used as the entry marker are known (e.g., YOS 10 62), but not collected. This use for the UD sign is suggested in the sign list Aa 18, BM 93037 obv i 19: (*u*$_3$ UD *ut-tu-ú* = *šum-ma*) and other lexical and bilingual texts equating UD and *šumma* (see CAD Š/3 s.v. *šumma* lex. sec.), which parallels how DIŠ is understood (e.g., the Principal Commentary to Šumma Izbu tablet 3: $^{di\text{-}eš}$diš = *šum*^{mu}*-ma*, see Leichty 1970: 216; De Zorzi 2014: 409).

13′. ⸢^{mul}⸣ ^{giš}GIGIR MUL.MEŠ-*šu* ⸢x⸣ […]

14′. […]

15′. ⸢^{mul}⸣ ^{giš}⸢gigir mul-bi⸣ […]

16′. ^{mul} ^{giš}GIGIR MUL.MEŠ-⸢*šu*⸣ […]

17′. ud ^{mul}gir$_2$-tab ud si-bi igi-⸢du$_8$⸣ a-⸢šub$^?$⸣ […]

18′. mu-e-ne ⸢saĝ⸣-bi-ta a-šub ⸢mu$^?$-x⸣-[…]

19′. ^{mul}GIR$_2$.TAB U$_4^{um}$ *qa-ra-an-šu in-nam-ru* ⸢x⸣ […]

20′. MU *ina re-ši-šá i-ma-qut ina re-iš šat*$^?$*-ti* ⸢x⸣ […]

21′. ud ^{mul}gir$_2$-tab ud ⸢si⸣-bi igi-du$_8$ a-⸢šub⸣ im a ⸢x⸣ […]

22′. mu-e-ne murub$_4$-bi-ta a-šub mu ⸢murub$_4$-bi-ta⸣ […]

23′. ^{mul}GIR$_2$.TAB U$_4^{um}$ ⸢*qa*⸣*-ra-an-šu in-nam-*⸢*ru*⸣ […]

24′. MU *ina qab-li-šá i-ma-qut ina* MURUB$_4$ ⸢MU⸣ […]

25′. ud ^{mul}apin mul-bi lul-aš al-su$_4$-su$_4$ ⸢x⸣ […]

26′. en-te-na an-sag$_3$ ud kum$_2$ ⸢x⸣ […]

27′. ^{mul}APIN MUL-*šú ma-ʾ-diš sa-a-mu ti-*⸢x⸣-[…]

28′. *ku-uṣ-ṣu ri-ih-ṣu* ⸢U$_4$⸣[^{um} …]

29′. ud ^{mul}ka-muš-i$_3$-gu$_7$-e mul-[bi …]

30′. ^{mul}KA.MUŠ.I$_3$.⸢GU$_7$.E⸣ […]

31′. ud ^{mul} ^{d}anzud⸢^{mušen}⸣ [mul-bi lul-aš al-su$_4$-su$_4$]

32′. ud-da e$_2$-[meš …]

33′. ^{mul}⸢ANZU⸣[^{mušen} MUL.MEŠ-*šu ma-ʾ-diš sa-a-mu*]

34′. ⸢BE⸣-*ma um-*[*ma-tum* …]

Reverse

1. ud ^{d}gu$_4$-u$_3$ ⸢x⸣ […]
2. gaz-e-ne […]
3. bara$_2$ suh-bi […]
4. ^{d}GU$_4$.U$_3$ *da-nu-*[…]
5. *di-ik-tu* KA^{ú} *tu-n*[*u*$^?$ …]
6. ud ^{mul}en-te-na-bar-hum igi mul-bi […]
7. ki-bal-a kalag-ga ⸢x⸣-[…]
8. ^{mul}EN.TE.NA.BAR.HUM *pa-an* MUL.MEŠ-*šu* AL$^{??}$ A […]
9. *na-bal-kàt-tum da-na-tum* ⸢x⸣ […]
10. ud ^{mul}an-ta-sur-ra mul-bi lul-aš al-⸢su$_4$⸣-[su$_4$ …]
11. ^{mul}AN.TA.SUR.RA MUL.MEŠ-*šú ma-ʾ-diš sa-a-*⸢*mu*⸣ […]
12. ud ^{mul}uz$_3$ igi-bi lul-aš al-su$_4$-su$_4$ ^{d}ir$_3$-ra kalam-ma ⸢x⸣ […] ⸢x⸣ […]
13. ^{mul}UZ$_3$ *pa-nu-ša ma-ʾ-diš sa-a-mu mu-ta-nu iš-*⸢x-x-x⸣
14. ud ^{mul}aš-gan$_2$ igi-bi lul-aš al-su$_4$-su$_4$ im rib-[ba$^?$ …]
15. ^{mul}AŠ.GAN$_2$ *pa-nu-šu ma-ʾ-diš sa-a-mu šá-a-r*[*u*$^?$ …]

16. [... n]am-ki-us$_2$ gi-na ⸢x⸣ NI an-na al-x-[...]
17. [...] ⸢x x si$^?$⸣ sa$_2$ buru$_{14}$ kalam-⸢ma⸣ [...]
18. [...] ⸢x⸣ *ina*$^?$ EN PA *šá iz-za-*⸢*as*⸣*-su šá* [...]
19. [...] ⸢x⸣-*ú ina* MU.BI *iš-šìr e-bu-*[...]

Colophon

20. LIBIR.RA.BI GIN$_7$ AB.SAR.AM$_3$
21. DUB^{pi} ^{md}AG-*zu-qu-up*-GI.NA DUMU ^{md}[AMAR.UTU-MU-BA^{šá}]
22. *traces*[22]

Translation

Obverse

1′–2′. [If] the Raven star [...]
3′. (If) the Raven star [...]
4′–8′. (If) the Raven star [...] very red [...] the borders of the city. The Raven star in$^?$ [...]. If the stars of the Raven [...] is changed, [...]
9′–10′. If the Chariot [...] Enmešara [...]
11′–14′. If the front star of the Chariot [... there will be] severe heat
15′–16′. The stars of the Chariot [...]
17′–20′. If when the pincers of the Scorpion are visible, *collapse*$^?$ [...] the beginning of the year [...] will *collapse*; in the beginning of the year [...]
21′–24′. If when the pincers of the Scorpion are visible, *collapse* of the ... [...] in the middle of the year [...] will *collapse*; in the middle of the year [...]
25′–28′. If the stars of the Plow are very red [...] devastating winter, hot days [...]
29′–30′. If the stars of KA.MUŠ.I$_3$.GU$_7$.E [...]
31′–34′. If the stars of the Anzu bird [are very red ...], summer [...]

Reverse

1–5. If the Ox ... [...] defeat ... [...] the foundations of the dais [...]
6–9. If the front stars of Enmebarhum ... [...] massive revolt [...]
10–11. If the stars of Antasura are very red [...]
12–13. If the front stars of the Doe Goat are very red, pestilence [...] in the land
14–15. If the front stars of the Field are very red, extensive wind [...]
16–17. (Sum.) [...] firm foundations ... in heaven ... [...] ... harvest in the land [...]
18–19. (Akk.) [...] who will stand with the lord of ... which [...] ... in that year will prosper; the harvest [...]

22 Compare other Nabû-zuqup-kēnu colophons (BAK 293–311). Note especially the colophon from BBR Nr. 7 (K.3242+; BAK 301): *ki-i pi-i* 2.DUB.MEŠ LIBIR.RA.MEŠ AB.SA[R.AM$_3$ ...] DUB^{pi} ^{md}AG-*zu-q*[*u*]-*up*-GI.NA DUMU ^{md}AMAR.UTU-MU-BA^{šá} ^{lu_2}DUB.SAR [ŠA$_3$.B]AL.BAL ^{m}gab-bi-DINGIR.MEŠ ^{ni}KAM^{eš} ^{lu_2}GAL.DUB.SAR.MEŠ.

Colophon

20–22. Written [and checked] according to its original. Tablet of Nabû-zuqup-kênu, son of [Marduk-šuma-iqīša ...]

Commentary

obv. 1′–8′. These opening lines are difficult to reconstruct due to their fragmentary state. It is difficult to ascertain which of the opening 2 lines are Sumerian and which are Akkadian; it is, however, likely that neither lines 3′ or 4′ are Sumerian since neither begins with ud. Moreover, it seems as though the beginning of 7′ should be considered part of the preceding Akkadian omen. However, such would entail that 4′–8′ constitutes a single omen and therefore does not appear to translate what occurs in the preceding lines.

Protases dealing with the Raven star may be found in BPO 2 XVII obv. i 11′–17′. 82-3-23, 120 1–3, another text in this group of bilingual omens, also gives protases with the Raven star.

If the beginning of 6′, *pāṭu ša āli* represents (part of) an apodosis, I can find no parallels among the published fixed star omens.

The verb *nakāru* in 8′ is often used to indicate the change of positions (*manzāzu*) of fixed stars and planets in the sky (see Reiner and Pingree 1981: 18). BPO 2 XVII 9–10 give protases that discuss the *manzāzu* of the Raven star.

obv. 9′–10′. The Akkadian line does not begin like the Sumerian, which suggests an equation of $^{mul\ ĝiš}$gigir and ^{mul}en.me.šar$_2$.ra. This association may be explicit in the Sumerian line, which juxtaposes the two star names. This identification is overt in the standard version of Ura = Hubullu 22 290′ (following Bloch and Horowitz 2015).

obv. 11′–14′. Cited in Reiner and Pingree 1981: 75, mentioned in reference to *narkabtu* as a constellation in CAD N/1 s.v. *narkabtu* mng. 2. See BPO 2 XVI 8′: [*] ^{mul}EN.ME.ŠAR$_2$.RA MUL IGI.BI GIN$_7$ *ka-ra-re-e ina* MURUB$_4$-*šú* 1 MUL *ma-diš* SA$_5$ *um-šum dan-nu* [GA]L$_2$ “If the front star of Enmešara is like the noon-day sun and in its middle one star is very red, there will be severe heat.” Based on this parallel, it is again likely that $^{mul\ ĝiš}$gigir and ^{mul}en-me-šar$_2$-ra are equated here. Following kum$_2$ kalag-ga, there is a large gap before a break; some form of the verb ĝal$_2$ is expected based on the parallel.

obv. 17′–20′. Partially cited in CAD Q s.v. *qarnu* mng. 2b. Due to the broken context, the nuance of šub = *maqātu* in this and the following omen is unclear.[23] It is possible that *maqātu* does not refer to the year, and that the actual subject is not preserved. The fact that both apodoses, however, deal with different times of the year suggests that the

23 The citation in CAD Q s.v. *qablu* A mng. 1f offers no translation for *maqātu*.

year is indeed the subject. The plural mu-e-ne in the Sumerian line is not reflected in the Akkadian and should probably not be considered.

obv. 21′–24′. Partially cited in CAD Q s.v. *qablu* A mng. 1f.

obv. 25′–28′. Cited in Reiner and Pingree 1981: 75. Partially cited in CAD S s.v. *sāmu* bil. section and CAD R s.v. *rihṣu* A mng. a. To my knowledge, the equation an-PA and *rihṣu* is unique. Sumerian an-pa typically refers to the zenith, Akkadian *elat šamê*. The lexical list Antagal, however, offers ĝir$_3$ sag$_2$$^{\text{sa-ag}}$ = MIN(*ra-ha-ṣu*) *šá* [GIRI$_3$] "to strike with the feet = to rush, said of [feet]" (Antagal N ii′ 13′, see MSL 17 = Roth 1985: 240),[24] which suggests a correspondence sag$_3$ "to hit, strike" = *rahāṣu* "to trample, destroy," a reasonable semantic translation.[25] The an in this word should be taken as perhaps an attempt to attach a verbal prefix.

obv. 29′–30′. Cited in Reiner and Pingree 1981: 72 with comparison to BPO 2 XV 5′.

obv. 31′–33′. Restoration suggested by Reiner (Reiner and Pingree 1981: 41) on the basis of BPO III 11b. See also BPO 2 XVI 9′: ⸢*⸣ ^{mul}ANZU^{mušen} MUL.BI IGI.BI *ma-diš* SA$_5$ BE-*ma* EN.TE.NA *šur-bu-ú* BE-*ma* E$_2$.MEŠ *u*[*m-šum* GAL$_2$] "If the front star of the Anzu bird is very red, if it is winter, there will be frost; if it is summer there will be heat."

rev. 1–5. The writing ^{d}gu$_4$-u$_3$—presumably reflecting a phonetic rendering for ^{d}gu$_4$-utu, usually identified as Mercury—is, to my knowledge, completely unique. The phrase following *dīktu* is unclear.

rev. 6–9. Partially cited in CAD N/1 s.v. *nabalkattu* bil. section and mng. 5.[26]

rev. 10–11. Cited in Reiner and Pingree 1981: 80.

rev. 12–13. Partially cited in CAD M/1 s.v. *mādiš* bil. section and CAD P s.v. *pānu* bil. section. To my knowledge, the correspondence ^{d}Erra = *mūtānu* (nor ^{d}Nergal = *mūtānu*) is not given anywhere else, but the cosmic domain with which Erra/Nergal is associated makes the equation reasonable.[27]

rev. 14–15. Referenced in CAD P s.v. *pānu* bil. section. Cited in Reiner and Pingree 1981: 40 with additional notation there that this same omen likely occurs in another text in the group under discussion, K.3094 ii 5. The reading and restoration of rib-ba is

---

24 The section offers three different correspondences for *rahāṣu*: ib$_2$-bi$_2$-ra = *ra-ha-ṣu šá* ⸢*e-de*⸣-[*e*] "… = to flood, said of a wave" (line 12′) as well as ma-da lu$_3$-lu$_3$ = MIN(*ra-ha-ṣu*) *šá ma-*⸢*ti*?⸣ "disturbed land = to flood, said of a land" (line 14′). Thus, in the section, it would seem that ra, sag$_2$, and lu$_3$ are equated to *rahāṣu*.

25 Note that sag$_2$(PA.GAN) must be understood as an alternative for sag$_3$(PA), used exclusively in lexical texts.

26 The CAD citation in mng. 5 cites K.3094, another text in the present group, but mentions K.2241+.

27 Note the implied correlation between Nergal and *mūtānu* in AbB 2, 118:9–11 and the apodosis in an Old Babylonian extispicy (Jeyes 1989 no. 3 iv 11′), which mentions Nergal which U. Jeyes rightly interprets as plague (i.e., *mūtānu*, Jeyes 1989: 121; see further Jeyes 1980: 109).

based on this latter text, which gives im rib-ba; BPO 2 III obv. 17 is ambiguous: IM KAL BA.AB.ZI.ZI.

rev. 16–19. These lines are slightly opaque as little appears to correspond between the Sumerian and the Akkadian. I have therefore given separate translations above to account for the discrepancies. It seems as though gi-na "to be firm" corresponds to *uzuzzu* "to stand." The equivalence may be semantically based as "to be firm" could conceivably be a near equivalent to "to stand." Alternatively, the correspondence may rely on the near homophone of Sumerian ĝen "to walk," but written with the same sign (DU) as gub "to stand." More difficult is nam-ki-us$_2$ apparently equated to *bēl* PA

FIGURE 8.1 *K.2241+ obverse.*
COURTESY THE TRUSTEES OF THE BRITISH MUSEUM.

FIGURE 8.2 *K.2241+ reverse.*
COURTESY THE TRUSTEES OF THE BRITISH MUSEUM.

(or similar). Perhaps the Akkadian should be read *bēl haṭṭi* (EN GIDRU) "lord of the scepter" as a metaphor for the king; nam-ki-us$_2$ "foundations, dais?" could then be understood as synecdoche for the king. The second line is slightly more transparent as si sa$_2$ = *ešēru* and buru$_{14}$ = *ebūru* would be expected. One would expect the traces preceding si sa$_2$ to correspond to mu-bi in some way, but such does not appear to be the case.

Bibliography

Baker, Heather D., and Laurie E. Pearce
2001 "Nabû-zuqup-kēnu." Pp. 912–913 in *A Prosopography of the Neo-Assyrian Empire 2*, ed. Heather D. Baker. Helsinki: The Neo-Assyrian Text Corpus Project.
Bloch, Yigal, and Wayne Horowitz

2015 "Ura = Ḫubullu XXII: The Standard Recension." *JCS* 67: 71–125.

Cavigneaux, A.

1988 "Un texte médical bilingue." *NABU*: no. 24.

De Nicla, Zorzi

2014 La Serie Teratomantica Šumma Izbu: Testo, Tradizione, Orizzonti Culturali. 2 vols, HANEM 15. Padova: S.A.R.G.O.N. Editrice e Libreria.

Falkenstein, Adam

1966 "«Wahrsagung» in der sumerischen Überlieferung." Pp. 45–68 in *La divination en Mésopotamie ancienne et dans les régions voisines. XIVe Rencontre Assyriologique Internationale (Strasbourg, 2–6 juillet 1965)*. Paris: Presses Universitaires de France.

Finkel, Irving L.

1986 "On the Series of Sidu." *ZA* 76: 250–253.

Frahm, Eckart

1999 "Nabû-zuqup-kēnu, das Gilgameš-Epos und der Tod Sargons. II." *JCS* 51: 73–90.

2010 "The Latest Sumerian Proverbs." Pp. 155–184 in *Opening the Tablet Box: Near Eastern Studies in Honor of Benjamin R. Foster*, ed. Sarah C. Melville and Alice L. Slotsky. CHANE 42. Leiden: Brill.

2011 *Babylonian and Assyrian Text Commentaries: Origins of Interpretation*, GMTR 5. Münster: Ugarit-Verlag.

Geller, M.J.

1991–92 "Rv. E. von Weiher, *Spätbabylonische Texte aus Uruk, Teil III*." *AfO* 38–39: 179–182.

George, A.R.

2016 *Mesopotamian Incantations and Related Texts in the Schøyen Collection*. CUSAS 32. Bethesda: CDL Press.

Goetze, Albrecht

1945 "The Vocabulary of the Princeton Theological Seminary." *JAOS* 65: 223–237.

Gössmann, P.F.

1950 *Planetarium Babylonicum oder die sumerisch-babylonischen Stern-Namen*. Šumerisches Lexikon IV 1. Roma: Päpstliches Bibelinstitut.

Horowitz, Wayne

2007 "The Astrolabes: Astronomy, Theology, and Chronology." Pp. 101–114 in *Calendars and Years: Astronomy and Time in the Ancient Near East*, ed. John M. Steele Oxford: Oxbow Books.

2010 "VAT 17081: A Forerunner to the Menology of Astrolabe B." Pp. 183–188 in *DUB.SAR É.DUB.BA.A. Studies Presented in Honour of Veysel Donbaz*, ed. Şevket Dönmez. Istanbul: Ege Publications.

2013 "The Astrolabes: An Exercise in Transmission, Canonicity, and Para-Canonicity." Pp. 273–287 in *Between Text and Text: The Hermeneutics of*

Intertextuality in Ancient Cultures and Their Afterlife in Medieval and Modern Times, ed. M. Bauks, W. Horowitz and A. Lange. JAJSup 6. Göttigen: Vandenhoeck & Ruprecht.

2014 *The Three Stars Each: The Astrolabes and Related Texts*. AfO Bei. 33. Vienna: Institut für Orientalistik der Universität Wien.

Hunger, Hermann

1972 "Neues von Nabû-zuqup-kēna." *ZA* 62: 99–101.

1976 *Spätbabylonische Texte aus Uruk*. ADFU 9. Berlin: Gebr. Mann.

Hunger, Hermann, and David Pingree

1989 *MUL.APIN: An Astronomical Compendium in Cuneiform*. AfO Bei. 24. Horn: Ferdinand Berger & Söhne Gesellschaft M.B.H.

1999 *Astral Sciences in Mesopotamia*. HdO 44. Leiden: Brill.

Jeyes, Ulla

1980 "Death and Divination in the Old Babylonian Period." Pp. 107–121 in *Death in Mesopotamia. Papers read at the XXVI*[e] *Rencontre assyriologique internationale*, ed. Bendt Alster. Mesopotamia 8. Copenhagen: Akademisk Forlag.

1989 *Old Babylonian Extispicy: Omen Texts in the British Museum*. PIHANS 64. Istanbul: Nederlands Historisch-Archaeologisch Instituut te Istanbul.

Koch, Johannes

1989 *Neue Untersuchungen zur Topographie des babylonischen Fixsternhimmels*. Wiesbaden: Harrassowitz.

Koch, Ulla Susanne

2015 *Mesopotamian Divination Texts: Conversing with the Gods. Sources from the First Millennium BCE*. GMTR 7. Münster: Ugarit-Verlag.

Koch-Westenholz, Ulla

1995 *Mesopotamian Astrology: An Introduction to Babylonian and Assyrian Celestial Divination*. CNI Publications 19. Copenhagen: Museum Tusculanum.

Lambert, W.G.

2013 *Babylonian Creation Myths*. MC 16. Winona Lake, IN: Eisenbrauns.

Leichty, Erle

1970 *The Omen Series Šumma Izbu*. TCS 4. Locust Valley, NY: J.J. Augustin.

Lieberman, Stephen J.

1987 "A Mesopotamian background for the so-called *Aggadic* 'measures' of biblical hermeneutics?" *HUCA* 58: 157–225.

Michalowski, Piotr

2006 "How to Read the Liver—in Sumerian." Pp. 247–257 in *If a Man Builds a Joyful House: Assyriological Studies in Honor of Erle Verdun Leichty*, ed. Ann K. Guinan, Maria de Jong Ellis, A.J. Ferrara, Sally M. Freedman, Matthew T. Rutz, Leonhard Sassmanshausen, Steve Tinney and M.W. Waters. CM 31. Leiden: Brill.

2011 *The Correspondence of the Kings of Ur: An Epistolary History of an Ancient Mesopotamian Kingdom*, ed. Jerrold S. Cooper. MC 15. Winona Lake, IN: Eisenbrauns.

Oppenheim, A. Leo

1974 "A Babylonian Diviner's Manual." *JNES* 33: 197–220.

Reiner, Erica

1995 *Astral magic in Babylonia*. TAPS 85. Philadelphia: American Philosophical Society.

1998 "Celestial Omen Tablets and Fragments in the British Museum." Pp. 215–302 in *Festschrift für Rykle Borger zu seinem 65. Geburtstag am 24. Mai 1994. tikip santakki mala bašmu…*, ed. Stefan M. Maul. CM 10. Groningen: Styx.

1999 "Babylonian Celestial Divination." Pp. 21–37 in *Ancient Astronomy and Celestial Divination*, ed. N.M. Swerdlow. Cambridge: MIT Press.

Reiner, Erica, and David Pingree

1981 *Babylonian Planetary Omens Part Two: Enuma Anu Enlil, Tablets 50–51*. BiMes 2. Malibu: Undena Publications.

Richardson, Seth

2010 "On Seeing and Believing: Liver Divination and the Era of Warring States (II)." Pp. 225–266 in *Divination and the Interpretation of Signs in the Ancient World*, ed. Amar Annus. OIS 6. Chicago: The Oriental Institute of the University of Chicago.

Rochberg, Francesca

1999a "Continuity and Change in Omen Literature." Pp. 415–425 in *Munuscula Mesopotamica. Festschrift für Johannes Renger*, ed. Barbara Böck, Eva Cancik-Kirschbaum and Thomas Richter. AOAT 267. Münster: Ugarit-Verlag.

1999b "Empiricism in Babylonian Omen Texts and the Classification of Mesopotamian Divination as Science." *JAOS* 119: 559–569.

2000 "Scribes and Scholars: The *ṭupšar Enūma Anu Enlil*." Pp. 359–375 in *Assyriologica et Semitica*, ed. Joachim Marzahn and Hans Neumann. AOAT 252. Münster: Ugarit-Verlag.

2004 *The Heavenly Writing: Divination, Horoscopy, and Astronomy in Mesopotamian Culture*. Cambridge; New York: Cambridge University Press.

2006 "Old Babylonian Celestial Divination." Pp. 337–348 in *If a Man Builds a Joyful House: Assyriological Studies in Honor of Erle Verdun Leichty*, ed. Ann K. Guinan, Maria de Jong Ellis, A.J. Ferrara, Sally M. Freedman, Matthew T. Rutz, Leonhard Sassmanshausen, Steve Tinney and M.W. Waters. CM 31. Leiden: Brill.

2009 "Conditionals, Inference, and Possibility in Ancient Mesopotamian Science." *Sci Context* 22: 5–25.

2010a “Beyond Binarism in Babylon.” *Interdisciplinary Science Reviews* 35: 253–265.

2010b “‘If P, Then Q’: Form and Reasoning in Babylonian Divination.” Pp. 19–27 in *Divination and Interpretation of Signs in the Ancient World*, ed. Amar Annus. OIS 6. Chicago: The Oriental Institute of the University of Chicago.

2011a “Divine Causality and Babylonian Divination.” Pp. 277–294 in *A Common Cultural Heritage: Studies on Mesopotamia and the Biblical World in Honor of Barry L. Eichler*, ed. Grant Frame, Erle Leichty, Karen Sonik, Jeffrey Tigay and Steve Tinney. Bethesda: CDL Press.

2011b “Observing and Describing the World through Divination and Astronomy.” Pp. 618–636 in *Oxford Handbook of Cuneiform Culture*, ed. Karen Radner and Eleanor Robson. Oxford: Oxford University Press.

2014 “The History of Science and Ancient Mesopotamia.” *JANEH* 1: 37–60.

2015 “The Babylonians and the Rational: Reasoning in Cuneiform Scribal Scholarship.” Pp. 209–248 in *In the Wake of the Compendia: How Technical Handbooks and Encyclopedia Reshape Early Mesopotamian Empiricism*, ed. J. Cale Johnson. Berlin: De Gruyter.

2016 *Before Nature: Cuneiform Knowledge and the History of Science.* Chicago: University of Chicago.

Roth, Martha T.

1985 “The Series An-ta-gál = *šaqû*.: Pp. 129–258 in *The Series Erim-ḫuš = anantu and An-ta-gál = šaqû*, ed. Antoine Cavigneaux, Hans G. Güterbock and Martha T. Roth. MSL 17. Rome: Pontificium Institutum Biblicum.

Rutz, Matthew T.

2006 “Textual Transmission between Babylonia and Susa: A New Solar Omen Compendium.” *JCS* 58: 63–96.

Veldhuis, Niek

2000 “Kassite Exercises: Literary and Lexical Extracts.” *JCS* 52: 67–94.

2014 *History of the Cuneiform Lexical Tradition*. GMTR 6. Münster: Ugarit-Verlag.

Verderame, Lorenzo

2002 *Le Tavole I–VI della serie astrologica Enūma Anu Enlil.* Nisaba 2. Rome: Di. Sc.A.M.

Westenholz, Joan Goodnick

2005 “Sing a Song for Šulgi.” Pp. 343–373 in *“An Experienced Scribe who Neglects Nothing”: Ancient Near Eastern Studies in Honor of Jacob Klein*, ed. Yitschak Sefati, Pinhas Artzi, Chaim Cohen, Barry L. Eichler and Victor A. Hurowitz. Bethesda: CDL Press.

CHAPTER 9

The Form and History of a Babylonian Prayer to Nabû

Tzvi Abusch

In an earlier study, I examined the *šuilla* Marduk, no. 2.[1] I demonstrated there, I believe, that this prayer did not follow the standard form of the *šuilla* type (Introductory Hymn, Prayer, Concluding Benediction[2]) and that it displayed a new pattern:

Introductory Hymn (1–9)
Capsule *šuilla* (10–12)
Prayer for Success (13–21)
Capsule *šuilla* (22–24)
Concluding Benediction (25–27).

Thus, the form of this *šuilla* was the result of innovation. But this Marduk *šuilla* does not stand in isolation. Some of its compositional features are found elsewhere.[3] Moreover, it is genetically related to Nabû, no. 3.[4] The purpose of the present article is to examine this Nabû *šuilla* in light of this genetic relationship in order to work out its structure, history of composition, and relationship to Marduk, no. 2.

---

* Professor Francesca Rochberg has devoted much of her scholarly attention and gifts to the study of Mesopotamian science, astronomy, and astrology. These fields border and often overlap those of Mesopotamian religion, and thus I hope that Chessie will receive this study of an Akkadian prayer in the spirit of respect and friendship in which it is offered.

1 Abusch 1983.

2 For this standard form, see Kunstmann 1932: 7–42; cf. Mayer 1976: 34–37.

3 See Abusch 1983: 14 and n. 38.

4 See already Kunstmann 1932: 99 s. Nabû 2: "Stilistisch auf das engste verwandt mit dem Gebete Marduk 2." Note that Kunstmann's Nabû 2 = Nabû, no 3 in W.R. Mayer's system of enumeration, which I follow here; see Mayer 1976: 400–401. For an edition of Nabû, no 3, see Mayer 1976: 473–475. For copies of MSS A–D, see Loretz and Mayer 1978: nos. 54–57. For a new copy and edition of Mayer's MS E (=*LKA* 56), see Maul and Strauss 2011: no. 64 (transliteration and translation, pp. 120–122; copy, p. 228). For a recent English translation, see Foster 2005: 695–696.

 | DOI 10.1163/9789004363380_010

Let us begin with a presentation of the two compositions in parallel columns:

Marduk, no. 2[a]	Nabû, no. 3
1. ÉN *gašru šūpû etel Eridug*	1. ÉN *rubû ašarēdu bukur* ^{d}*Tutu*
2. *rubû tizqāru bukur* ^{d}*Nudimmud*	2. *massû itepšu ilitti* ^{d}*Papnunanki*
3. ^{d}*Marduk šalbābu murīš E'engura*	3. ^{d}*Nabû nāš ṭup šīmāt ilī āšir Esagila*
4. *bēl Esaĝila tukulti Bābili*	4. *bēl Ezida ṣulūl Barsipa*
5. *rā'im Ezida mušallim napišti*	5. *narām* ^{d}*Nudimmud* (var. ^{d}*Papnunanki*) *qā'išu balāṭi*
6. *ašarēd Emaḫtila mudeššû balāṭi*	
7. *ṣulūl māti gāmil nišī rapšāti*	6. *ašarēd Bābili nāṣiru napišti*
8. *ušumgal kalîš parakkī*	7. *andul dadmi ēṭir nišī bēl ešrēti*
9. *šumka kalîš ina pī nišī ṭāb*	8. *zikirka ina pī nišī šuṭubba* ^{d}*lamassi*
10. ^{d}*Marduk bēlu rabû ilu rēmēnû* (var. Ø)	9. *mār rubê rabî* ^{d}*Marduk ina amātika kītti*
11. *ina qibītika kītti* (vars. *ṣērti* / *kabitti*) *lubluṭ lušlimma*	10. *ina siqrika kabti ina qibīt ilūtika rabīti* (var. *ina qibītika* [*rabīti*])
12. *luštammar ilūtka*	11. *anāku annanna mār annanna marṣu šumruṣu aradka* (var. lines 11–12 absent)
	12. *ša qāt eṭemmi ušburrudâ*[(sic)] *māmītu iṣbatūnima irteneddûni*
	13. *lubluṭ lušlimma* →
13. *ēma uṣammaru lukšud*	*ēma akappudu lukšud*
14. *šuškin kīttu ina pîya*	14. *šuškin kīttu ina pîya*
15. *šubši amāt damiqti ina libbiya*	15. *šubši amāt damiqti ina libbiya*
16. *tīru u nanzāzu liqbû damiqtī*	16. *tīru u manzāzu liqbû damiqtī*
17. *ilī lizziz ina imniya*	17. *lizziz ilī ina imniya*
18. *ištarī lizziz ina šumēliya*	18. *lizziz ištarī ina šumēliya*
19. *ilu mušallimu idāya* (var. *ina idīya*) *lū kayān*	19. *šēdu damqu lamassu damiqtu lirrakis*[(sic)] *ittiya*
20. *šurqamma qabâ šemâ u magāru*	
21. *amāt aqabbû ēma* (var. *kīma*) *aqabbû lū magrat*	20. *šutlimamma tašmâ u magāra*
	21. *zikir atammû qibītī lišl*[*im*]
22. *Marduk bēlu rabû napištī qīša*	22. *mār rubê rabî* ^{d}*Marduk* [*bal*]*āṭa qīša*
23. *balāṭ napištiya qibi*	23. *maḫarka kīniš atalluka lušbi*
24. *maḫarka namriš atalluka lušbi*	

Marduk, no. 2[a]	Nabû, no. 3
25. ^{d}*Enlil liḫdūka* ^{d}*Ea lirīška*	24. ^{d}*Šazu liḫdūka* [^{d}*Nudimmu*]*d lirīška*
26. *ilū ša kiššati likrubūka*	25. *ilū ša šamê u erṣeti likrubūka*
27. *ilū rabûtu libbaka liṭibbū*	26. *šamû* [(*u*) *erṣetu* … *l*]*išālilūka*
	27. *ilū rabûtu libbaka liṭibbū*
	28. ^{d}*Taš*[*mētu ḫērtu narāmtaka*]
	29. *ina Ez*[*ida bīt ši*]*kin napišti š*[*a ilī* (*rabûti*) *damiqtī*[?]] *liqb*[*īka*].

a Variants deemed significant for our discussion are given on the line in round brackets.

We may now turn to an analysis of the two texts. We first set out the textual segments and structures that the two *šuilla*s share, citing for each corresponding section the Akkadian text and its English translation.[5] Only after establishing the parallel structures will we point to some imperfections in the Nabû text and consider some of their implications.

1) **Hymnic Introduction**: Both *šuilla*s begin with hymnic introductions. The hymnic introduction of Marduk, no. 2 is made up of three stanzas.

1. ÉN *gašru šūpû etel Eridug*
2. *rubû tizqāru bukur* ^{d}*Nudimmud*
3. ^{d}*Marduk šalbābu murīš E'engura*

4. *bēl Esaĝila tukulti Bābili*
5. *rā'im Ezida mušallim napišti*
6. *ašarēd Emaḫtila mudeššû balāṭi*

7. *ṣulūl māti gāmil nišī rapšāti*
8. *ušumgal kalîš parakkī*
9. *šumka kalîš ina pī nišī ṭāb*

1. Famed mighty one, chieftain of Eridu,
2. Exalted prince, first-born of Nudimmud,
3. Raging Marduk, restorer of rejoicing to E'engura.

5 I assume that the reader is familiar with my study of the text of Marduk, no. 2; I will repeat as little as possible from that earlier study.

4. Lord of Esagila, hope of Babylon,
5. Lover of Ezida, preserver of life,
6. Lone one of Emahtila, multiplier of living.

7. Protection of the land, savior of the multitudes of people,
8. The single great one of chapels everywhere,
9. Your name is sweetly hymned by the people everywhere.

The first stanza (1–3) presents Marduk as the young hero of Eridu. The focus is upon his relationship to his parents and his actions on their behalf (*murīš E'engura*). The second stanza (4–6) presents Marduk as the god of Babylon and its environs: in line 4, he is the lord of Esagilla and the trust of Babylon (note the order temple–city); in lines 5–6, epithets describing Marduk's relation to temples in Borsippa are juxtaposed to those describing his care for human life. The third stanza (7–9) presents Marduk in relationship to land, people, and sanctuaries, and ends with a declaration that his praise is sung by people everywhere.

We find a similar structure in Nabû, no. 3, for here too the hymnic introduction is made up of three stanzas that seem to parallel those of the Marduk text:

1. ÉN *rubû ašarēdu bukur* ^{d}*Tutu*[6]
2. *massû itepšu ilitti* ^{d}*Papnunanki*
3. ^{d}*Nabû nāš ṭup šīmāt ilī āšir Esagila*

4. *bēl Ezida ṣulūl Barsipa*
5. *narām* ^{d}*Nudimmud* (var. ms E: ^{d}*Papnunanki*) *qā'išu balāṭi*
6. *ašarēd Bābili nāṣiru napišti*

7. *andul dadmi ēṭir nišī bēl ešrēti*
8. *zikirka ina pī nišī šuṭubba* ^{d}*lamassi*

1. O Pre-eminent prince, scion of Tutu,
2. Accomplished leader, offspring of Papnunanki,
3. Nabû, bearer of the tablet of destinies of the gods, supervisor of Esagila.

4. Lord of Ezida, protection of Borsippa,
5. Beloved of Nudimmud (var.: Papnunanki), bestower of life,

6 My transcription generally follows the main text in Mayer's edition, which is usually based on the Nineveh MSS. Note that the Aššur MS *LKA* 56 = KALI 4, no. 64 (= Mayer's MS E) has a number of earlier and better readings than the Nineveh MSS.

6. Pre-eminent one of Babylon, protector of life.

7. Shelter of settlements, savior of the people, lord of shrines,
8. Your mention is sweeter in the mouth of the people than that of their own protective spirit.[7]

The first stanza (1–3) places Nabû in relationship to his parents and sets out his activities in his native home: He is the son of Marduk (under the name Tutu, the original god of Borsippa) and Ṣarpanītu (under the name Papnunanki, originally the wife of Asalluḫi) and administrator of Esagila (the temple of Marduk in Babylon). The second stanza (4–6) presents Nabû as the god of Borsippa and its environs: in line 4, he is the lord of Ezida and the protector of Borsippa (note the order temple—city); in lines 5–6, epithets describing his relationships are juxtaposed to those describing his care for human life. The third stanza (7–8) presents Nabû in relationship to land, people, and sanctuaries, and ends with a declaration that his praise is sung by people everywhere.

2) **Capsule *šuilla***: Marduk, no. 2 has a capsule *šuilla* in lines 10–12, immediately following the hymnic introduction:

10. ^{d}*Marduk bēlu rabû*
11. *ina qibītika kītti* (vars.: *ṣērti* / *kabitti*) *lubluṭ lušlimma*
12. *luštammar ilūtka*

10. O Marduk, great lord,
11. By your affirmative decree, may I live and be well,
12. I will then constantly praise your godhead.

Marduk, no. 2 has a second capsule *šuilla* in lines 22–24, immediately preceding the concluding benediction:

22. ^{d}*Marduk bēlu rabû napištī qīša*
23. *balāṭ napištiya qibi*
24. *maḫarka namriš atalluka lušbi*

22. O Marduk, great lord, grant me my life,
23. Decree for me a healthy life,
24. In joyfully serving you regularly will I (then) find satisfaction.

7 Cf. *CAD* L, 62b.

Each of these summaries contains (a) an invocation, (b) a prayer for life, (c) a promise of praise or service. The two summary *šuilla*s in lines 10–12 and 22–24 serve as a frame for the prayer for success with Marduk in lines 13–21.

Similarly, Nabû, no. 3 contains two summary *šuilla*s, one after the hymnic introduction (9–13a), the other before the concluding benediction (22–23):

9. *mār rubê rabî* ^{d}*Marduk ina amātika kītti*
10. *ina siqrika kabti ina qibīt ilūtika rabīti* (var. MS E: *ina qibītika* [*rabīti*])
11. *anāku annanna mār annanna marṣu šumruṣu aradka* (var. MS E: lines 11–12 absent)
12. *ša qāt eṭemmi*[8] *ušburrudâ*[(sic)9] *māmītu iṣbatūnima irteneddûni*
13a. *lubluṭ lušlimma* (13b *ēma akappudu lukšud*)

……

22. *mār rubê rabî* ^{d}*Marduk* [*bal*]*āṭa qīša*
23. *maḫarka kīniš atalluka lušbi*

9. O Son of the great prince Marduk, by your affirmative decree,
10. By your honored command, by the command of your great godhead (var.: by your great command)—
11. I, so-and-so, the son of so-and-so, your sick, suffering servant
12. Whom the Hand of a Ghost, Witchcraft, Hand of a Curse have seized and constantly persecute (var.: lines 11–12 absent)—
13a. May I live and be well.

……

22. O Son of the great prince Marduk, grant me life,
23. In faithfully serving you regularly will I (then) find satisfaction.

Each of the capsule *šuilla*s in Nabû, no. 3 contains an invocation of the god and a prayer for life; the latter one concludes with a promise of service.

3) **Prayer for Success:** The prayer for success in Marduk, no. 2 reads:

13. *ēma uṣammaru lukšud*
14. *šuškin kīttu ina pîya*
15. *šubši amāt damiqti ina libbiya*
16. *tīru u nanzāzu liqbû damiqtī*
17. *ilī lizziz ina imniya*

---

8 Or *šugidimmakku.*

9 UŠ$_{11}$.BÚR.RU.DA is a mistake for "witchcraft" = *kišpū* (UŠ$_{11}$ or UŠ$_{11}$.ZU). UŠ$_{11}$.BÚR.RU.DA refers to a ritual to undo witchcraft.

18. *ištarī lizziz ina šumēliya*
19. *ilu mušallimu idāya* (var. *ina idīya*) *lū kayān*
20. *šurqamma qabâ šemâ u magāru*
21. *amāt aqabbû ēma* (var. *kīma*) *aqabbû lū magrat*

13. Whatsoever I seek may I attain,
14. Place effective speech in my mouth,
15. Fashion an agreeable word in my mind,
16. May courtier and attendant seek agreement on my behalf,
17. May my god stand at my right,
18. May my goddess stand at my left,
19. May the guardian-deity be constantly at my side,
20. Grant me (the power) to speak, to be hear, and to meet with consent (so that)
21. Whatsoever words I utter may meet with consent.

As noted above, the capsule *šuilla*s in Marduk, no. 2 serve as a frame for the prayer for success in the audience before the deity (13–21). Similarly, the capsule *šuilla*s in Nabû, no. 3 serve as a frame for the prayer for success (13b–21) in that *šuilla*:

13b. ... *ēma akappudu lukšud*
14. *šuškin kīttu ina pîya*
15. *šubši amāt damiqti*[10] *ina libbiya*
16. *tīru u manzāzu liqbû damiqtī*
17. *lizziz ilī ina imniya*
18. *lizziz ištarī ina šumēliya*
19. *šēdu damqu*[11] *lamassu damiqtu lirrakis*[12] *ittiya*
20. *šutlimamma tašmâ u magāra*
21. *zikir atammû qibītī lišl*[*im*]

13b. ... Whatsoever I plan, may I attain,
14. Place effective speech in my mouth,

10 Text: *damiqtu*.

11 Perhaps, rather, *šēd dumqi lamassi dumqi*. See Mayer 1976: 245 and n. 43, and p. 247. I choose *damqu* and *damiqtu* because of a writing cited there in n. 43: ^{d}LAMMA SIG$_5$-*tú*/*tu*$_4$—see *BMS* 12:110 = now *CMAWR* 2, 8.28: 110, and *KAR* 298: 36, cited *CAD* L, 64a. But note that Mayer 1976: 247 reads our passage as: *šēd dumqi lamassi dumqi*.

12 We expect a plural verb; but see Köcher, Oppenheim, and Güterbock 1957–1958: 67 iii 30, cited *CAD* Š/2, 258b.

15. Fashion an agreeable word in my mind,
16. May courtier and attendant seek agreement on my behalf,
17. May my god stand at my right,
18. May my goddess stand at my left,
19. May a protective spirit and a spirit of vitality be bound to my side.
20. Accord me (the power) to speak and to meet with consent (so that)
21. My command be fulfilled as I pronounce it.

Even a cursory examination of the prayers for success in the two *šuilla*s will indicate their likeness. Note the virtual identity of the central part of the prayers for success: Marduk, no. 2, lines 14–18 = Nabû, no. 3, lines 14–18; note also that while the opening and closing lines (Marduk, no. 2, lines 13, 20–21 // Nabû, no. 3, lines 13, 20–21) of the prayers for success are not identical, they are structurally parallel and quite similar:

Marduk:

13. *ēma uṣammaru lukšud*
...
20. *šurkamma qabâ šemâ u magāru*
21. *amāt aqabbû ēma aqabbû lū magrat.*

13. Whatsoever I seek, may I attain,
...
20. Grant me (the power) to speak, to be heard, and to meet with consent (so that)
21. Whatsoever words I utter may meet with consent.

Nabû:

13. *ēma akappudu lukšud*
...
20. *šutlimamma tašmâ u magāra*
21. *zikir atammû qibītī lišlim.*

13. ... Whatsoever I plan, may I attain,
...
20. Accord me (the power) to speak and to meet with consent (so that)
21. My command be fulfilled as I pronounce it.

4) **Concluding Benediction**: Finally, both *šuilla*s (25–27) end with similar benedictions (Marduk: lines 25–27; Nabû: lines 24–29). The benediction in Marduk, no.2 reads:

25. ^{d}*Enlil liḫdūka* ^{d}*Ea lirīška*
26. *ilū ša kiššati likrubūka*
27. *ilū rabûtu libbaka liṭibbū*

25. May Enlil rejoice over you, may Ea exult over you,
26. May the gods of the universe bless you,
27. May the great gods make you happy.

The benediction in Nabû, no.3 reads:

24. ^{d}*Šazu*[13] *liḫdūka* [^{d}*Nudimmu*]*d*[14] *lirīška*
25. *ilū ša šamê u erṣeti likrubūka*
26. *šamû* [(*u*) *erṣetu* … *l*]*išālilūka*
27. *ilū rabûtu libbaka liṭibbū*
28. ^{d}*Taš*[*mētu hērtu narāmtaka*]
29. *ina Ez*[*ida bīt ši*]*kin napišti š*[*a ilī* (*rabûti*) *damiqtī*$^{?}$] *liqb*[*īka*].[15]

24. May Šazu rejoice over you, may [Nudimmu]d exult over you.
25. May the gods of heaven and the netherworld bless you.
26. May heaven [and the netherworld …] celebrate you.
27. May the great gods make you happy.
28. May Taš[mētu, your beloved wife]
29. Speak [well of me to you] in Ez[ida, the house of the Grant]ing of life o[f the (great) gods].

The first part of the Nabû benediction is quite similar to the benediction in Marduk, no. 2;[16] the last two lines of the Nabû benediction are a special addition to the Nabû composition.[17]

Overall, the two *šuilla*s parallel each other. But while the Marduk one seems to have attained a new form that is complete and as free as possible from flaws, this cannot be said of the Nabû one. There are a number of imperfections in our Nabû prayer, the most significant of which is found in line 13. For there, the expected promise of praise is absent, and the present text juxtaposes *lubluṭ*

13 A name of Marduk.
14 Originally probably ^{d}Papnunanki or the like, a reading that I imagine MS E had; unfortunately, MS E is broken at this point.
15 For the reading of the last two lines, see Beaulieu and Mayer 1997: 174; for a different reading, see Maul and Strauss 2010: 121–22.
16 Marduk 25 // Nabû 24; Marduk 26 // Nabû 25; Marduk 27 = Nabû 27.
17 Cf. Mayer 1976: 348.

luššlimma, the request for life, with the beginning of the prayer for success.[18] Contrast the absence here of the promise of praise with the closing summary *šuilla* in this text (22–23), but especially with the first summary *šuilla* in Marduk, no. 2, which ends with *luštammar ilūtka*. There, this phrase occurs immediately after *lubluṭ lušlimma* and just before the beginning of the prayer for success. In our Nabû *šuilla*, the phrase *luštammar ilūtka* is missing in line 13 between *lubluṭ lušlimma* and *ēma akappudu lukšud*.[19]

The absence—which reflects either a textual omission or a failure to provide the correct form in the process of composition—is not unrelated to other problems in this part of the text, difficulties which may be at least partially responsible for the absence:

(1) Lines 9–10 and 13a are separated from each other by 11–12, where the speaker introduces himself and asserts that various evils have seized him (*anāku annanna mār annanna marṣu šumruṣu aradka* / *ša qāt eṭemmi ušburrudâ*(sic) *māmītu iṣbatūnima irteneddûni*). Lines 11–12 have no correspondence in Marduk, no. 2,[20] and should be treated as an insertion that breaks up the segment 9–10 + 13a; in fact, it is absent in MS E.[21]

(2) Lines 9b–10 essentially say the same thing three times and parallel the variants in Marduk, no. 2: 11: *ina qibītika kītti* / *ṣērti* / *kabitti*. That is to say, the three variants in different manuscripts of Marduk, no. 2 are preserved in Nabû, no. 3 as three parallel phrases.[22] Leaving for later the question as to which of our two compositions preserves the more original text, it is clear that *ina qibīt ilūtika rabīti* in Nabû, no. 3, line 10b is a corrupt form of the expected *ina qibītika rabīti*,[23] which, in fact, is preserved in MS E. *ilūtka* belongs in line 13. This corruption is perhaps due to the insertion of lines 11–12, which caused the displacement of *ilūtika*[24] from line 13 to line 10. (We should probably move *ilūtka* down from line 10 to line 13 and supply *luštammar*, thus reconstructing the expected <*ilūtka luštammar*>.)

---

18 This has led translators to erroneously connect the two; see, for example, Falkenstein and von Soden 1953: 312: "möge ich ... zu Leben und Gesundheit kommen und es so, wie ich es erstrebe, (auch) erreichen!"; Seux 1976: 298: "Que je vive, que je sois sauf et que je réalise tout ce que je projette."

19 Further support that the two half lines 13a and 13b are separate and are not to be connected is provided by MS E, which "verteilt 13 auf 2 Zeilen" (Mayer 1976: 473).

20 Marduk, no. 2 does have the inserted *attalû* formula before line 11 in *BMS* 54: 1′–4′, Loretz and Mayer 2011: no. 27, obv. 11–14, and PBS 1/2, no. 108, obv. 1′–8′.

21 Cf. Mayer 1976: 473: "E: lässt den Einschub 11f weg."

22 See Abusch 1983: 9–10, n. 23. In place of *ṣērti*, Nabû, no. 3 has *rabīti*.

23 But note *ina qibīt ilūtika rabīti* in Nabû, no. 4, rev. 25, edited Mayer 1976: 476–477 and cited ibid., 304.

24 And the change of form from DINGIR(-*ut*)-*ka* = *ilūtka* to DINGIR-*ti*-*ka* = *ilūtika*.

Other imperfections in our text agree with the inelegant, actually clumsy or maladroit, form of the summary *šuilla* in lines 9–10 + 13a of our main text. Here I would simply turn back to the hymnic introduction and take note of the text of lines 5–6: the Marduk *šuilla* suggests that there should be a connection between 5a and 5b and between 6a and 6b—but I, for one, do not see a connection here in the Nabû composition. In 5a (and 6a) we expect a temple (or city) name, instead we have the name Nudimmud (= Ea); Nudimmud is Nabû's grandfather (according to the Eridu-Babylon lineage),[25] but following *narām* we expect a female name, a reading that is actually found in MS E.[26]

There can be no doubt that a relationship exists between Nabû, no. 3 and Marduk, no. 2, though it should be clear by now that while the Marduk composition is a consummately executed example of a new form, the Nabû composition is an imperfect, perhaps second-rate, example of the form. But now we must face a difficult question: which of the two is earlier and which is constructed using the other as a model? To rephrase the question: is the Nabû composition dependent on the Marduk one or is the Marduk composition dependent on the Nabû one? We can simplify the problem by noting that MS E, the Aššur version of Nabû, no. 3 represents an earlier version of the text than do the Nineveh MSS A–D; thus, in determining dependency, we may ignore the textual problems and corruptions in A–D and compare only MS E. Nevertheless, even without the textual problems of Nabû, no. 3, MSS. A–D, the text preserved in Nabû, no. 3 MS E is still not as well formed as that in Marduk, no. 2.

I imagine that ancient composers/scribes would appreciate a newly worked out text form and, without good reason, would hesitate to alter and damage the form. This being the case, we may well conclude that a new literary form would not immediately degenerate[27] and that an imperfect example of a new form might well be earlier than a fully formed example. Accordingly, we may treat Nabû, no. 3 (MS E) as the precursor—a trial run or an early draft—of a new form and Marduk, no. 2 as the more developed creation.

Against this conclusion is the observation that the Nabû composition seems to reflect a later stage of history than Marduk, no. 2. For while in Marduk, no. 2, Marduk is associated with the main temples of Borsippa (ll. 5–6), in Nabû, no. 3, Nabû is the protector of Borsippa (l. 4) and the master of Ezida (l. 4; cf. l. 29).[28] Since historically Nabû replaces Marduk as the god of Borsippa

25 For Nabû in association with Ea, see Pomponio 1978: 165–168.

26 Thus, the reading ^{d}Papnunanki in E is to be preferred.

27 I can see no reason why the composer of Nabû, no. 3 would wish to alter the form.

28 This is not to deny that Nabû has several important functions in Babylon (cf. ll. 3 and 6) even after he took over Borsippa; compare, for example, his roles in the *Akitu* festival in Babylon and statements about him in other texts such as *kudurrus*.

(as earlier Marduk took over this role from Tutu),[29] Marduk's connections with Borsippa in Marduk, no. 2, but Nabû's connections with it in Nabû, no. 3 suggest that the Nabû *šuilla* is later than the Marduk one. Thus, given the shared forms, we should conclude that the Nabû composition is dependent on the Marduk one, that it is inferior, and that it does represent a deterioration or degeneration in form.[30]

We thus have arguments in support of the priority of each of the texts, though the latter argument in favor of the priority of Marduk, no. 2 and the dependence thereupon of Nabû, no. 3 undoubtedly carries much more weight. But it would be best if we could decide which is earlier on the basis of a parallel passage in the two texts. I would suggest, therefore, that we turn back to a passage mentioned earlier, the invocation of the god's decree in the first summary *šuilla*.

In Nabû, no. 3 MS E, the invocation reads: *ina amātika kītti, ina siqrika kabti, ina qibītika* [*rabīti*]; in Marduk, no. 2, the equivalent passage reads: *ina qibītika kītti* (vars.: *ṣērti*; *kabitti*). If the Marduk composition is dependent on the Nabû one, the variant readings in the Marduk text would reflect the break-up of a ἓν διὰ τριῶν (*hendiatrion*) in the Nabû text and the preservation of synonymous variant readings in different manuscripts of the Marduk composition; if, on the other hand, the Nabû composition is dependent on the Marduk one, then the *hendiatrion* may be the result of the conflation of several variants and should be regarded as a triple reading or triplet.[31]

It is possible that it is easier to imagine the break-up of an expansive *hendiatrion* into three synonymous variant readings rather than the reverse (in which case our Nabû text would be the earlier of the texts); but this is a subjective (aesthetic) judgment. And, for purposes of a decision regarding priority, it would be better to ascertain whether or not there are other examples of similar *hendiatrion*s or triplets used as invocations of a god's decree. If there are, then this passage in the Nabû prayer need not be dependent on the Marduk prayer. But if such a usage is rare, this anomaly would suggest that the formulation in the Nabû text is the result of a particular circumstance and would support the proposition that the Nabû text is dependent on the Marduk one. A perusal

29 See Pomponio 1978: 63–65, and idem 1998: 19. In Abusch 1983: 14–15, I noted: "The Marduk šuilla was composed, I should guess, sometime before the reign of Adad-apla-iddina." In view of the inscription of Marduk-šāpik-zēri (Frame 1995: 47), cited by Pomponio 1998: 19, I would now change my estimate of the date of composition to "before the reign of Marduk-šāpik-zēri."

30 This conclusion would agree with my supposition that an original composition usually coheres better than its descendent. Of course, texts undergo change (additions and reformulations) because of ritual or conceptual changes (not to mention mistakes in transmission) and may become less coherent in the process of transmission.

31 See already my comment in Abusch 1983: 10, n. 23.

of the examples collected by Mayer (1976: 300–306) reveals that our passage (1976: 304; cf. 301–302) is the only one in this function that is made up of three elements (all others have one or two). Given the uniqueness or, at least, oddity of the repetition of a similar phrase three times in the invocation of the god's decree, I surmise that our Nabû prayer represents an attempt to preserve earlier synonymous variants (found in the Marduk *šuilla*) by conflating them into the present triplet.[32]

Accordingly, I feel compelled to conclude that our Nabû *šuilla* is dependent upon Marduk, no. 2 and made use of it.[33] To be sure, not only is Nabû, no. 3 not an improvement on Marduk, no. 2, but it also seems to be a degraded form. Why this should be the case is certainly worthy of further study.

Bibliography

Abusch, I. Tzvi

1983 "The Form and Meaning of a Babylonian Prayer to Marduk." *JAOS* 33: 3–15.

Abusch, I. Tzvi and Daniel Schwemer with Mikka Lrukko and Greta Van Buylacne

2016 Corpus of Mesopotamian Anti-Witchcraft Rituals. vol. 2 AMD 8/2. Leiden: Brill.

Beaulieu, Paul-Alain and Werner R. Mayer

1997 Review: Akkadische Lexikographie: CAD $Š_2$ und $Š_3$. *Or*NS 66: 157–180.

Falkenstein, Adam and Wolfram von Soden

1953 *Sumerische und akkadische Hymnen und Gebete*. Zürich: Artemis-Verlag.

Foster, Benjamin R.

32 I have modeled my understanding of triple readings and triplets on the phenomena of double readings and doublets. The phenomenon of preserving variant readings by creating doublets is well known in biblical text criticism. For double readings there, see especially Talmon 1960: 143–184 = 2010: 217–266, and Talmon 2010: 35–46. Cf. also Talmon 1976: 170–173; for purposes of the comparison of Marduk, no. 2 and Nabû, no. 3, note Talmon's comment in idem 1976: 171: "The recognition of a doublet in the MT or in a version usually arises from a comparison of two or more witnesses to the text. A simple or short reading in one indicates that an expanded reading in the other may constitute a doublet, if the assumedly redundant component is but a reiteration of what already is expressed in the other textual component."

For a concise treatment of doublets and synonymous readings in the Bible, see Tov 2012: 225–227 and 257–258, and for the preservation of variant readings in different manuscripts (as in Marduk, no. 2, line 11), see Talmon 1961: 335–383 = 2010: 171–216.

33 Here, then, I should mention that the corruptions in Nabû, no. 3 MSS A–D are later developments of the Nabû text, changes that are independent of Marduk, no. 2, and may have been introduced into the text either in Nineveh or whence the Nineveh tradition derived.

2005 *Before the Muses: An Anthology of Akkadian Literature*. 3rd ed. Bethesda: CDL Press.

Frame, Grant

1995 *Rulers of Babylonia from the Second Dynasty of Isin to the End of Assyrian Domination: 1157–612 BC*. RIMB 2. Toronto: University of Toronto Press.

King, Leonard W.

1896 *Babylonian Magic and Sorcery*. London: Luzac & Co.

Köcher, Franz, A. Leo Oppenheim, and Hans G. Güterbock

1957–1958 "The Old Babylonian Omen Text VAT 7525." *AfO* 18: 62–80.

Kunstmann, Walter G.

1932 *Die babylonische Gebetsbeschwörung*. LSS NF 2. Leipzig: Hinrichs.

Loretz, Oswald and Werner R. Mayer

1978 *Šu-ila-Gebete: Supplement zu L. W. King, Babylonian Magic and Sorcery*. AOAT 34. Kevelaer: Butzon & Bercker; Neukirchen-Vluyn: Neukirchener.

Maul, Stefan M. and Rita Strauß

2011 *Ritualbeschreibungen und Gebete I*. Mit Beiträgen von Daniel Scheme KALI 4 [= WVDOG 133]. Wiesbaden: Harrassowitz.

Mayer, Werner R.

1976 *Untersuchungen zur Formensprache der babylonischen "Gebetsbeschwörungen."* StPohl SM 5. Rome: Pontificio Istituto Biblico.

Pomponio, Francesco

1978 *Nabû. Il culto e la figura di un dio del Pantheon babilonese ed assiro*, StSem 51. Rome: Istituto di Studi del Vicino Oriente. Università di Roma.

1998 "Nabû. A. Philologisch." *RLA* 9: 16–24.

Seux, Marie-Joseph

1976 *Hymnes et prières aux dieux de Babylonie et d'Assyrie*. Paris: Éditions du Cerf.

Talmon, Shemaryahu

1960 "Double Readings in the Massoretic Text." Textus 1: 143–184 [= 2010 Pp. 217–266 in Talmon, *Text and Canon of the Hebrew Bible: Collected Studies*. Winona Lake, IN: Eisenbrauns].

1961 "Synonymous Readings in the Textual Traditions of the Old Testament." *SH* 8: 335–383 [= 2010 Pp. 171–216 in Talmon, *Text and Canon of the Hebrew Bible: Collected Studies*. Winona Lake, IN: Eisenbrauns].

1976 "Conflate Readings (OT)." Pp. 170–173 in *Interpreter's Dictionary of the Bible: Supplementary Volume*. New York: Abingdon Press.

2010 "The Textual Study of the Bible: A New Outlook." Pp. 19–84 in Talmon, *Text and Canon of the Hebrew Bible: Collected Studies*. Winona Lake, IN: Eisenbrauns.

Tov, Emmanuel

2012 *Textual Criticism in the Hebrew Bible*. 3rd ed. Minneapolis: Fortress Press.

CHAPTER 10

Translation in The Elevation of Ištar

Niek Veldhuis

The bilingual poem "The Elevation of Ištar" tells the story of how, at the behest of the great gods, Anu decides to give Ištar power over heaven and earth and to elevate her to his own majesty. This is followed by a speech of praise by Enlil, confirming that Ištar is now equal in power to himself (tablet 4) and, presumably, by another speech by Ea in tablet 5 of which only the introductory line is preserved. The poem, which was edited by Hruška 1969, is primarily attested in Neo-Assyrian Nineveh and Late Babylonian Uruk. Since Hruška's publication a few more tablets and fragments have been recognized as probably or certainly containing parts of the poem.[1] The series consisted of at least five tablets; what remains today mostly belongs to tablets three and four. A new edition and translation by Daniel Foxvog (2013) was not formally published, but is available on his academia.edu page.[2] In the context of the NEH-supported ORACC project Bilinguals in Late Mesopotamian Scholarship (BLMS) all tablets have been re-edited and translated. The initial work (edition, translation, and lemmatization) was done by Jeremiah Peterson; an editorial team consisting of Jamie Novotny, Josh Jeffers and myself has prepared Peterson's files for online publication at http://oracc.org/blms.

The goal of the present paper is to place The Elevation of Ištar among the various groups of first millennium bilingual compositions. In the process I will discuss the likely date and place of the composition and consider its intellectual context in Neo-Assyrian Nineveh and Late Babylonian Uruk.

The Manuscripts

The Elevation of Ištar is known in 12 or 13 manuscripts from Nineveh and Late Babylonian Uruk, plus one fragment from Neo-Babylonian Babylon.

1 Primarily by Lambert 1971.

2 Daniel Foxvog: "The Late Bilingual Exaltation of Ištar (Inannas Erhöhung)" on http://berkeley.academia.edu/DanielFoxvog. The transliterations of individual tablets on CDLI (http://cdli.ucla.edu) are also by Daniel Foxvog.

 | DOI 10.1163/9789004363380_011

Neo-Assyrian:

Rm 953	unplaced	P424891
K 8482	unplaced	P397648
1880-07-19,281	IV	P452027
K 13459	IV	P357130
K 15340	IV	P357142
K 4940+	attribution uncertain	P395798

Neo-/Late Babylonian Uruk:

SpTU 2, 28	I	P348633
TCL 6, 51	III	P363723
TCL 6, 52	III	P363724
LKU 12	III	P414154
LKU 135	III	P414266[a]
Bod S 302 (RA 12, 73–84)	IV	P368468

a This tablet, kept in the Vorderasiatisches Museum in Berlin, remains unpublished. Adam Falkenstein cataloged (but did not copy) the tablet as LKU 135; he provided some variant readings (comparing it to TCL 6, 51 and 52), but no museum number. The variants are included in the editions by Hruška and Foxvog but the text is not included in BLMS and is not considered in this article.

Neo-Babylonian Babylon:

VS 24, 37[a]	IV	P347156

a Edition by Falkenstein 1952; see Pedersén 2005: 192 N8:40.

The Neo-Assyrian tablet K 4940 + K 5118 + K 6020[3] is mentioned in the W.G. Lambert Notebooks as possibly belonging to our composition (Lambert Folios 8992 and 9001).[4] The tablet contains a prayer or speech to the spouse of Sîn, presumably Ningal. If it belongs to our composition (which is by no means certain) it may be an attempt to convince Ningal to intercede with Sîn on behalf

3 K 5118 was published by MacMillan 1906, no. 26 and Meek 1913, no. 23; see also Delitzsch 1913: 141. An edition and translation of the joined fragments by John Carnahan is available on BLMS (P395798).

4 The Notebooks were made available by Andrew George in http://oracc.org/contrib/lambert.

of Inana/Ištar. The style of the text and the odd mixing of Emesal and regular Sumerian (see §3.2) is entirely in line with the style of the known fragments of The Elevation of Ištar.

The exemplar from Babylon (VS 24, 37) is associated with the Ištar of Akkade temple in the Merkes district (see Pedersén 2005: 188–192).

First Millennium Bilinguals

First millennium bilinguals may roughly be divided into four groups: Old Babylonian Sumerian literary compositions to which an Akkadian translation had been added later on; liturgical texts in Emesal; incantations or incantation series; and finally bilingual-born compositions. The Sumerian used in each of these groups has its own peculiarities and characteristics.[5] The first group, best exemplified by Lugale (also known as Ninurta's Exploits) and Angim (or The Return of Ninurta) preserved Old Babylonian literary Sumerian—with occasional variants, adjustments or reinterpretations. The Akkadian usually follows the Sumerian fairly closely but may add an interpretative layer (Seminara 2002). Liturgical texts in Emesal (Eršahunga's, Balaĝs, Eršemmas, and Šu'ilas) tend to use more or less standardized turns of phrases and build upon a continuous tradition of temple practice. The texts of these compositions was in flux for a long time and could be adapted to new situations (see Delnero 2015; Maul 1998). The Akkadian translation often follows the Sumerian, but there are many attestations of interpretative translations, sometimes using complex exegetical methods (Maul 1997). The Sumerian of first millennium incantations and incantation series has, as far as I know, hardly been studied in detail and the sheer amount of evidence may hamper such a study. Some incantations went back to Old Babylonian examples and partly preserved an Old Babylonian understanding of Sumerian grammar and vocabulary; others were clearly composed at a later date and entirely ignore Sumerian syntax.[6] The use

5 See Falkenstein 1953; and the brief but thoughtful discussion of translation in Frahm 2011: 17–19.

6 See, for instance, KAR 50, edited and discussed by Linssen 2003: 96 and 267–268; Gabbay 2015: 194–195. A new edition with a new duplicate appears in the article by Alan Lenzi in this volume. The last line of the incantation in KAR 50 (eme hul-ĝal$_2$ bar-še$_3$ he$_2$-en-ta-gub) stands out because it follows acceptable Sumerian syntax and is not translated into Akkadian. This is a stock line that is found in numerous incantations and was thus readily available to the scribes of the time. For examples of this phrase, see PSD B: 95 (bar A 1.4.1.2) and 103 (bar A Bilingual 4).

of Emesal in incantations is minimal.[7] Bilingual-born compositions constitute a mixed bag of prayers and hymns, royal inscriptions, and Eduba texts.[8] It is in this category (which hardly constitutes a category in the sense of a homogeneous group) that the Elevation of Ištar is to be placed. Like the incantations, the Sumerian of the bilingual-born group may use morphology and syntax that deviates considerably from classical Sumerian. In addition, these texts tend to use rare and artificial Sumerian lexemes, apparently taken from lexical lists or produced by analogy (see below). Moreover, a number of these texts use Emesal at places where such forms are not expected.[9]

Emesal and artificial Sumerian words thus belong to the practice of producing new Sumerian texts that are not based directly on earlier examples. It is against this background that we may appreciate the Sumerian of the Elevation of Ištar.

Sumerian in The Elevation of Ištar

In addition to regular, run-of-the-mill Sumerian words and their Akkadian equivalences (§3.1), The Elevation of Ištar also includes a fair number of Emesal words (§3.2) and a couple of very rare Sumerian words and expressions (§3.3). Although each of those types is well known from lexical and bilingual texts, the particular mix found in the Elevation of Ištar is somewhat unusual.

Regular Equations

The largest group of Sumerian–Akkadian equations in the Elevation of Ištar corresponds more or less to expectations. The first lines of tablet III may serve as an example:[10]

an kug-ga gal-bi inim ka-na nu-mu-un-$^{\text{ti-il}}$til-le-e-de$_3$
^{d}*a-nu el-lu zi-kir pi-i-šu*$_2$ *la ga-ma-ru*
diĝir gal-gal-e-ne ul-la a-ra-zu-ĝa$_2$ ^{zu}zub-bi-gin$_7$ ba-an-gur$_2$-ru-uš
DINGIR.MEŠ GAL.MEŠ *ina an-nim tes*$_2$*-li-tum gam-li-iš ik-tan-šu*$_2$*-uš*

7 Falkenstein 1953: 10 nt. 58 came up with one single example. More than 60 years later there may well be more, but the occurrence remains very exceptional.

8 In Praise of Scribal Art (Hurowitz 2000); "Examenstext A" (Sjöberg 1974), and the Letter of Lugal-ibila to Lugal-nesaŋ (Civil 2000). The letter of Lugal-ibila is known in copies from Hattuša, and Ugarit, as well as first millennium sources. For its history see Civil 2000: 113.

9 Numerous examples in Falkenstein 1953; see below (§3.2).

10 TCL 6, 51 (P363723) o 1–2.

The great words of holy An are not to be nullified.
The great gods bow down to him like a curved staff with: "Please, be my guest."

In these lines the following unsurprising correspondences may be observed:

an	*anu*	Anu
kug	*ellu*	pure
inim	*zikru*	utterance
ka	*pû*	mouth
til	*gamāru*	to end
diĝir	*ilu*	god
gal	*rabû*	great
zubi	*gamlu*	bent stick
gur_2	*kanāšu*	to bow

A competent scribe would know these Sumerian words (and their Akkadian equivalents) from lexical lists, from other bilinguals, or from both.

In addition, the passage displays several well-known Sumerian–Akkadian morphological correspondences, such as nu- corresponding to *lā*; the ba- prefix corresponding to a /t/ infix in Akkadian, etc.[11] Again, a competent scribe would have known those and would have no problem applying such rules.

One may also notice that the Sumerian verbs are conjugated and that the Sumerian sentences are more or less syntactically sound. Producing such sentences goes beyond mere plugging of lexemes from a lexical list and presumes an active knowledge of how Sumerian works. That is not to say that there are no oddities when compared to classical Sumerian. Instead of ka-na one would expect ka-ga-na (or ka-ka-na) and gal-bi ("greatly?") has no corresponding lexeme in the Akkadian and its syntactic position in the Sumerian is unclear.

Most intriguing in this section is the expression ul-la a-ra-zu-$ĝa_2$, translated as *ina annim teslītum* ("in approval and prayer"). The expression is known from a handful of Old Babylonian literary passages where it represents a polite phrase (my translation "Please, be my guest" intends to render that).[12]

---

11 Black 1991: 24; 27.

12 The expression was discussed most recently by Koslova 2014: 314–317. The phrase is found in Enlil and Ninlil 28 (Civil 1983: 52); Heron and Turtle 107 (Peterson 2007: 395); Inana and Enki (Farber-Flügge 1973: 253) and in the Late Babylonian ritual text TCL 6, 43 rev. 9 (RAcc: 109 and Linssen 2003: 197). Note that the Old Babylonian attestations all appear in literary texts with a rather restricted number of duplicates, falling outside the mainstream of the

Emesal

The use of Emesal in the Elevation of Ištar is unusual in two respects.[13] First, Emesal forms are found in narrative passages and in speeches of male gods (no speeches of female deities are preserved). Second, the use of Emesal forms is very inconsistent; the same word may appear in regular Sumerian and in Emesal in adjacent lines. These phenomena can be seen in TCL 6, 51: o 4–5 with duplicate TCL 6, 52 o 1′–2′:

> an dug$_4$-ga mah-zu saĝ ba-du ul-la mu-lu im-me
> ^{d}*a-nu qi$_2$-bit-ka ṣir-tu$_4$ ina mah-ri il-lak ul-la man-nu i-qab-bi*
> a-a dim$_3$-me-er-e-ne-ke$_4$ ^{en}inim-zu an-ki-a te-me-en-bi diĝir na-me nu-še
> *a-bi* DINGIR.MEŠ *a-mat-ka te-me-en* AN^{e} *u* KI^{tim} *a-a-u$_2$* DINGIR *le-em-ma*

> An, your exalted command goes in the vanguard, who would say “no”? Father of the gods, your word is the foundation of heaven and earth, which god would disobey?

The forms mu-lu and dim$_3$-me-er-e-ne-ke$_4$ are in Emesal, even though this is a group of male gods speaking. Similarly, in An’s response (TCL 6, 51 o 21–22):

> li-bi-ir zid-da-ĝu$_{10}$ šu-um-du-um kal-kal-la-bi ka dib-ba-ĝu$_{10}$ mu-un-zu-a
> *suk-kal-lum ki-nu šap-tan šu-qu-ra-a-tuš mu-du-u$_2$ pi-riš-ti-ia*
> ^{d}nin-šubur sukkal me-te-ĝu$_{10}$ za-e-da nam-ab-ba ak-a he$_2$-na-nam
> ^{d}*i$_3$-li$_2$-ab-rat suk-kal-lum si-ma-ti-ia lu-u$_2$ e-piš a-bu-tu$_2$ it-ti-ka*

> May my loyal vizier of precious lips, who knows my secrets,
> may Ninšubur, my appropriate vizier intercede before you.

In these two passages, we see the same word appearing in Emesal and in the regular form side by side (dim$_3$-me-er vs. diĝir in TCL 6, 51 o 5 and li-bi-ir vs. sukkal in TCL 6, 51 o 21–22). Emesal forms are not restricted to speeches; they also appear in narrative passages, as in (TCL 6, 51 o 13):[14]

literary tradition of the time. TCL 6, 43 comes from Uruk and may easily have borrowed the phrase from The Elevation of Ištar.

13 For the use of Emesal in the Elevation of Ištar see Falkenstein 1952: 91, Falkenstein 1953: 11, and Hruška 1969: 478–479. The writing ur for ur$_5$ (corresponding to *šī*; TCL 6, 51 o 12 and TCL 6, 52 r 1) is classified as Emesal by Hruška 1969: 499, but there is no other evidence for such a form.

14 Duplicates LKU 12: o 9′ and TCL 6, 52: r 2.

an-na-ra inim bal bar ze$_2$-eb-ba-ke$_4$ hul$_2$-le-eš nam-mi-in-ĝar
ana ^{d}*a-nu na-pa-le-e ṭu-ub ka-bat-ti ha-diš iš-ša*$_2$*-kin-šum-ma*

As for An, this heart-warming speech made him very happy.

All in all, the distribution of Emesal forms seems random. Clearly, the use of Emesal has nothing to do here with gender or performance.

In the current edition of the Elevation of Ištar in BLMS there are 898 Sumerian tokens representing 347 individual words, against 46 Emesal tokens representing 20 words (plus 975 Akkadian tokens representing 413 words). In other words the Sumerian text has approximately 5% explicit Emesal words. This is a low percentage when compared to ritual Emesal texts such as the Eršahungas, where approximately 15% of the Sumerian words are explicitly in Emesal.[15]

This usage of Emesal in the Elevation of Ištar may be compared to a few other first (or late-second) millennium bilingual-born compositions, such as the Hymn to Nabû from the House of the Incantation Priest[16] at Assur, the hymn to Ninurta KAR 97, the bilingual royal inscriptions of Adad-apla-idinna, Šamaš-šum-ukkin, and Nebukadnezzar I, the bilinguals associated with Tukulti-Ninurta (Prayer of Tukulti-Ninurta KAR 128 + 129 and the historical

	Bab/ Assyr	Mainly Emesal	Some Emesal	Mainly Emegir	No Emesal
Examenstext A[a]					x
In Praise of Scribal Art[b]					x
Elevation of Ištar			x		
Hymn to Nabû[c]	Assyr		x		
KAR 97[d]	Assyr	x			

15 These numbers are computed by analyzing the editions in BLMS. For these computations, a word is considered to be in Emesal when it has at least one explicit Emesal morpheme. Thus a form like de$_3$-en-sed is considered Emesal, but mu-un-sed is Emegir. The current computation ignores the fact that many words do not have a separate Emesal form and it should therefore be taken only as a rough measurement. The group of Eršahungas used for this computation corresponds to Eršahunga 1–109 in Maul 1988; other Ershahungas have been edited in BLMS, but have not (yet) been lemmatized (as of May 2016) and therefore cannot be analyzed computationally.

16 For the House of the Incantation Priest see Maul 2010. Maul's analysis refers to the finds of the German excavations in the early twentieth century. The Prayer to Nabû belongs to a group of texts excavated by an Iraqi team in the same area of Assur.

	Bab/ Assyr	Mainly Emesal	Some Emesal	Mainly Emegir	No Emesal
KAR 128–129[e]	Assyr	x			
BM 98496[f]	Assyr				x
CBS 1642[g]	Assyr				x
Hymn to Nergal[h]	Assyr			x	
Hymn to Gula[i]	Assyr			x	
KAR 119/N 3462[j]	Bab				x
Hymn to Erra[k]	Bab				x
Hymn to Marduk[l]	Bab				x
Hymn to Utu[m]	Bab				x
Nebuchadnezzar I[n]	Bab			x	
Adad-apla-idinna[o]	Bab			x	
Šamaš-šum-ukin[p]	Bab			x	

a Sjöberg 1974, with a few additional exemplars.

b See most recently Hurowitz 2000, with further literature.

c Cavigneaux and Ismail 1998; this text has approximately 5.5% Emesal.

d Hymn to Ninurta; edited by Ebeling 1918: 75–78; see Ebeling 1919: 78–79.

e Prayer of Tukulti-Ninurta (P282608); Chang 1979; see the translation by Foster 1993: III.2, and the discussion in Veldhuis 2014: 347–351.

f Middle Assyrian historical text; Lambert 1976.

g Cavigneaux 1996: 11–13. The obverse (of which little is preserved) has an unidentified bilingual text; the reverse has a copy of the Letter of Lugal-ibila to Lugal-nesaŋ (Civil 2000).

h 4R 42: 1 (K 4809+). The text has the Emesal form i-bi$_2$ (eye) twice.

i KAR 100; only the refrain (ĝešgiĝal) has Emesal forms.

j The hymn to Ninurta KAR 119 (P369099; in Babylonian script) was edited by Lambert 1960: 118–120. That the text is indeed Kassite in origin (as argued by Lambert) is now supported by the duplicate N 3462 (P278493) from Nippur. See now Michalowski 2017.

k K 5268+ (BA 5, 642–643); in Babylonian script.

l CTMMA 2, 18.

m K 5970; Rm 2, 290; and W 18828 (Falkenstein 1959). I wish to thank Henry Stadhouders for drawing my attention to these texts and providing me with editions of the Kuyunjik fragments.

n Grant Frame, RIMB2 2.4.1 and 2.4.8–9.

o Grant Frame, RIMB2 2.8.5.

p Grant Frame, RIMB2 6.33.1.

text BM 98496), and several texts known from the Kuyunjik collection and elsewhere:[17]

An inspection of this table yields some interesting trends. Apart from the royal inscriptions (where Emesal forms are rare), the use of Emesal in bilingual-born texts is mostly Assyrian. One may recall that the Emesal Vocabulary is primarily an Assyrian compilation with only two exemplars from Babylonia.[18] The Emesal inventory of words becomes one more resource to draw on when attempting to write Emegir. As the table shows, this happens less in Babylonia and more in Assyria.

One more observation, somewhat contradicting the previous one, may be added here. The Prayer of Tukulti Ninurta (KAR 128+) is styled as an Eršahunga prayer and is therefore primarily in Emesal. By contrast, the Middle Assyrian bilingual historical text BM 98496 (Lambert 1976), which is closely related to the Prayer of Tukulti-Ninurta in other respects, does not use Emesal forms at all. If these two texts were indeed (as Lambert suspects) authored by the same scribe, this Middle Assyrian scholar was very well aware of the differential usage of Emesal and Emegir.

Rare Sumerian Words and Expressions

The Elevation of Ištar uses a number of rare expressions that may have been taken either from lexical compositions or from other bilinguals that were accessible at the time. The line between "rare" and "regular" Sumerian expressions is, of course, arbitrary. Post-Old Babylonian usages such as the prepositional use of ki (correlating with Akkadian *itti*; TCL 6, 51: r 9) may strike a modern Sumerologist as odd, but are regular in the first millennium. The following contains a discussion of a selection of rare or somewhat rare Sumerian words and Sumerian–Akkadian equivalencies from the Elevation of Ištar and the places where these equivalencies may have been found in lexical and/or in other bilingual texts.

te-e = *kannu* "loincloth"

This equation appears in the following context in the Neo-Assyrian exemplar 1880-07-19, 281: o 10′ (tablet IV):[19]

17 The majority of the texts listed here are edited in some form in BLMS and (for the royal inscriptions) in the forthcoming ORACC project The Royal Inscriptions of Babylonia online, directed by Jamie Novotny. The table includes only "bilingual-born" compositions; incantations, Emesal rituals, and first millennium copies of Old Babylonian literary texts are left out.

18 See Veldhuis 2014: 318–320.

19 Published in photograph on CDLI (P452027).

[t]e-e sig-ga-gin$_7$ ^{d}in-nin hul-d[u ...]

ki-ma nadi-i kan-ni ^{d}MIN *ana za-*[*ap-ri* ...]
Lady, like [throwing] a discarded loincloth at a wicked one, ...

The verb sig, which represents Akkadian *nadû* is an unorthographic spelling for sig$_9$/sig$_{10}$ (or si.g/si$_3$.g), "to put" or "to cast away." This specific usage of the verb is dealt with in Aa III/4: 218 (BM 92692; MSL 14: 339): [si-ig] [SI] *nadû ša karrum* "to discard, said of dirty cloths." The equation te-e = *kannu* is more problematic. The entry is found in the section Aa VIII/1, 185–205[20] which presents more than 20 Akkadivan equivalents for the Sumerian word te (as well as ten and teĝ), many of them repeated in the TE section in Izi E (MSL 13, 187). The Aa passage begins with te = *simtum* (185; "sign, symbol") an equivalence that is found elsewhere and is based on the Sumerian word me-te ("ornament, or fitting attribute").[21] Other entries are based on te = *lētum* (193; "cheek"); teĝ$_3$(TE) = *ṭehû* (186; "to approach"); ten(TE) = *bullû* (196; "to extinguish"), or explain that the sign TE may be used for its syllabic value /te/ (190; *ša tēltim*). All of these may be found elsewhere in the lexical and bilingual tradition. Several other entries in this section remain hard to explain or may be based on complex reasoning. Thus te = *duppurum* (191; "to go away") may provide an antonym of the regular equation *ṭehû*; or, perhaps more likely, may be based on kar(TE.A) "to leave; chase away." Whatever the proper solution, the entry *nesû* (188; "far away") is no doubt based on the same principle and the following entry *nešû* (189; "to vomit"; not found in Izi E) is based on a phonological association with *nesû*. The entry te = *salāhu* (195; "to sprinkle") is a mistake for *sahālu* ("to pierce"; see Izi E 101), well attested in lexical and bilingual texts.

None of these entries has anything to do with te-e = *kannu*, but they are useful to show the extent of speculation and association that goes into Aa (and, by extension, the commentaries to Aa). A fairly rare equivalence of Sumerian te is *šišītum* (199; "membrane"), which already appears in one exemplar of Old Babylonian Izi II (HS 1802; see Crisostomo 2014: 428) but is not, as far as I know, ever attested in non-lexical texts. In Aa VIII/1 it is found with *ipu* (197; "membrane" or "afterbirth") and *ruššû* (198; "red"). This range of associations may well include te = *ishunnatu* (200; "cluster of grapes"; appears in similar context in Izi E 108) which might have a secondary meaning "ovaries"

20 Aa VIII/1 is known from a single Late Babylonian exemplar BM 92691 (CT 12 10f); MSL 14: 489.

21 Izi E has me-te = *simtum*; te = *simtum* in sequence (92–93).

(see CAD I 190b).[22] If we are correct to include these associations and extensions around te = *šišītum* then the association with *kannu* as "loincloth" or "sanitary cloth" makes some (remote) sense.

The relevant section in the commentary to Aa VIII/1–2 (AO 3555 o 18–19)[23] has

> [… *ši-ši*]*-tum* : *ši-i-ri* : *šu-u* : *ka-an-nu* : *kan-nu ša*$_2$ *qe*$_2$*-e* […]
> [… te si]-ga : *kan-nu na-du-tu* (continues with KAR section)

It is very likely that line 19 of the commentary uses the same verb si.g = *nadû* for "to discard (a loincloth)."[24] This may represent an actual quote from the Elevation of Ištar or refer to some very similar passage.

The motive of taunting enemies by throwing(?) women's underwear at them, which is the subject of our line in The Elevation of Ištar, is probably also found in Samsuiluna A line 28, an Old Babylonian victory song that features Inana. That text uses Sumerian zi-in-bu (> Akkadian *sinbu*)[25] for underwear.

ti-sah$_4$ = *anantu* and Inana's Games

Sumerian ti-sah$_4$ is a rare word for "fight," or "battle"; it appears in a passage in The Elevation of Ištar that extols Inana's joy in battle:[26]

> ^{d}inana ti-sah$_4$ ĝeš-la$_2$ ešemen$_2$$^{eš\text{-}me\text{-}en\text{-}gin_7}$-gin$_7$ u$_3$-mi-ni-ib$_2$-sar-sar
> *^{d}iš-tar a-na-an-ti u tu-qu-un-ta ki-ma kip-pe-e šu-tak-pi-ma*
>
> Inana, make war and battle whirl around like a skipping rope!

---

22 Note that the reference to KAR 40: 8 in the discussion section is not relevant for the presumed meaning "ovaries." The duplicate VAT 14255 (P381848; Cavigneaux 1980: 261) reads [ellaĝ$_2$ tun$_3$-na] = [MIN] *iṣ-ṣa-ba-tu*[*m* x x]; the Akkadian derives from the word for ring (*anṣabtum*). See MSL 17: 40 255.

23 The section is now lost. See the introduction and edition in MSL 14, 495–496 with further bibliography.

24 This terminology is also found in Ura 19, 312 (MSL 10: 136), which should be corrected to read: 311 [^{tug_2}]te = *kan-nu*; 312 [^{tug_2}te si]-ga = MIN *na-du-u*$_2$ (BM 40491 + 40724; collated by Jay Crisostomo).

25 See the interpretation in Durand 1984, 155 and the discussion by Nicolet 2015, vol. 2, 217. The Samsuiluna text was edited (without translation) by Farber-Flügge 1976; a tentative translation was offered in ETCSL (text 2.8.3.1). For Akkadian *sinbu* see now SpTU 3, 116 (Murgud; P348714), o i 45: ^{tug_2}niĝ$_2$-dara$_2$ gal$_4$-la sir$_2$-ra = *sinbu* = MIN (*ulāpi*) MIN (*dāmu*). This passage also includes the word *kannu* (o i 46), confirming its meaning "sanitary towel" (see also Stol 2012: 274).

26 Bod S 302 (RA 12, 73–84): o 2 with duplicate 1880-07-19, 281: r 6.

As far as I know ti-sah$_4$ is not attested in Old Babylonian (or earlier) Sumerian, but it appears in Erimhuš, where it has a section of its own (Erimhuš II, 241–243; MSL 17: 40):

> ti-sah$_4$ = *anantu*
> ti-sah$_4$ = *tuquntu*
> ti-sah$_4$-sah$_4$ = *ašgagu*

Sjöberg 1970: 83 and Black 2003: 42 have both argued that ti-sah$_4$ is related to suh$_3$-sah$_4$, used in suh$_3$-sah$_4$ … za ("to thud") and in (igi)-suh$_3$-sah$_4$ (*sahmaštu*: "chaos, confusion"—commonly used for battle scenes). The latter word is used in this same passage in the Elevation of Ištar (line 5). The only other context usage of ti-sah$_4$ = *anantu* that I am aware of is the bilingual royal inscription of Nebuchadnezzar I (RIMB 2.02.04.01 rev. ii 1).

The word ti-sah$_4$ in the Elevation of Ištar appears in a context that compares Inana's joy in battle with playing games:

> Inana, make war and battle whirl like a skipping rope (ešemen$_2$-gin$_7$ u$_3$-mi-ni-ib$_2$-sar-sar)
> Lady of combat, let battles clash like ball and mallet!
> Goddess of strife, pursue fighting like playing with dolls! (bi-za šu-tag-ga-gin$_7$)
> Lady, where weapons clash, play with the chaos like knucklebones! (zi-in-gi ra-ra-da-gin$_7$)

Three of the four games mentioned here appear in a four-line section in Antagal F (243–246; MSL 17: 219):

TUR.DIŠ$^{hi-bi-iz}$.KAR	*mēlulu*	to play
EŠ$_2$.HUL$_2$$^{e-še-min}$ sar-ra	MIN *ša kippê*	to play with a skipping rope
^{ĝeš}bi-za šu tag-ga	MIN *ša passi*	to play with dolls
zi-in-gi ĝiri$_3$ ra-ra	MIN *ša$_2$ ta-x-x*	to play knucklebones

Skipping rope, dolls, and knucklebones appear in the Elevation of Ištar in the same or almost the same expressions, suggesting that there is a direct relationship between Antagal F and The Elevation of Ištar.

a-ga-zu = ***ēdiššiki***

Anu, after receiving advice from the great gods, admonishes Ištar to exercise her full power in the following words (TCL 6, 51: o 18):

> zu_2-$keš_2$ ki-sur-bi zag-zu u_3-bi_2-gi a-ga-zu nam-en-bi ak-ab
> *kip-pat ki-sur-ri-šu-nu ana i-di-ka ter-ri-ma e-diš-ši-ka be-el*[27]
>
> Bring the entirety of their area to your side, you alone are lord!

This reflects a rare usage of Sumerian a-ga. The word a-ga usually means "rear" or "back room" and frequently stands in opposition to igi "front."[28] Unlike igi, the word a-ga is not used for a body part and thus does not appear in the list of body parts Ugumu. The equation with *ēdišši-* + Pronominal Suffix, meaning "on his/her own (accord)" is known from Ninurta's Exploits (or Lugal-e) line 2:[29]

> ^{d}nin-urta saĝ-kal usu mah tuku kur a-ga-na DU.DU
> *Ninurta ašarēdu rāš emuqān ṣīrāti šālil šadî ēdiššišu*
>
> Ninurta, foremost, who has supreme power, the one who plunders the mountain on his own.

This usage is also found in Ninurta's Exploits 466 and as a variant of an-ga-am_3 in Father and Son 93.[30] In the lexical tradition a-ga is usually rendered *arkatu* "rear side," but the Neo-Assyrian group vocabulary Šarru offers a-ga-ba = *ahamma* ("separately") and *īdiššišu* ("he alone").[31]

The source for this usage of a-ga in The Elevation of Ištar may well have been Lugale, which was well known in first millennium learned circles. One may also note that this particular use of a-ga-zu in The Elevation of Ištar is remarkably consistent with the (rare) Old Babylonian occurrences, which seem to mean "by his own power" (or, with Heimpel and Salgues 2015: "auf eigene Faust").[32]

šu-ta-ta = *malmališ*

Anu, in his speech, describes the beginning of the universe, which includes the line (TCL 6, 51: r 2):

---

27 Variant (TCL 6, 52: r 7): *ēdiššiki bēli*.

28 See the references in PSD A/1: 68–70 and the discussions in Gragg 1973: 17–18 and Balke 2006: 63.

29 See Van Dijk 1983: 51; Al-Rawi 1995: 199; and Seminara 2001: 42.

30 Sjöberg 1973: 126 and Wilcke 1968: 240–241.

31 The text is known in two duplicates: K 2054 (CT 18, 29–30; P346055) r ii 27–28′ and AO 7092 (Thureau-Dangin 1919) r ii 31–32 (only the ends of the Akkadian words preserved). The Ugarit Diri entry (RS 17.154) a-ga-ba = *e-de-eš-š*[*u*] (quoted in PSD A/1: 72 from CAD E) is heavily restored (see now MSL 15, 76); the basis of its reconstruction remains unclear.

32 Seminara 2001: 218 prefers a more literal translation, ("on his back") and takes the Akkadian as a reinterpretation of the Sumerian original.

^{d}nanna ^{d}utu-ra ĝi$_{6}$^{gi} ud-da šu-ta-ta an-ni-ši-ib$_{2}$-si
ana ^{d}30 *u* ^{d}UTU *u$_{4}$-mu u mu-ši ma-al-ma-liš ba-šim-ma*

For Sîn and Šamaš night and day were created in equal measures.

The verb si/sig$_{9}$ probably corresponds to si$_{3}$/sig$_{10}$, which is equated with *bašāmu* in Nabnitu 7 111 and 114 in broken context.[33] This usage may derive from si$_{3}$.g = *šapāku*, "to cast" (said of metals), which is occasionally used for "to fashion."[34] More puzzling is šu-ta-ta, which is rendered *malmališ* in Akkadian. Double ablatives may occur if the first is adnominal (distributive) and thus bound into a larger noun phrase.[35] One might imagine that it could mean "by means of a hand each," but the construction seems strained and one may wonder whether a late author would even have access to this use of the ablative. Alternatively, one may see here a Sumerian-looking rendering of *šutātû* = "to face each other" (*atû* Š), which is used as the technical term for "to be in opposition" in astronomical texts. The commentary on Venus omens K 148+ includes the entry *šutātû* = *malmališ* (obv. 38), which is uncannily close to our line.[36] One may thus treat šu-ta-ta in The Elevation of Ištar as an artificial creation or a loan derived from Akkadian *šutātû*.

ki-la$_{2}$ = *sāniqu*

Describing Ištar's exalted position, Anu emphasizes that she will have no one above her (TCL 6, 51: r 11):

diĝir-re-e-ne-gin$_{7}$ lu$_{2}$ ki-la$_{2}$-zu nu-ĝal$_{2}$-la-ar saĝ u$_{6}$ he$_{2}$-ri-in-e-še
ki-ma ina DINGIR-MEŠ *sa-ni-qa la te-ši-i* UN-MEŠ *lib-ra-ki*

May the people admire you as one who has no one among the gods to supervise you.

33 The lines in VAT 8755 (P368989) o ii 13′ and 16′ read […] si$_{3}$ = *bašāmu* and […] si$_{3}$ = *bašāmu ša napišti*, respectively. The note in MSL 16: 108 suggests that the Sumerian in o ii 16′ may be read [zi] si$_{3}$ and that nothing may be missing in o ii 13′—which would work well with the current line. Note that line 110 may be restored [saĝ]-sur = *bašāmu* (see MSL SS 1: 18 34); similarly in Antagal E 26′ (MSL 17: 210).

34 See Veldhuis 2004: 137.

35 See Wilcke 1990: 471–476; Balke 2006: 127–129; and Jagersma 2010: 191–196.

36 The text is edited in Reiner and Pingree 1998: 56–59 (and see 36–38); see further Frahm 2011: 151–152 and http://ccp.yale.edu/P237789. At the moment of writing this article, the CCP page for K 148+ offered bibliography only; eventually it will include a full edition.

The equivalence saĝ = *nišū* is somewhat unusual, but not unparalleled (see CAD *nišū* lexical section) and readily understandable. Less straightforward is ki-la$_2$ = *sāniqu* ("controller, supervisor"). The verb *sanāqu* in the meaning "to check" is usually the equivalent of Sumerian dim$_4$. Sumerian ki-la$_2$ means "weight," derived from ki—la$_2$ "to weigh." The equation ki-la$_2$ = *sāniqu* is based on Nabnitu 28: 100–102:

ki-la$_2$ = *sanāqu ša awāti*	to check, said of statements
ki-la$_2$ = *sanāqu ša šuqulti*	to check, said of a weight
ki-la$_2$ = *sanāqu ša kaspi*	to check, said of silver

Lines 101 and 102 are fairly clear: in order to calibrate a weight or to check silver one may weigh it. Line 100 and the author of The Elevation of Ištar expand the usage of ki-la$_2$ to cover further usages of *sanāqu* and derived words.

un-gi = *nāpalû*

Akkadian *nāpalû* ("conversation") is a fairly rare word, derived from *apālu* "to answer." The word appears twice in The Elevation of Ištar, once as the equivalent of inim bal ("to converse");[37] and once as the equivalent of un-gi.[38] The latter equivalence is presumably based on gi$_4$ = *apālu*, creating a pseudo-verbal form.

ki mar-ra = *tanēhtu*

This equation appears in a speech by Enlil (Bod S 302: r 4):[39]

> barag dur$_2$-ĝar-ra dag ki mar-ra niĝ$_2$ mu še$_{21}$-ĝu$_{10}$ in-pad$_3$-da
> MIN *šu-bat ta-ne$_2$-eh-tu ša$_2$ ana ni-bit šu-mi-ia zak-ru*
>
> at the dais, the tranquil dwelling that is called after my name,

The equation ki mar-ra/ĝar-ra = *tanēhtu* is not otherwise found; it derives from ĝar = *pašāhu* or *nâhu* that is found in incantations such as Udughul and at least once in an Emesal lament (SBH 44 line 10–11; see Schramm 2008: 188).[40]

37 TCL 6, 51: o 13; duplicate TCL 6, 52: r 2.

38 Bod S 302 (RA 12, 73–84): r 9; duplicate VS 24, 37: r 13′.

39 Duplicate VS 24, 37: r 8′.

40 Also see Geller 2007: 280 (under ki—gar-gar) and 287 (under níg—gar) and Geller and Vacin 2015: 125–126 lines 165–167 with commentary.

The earliest attestation of this usage of ĝar that I am aware of is in a unilingual Sumerian Udughul incantation from Emar (Arnaud 1987: no.729 line 1). It is remarkable that the equation ĝar = *nâhu* is nowhere to be found in the lexical tradition—or perhaps not so remarkable, given that the relevant section in first millennium Ea 3:220–224 (MSL 14: 312–13) and the corresponding passage in Aa are largely broken.

Rare Words: Summary

The Sumerian of The Elevation of Ištar includes a good number of rare words and unusual correspondences between Sumerian and Akkadian; the above does not attempt to be exhaustive, but is rather meant to give an impression of the range. Occasionally, such words may have been made up or derived from Akkadian words (un-gi; šu-ta-ta); more commonly they are based directly on lexical texts (te = *kannu*) or are developed from such an entry (ki-la_2 = *sāniqu*). The boundary between "regular equations" (discussed in section "Regular Equations") and rare equations, discussed in the present section, is, of course, artificial. The use of a-ga for "on his/her own" is rare but is clearly idiomatic in Old Babylonian Sumerian. It is not more or less unusual than ul-la a-ra-zu-$ĝa_2$ for *annim teslītum* (see section "Regular Equations"). The use of ĝar for *nâhu* (or rather ki mar-ra for *tanēhtu*) may be based on similar usage in incantation texts—or perhaps on a lexical text now lost.

The Sumerian of The Elevation of Ištar demonstrates a broad knowledge of the lexical and bilingual tradition of the time, using a mix of common vocabulary, artificial Sumerian, and rare but idiomatic phraseology.

The Elevation of Ištar—An Assyrian Poem?

The analysis of the particularities of the Sumerian in The Elevation of Ištar makes it likely that the text originates in an Assyrian learned context. Several aspects of the contents of the composition support this conclusion. Although none of the indicators discussed here is conclusive by itself, the accumulated evidence favors a first millennium Assyrian origin.[41]

41 Earlier discussions of the date of composition of the Elevation of Ištar mostly favored a Kassite date (Falkenstein 1952: 91–92 with a discussion of previous scholarship; Falkenstein 1953: 11; Lambert 1971: 91–93), a theory that cannot be excluded but that does not offer a plausible context for the elevation of Ištar's power above the level of traditional Babylonian gods such as Enlil and Ea. Wilcke 1993: 46 has suggested the possibility of an origin in the Sargonic period (see also Zgoll 1997: 169).

First and foremost, the somewhat indiscriminate usage of Emesal is more likely in an Assyrian context than in Babylonia (see section "Emesal"). In the Old Babylonian period Sumerian literature, as taught at the Eduba, and Emesal laments, used by gala-priests, were two very different corpora that lived under different circumstances in different social environments and were transmitted in different ways.[42] In the Old Babylonian period the overlap between the learned corpus of Sumerian literature at the Eduba and liturgical literature at the temples was minimal and is to be found primarily in the city laments. In first millennium Assyria the situation was obviously very different. The *kalû* priest who was mainly responsible for the performance and maintenance of the Emesal laments belonged to the category of learned scholars in the circle of advisors to the king.[43] Sumerian literature had become marginal; the maintenance of the knowledge of Sumerian was now in the hands of those who perform Sumerian in laments and incantations. With the introduction of the Emesal Vocabulary in the Middle Assyrian period the language of the laments had received its proper place in the learned (lexical) tradition—and one may note that this composition is almost exclusively Assyrian.[44] Gabbay 2014 has emphasized the foreignness of *kalûtu* literature and cultic performance (in particular of Balaĝs and Eršemmas) in Assyria. Outside of Nineveh, *kalûtu* was fairly rare in Assyria and the Nineveh corpus depended in many ways on Babylonian examples. Thus displaced from and deprived of its original intellectual and ritual context, it may have been easier in Assyrian circles to basically obliterate the line between Emesal and regular Sumerian.

Secondly, the lexical text Antagal, which provides the most extensive parallel to The Elevation of Ištar (see section "ti-sah$_4$ = *anantu* and Inana's Games"), is an Assyrian composition (the only exception is VS 24, 13 from Neo-Babylonian Babylon).[45] Some of the uncommon Sumerian expressions that appear in The Elevation of Ištar are otherwise attested in the Old Babylonian compositions Enlil and Sud (see section "Regular Expressions") and Ninurta's Exploits (see section "a-ga-zu = *ēdiššiki*"). Both texts are known in first millennium copies; Enlil and Sud is attested in Nineveh and Huzirina (Sultantepe), The Exploits of Ninurta is found at multiple sites in Babylonia and Assyria. A scholar working at Nineveh, or at some other well-endowed scholarly library, may have had

42 See, for instance, Delnero 2015.

43 For the role of the *kalû* in first millennium Assyria, see Gabbay 2014.

44 Today only two fragments of Neo-Babylonian origin are known, one from Nippur and one from Babylon (Veldhuis 2014: 318–320).

45 See Veldhuis 2014: 361–363.

access to all of these resources, as well as to the numerous lexical texts, incantations, and other bilinguals that were collected there.

Several aspects of The Elevation of Ištar make sense when read in an Assyrian context. The high position of Ištar in the Assyrian pantheon, in particular in relation to the royal court, is not in need of further elaboration.[46] It is worth mentioning, though, that the composition, as far as preserved, avoids praising Ištar for her beauty or sex appeal. The only aspects of her personality that are elaborated are her power, her violence, and her joy in battle. Most significantly, at those places where her femininity (in connection with weaving) and her character as a young girl are introduced, these aspects are immediately related to violence on the battle field:[47]

> Lady, [may] the fierce dagger [...] like a pointed needle,
> May your heddle [...] the tendon like a ŠID-bow,
> May the arrowhead, piercer of heart and lungs, fly like a shuttle!
> Ištar, [make] war and battle [whirl around] like a skipping rope!
> Like ball and mallet, lady of battle, let battles clash!
> Goddess of strife, pursue battle like playing with dolls!
> Lady, where weapons clash, play with the chaos like knucklebones!

This exclusive emphasis on Ištar's role in battle is understandable in an Assyrian context and so is the idea of several high-ranking Babylonian gods (Anu, Enlil, and Ea) addressing Ištar as their equal or superior. The text, written in Standard Babylonian and provided with the sheen of antiquity through its Sumerian version, may thus be read as a reflection on the relationship between Assyrian and Babylonian intellectual culture.

When read in an Assyrian context we have to assume that Anu plays the role of Aššur. There is at least one place that supports such a reading, namely where Anu refers to himself as An-šar$_2$ (Akkadian: *Anu*) while Ištar is named Ki-šar$_2$ (Akkadian: *Antu*).[48] This move not only placed Anu and Ištar at the hoary beginning of time, among the earliest divine ancestors, but it also created the possibility of reading the text as one about the two main gods of Assyria: Aššur[49] and Ištar.

---

46 Porter 2004; Zsolnay 2010.

47 Bod S 302 (RA 12, 73–84): o 1–5 and 1880-07-19, 281 r 3–9.

48 TCL 6, 51: o 10 and 17. For this passage and the various divine identifications involved, see Beaulieu 1995: 204 and Lambert 2013: 422.

49 For An-šar$_2$ as a writing for Aššur, see Beaulieu 1997; and Lambert 2013: 5.

The Elevation of Ištar found its way to Late Babylonian Uruk and it is not hard to see why the composition was attractive to scholars there. The two main and most powerful characters in the composition are the city gods of Uruk: Anu and Ištar. Beaulieu 1997 has discussed the association between Aššur, Anšar, and Anu that may have been due, in part, to Assyrians who established the cult of Anšar/Aššur in the city of Uruk. The Assyrian influence on Uruk's intellectual life is most dramatically illustrated by a tablet with an Assurbanipal colophon found at Uruk (SpTU II 46) but has also been demonstrated in numerous other details of Urukean scholarship (Beaulieu 1997 and Beaulieu 2010).

In conclusion, I suggest that The Elevation of Ištar may be a product of a Neo-Assyrian scholarly milieu. The composition was then transplanted to Late Babylonian Uruk and to the Ištar of Akkade temple in Babylon (see section "The Manuscripts"). The Neo-Assyrian and Late Babylonian scribes who were responsible for the copies of The Elevation of Ištar that we have today would probably beg to differ. The Catalogue of Texts and Authors[50] assigns our texts to Taqiša-Gula the lamentation priest who, according to the Uruk List of Kings and Sages, was a scholar at the court of the Old Babylonian ruler Abi-ešuh (Lenzi 2008).

Bibliography

Al-Rawi, Farouk N.H.

1995 "Tablets from the Sippar Library IV. 'Lugale.'" *Iraq* 57: 199–223.

Arnaud, Daniel

1987 *Textes de la bibliothèque: transcriptions et traductions.* Recherches au Pays d'Aštata. Emar 6/4. Paris: Editions Recherche sur les Civilisations.

Balke, T.E.

2006 *Das sumerische Dimensionalkasussystem.* AOAT 331. Münster: Ugarit-Verlag.

Beaulieu, Paul-Alain

1995 "Theological and Philological Speculations on the Names of the Goddess Antu." *Or*NS 64: 187–213.

1997 "The Cult of AN.ŠÁR/Aššur in Babylonia after the Fall of the Assyrian Empire." *SAAB* 11: 55–73.

2010 "The Afterlife of Assyrian Scholarship in Hellenistic Babylonia." Pp. 1–18 in: *Gazing on the Deep. Ancient Near Eastern and Other Studies in Honor of Tzvi Abusch*, eds. J. Stackert, B.N. Porter, and D.P. Wright. Bethesda: CDL Press.

50 Lambert 1962: 75–76.

Black, Jeremy A.

1991 *Sumerian Grammar in Babylonian Theory*. StPohl SM 12. Rome: Editrice Pontificio Istituto Biblico.

2003 "Sumerian Noises: Ideophones in Context." Pp. 35–52 in: *Literatur, Politik und Recht in Mesopotamien. Festschrift für Claus Wilcke*, eds. W. Sallaberger, K. Volk, and A. Zgoll. OBCh. Wiesbaden: Harrassowitz.

Cavigneaux, Antoine

1980 "Die Fragmente der sumerisch-akkadischen Wortliste 'ERIMHUŠ' im Vorderasiatischen Museum zu Berlin." *FuB* 20/21: 259–270.

1996 "Miettes de l'edubbâ." Pp. 11–26 in: *Tablettes et images aux pays de Sumer et d'Akkad: mélanges offerts à Monsieur H. Limet*, eds. Ö. Tunca and D. Deheselle. Liège: Université de Liege.

Cavigneaux, Antoine, and Bahija Khalil Ismail

1998 "Eine zweisprachige Hymne aus dem Haus des Beschwörungspriesters." *ASJ* 20: 1–11.

Chang, K.W.

1979 *Dichtungen der Zeit Tukulti-Ninurtas I. von Assyrien*. Frankfurt am Main: Peter Lang.

Civil, Miguel

1983 "Enlil and Ninlil: The Marriage of Sud." *JAOS* 103: 43–66.

2000 "From the Epistolary of the Edubba." Pp. 105–118 in: *Wisdom, Gods and Literature. Studies in Assyriology in Honour of W.G. Lambert*, eds. A.R. George and I.L. Finkel. Winona Lake, IN: Eisenbrauns.

Crisostomo, Jay

2014 "Bilingual Education and Innovations in Scholarship: The Old Babylonian Word List Izi." Doctoral Dissertation, Berkeley, CA: University of California at Berkeley.

Delitzsch, Friedrich

1913 "Bemerkungen zu Prof. Meek's zweisprachigen Fragmenten." *BA* 10: 129–146.

Delnero, Paul

2015 "Texts and Performance. The Materiality and Function of the Sumerian Liturgical Corpus." Pp. 87–118 in: *Texts and Contexts: The Circulation and Transmission of Cuneiform Texts in Social Space*, eds. P. Delnero and J. Lauinger. SANER 9. Berlin: De Gruyter.

Durand, Jean-Marie

1984 "Trois études sur Mari." *MARI* 3: 127–180.

Ebeling, Erich

1918 *Quellen zur Kenntnis der babylonischen Religion I*. MVAG 23/I. Leipzig: Hinrichs'sche Buchhandlung.

1919 *Quellen zur Kenntnis der babylonischen Religion II.* MVAG 23/II. Leipzig: Hinrichs'sche Buchhandlung.

Falkenstein, Adam

1952 "Inannas Erhöhung (Tafel IV)." *BiOr* 9: 88–92.

1953 "Zur Chronologie der sumerischen Literatur. Die nachaltbabylonische Stufe." *MDOG* 85: 1–13.

1959 "Zwei Rituale aus seleukidischer Zeit." in: *XV. Vorläufiger Bericht über die von dem Deutschen Archäologischen Institut und der Deutschen Orient-Gesellschaft aus Mitteln der Deutschen Forschungsgemeinschaft unternommenen Ausgrabungen in Uruk-Warka Winter 1956/57.* ADOG 4. Berlin: Gebr. Mann.

Farber-Flügge, Gertrud

1973 *Der Mythos "Inanna und Enki" unter besonderer Berücksichtigung der Liste der m e.* StPohl 10. Rome: Biblical Institute Press.

1976 "Zur sogenannten Samsuilunahymne PBS 10/2 Nr. 11." Pp. 177–181 in: *Kramer Anniversary Volume. Cuneiform Studies in Honor of Samuel Noah Kramer.* AOAT 25. Neukirchen-Vluyn: Neukirchener Verlag.

Foster, B.R.

1993 *Before the Muses. An Anthology of Akkadian Literature.* Bethesda, Maryland: CDL Press.

Frahm, Eckart

2011 *Babylonian and Assyrian Text Commentaries. Origins of Interpretation.* GMTR 5. Münster: Ugarit-Verlag.

Gabbay, Uri

2014 "The kalû Priest and kalûtu Literature in Assyria." *Orient* 49: 115–144.

2015 "Ancient Mesopotamian Whispering into the Ears." Pp. 185–220 in: *Marbeh Ḥokmah. Studies in the Bible and the Ancient Near East in Loving Memory of Victor Avigdor Hurowitz*, eds. S. Yona, E.L. Greenstein, M.I. Gruber, et al. Winona Lake, IN: Eisenbrauns.

Geller, Markham J.

2007 *Evil Demons: Canonical Utukkū Lemnūtu Incantations.* SAACT 5. Helsinki: Neo-Assyrian Text Corpus Project, Institute for Asian and African Studies, University of Helsinki.

Geller, Markham J., and Ludek Vacin

2015 *Healing Magic and Evil Demons: Canonical Udug-Hul Incantations.* BAM 8. Boston: De Gruyter.

Gragg, G.

1973 *Sumerian Dimensional Infixes.* AOATS 5. Neukirchen-Vluyn: Neukirchener Verlag.

Heimpel, Wolfgang, and Emmanuelle Salgues

2015 "Lugale oder wie Ninurta dem Tigris mehr Wasser Schuf." Pp. 33–67 in: *Erzählungen aus dem Land Sumer*, ed. K. Volk. Wiesbaden: Harrassowitz Verlag.

Hruška, Blahoslav

1969 "Das spätbabylonische Lehrgedicht 'Inannas Erhöhung.'" *ArOr* 37: 473–522.

Hurowitz, Victor A.

2000 "Literary Observations on 'In Praise of the Scribal Art.'" *JANES* 27: 49–56.

Koslova, Natalia V.

2014 "Zu den Anfangszeilen einiger Edubba'a Kompositionen." Pp. 305–326 in *Studies in Sumerian Language and Literature. Festschrift für Joachim Krecher*, eds. N.V. Koslova, E. Vizirova, and G. Zólyomi. Babel und Bible 8. Winona Lake, IN: Eisenbrauns.

Lambert, W.G.

1960 *Babylonian Wisdom Literature*. Oxford: Clarendon Press.

1962 "A Catalogue of Texts and Authors." *JCS* 16: 59–77.

1971 "Critical Notes on Recent Publications." *Or*NS 40: 90–98.

1976 "Tukulti-Ninurta I and the Assyrian King List." *Iraq* 38: 85–94.

2013 *Babylonian Creation Myths*. MC 16. Winona Lake, IN: Eisenbrauns.

Lenzi, Alan

2008 "The Uruk List of Kings and Sages and late Mesopotamian Scholarship." *JANER* 8: 137–169.

Linssen, Marc J.H.

2003 *The Cults of Uruk and Babylon: The Temple Ritual Texts As Evidence for Hellenistic Cult Practice*. CM 25. Leiden: Brill.

MacMillan, Kerr Duncan

1906 "Some Cuneiform Tablets Bearing on the Religion of Babylonia and Assyria." *BA* 5: 531–712.

Maul, Stefan M.

1988 *"Herzberuhigungsklagen" Die sumerisch-akkadischen Eršahunga-Gebete*. Wiesbaden: Harrassowitz.

1997 "Küchensumerisch oder hohe Kunst der Exegese? Überlegungen zur Bewertung akkadischer interlinearübersetsungen von Emesal-Texten." Pp. 253–267 in: *Ana šadî Labnani lu allik. Beiträge zu altorientalischen und mittelmeerischen Kulturen. Festschrift für Wolfgang Röllig*, eds. B. Pongratz-Leisten, H. Kühne, and P. Xella. AOAT 247. Neukirchen: Neukirchener Verlag.

1998 "Marduk, Nabû und der assyrische Enlil. Die Geschichte eines sumerischen Šu'ilas." Pp. 159–197 in: *Festschrift für Rykle Borger zu seinem 65. Geburtstag am 24. Mai 1994. tikip santakki mala bašmu.*, ed. S.M. Maul. CM 10. Groningen: Styx.

2010 "Die Tontafelbibliothek aus dem sogenannten 'Haus des Beschwörungspriesters.'" Pp. 189–228 in: *Assur-Forschungen*, eds. S.M. Maul and N. Heeßel. Wiesbaden: Harrassowitz.

Meek, Theophile James

1913 "Cuneiform Bilingual Hymns, Prayers and Penitential Psalms." *BA* 10: 1–127.

Michalowski, Piotr

2017 "Literary Journeys from Babylonia to Assyria: Second Millennium Copies of a Bilingual Poem Concerning Ninurta." Pp. 205–230 in: *The First Ninety Years: A Sumerian Celebration in Honor of Miguel Civil*, eds. L. Feliu, F. Karahashi, and G. Rubio. SANER 12. Boston: De Gruyter.

Nicolet, Grégoire

2015 "La 'Maison aux tablettes' et l'enseignement à Mari à l'époque paleobabylonienne (ca 1800 av. J.-C.)." Doctoral dissertation, Geneva: Université de Genève.

Pedersén, Olof

2005 *Archive und Bibliotheken in Babylon. Die Tontafeln der Grabung Robert Koldeweys 1899–1917*. ADOG 25. Saarländische Druckerei und Verlag.

Peterson, Jeremiah

2007 "A Study of Sumerian Faunal Conception with a Focus on the Terms Pertaining to the Order Testudines." Doctoral dissertation, Philadelphia: University of Pennsylvania. http://search.proquest.com/docview/304836156.

Porter, Barbara Nevling

2004 "Ishtar of Nineveh and Her Collaborator, Ishtar of Arbela, in the Reign of Assurbanipal." *Iraq* 66: 41–44. doi:10.2307/4200556.

Reiner, Erica, and David Edwin Pingree

1998 *Babylonian planetary omens. Part Three.* CM 11. Groningen: Styx.

Schramm, Wolfgang

2008 *Ein Compendium sumerisch-akkadischer Beschwörungen.* GBAO 2. Göttingen: Univ.-Verl. Göttingen.

Seminara, S.

2001 *La versione accadica del Lugal-e. La tecnica babilonese della traduzione dal sumerico e le sue regole.* MVS 8. Rome: Università degli studi di Roma La Sapienza, Dipartimento di Studi Orientali.

2002 "The Babylonian Science of the Translation and the Ideological Adjustment of the Sumerian Text to the 'Target Culture.'" Pp. 245–255 in: *Ideologies as Intercultural Phenomena. Proceedings of the Third Annual Symposium of the Assyrian and Babylonian Intellectual Heritage Project. Held in Chicago, USA, October 27–31, 2000*, eds. A. Panaino and G. Pettinato. Melammu Symposia. Milan: Università di Bologna & IsIao.

Sjöberg, Åke W.

1970 "Beiträge zum sumerischen Wörterbuch." *Or*NS 39: 75–98.

1973 "Der Vater und sein missratener Sohn." *JCS* 25: 105–169.

1974 "Der Examenstext A." *ZA* 64: 137–176. doi:10.1515/zava.1974.64.2.137.

Stol, M.

2012 *Vrouwen van Babylon: prinsessen, priesteressen, prostituees in de bakermat van de cultuur*. Utrecht: Kok.

Thureau-Dangin, F.

1919 "Un vocabulaire de Kouyoundjik." *RA* 16: 165–171.

Van Dijk, J.J.A.

1983 *LUGAL UD ME-LÁM-bi NIR*-GÁL. *Le récit épique et didactique des Travaux de Ninurta, du Déluge et de la Nouvelle Création. Texte, traduction et introduction*. Leiden: Brill.

Veldhuis, Niek C.

2004 *Religion, Literature, and Scholarship: The Sumerian Composition "Nanše and the Birds," With a Catalogue of Sumerian Bird Names*. CM 22. Leiden: Brill/Styx.

2014 *History of the Cuneiform Lexical Tradition*. GMTR 6. Münster: Ugarit-Verlag.

Wilcke, Claus

1968 "Das modale Adverb i-gi$_4$-in-zu im Sumerischen." *JNES* 27: 229–242.

1993 "Politik im Spiegel der Literatur, Literatur als Mittel der Politik im älteren Babylonien." Pp. 29–75 in: *Anfänge politischen Denkens in der Antike. Die nahöstlichen Kulturen und die Griechen.*, ed. K. Raablauf. Munich: R. Oldenbourg.

Zgoll, A.

1997 *Der Rechtsfall der En-hedu-Ana im Lied nin-me-šara*. AOAT 246. Münster: Ugarit-Verlag.

Zsolnay, Ilona

2010 "Ishtar, Goddess of War, Pacifier of Kings." Pp. 389–402 in: *Language in the Ancient Near East. Proceedings of the 53e Rencontre Assyriologique Internationale Vol. 1*, eds. L. Kogan, N. Koslova, S. Loesov, et al. BabBib 4/1. Winona Lake, IN: Eisenbrauns.

CHAPTER 11

Judean Scribalism, Documentary Epistemology, and the Name ישׂראל

Jeffrey L. Cooley

Introduction

The study of the epistemology of first millennium Mesopotamian scribes has taken a decidedly constructivist turn in the last few decades. This reorientation to the issue of how the intelligentsia of ancient Iraq knew their world has proven to be a powerful explanatory model for understanding Babylonian mantics, textual hermeneutics, and historiography, etc.

In this paper I argue that a constructivist perspective on Judean scribalism is similarly productive and that Judean scribes, too, constituted a distinctive epistemic culture. This sub-group of educated elites privileged the manufacture and visual inspection of documents in their knowledge-building process. Thus, the primary extant product of their craft, the Hebrew Bible, should be understood through the lens of that group's specific epistemology.

Their epistemic processes, I maintain, had prominent graphic and material characteristics. Recognizing this has the potential to solve a number of interpretive difficulties, especially those that revolve around the biblical interpretation of names. To demonstrate the effectiveness of this approach, I examine the rationalization of the name "Israel" in Genesis 32:28–29 and 35:9–11.

* Though I had not met her at the time, it is no exaggeration to say that my first monograph (Cooley 2013), could not have been written without the documentary presence of Francesca Rochberg. That I continue to be inspired by her starry messages is manifest in this paper (even with its mundane topic!), which draws on her ideas as well as those that she introduced to me through her science. I am truly honored to participate as a small twinkling speck in this august constellation of studies that celebrates her radiant achievements, and I am grateful to the editors for inviting me to do so.

I also thank my colleague David Vanderhooft, the members of the Boston College biblical studies doctoral colloquium, as well as Alan Lenzi (University of the Pacific) and C. Jay Crisostomo (University of Michigan), for generously reading previous drafts of this paper and providing many critical comments.

Note: abbreviations that do not appear in the *CAD* or *OCD* are from SBL *Handbook of Style, 2nd Edition* (Atlanta, SBL Press, 2014).

 | DOI 10.1163/9789004363380_012

Epistemology in Ancient Mesopotamia: Previous Approaches

In the early days of the study of ancient Near East, Assyriologists had an especially high regard for the knowledge-building of their objects of research.[1] The publication of sophisticated astronomical texts that featured detailed observations, together with mathematical models for planetary movements by scholars such as Epping and Strassmeier, demonstrated both unexpected empirical and rational dimensions of Babylonian science.[2] It is not an understatement to say that Victorian scholars were dazzled by the complexity of the thought displayed by these materials. Typical for the period, Jastrow, in his monographic juggernaut *The Religion of Babylonia and Assyria*, would lay out an essentially positivist characterization for Babylonian epistemology: the empirical recording of celestial phenomena eventually allowed for predictive models.[3] These models were, in turn, systematized within a pre-existent religious system. Summarizing the ancient scientific outlook and its religious framing, Jastrow would write:

> ...the point of importance for our purposes is to detect the extension of religious ideas into the domain of science, and, on the other hand, to note the reaction of scientific theories on the development of religious thought. The cosmology of the Babylonians results from the continued play of these two factors. Hence the strange mixture of popular notions and fancies with comparatively advanced theological speculations and still more advanced scientific theories that is found in the cosmological

1 For my purposes, "knowledge" is a communally recognized "set of organized statements of [perceived] fact or ideas" (Bell 1973: 41, cited in Knorr Cetina 1999: 6; bracketed insertion mine). This set functions as "a tool for defining one's place in society; its validity and usefulness are inextricably linked to social structure and to the place within that structure where knowledge is produced" (Veldhuis 2014: 21). The last point, that knowledge is "produced," that is to say, *socially constructed*, is the underlying observation of constructivist sociology of knowledge.

This is undoubtedly a more controversial statement, at least in regard to the modern scientific enterprise. That debate, between objectivism and constructivism, need not concern this study since there is little discord regarding the constructed character of knowledge in pre-modern societies. Thus, "knowledge" here has a sociological definition that needs to concede to cultural relativity, even if I need not deny the genuinely effective understanding of the cosmos that modern scientific processes demonstrably generate. In this regard, my modernist proclivities are comforted by the "model-dependent realism" employed by Hawking and Mlodinow (2011: 8).

2 Strassmeier and Epping 1889.

3 Jastrow 1898: 362, 370, 457, 466.

> system. Even mysticism is given a scientific aspect in Babylonia. The identification of the gods with the stars arises, as we have seen, from a scientific impulse, and it is a scientific spirit again that leads to the introduction of the gods into the mathematics of the day.[4]

Jastrow's characterization of Babylonian knowledge-making was based in part on his high regard for what he regarded as Babylonian "scientists" and his dramatically lesser consideration of their popularly pious contemporaries.

A different take on Mesopotamian epistemology was offered by scholars of Pan-Babylonian school of thought. They marveled, as well, at the surprising mathematical and observational erudition demonstrated in cuneiform sources. This group of scholars, most prominently represented by Winckler and especially Jeremias, portrayed the Babylonian priesthood as nearly modern scientists whose epistemology was thoroughly grounded in empiricism.[5] The Pan-Babylonianists were impressed by the manifest mathematical competency of the ancient documents revealed by Epping and Strassmeier. Additionally, they were struck by the obvious reliance of the Greeks and biblical authors on Mesopotamian sources, the former for astronomical knowledge, the latter for narrative material, such as the flood account. The Babylonian priority of knowledge, tethered to 19th century trends in the absolute dating of Mesopotamian history, led the Pan-Babylonianists to maintain that Babylonian celestial science was not only advanced but was also remarkably ancient. Furthermore, Winckler and Jeremias posited that Babylonian celestial science was the underlying inspiration for the *content* of Babylonian literature and culture, and was deliberately woven into mythic narratives by the religio-scientific elite so that the general populace might glean a modest intellectual morsel of their empirical erudition. Even more radically, they argued, that literature itself became the germ of much of the world's mythic traditions. Later interpreters of the Babylonian tradition, Hebrew, Greek or otherwise, typically did not fully comprehend the underlying scientific agenda of the narratives.

Both Jastrow and the Pan-Babylonianists largely restricted their discussions of Mesopotamian epistemology to celestial science material for a number of reasons.[6] First, and most obviously, the availability and prominence of the tablets brought to light by Epping and Strassmeier demanded intellectual

4 Jastrow 1898: 465.

5 E.g., Winckler 1904; Jeremias 1913. For a summary of Pan-Babylonianism and bibliography, see Cooley 2013: 2–6.

6 They were aware, of course, that the emic Mesopotamian perspective on knowledge was that it was something that was divinely revealed to humanity. See, e.g., Jastrow 1898: 62, 122, 131, 276, 552, 553.

attention. As well, these texts presented an undertaking that was considered commensurate with a modern scientific activity. Except for its popular adulteration, the ancient epistemic engine and those ancient "scientists" who operated it mirrored their Victorian analogs.

Nonetheless, late 19th–early 20th century Assyriologists were at a loss, it seems, to explain why the Babylonians, who were supposed to be such acute observers of the natural world, had not abstracted their models in an Ionian (i.e., modern) way, despite the empirical corset that should have starkly delineated their conclusions. The Babylonians, Greeks—and nineteenth century astronomers—were all observing the selfsame sky, were they not? Thus, Jastrow and the Pan-Babylonianists rationalized this empirically anchored epistemic discord by appealing to popularism. For Jastrow, the doctrinal, scientific scholarship of the Babylonian intelligentsia was ultimately trumped by popular insipidness and simplistic religiosity.[7] In contrast, the Pan-Babylonianists argued that the ancient scholars deliberately popularized their science by means of mythic narrative so that the unwashed masses could receive, at least, a modicum of scientific knowledge.

Much of the obsession with Babylonian celestial science within its cultural context collapsed with the publication of Kugler's *Im Bannkreis Babels*.[8] The Jesuit's work dismantled the erroneous chronology of scientific discovery that was at the core of the Pan-Babylonianist thesis and cast a dark, velvet pallor over the conversation. Consequently, the issue of the place of celestial science within Mesopotamian culture through much of the 20th century was dominated by science historians in the pre-Kuhnian vein, who were chiefly concerned with the intellectual achievements of the ancients inasmuch as they provided links in the stainless steel chain that led to the *telos* of modern science.

This shunning of cultural-context questions sequestered Babylonian celestial science from the evaluation of ancient epistemology. Thus, the next major contribution to the topic came in von Soden's influential article "Leistung und Grenze sumerischer und babylonischer Wissenschaft."[9] Von Soden's analysis began with—and in many ways was delimited by—the examination of the Sumerian list texts. Because these lists covered almost all conceivable material facets of life in ancient Mesopotamian rather than a singular enterprise (such as celestial science), von Soden seemingly approached the issue

7 For his general disdain for the ancient Babylonian popular perspective, see, e.g., Jastrow 1898: 150, 153, 465, 447.

8 Kugler 1910.

9 von Soden 1936, revised in Landsberger and von Soden 1965. See also briefly, highlighting the longevity of his thoughts on this issue, von Soden 1994: 145–148.

of epistemology in ancient Iraq rather broadly, unlike his predecessors. Notably, he acknowledged the lists' original role as part of scribal formation and recognized that their organization appeared to be based on philological relationships rather than ones grounded in observable physical reality. But von Soden furthermore asserted that the lists needed to be understood not merely as philological constructs but as religio-ideological documents that, in their essence, reflected the Sumerian race's desire for an orderly cosmos. The lists were indeed a kind of concretization of a philological worldview, but in their intellectually slavish and repetitive copying, they ultimately expressed a simplistically pious epistemology. For von Soden, this meant that the Sumerians' attempt to understand the world around them was palsied at its conception and was sessile in the long term. Knowing apparently meant organizing, and that was all. Their Semitic intellectual successors, the Akkadians, Babylonians, and Assyrians, adopted the list form but, not comprehending its deep intellectual significance, transformed the original lists into learning tools by augmenting the original columns of Sumerian text with additional ones in Akkadian. Additionally, von Soden maintained, the Babylonians expanded the use of the list form to other fields of scholarship. The Semitic reception of the Sumerian *Ordnungswissenschaft* was a kind of bastardization of the *Ordnungswille*, both an unsuccessful endeavor to understand the Sumerian written record as well as a radical reorientation of the scribal craft towards the pragmatic. Von Soden, characteristic for his time and place, attributed the differences between Sumerian and Babylonian *Wissenschaft* on the "differences of the racial aptitudes of both peoples."[10]

In addition to the thoughtful critiques offered by scholars such as Oppenheim, Veldhuis, Hilgert, and most recently, van de Mieroop, it is difficult not to view von Soden's understanding of Mesopotamian epistemology as brightly colored by his own experience and method of creating knowledge.[11] Von Soden, after all, was a lexicographer and was neck-deep in the Mesopotamian lexical lists as part of the lexicographical process. He was the scholarly inheritor of Bruno Meissner's (1868–1947) archive that ultimately was the basis of the

10 "Verschiedenheit der rassischen Anlagen beider Völker" (von Soden 1936: 552). Racial characterizations dominate the end of von Soden's study, perhaps culminating in his statement that "science, in the strict sense of the word, is something that could only be created by the northern race of certain Indo-Europeans" ("Wissenschaft im strengen Sinn des Wortes etwas ist, das nur von den durch die nordische Rasse bestimmten Indogermanen geschaffen werden konnte") (von Soden 1936: 556).

11 Oppenheim 1977: 248–249; Veldhuis 1997: 137–139; Hilgert 2009; van de Mieroop 2015: 64–70.

Akkadisches Handwörterbuch.[12] Notably, assembling the *AHw* was contemporary to the inception and fulfillment of the *Assyrian Dictionary of the University of Chicago*, begun in 1921.[13] In the mid-twentieth century, the process of Assyriology was one of morphologically splitting the atoms of words, breaking and reforming the molecular bonds that spanned the parallel columns of the lexical texts to create the scientific knowledge of ancient Mesopotamian culture through its language. That von Soden, one of the great scholarly personalities and contributors to the enterprise of reconstructing the *Wortschatz* of ancient Iraq, would characterize his ancient scholarly counterparts as possessing a list-based, philologically oriented and organized epistemology should come as little surprise.

Regardless of the shortcomings of von Soden's evaluation, at least he considered the Sumerians as systematic and orderly and the Babylonians as practical and innovative. In the heart of the twentieth century, much of the scholarship on the ancient Near East, perhaps overly influenced by the protestant Biblical Theology movement, was far less generous. In the wildly influential collection of essays *The Intellectual Adventure of Ancient Man: An Essay on Speculative Thought in the Ancient Near East*, the Frankforts considered ancient Near Eastern epistemology—with the exception of that of ancient Israel—as an essentially emotional, rather than intellectual process.[14] Necessary to the scientific episteme, for the Frankforts, was the clear and soulless objectification of the phenomenon under consideration. The peoples of the ancient Near East, however, were never able to truly objectify the world around them—it was always a sentient entity—and therefore their evaluation of it was inevitably a relational and emotionally charged muddle.[15] Knowing was feeling, and that was all.[16]

Perhaps in response to the mythopoetic characterization of Mesopotamian thought by mid-twentieth century biblical scholars, Assyriologists beginning in the 1970s increasingly pointed, like Jastrow and the Pan-Babylonianist at the end of the nineteenth century, to the genuinely empirical dimensions of ancient Mesopotamian scholarship. This time, however, mantic texts in

---

12 Reiner 2002: 19.

13 Reiner 2002: 19.

14 H. and H.A. Frankfort 1946a; see also H. and H.A. Frankfort 1946b in the same volume. For my notes on William A. Irwin's contribution (1946), see below. It is striking that, in his contribution to the collection, Jacobsen (1946) demurs from saying anything regarding Mesopotamian "knowledge."

15 H. and H.A. Frankfort 1946a: 5, 11.

16 The Frankforts' analysis has had a profound longevity; see, e.g., Wiggerman 1993 (esp. 297, note 4). For evaluations, see Rochberg 2004: 37–42.

particular played an important role in the argumentation.[17] To be sure, von Soden had previously argued that the relationships between protases and their respective apodoses were considered on some level empirically derived: someone in the past had noticed a correlation between an observable phenomenon and an event, and that observation became the basis for future prognostication.[18] As Trolle Larsen noted in his remarkable study, building on the work of Bottéro and Rochberg, "We may be dubious about the historical validity of the information in the omens, but the empirical principle cannot be argued away."[19] (Just what precisely was the actual object of the observation that made Mesopotamian mantic epistemology empirical is a question I will shortly address below.) With respect to the observation of natural phenomena, the culmination of the empiricist approach to Mesopotamian epistemology has been Brown's *Mesopotamian Planetary Astronomy-Astrology*.[20] Brown's thesis is quite deliberately situated in the Kuhnian model and argues for a paradigmatic shift in ancient Babylonian celestial science from an observational practice to a predictive one based on the accumulation of decades of empirically derived data.

Mesopotamian Epistemology and Epistemic Cultures

Repeatedly, scholars have pointed to empirical dimensions of Mesopotamian thought, at least in principle. Still, while observational facets of Mesopotamian omen scholarship are hard to deny, it is also clear that there was a robust, even dominant, deductive component to the undertaking, characterized by culturally based schema (e.g., right = good, left = bad), as Koch-Westenholz has amply demonstrated.[21] When it comes to the question of epistemology in ancient Babylonian scholarship, how should we frame the interplay between deductivism on the one hand, and empiricism on the other? Is it enough to characterize

17 Jastrow actually gave significant shrift to mantic texts in his nineteenth century synthesis, and frequently noted its observational and systematic character. However, he sharply distinguished in his discussion between the mantic, understood as magic, and that which was to be considered scientific (Jastrow 1898: 328–406 [omens], and 454–466 ["Zodiacal System"]).

18 Von Soden 1936: 536–538.

19 Trolle Larsen 1987: 213; Bottéro 1974, especially 144–168; Rochberg 2010, especially 26–27 (originally published Rochberg 1982).

20 Brown 1999.

21 Koch-Westenholz 1995: 15–19.

the former as a kind of mysticism, and the latter as a mere augment of the former thereby largely dismissing it?

At the nexus of the Babylonian intellectual enterprise's empirical and deductive characters is the technology of writing itself. Jastrow, for instance, thoroughly acknowledged its patent sophistication was inextricably linked to the scribal craft.[22] Von Soden, too, understood that the organization of the lists was philological in nature, even if for him the underlying *Ordnungswille* was a motivation broadly shared by all Sumerians as a hallmark of their "race."[23] Only in the later twentieth century had scholars really begun to articulate writing as an explicitly *technological* facet of the epistemic undertaking; Trolle Larsen, citing the anthropological work of Goody and Ong, was already keenly aware of the impact of the technology and tradition of writing on the mantic enterprise.[24] A number of scholars have further contributed to this discussion, especially Veldhuis, Selz, Rochberg, and most recently van de Mieroop in his sweeping monograph.[25] Rochberg writes that "the crux for a cuneiform epistemology is in what was considered the object of knowledge, especially when words and signs were of central interest."[26] This is a keen insight. Scribes observed the world around them to construct knowledge, no doubt, but they also observed the documents in their hands. Indeed, van de Mieroop notes that "The Babylonian theory of knowledge was to an extent empirical—observation was crucial. It was also fundamentally rooted in a rationality that depended on informed reading. Reality had to be read and interpreted as if it were a text."[27] Whether they prioritized observed documentary knowledge over observed

22 Jastrow 1898: 129–134.

23 Von Soden 1936: 552.

24 Trolle Larsen 1987: 213, 220, 223.

25 Veldhuis 1997; Selz 2011; Rochberg 1999; van de Mieroop 2015. See also Winitzer 2011, especially 94.

26 Rochberg 2014: 50.

27 Van de Mieroop 2015: 10. Though this is not the place for a full review of van de Mieroop's significant contribution to the question at hand, suffice it to say that he brings together many of the contemporary strands of thought on the issue of Babylonian epistemology. Among the book's many merits, I will highlight only a couple that I find particularly helpful. First he (re-)introduces the term "grammatological" (as understood by Derrida) to the discussion to highlight how "writing creates its own meaning" (78, for his full presentation 77–84). Second, van de Mieroop does a fine job reasserting of the place of the concept and execution of lists in Babylonian thought, tempered by the recent studies of these crucial documents; this important move allows him to extend the consideration of the epistemological question to the law codes (143–181), and thus, he is able to consider in his survey the entire scribal corpus essentially comprehensively.

non-documentary knowledge is, for the moment, beside the point. What *is* important is that the distinctively scribal manner of seeing things necessarily colored all of their empirical experiences and, consequently, their understanding of the observable world around them. We must situate Mesopotamian empiricism—and the epistemology of which it is a key facet—firmly *within* the scribal tradition. Thus, we need a method that acknowledges this tightly professional social context of knowing.[28]

Exceptionally productive in trying to understand the intersection of knowledge and writing in ancient Mesopotamia has been the application of constructivist methodologies, developed within the field of the sociology of knowledge, to ancient texts and the literate sub-cultures that created them. The obvious attention in this regard has been on those fields of knowledge that have been most easily identified with what moderns consider approximations of our sciences, such as astronomy (i.e., celestial divination) and mathematics. Here, Rochberg's *The Heavenly Writing* and Robson's *Mathematics in Ancient Iraq* should be considered genuinely pioneering.[29]

Constructivist studies characterize the epistemic process very nearly as one of manufacturing. Briefly stated, a knowledge (or epistemic) subject, in the modern context, a scientist or technician, etc., operates within a laboratory.[30] The lab is a kind of cosmos in a box, a place where practice (see below) can take place in a controlled setting; it is filled with tools and materials, some of which are manufactured on site, others are procured from specialized manufacturers.[31] The knowledge subjects, utilizing their tools in the lab space, apply their practice ("recurrent processes governed by specifiable schemata of preferences and prescriptions")[32] to "knowledge objects" or "epistemic things." In Rheinberger's words, epistemic things are

> material entities or processes ... that constitute the objects of inquiry. As epistemic objects they present themselves in a characteristic, irreducible vagueness. This vagueness is inevitable because, paradoxically, epistemic things embody what one does not yet know.[33]

28 To be sure, since at least the time of Durkheim and Mannheim it has been recognized that knowledge is socially produced (Durkheim 1915 and Mannheim 1936).

29 Robson 2008.

30 Knorr Cetina 1999: 11.

31 Knorr Cetina 1981: 3–4. See also Latour and Woolgar 1986, especially 43–90 (Chapter 2, "An Anthropologist Visits the Laboratory").

32 Knorr Cetina 2001: 175.

33 Rheinberger 1997: 28. See also, Knorr Cetina 2001: 181.

Their process-embeddedness means that epistemic things are constantly taking on "new properties and changing the ones they have."[34] Furthermore, they often "exist in a variety of forms" and have "multiple instantiations, which range from figural, mathematical, and other representations to material realizations."[35] Epistemic things are also signifying in nature, in that they point "to possible further explorations."[36] The outcome of applying practice to epistemic objects is knowledge, which epistemic subjects then "enact in concrete situations."[37] While the constructivist terminology is perhaps hopelessly jargony, the model that is buried under that thick terminological dermis is relatively simple and describes a fundamentally *syntactical* relationship, i.e., subject—verb/process—object—result.

The actual manifestation of these subjects, processes, and objects—and the knowledge they produce—is by no means universal. Galison shows that the material cultures (i.e., the laboratory tools) of the various branches of microphysics vary by discipline, that is, by *sub-culture*.[38] These sub-cultures have their own epistemic subjects, processes, objects, and, indeed, even their own languages that need to be deliberately translated in order for the sub-cultures to effectively exchange and utilize their respective "knowledges." Knorr Cetina has dubbed such respective groups of technical experts "epistemic cultures." As she defines the term, an "[epistemic or knowledge] culture refers to the aggregate patterns and dynamics that are on display in expert practice and that vary in different settings of expertise."[39]

To be sure, Knorr Cetina developed the concept to address epistemology in contemporary scientific cultures, and her study placed two specific groups under the microscope: molecular biologists and high-energy physicists. In this study I am discussing two groups of experts from antiquity, namely Mesopotamian and Judean scribes of the middle of the first millennium. The differences between these sets of experts, ancient and modern, might appear to be irresponsibly vast at first glance. In terms of sophistication, particle accelerators and autoradiographs are comically distant from the stylus, tablet, reed-pen, papyrus, logo-syllabary and alphabet. Nonetheless, like the devices

34 Knorr Cetina 2001: 181.

35 Knorr Cetina 2001: 182.

36 Knorr Cetina 2001: 183. See also Latour 1987: 87–88.

37 Knorr Cetina 2001: 185. My largely syntagmatic understanding is, in fact, *contra* Knorr Cetina's, who sees the result of practice as relational. For the enactment of knowledge, see, e.g., Bourdieu 1977 and Lynch 1993.

38 Galison 2001.

39 Knorr Cetina 1999.

of the modern laboratory, they are all tools that require specific training to use expertly and meaningfully. And all of these apparatus, whether ancient or modern, have little use outside those expert circles and are, most importantly, fundamentally employed in the process of constructing knowledge. By definition, the tools of the scribe are instrumental in that process.

If we apply the basic model of epistemic constructivism to the ancient scribes, we can say that, as a result of their education, livelihood, tools, skill, etc., scribes situated sight, sound, and tradition into their specific sub-cultural episteme in a manner different than those without these things. Scribes *ex officio* saw documents and made new copies of them. Sometimes, in various commentaries, they provided explicit exegesis of the writing on the document, elucidating obscure words and signs by means of other words and signs. When scribes lifted their squinting eyes up from the slabs of clay or scraps of papyrus and trained their gazes on the world around them, they did not compartmentalize these habits of seeing. For scribes, the point of seeing is to acquire knowledge by means of written signs.[40] This scribal vision of the world can be observed in the sign-based categorization of objects we see in the great lexical lists, originating in the third millennium but utilized throughout the history of cuneiform scribalism. These provided an organization of contents of the world—an organization whose detailed principles are based on the careful observation of written form of the names of the objects, not on a carefully observed physical or functional taxonomy of the objects themselves.[41] The lists, employed at the very beginning of scribal inculcation, inadvertently trained the scribe to order the world around himself as a collection of written words.

In Mesopotamian documents from the first millennium this can be seen in a number of places. The basic idea that the written form of a word revealed something more than its mere utterance could reveal was a crucial principle in *Enūma Eliš*. There, Marduk's fifty names (Tablets VI–VII) are exegeted based entirely on their visual properties and bilingual polyvalences.[42] But this scribal understanding of the world is perhaps most clear in a number of mantic practices. It goes without saying, for example, that for the average (i.e., illiterate) first-millennium inhabitant of Babylon the sky was very much populated by visible celestial deities. This is true for the contemporaneous scribe as well, but for him additionally, the sky was covered with a kind of writing, *šiṭir šamê* ("writing of the sky"), the celestial surface itself being a gargantuan lapis

40 Rochberg 2014: 50.

41 Veldhuis 1997: 139; see also Veldhuis 2012: 16.

42 Bottéro 1977.

writing tablet.[43] The sky was a vast and ever-changing document whose true significance was only deeply discernable to those with specific scribal training. Similarly, the gods used the exta of sheep as documentary surfaces to convey their judgments to humanity. The liver, lung, intestines, etc., were "tablets of the gods," *ṭuppū ša ilī*, on which messages were composed.[44] The living animal itself would carry the message to the haruspex who could, because of his scribal training, interpret the written document of the exta. In this sense, the sky and the animal's organs were no more—or less—a "natural" phenomenon than the scribe's clay tablet.[45] Inasmuch as the traditional texts of the Mesopotamian scribal craft (e.g., series of literary, mantic, exorcistic, etc., documents) were considered to have divine origins, the scribally manufactured documents, the documents displayed in the sky, or on a sheep's liver or lung, ultimately had the same source. Indeed, they shared the same documentary function.

A striking example of scribal outlook as it was projected on the cosmic scale can also be seen in one of Esarhaddon's prisms describing the restoration of the Esagil in Babylon. The king recounts how previously the inhabitants of the land had angered the gods with their impiety; the latter struck the land with flooding and had abandoned it—a sentence Marduk determined should last seventy years. The god, however, was merciful and gracious, and he lessened the sentence, literally by inverting the signs used to write the numeral 70 (a vertical wedge followed by a *Winkelhacken*) into the numeral 11 (a *Winkelhacken* followed by a vertical wedge):

mar-qí-ti 70 MU.AN.NA.MEŠ	The merciful god Marduk wrote
mi-nu-ut ni-du-ti-šu	that the calculated time of its
iš-ṭur-ma re-me-nu-ú	abandonment (should last) 70 years,
^{d}AMAR.UTU *sur-riš lìb-ba-šu*	(but) his heart was quickly soothed,
i-nu-uḫ-ma e-liš a-na	and he reversed the numbers and
šap-liš uš-bal-kit-ma	(thus) ordered its (re)occupation to be
a-na 11 MU.AN.NA.MEŠ	(after) 11 years.[a]
a-šab-šú iq-bi	

a Esarhaddon 104 ii 2–9a (RINAP 4, 196). See also Beaulieu 1995: 4–5; Robson 2008: 148–149.

43 CAD Š/3, 146; see also Rochberg 2004: 1; Frahm 2010: 98; Cooley 2013: 36–37.

44 Starr 1983: 53–57.

45 For the problem of talking about "nature" in Babylonian thought, see Rochberg 2016: 21–39.

The god's judgment was cut short and this unexpected outcome merited a rationalization: clearly, Marduk had been merciful. The divine act of mercy is affected by means of the scribal craft. The written documentation of a sentence, perhaps recorded on the so-called "Tablet of Destinies," is physically altered by the god pulling a graphic switcheroo, such that when the sentence was crafted the actual reality of the land's punishment was altered. Like the fifty names of Marduk in *Enūma Eliš*, the actual written documentation of something, rather than its mere oral verbalization, is what is ontologically tethered to physical reality. And the god is not merely using a simple knowledge of cuneiform to effect this change; rather, he is utilizing a sophisticated, scholarly hermeneutic involving the twisting of signs. For scribes, meaningful phenomena worthy of observation included not only the positions of planets, deformed fetuses, the urban appearance of wayward foxes, but also dockets, contracts, word lists and literature. For the scribes and the scribally-literate god alike, writing was as much a reality as anything else.[46]

Epistemology in Ancient Judah: Previous Approaches

In the mid-twentieth century, biblical theologians addressed the epistemology of ancient Israel, and characteristically for the period, their analysis was an argument for a sort of Israelite intellectual exceptionalism. Boman, believing the Hebrew episteme to be aurally derived, considered the Israelites masters of psychological thought, forerunners of the mid-twentieth century psychoanalyst.[47] For both Irwin and Wright on the other hand, the Hebrews were unique in the ancient Near East for their experiential/empirical basis of knowing.[48] Ultimately, the marriage of Hebrew experiential knowledge and Greek analytical thought created the wonder of the Western intellectual tradition.[49]

46 As Liverani wrote almost presciently more than forty years ago, "The thing to do should be to view the document not as a 'source of information,' but as information in itself; not as an opening on a reality laying beyond, but as an element which makes up that reality" (Liverani 1973: 179). Many thanks to my colleague David Vanderhooft for bringing this quotation to my attention.

47 Boman 1960: 201–204.

48 Irwin 1946; Wright 1952: 42–44.

49 H. and H.A. Frankfort 1946b, especially 373. Of course, the subject has also been the focus of a considerable amount of theological and philosophical work; see, e.g., Healy and Parry 2007; Dowd 2009; Gericke: 2012: 371–404; Hazony 2012: 161–192; Johnson 2013. While these have their value, my approach here is neither theologically nor philosophically oriented, and so I will not engage with this scholarship to any large degree.

Within strictly biblical studies, epistemology has long been a focus of those working with Qohelet for obvious reasons.[50] More recently, the anthropological approaches of Malul and Avrahami, as well as the philological approach of Carasik, have sought to examine the question of the epistemology of ancient Israel from a consciously emic perspective.[51] While I maintain that their methods are problematic due to the fact that they do not explicitly front the scribal context in which the biblical text was fashioned, their contributions are nonetheless quite valuable. Still, in stark contrast to the current situation in Assyriology (see above), very little work has been done on the relationship between epistemology and ancient Judean scribalism.[52] This fact is all the more surprising given the explosion of studies regarding scribalism, literacy, and literate education that have been produced in the last couple of decades, most recently by Carr, van der Toorn, Rollston, and Schniedewind, among others.[53]

One of the few contributors who has explicitly discussed the impact and relationship between epistemology and education has been James Crenshaw, who broached the subject in his monograph *Education in Ancient Israel: Across the Deadening Silence*. His focus is not on scribal training, per se, but on education broadly speaking. But Crenshaw's insights on Judean epistemology are derived from an inspection of the texts that tradition and modern scholarship have labeled "wisdom." Since it is commonly assumed that these documents were central in the Judean scribal training process as they were at times in neighboring Mesopotamia and Egypt, his contribution is worth addressing.[54]

---

50 E.g., Fox 1987; Schellenberg 2002. The date of Qohelet's composition and its overall intellectual novelty limits its utility in this study, which focuses on the mid-first millennium.

51 Carasik 2006; Malul 2002 (for method, see 3–7); Avrahami 2012 (for method, 4–64). Malul, in particular, is problematic since he regards the biblical text as the product of an oral culture rather than a literate professional culture; in his own words (52): "Needless to say, the methods applied [to the biblical text], being geared to the analysis of *specimens of writing*, must be in general erroneous, or at least grossly inappropriate for the analysis of the type of source-material that is at our disposal and for carrying out the task of the present book."

52 Explaining this situation is difficult, but it seems to be related to the desire by biblical scholars to find in ancient Israel a kind of democratization of reading/writing that is otherwise not evinced in the broader ancient Near East and which in antiquity is only seen, perhaps, in certain parts of the Greek world in certain periods. For this bent, see Schniedewind 2004 and Sanders 2009.

53 For education, see Lemaire 1981; Puech 1988; Crenshaw 1998. For scribalism and literacy in particular, see Carr 2004; van der Toorn 2007; Sanders 2010; Schniedewind 2013.

54 See discussions of Carr 2004 and van der Toorn 2007, below. For a serious critique of the assumption that Proverbs was actually employed pedagogically in ancient Israel based on form-critical criteria, see Vayntrub 2016 (for her comments on Crenshaw, see 97 note 5).

Crenshaw's basic conclusion is that knowledge in ancient Israel was considered to derive from three sources: observation, analogy, and tradition. Crenshaw maintains that some of the Bible's wisdom material, such as Proverbs, is a product of family life and "parental instruction."[55] According to his reconstruction of the epistemology of the wisdom literature, the thoughtful Israelite acquires knowledge, first and foremost, by observation of the human and natural world. The ancient sage has a creed, but dynamically reinterprets that creed to fit with observed reality. The Israelite God himself does reveal knowledge, but only as a divine watchmaker, who places it all in nature to be revealed, it seems, by means of careful human observation. Revelation as a source for knowledge, according to Crenshaw's understanding of the wisdom tradition, had a place only at the creation—humanity rediscovers this cosmically embedded knowledge through careful observation.[56] Overall his formulation sounds suspiciously like the epistemic process of a mid-twentieth century liberal biblical theologian.[57] More positively, Crenshaw is undoubtedly correct that tradition, analogical reasoning, and observation were all part of the epistemic process in ancient Israel and Judah—as they are in any ancient or modern culture.

In part, Crenshaw's selection of texts leads him to these conclusions regarding ancient Judean epistemology. Focusing solely on (modernly) labeled "wisdom literature" (e.g., Proverbs, Job, Qohelet), tacitly assuming that these works stood in a kind of intellectual and literary isolation from others, provides a profoundly skewed perspective.[58] If texts such as Proverbs (whether it be an earlier edition or something similar) were involved in the scribal education process, as were similar collections from Egypt and Mesopotamia, to divorce their pedagogy from the acquisition of the craft of writing is difficult, in spite of the dearth of explicit references to writing in that biblical book.[59] The "sages" (חכמים) of the "wisdom" material, Proverbs in particular, are first and foremost to be identified with the scribes themselves or at the very least those scribes involved with the formulation of Judah's literature.[60]

Somewhat more valuable to my questions here is Carr's study of education in the ancient Near East and Greece. Elaborating on the intricate interplay in the pedagogic process between orality and textuality, memorization

55 Crenshaw 1998: 230.

56 Crenshaw 1998: 130.

57 For a discussion of a similar twentieth-century characterization of Hebrew thought, including bibliography, see Cooley 2013: 13–16.

58 For "wisdom literature" as a problematic grouping, see Carr 2004: 132.

59 See especially Carr 2004: 128.

60 E.g., Grabbe 1995: 155; Schniedewind 2004: 206–207; van der Toorn 2007: 77; Dell 2008: 130.

and documentation, Carr argues that educating a scribally literate class was a process of enculturation. Though he does not address the issue of epistemology explicitly, his recognition that scribal education was a path to enculturation emphasizes the constructive nature of ancient intellectualism as well as its concentration within a distinctive subgroup within a larger society. He does, of course, obliquely address the scribal episteme, but the primary kind of knowledge that is germane for his discussion is by definition the reification of traditional knowledge rather than the generation of the novel. Part of that tradition was, to be sure, the texts that were operative in the scribal formation process (e.g., Proverbs) and, by extension, in the training of literate elites.[61] Additionally, the tradition encompassed the means to produce such texts physically (i.e., pen and papyrus—though Carr does not dwell on the material angle), as well as intellectually, including the reapplication of traditional, internalized/memorized texts to novel problems.[62] Still, Carr's emphasis on the role of memorization in the process of enculturation means that he, by definition, downplays its material aspect and the impact of that materiality on Judean scribal sub-culture.

Van der Toorn has provided a more helpful contribution to the object of my inquiry, Judean epistemology as it intersects with the scribal craft. Indeed, his work in *Scribal Culture and the Making of the Hebrew Bible* is close to a constructivist model. In his discussion of the nativity of the biblical prophetic corpus, van der Toorn argues that scribes coopted the idea of divine revelation, originally the purview of non-literate prophets, and embodied it in documentation thereby imbuing the written text with oracular authority:[63]

> ...I call revelation a "scribal construct." It will be clear by now that this expression was not intended to mean that scribes invented the notion of revelation as such; their invention was rather in the nature of a radical transformation. They used the concept of revelation as an epistemological category to qualify a body of literature. By identifying revelation with a circumscribed group of texts, the scribes shifted the focus of the concept. Until then revelation had been understood as an interaction between superhuman beings and human individuals in which the former imparted

61 Carr 2004: 126–156.

62 Carr 2004: 161–173.

63 van der Toorn 2007: 205–232. Lenzi, too, argues that Judean scribes, specifically those involved with the composition and promulgation of the ritual legislations in Leviticus, Ezekiel, and Deuteronomy, deliberately assumed the prophetic character for the purpose of mantically grounding their authority (Lenzi 2007: 273–305; see also, Lenzi 2014: 65–86).

> knowledge to the latter; in the concept developed by the scribes, revelation became an object rather than an interaction: it was coterminous with a set of texts.[64]

I am rather sympathetic to van der Toorn's take on the textualization of prophecy. Still, his approach is less about *how* Judean scribes created knowledge and more about how they added a novel (mantic) dimension to texts they manufactured as scribes. What is only partially present in his work and missing entirely from the work of Crenshaw (as well as other recent interlocutors on the subject of Judean epistemology, such as Malul and Avrahami), is an adequate recognition that the biblical text does not reveal a view of Israelite epistemology *per se*, but rather, that of the particular sub-group of individuals who created those documents, i.e., Judean scribes.[65]

Scribal Epistemology among Other Epistemologies

Like that of their Mesopotamian counterparts, the job of the Judean scribe was first and foremost to manufacture documents.[66] Though we know that scribes served in a number of loci (temple, palace, army, etc.), the produce of their specific training was not *texts* per se (here defined as the content of a physical document, which could be dictated to them by a second party, e.g., the

64 van der Toorn 2007: 232. This is not to deny the orality of Judean culture as a whole, *a lá* Niditch, nor the high value that scribes must have placed on memorization as Carr, in particular, has highlighted (Niditch 1989; cf., Malul 2002: 25–54). But the end of the matter is that for a scribe to be a scribe he was expected to be able to manufacture a written document. It is also important here to mark the difference between literacy and scribalism. Schniedewind has made a case for widespread literacy in pre-exilic Judah, and I do not deny that the ability to read was not confined to those who were inculcated in the scribal craft. But, as both Carr and Veldhuis have emphasized, being able to verbalize a simple document is not the same as being able to read and interpret a piece of written literature, technical or otherwise (Carr 2004: 13; Veldhuis 2011). I would add, as well, that the ability to verbalize and understand something as basic as the name on a seal or even the contents and implications of a contract is a rather different skill set than that needed to craft a written document. For the elite status of writers and readers in the biblical text, see also Young 1998 and Rollston 2010: 129.

65 Carasik 2006 is an outlier among the recent contributors; the title of his monograph, *Theologies of the Mind in Biblical Israel*, is a recognition that it is not ancient Israel as a whole that he is examining, but rather the vision presented by the Bible's authors in particular.

66 van der Toorn 2007: 1–2.

contents of an official letter), but rather the physical documents themselves.[67] This physical materiality, and the control over the manufacture and distribution that goes hand-in-hand with it, is important to underscore. Documents are technological products that physically store knowledge. Their limited physicality allows religious and social elites (priests, kings, scribes, etc.) control over the production of and access to knowledge. This knowledge could be explicitly labeled as divine, as Stavrakopoulou has recently argued.[68]

As manufacturers of a product that stored knowledge, for scribes epistemic authority lay in documents. Documents were certainly not the sole source of epistemic authority for scribes, but, by definition, they were a privileged one. The numerous documentary references in Joshua, Judges, Samuel (ספר הישר and ספר מלחמות יהוה) and especially Kings (ספר דברי שלמה, דברי הימים למלכי יהודה/ישׂראל) and Chronicles (ספר מלכי ישׂראל ויהודה) attest to the idea that written sources contain knowledge about the past that is otherwise unobtainable. Only scribes have physical and intellectual access to these documents.

Other epistemologies were operative in ancient Judah, frequently overlapping with each other. Certainly scribes were partakers of the common cultural episteme (of the sort Malul and Avrahami seek to define). One could also point to a legal episteme in which eyewitnesses as well as contractual documentation are salient. Another obvious episteme is the mantic: Yahweh's knowledge revealed through the proper diviners was a potentially valid source of practical knowledge. As noted, these epistemologies could and did overlap; surely, for example, mantic and scribal epistemologies were not always mutually exclusive (e.g., Ezek 2:8–3:4).[69]

Nonetheless, tensions between these various epistemologies could exist. In 2 Kgs 22:11–20, for instance, we see the concern for the validity of a written *torah* supposedly uncovered in Yahweh's temple in Jerusalem by the high priest Hilkiah during the reign of Josiah. When the document's contents are revealed to the king, he is dismayed, but he nonetheless immediately orders the small entourage responsible for the document's delivery (which, notably include a named scribe שׁפן הספר and his son אחיקם, verse 12) that the document's

67 Here I am consciously de-emphasizing the idea of the "text"—what a modern media librarian would refer to as "content." This is in contrast to those scholars who have highlighted that idea for text-critical reasons, such as Fox (Fox 2015: 4, drawing on Tanselle 1989: 14–18). Similarly, Carr, though defining "textuality" *per se* as written phenomenon, nevertheless employs "(textual) performance" to refer to copying, reading, or reciting orally a text from memory (Carr 2004: 12, 38, 94, 236, 291).

68 Stavrakopoulou 2013 (especially 233).

69 van der Toorn 2007 and Lenzi 2014, as noted above, have convincingly argued for the scribal adoption of the prophetic epistemic mantle.

contents (דברי הספר, verse 13) be verified mantically. They approach the prophet Huldah, conveniently living in Jerusalem, and she confirms the newly discovered document's epistemic validity by means of prophetic divination.[70] Interestingly, Josiah's skepticism regarding the *document* is not punished; rather, Yahweh rewards Josiah for his immediate reaction of rending his clothes when he initially heard its contents:

יַעַן רַךְ־לְבָבְךָ וַתִּכָּנַע ׀ מִפְּנֵי יְהוָה בְּשָׁמְעֲךָ אֲשֶׁר דִּבַּרְתִּי עַל־הַמָּקוֹם הַזֶּה וְעַל־יֹשְׁבָיו לִהְיוֹת לְשַׁמָּה וְלִקְלָלָה וַתִּקְרַע אֶת־בְּגָדֶיךָ וַתִּבְכֶּה לְפָנָי וְגַם אָנֹכִי שָׁמַעְתִּי נְאֻם־יְהוָה׃ לָכֵן הִנְנִי אֹסִפְךָ עַל־אֲבֹתֶיךָ וְנֶאֱסַפְתָּ אֶל־ קִבְרֹתֶיךָ בְּשָׁלוֹם וְלֹא־תִרְאֶינָה עֵינֶיךָ בְּכֹל הָרָעָה אֲשֶׁר־אֲנִי מֵבִיא עַל־הַמָּקוֹם הַזֶּה וַיָּשִׁיבוּ אֶת־הַמֶּלֶךְ דָּבָר׃	"Because your heart was receptive and you were humble in front of Yahweh when you heard what I said concerning this place and its inhabitants becoming a waste and curse, you tore your garments [cf., verse 11] and wept in my presence—and I myself have heard, oracle of Yahweh—therefore, look: I am gathering you to your fathers, and you will be gathered to your grave intact. Your eyes will not see all the evil which I am bringing on this place." And they brought an oracle back to the king. (2 Kgs 22:19–20)

That is to say, the king is praised and rewarded for trusting the purportedly antique knowledge that the document recorded *before* it had been mantically verified. This is a powerful statement: trust the knowledge contained in the documents that scribes offer, and you will be rewarded.[71]

Jer 8:8–9, according to van der Toorn, offers a sharply contrasting perspective on the intersection between scribal and mantic epistemologies. He argues that this passage refers to Jeremiah's rejection of the same document whose purported discovery is described in 2 Kings 22:[72]

70 Inasmuch as the discovery of the *torah* was accidental, it could be considered an unprovoked oracle that needed to be verified by a provoked one. As such it conforms to the mantic stacking seen broadly in ancient Near Eastern divinatory practices. For discussion and bibliography, see Cooley 2011.

71 Ben Dov (2008) argues that the *torah* in this passage is not an early form of Deuteronomy, but is an oracle in itself which had possibly even been provoked at Josiah's behest.

72 van der Toorn 2007: 35. This is a difficult passage, and interpretations abound: for Schniedewind, Jer 8:7–9 is referring much more broadly of emerging tension between oral

אֵיכָה תֹאמְרוּ֙ חֲכָמִ֣ים אֲנַ֔חְנוּ וְתוֹרַ֥ת יְהוָ֖ה
אִתָּ֑נוּ אָכֵן֙ הִנֵּ֣ה לַשֶּׁ֣קֶר עָשָׂ֔ה עֵ֖ט שֶׁ֥קֶר
סֹפְרִֽים׃ הֹבִ֣שׁוּ חֲכָמִ֔ים חַ֖תּוּ וַיִּלָּכֵ֑דוּ הִנֵּ֤ה
בִדְבַר־יְהוָה֙ מָאָ֔סוּ וְחָכְמַ֥ת־מֶ֖ה לָהֶֽם׃

How can you say, “We are sages, and Yahweh’s *torah* is with us”? What a lie the lying pen of scribes has made! (The) sages are hereby shamed, shattered; they were captured! Look, they refused Yahweh’s oracle, and so what kind of sagacity do they have? (Jer 8:8–9)

Regardless of whether Jer 8:8–9 references Josiah’s book of the law (דבר־יהוה), what is evinced in both this and 2 Kings 22 is the tension between mantic (here: prophetic) and strictly scribal epistemology. For the scribes/sages, the document was authoritative by its virtue of being a written text; for the prophets it was Yahweh’s oracular word, i.e., דבר־יהוה. Nonetheless, by the exilic or early Second Temple period at least, the idea born, it seems, in 2 Kings 22, was common: scribes could and did assert that the documents they manufactured and the knowledge contained therein originated with God on some level.

Judean Scribal *Weltanschauung*: Documentary Epistemology

As Carr has convincingly argued, the process of scribal education not only trained these craftsmen how to produce documents, it also enculturated them into a specifically scribal mindset.[73] That the authors of many biblical texts understood the world around through a scribal lens is evident in a number of places. For example, scholars have compellingly argued that the materials from book of Proverbs played a role Judean scribal training.[74] Striking, however, is the dearth of references to the process of writing in that text. Only Prov 3:3; 7:3, and 22:20 mention writing, and in the former two cases, it seems, writing is used as an image for memorization:

authority (i.e., the משפט יהוה in vs. 7) and the written authority of the scribes (2004: 114–117). Friedman (1987: 188–206) sees the conflict as one of the Deuteronomic versus the Priestly school.

73 Carr 2004: 126–134.

74 Carr 2004: 126–134. The idea that Proverbs, a lengthy, sophisticated written document, does not reflect a literate mindset as Carasik and Kovacs argue, seems highly problematic at best (Kovacs 1974: 183–184, 186; Carasik 2006: 63). Again, for a caveat, see Vayntrub 2016.

חֶסֶד וֶאֱמֶת אַל־יַעַזְבֻךָ קָשְׁרֵם עַל־ גַּרְגְּרוֹתֶיךָ כָּתְבֵם עַל־לוּחַ לִבֶּךָ׃	Do not abandon loyalty and reliability. Bind them on your neck; write them on the tablet of your heart. (Prov 3:3)

Similarly, in what appears to be a plea for the memorization of a text,

שְׁמֹר מִצְוֺתַי וֶחְיֵה וְתוֹרָתִי כְּאִישׁוֹן קָשְׁרֵם עַל־אֶצְבְּעֹתֶיךָ כָּתְבֵם עַל־לוּחַ עֵינֶיךָ׃ לִבֶּךָ׃	Keep my commandments and live, and my torah like the pupil of your eye. Bind them on your fingers; write them on the tablet of your heart. (Prov 7:2–3)

A similar use of לוח לב is found in Jer 17:1:

חַטַּאת יְהוּדָה כְּתוּבָה בְּעֵט בַּרְזֶל בְּצִפֹּרֶן שָׁמִיר חֲרוּשָׁה עַל־לוּחַ לִבָּם וּלְקַרְנוֹת מִזְבְּחוֹתֵיכֶם׃	Judah's sin is written with an iron pen; with a flint stylus it is engraved on the tablet of their heart, and on the horns of their altars.

Here, the image serves not to indicate memorization *per se*, but as a perpetual documentation of Judah's sin. The image of the לוח לב is perhaps correctly taken as a metaphor, but the fact that organs were sometimes considered real writing surfaces in the ancient Near East should caution us against too flippantly labeling it as purely metaphorical.[75] It is quite possible that the Judean writer actually understood the process of intellectual internalization/memorization as having a text written on the surfaces of the physical organ.[76] Whatever the case, the author could have used an idiom that did not necessarily derive from scribal practice, such as שית לב (Prov 22:17; 24:32) or שים בלב (1 Sam 21:13; Job 22:22). The point is that the cognitive process is described in terms of the practice of written documentation and this, I maintain, shows the scribal orientation of the text. Moreover, if we are to identify the חכמים, "sages," that are frequently mentioned in Proverbs (ca. 40X) as the scribes (or at least

75 See brief discussion and bibliography in Shupak 2003: 425–426.

76 For wearing words on the body (e.g., Prov 1:9, 3:3; 6:21; 7:3), and its possible relationship to written amulet use, see Schniedewind 2013: 115.

the scribes from the past who were responsible for the composition of the works that comprise Proverbs), then we have in that book a document that is by definition highly scribally-oriented, regardless of the lack of overt references to writing.[77]

Avrahami highlights how the complete *sensorium* is involved with knowing the world and how its perception is culturally conditioned.[78] I will offer an example below (focusing on Genesis 32 and 35) of how sight in particular functioned among those Judeans with expert scribal training. For the moment, I will highlight that other senses, too, can be scribally oriented. For example, in Prov 22:17–20 the instructor expects his pupil to *hear* the contents of a written document:

הַט אָזְנְךָ וּשְׁמַע דִּבְרֵי חֲכָמִים וְלִבְּךָ תָּשִׁית לְדַעְתִּי׃	Incline your ear and hear the words of the sages; set your heart to my knowledge.
כִּי־נָעִים כִּי־תִשְׁמְרֵם בְּבִטְנֶךָ יִכֹּנוּ יַחְדָּו עַל־שְׂפָתֶיךָ׃	It is so pleasant that you should keep them in your innards; let them be established together on your lips.
לִהְיוֹת בַּיהוָה מִבְטַחֶךָ הוֹדַעְתִּיךָ הַיּוֹם אַף־אָתָּה׃	In order that your trust be in Yahweh, I hereby inform you of Amenemopet.[a]
הֲלֹא כָתַבְתִּי לְךָ שִׁלְשׁוֹם [שָׁלִישִׁים] בְּמוֹעֵצֹת וָדָעַת׃	Have I not written for you the Thirty Sayings with counsel and knowledge?

a Amending the mt's הַיּוֹם אַף־אָתָּה (which is gibberish as it stands) as ימנמאפת, following Rendsburg 2001: 192–195. See also the translation offered already in the original *New American Bible* (1970), "That your trust may be in the LORD, I make known to you the words of Amen-em-ope." Note, this rendering is eliminated in the *New American Bible, Revised Edition* (2011).

This passage is often understood to highlight the oral element of pedagogy in ancient Judah, and I do not mean to deny that this is the case.[79] However, while epistemology in the Hebrew Bible overall is dominated by a vocabulary related to seeing,[80] the educative process itself is most often described in terms of

77 For this identity of the חכמים, see van der Toorn 2007: 77, as well as Dell 2008: 130.

78 Avrahami 2012 (for Ezek 3:1–3, see 93–94).

79 The role of texts like Proverbs in ancient Judean didactism as well as its featuring of orality and writing is well treated in Carr 2004: 126–134.

80 Avrahami 2012: 223–275.

hearing regardless of the actual form it might have taken.[81] Indeed, it is worth noting that, so dominant is the auditory rhetoric for learning in Proverbs, the verb למד ("to learn") itself is used only once (Prov 30:3).[82] In this regard it is also notable that the passage from the Wisdom of Amenemopet on which this is based does not mention writing.[83] That is to say, the West Semitic translator (if that is what we want to call him), chose to *add* the element of written documentation to the didactic rhetoric: "hear my document!"

More unexpectedly, other senses can be scribally understood as well. Thus, in Ezek 2:8–3:4, Yahweh orders the prophet to consume a double-sided scroll (מגלה) that originates with the god; it bears a series of "lamentations, dirges, and woes" (קנים והגה והי), and Yahweh commands him to deliver its message to Israel. The document is "as sweet as honey" (כדבשׁ למתוק), an indication of its finest origin and manufacture. Furthermore, Yahweh notes that the scroll was written in the language of the people of Israel, not a foreign one that would be unintelligible (3:5–6, עמקי שׂפה וכבדי לשׁון, literally, "deep of lip and heavy of tongue"). Similarly, Ps 19:11, in which Yahweh's *torah* (with various epithets and undoubtedly written) is described as מתוקים מדבשׁ ונפת צופים, "sweeter than honey, and honey from the comb."[84]

Perhaps the best evidence for a scribal *Weltanschauung* in the biblical text, however, is the characterization of the Judean god as a skilled scribe. Yahweh himself writes documents:

וַיֹּאמֶר יְהוָה אֶל־מֹשֶׁה עֲלֵה אֵלַי הָהָרָה וֶהְיֵה־שָׁם וְאֶתְּנָה לְךָ אֶת־לֻחֹת הָאֶבֶן וְהַתּוֹרָה וְהַמִּצְוָה אֲשֶׁר כָּתַבְתִּי לְהוֹרֹתָם׃	Yahweh said to Moses, "Come up to me, to the mountain, and be there, so I can give you the stone tablets, and the *torah* and commandment that I have written to instruct them." (Exod 24:12)

81 Shupak 2003: 417. For hearing a document, see, e.g., Exod 17:14 and Deut 30:10.

82 See Shupak 2003: 420.

83 Amenemopet 3:9–10 (Wilson 1969: 421; see also Lichtheim 2006: 149).

84 Intriguing in this regard is Jer 15:16:

נִמְצְאוּ דְבָרֶיךָ וָאֹכְלֵם וַיְהִי־דבָרֶיךָ [דְּבָרְךָ] לִי לְשָׂשׂוֹן וּלְשִׂמְחַת לְבָבִי כִּי־נִקְרָא שִׁמְךָ עָלַי יְהוָה אֱלֹהֵי צְבָאוֹת׃	Your words were discovered, and I ate them. Your word was such a delight to me, and my heart's joy, for your name was read off of me [lit. "upon me"]: "Yahweh, the god of armies."

The biblical authors can be quite insistent that Yahweh himself physically manufactured the text:

וַיִּתֵּן אֶל־מֹשֶׁה כְּכַלֹּתוֹ לְדַבֵּר אִתּוֹ בְּהַר סִינַי שְׁנֵי לֻחֹת הָעֵדֻת לֻחֹת אֶבֶן כְּתֻבִים בְּאֶצְבַּע אֱלֹהִים׃	He gave Moses, when he finished speaking with him on Mount Sinai, the two tablets of the treaty, stone tablets written with the finger of God."(Exod 31:18; cf., Deut 9:10)

Also,

וַיִּפֶן וַיֵּרֶד מֹשֶׁה מִן־הָהָר וּשְׁנֵי לֻחֹת הָעֵדֻת בְּיָדוֹ לֻחֹת כְּתֻבִים מִשְּׁנֵי עֶבְרֵיהֶם מִזֶּה וּמִזֶּה הֵם כְּתֻבִים׃ וְהַלֻּחֹת מַעֲשֵׂה אֱלֹהִים הֵמָּה וְהַמִּכְתָּב מִכְתַּב אֱלֹהִים הוּא חָרוּת עַל־הַלֻּחֹת׃	He turned and Moses descended from the mountain with the two tablets of the treaty in his hand, tablets written on their two sides (they were inscribed on this side and on that side). Now, as for the tablets, they were the manufacture of God, and as for the writing, it was God's writing inscribed on the tablets. (Exod 32:15–16)

These documents, above, seem to have been novel compositions, but the deity, as a competent scribe, is capable of creating a new copy of a document from memory:

בָּעֵת הַהִוא אָמַר יְהוָה אֵלַי פְּסָל־לְךָ שְׁנֵי־לוּחֹת אֲבָנִים כָּרִאשֹׁנִים וַעֲלֵה אֵלַי הָהָרָה וְעָשִׂיתָ לְּךָ אֲרוֹן עֵץ׃ וְאֶכְתֹּב עַל־הַלֻּחֹת אֶת־הַדְּבָרִים אֲשֶׁר הָיוּ עַל־הַלֻּחֹת הָרִאשֹׁנִים אֲשֶׁר שִׁבַּרְתָּ וְשַׂמְתָּם בָּאָרוֹן׃... וַיִּכְתֹּב עַל־הַלֻּחֹת כַּמִּכְתָּב הָרִאשׁוֹן אֵת עֲשֶׂרֶת הַדְּבָרִים אֲשֶׁר דִּבֶּר יְהוָה אֲלֵיכֶם בָּהָר מִתּוֹךְ הָאֵשׁ בְּיוֹם הַקָּהָל וַיִּתְּנֵם יְהוָה אֵלָי׃	At that time, Yahweh said to me, "Fashion for yourself two stone tablets like the first ones, and come up to me, to the mountain, and make for yourself a wooden box, so I can write on the tablets the words that were on the first tablets that you broke, and you can place them in the box."... And he wrote on the tablets, like the first writing, the ten words that Yahweh told you on the mountain from the fire on the day of assembly. And Yahweh gave them to me. (Deut 10:1–2, 4)

Yahweh is also capable of reading physical documents presented to him for inspection, in this case a formal missive from a foreign monarch, Sennacherib:

וַיִּקַּח חִזְקִיָּהוּ אֶת־הַסְּפָרִים מִיַּד הַמַּלְאָכִים וַיִּקְרָאֵם וַיַּעַל בֵּית יְהוָה וַיִּפְרְשֵׂהוּ חִזְקִיָּהוּ לִפְנֵי יְהוָה׃	Hezekiah took the documents from the hand of the messengers and read them. He went up to the Temple of Yahweh and Hezekiah spread it out in front of Yahweh. (2 Kgs 19:14; cf. also Isa 37:14)

Furthermore, Yahweh has his own personal collection of documents at his disposal for reference and consultation. Some of these are administrative in nature, including primarily lists of names, families, etc. (Exod 32:32–34; Isa 4:3; Ezek 13:9; Pss 69:29; 87:6). These texts function as do human archives, to serve as legal witness, but also as a technologically-enabled mnemonic aid for the administratively-busy god-on-the-go:

וַיַּקְשֵׁב יְהוָה וַיִּשְׁמָע וַיִּכָּתֵב סֵפֶר זִכָּרוֹן לְפָנָיו לְיִרְאֵי יְהוָה וּלְחֹשְׁבֵי שְׁמוֹ׃	Yahweh noted it [the contents of the conversation between those who feared him], and heard it, and a book of remembrance was written in his presence, for those who fear Yahweh and those who consider his name. (Mal 3:16b)

The Judean god is depicted, as well, as using documents in the process of planning (Ps 139:16), and responds to documentation that requires an active response (Isa 65:6; Ps 149:8–9). Furthermore, Yahweh is depicted as textually aware and citing already existing documents:

וְהֵבֵאיתִי [וְהֵבֵאתִי] עַל־הָאָרֶץ הַהִיא אֶת־כָּל־דְּבָרַי אֲשֶׁר־דִּבַּרְתִּי עָלֶיהָ אֵת כָּל־הַכָּתוּב בַּסֵּפֶר הַזֶּה אֲשֶׁר־נִבָּא יִרְמְיָהוּ עַל־כָּל־הַגּוֹיִם׃	I will bring against that land all of my oracles that I spoke concerning it, all that is written in this book that Jeremiah prophesied concerning all the nations. (Jer 25:13)

The Chronicler's account of the Hilkiah's scroll-find during Josiah's reign yields a number of striking differences from its documentary source in 2 Kings 22–23.

For my discussion here, I will merely highlight that in Huldah's confirming oracle, Yahweh explicitly cites the copy of his own *torah* (described in 2 Chr 34:14 as ספר תורת־יהוה ביד־משה):

כֹּ֚ה אָמַ֣ר יְהוָ֔ה הִנְנִ֨י מֵבִ֥יא רָעָ֛ה עַל־הַמָּק֥וֹם הַזֶּ֖ה וְעַל־יוֹשְׁבָ֑יו אֵ֣ת כָּל־הָאָל֗וֹת הַכְּתוּב֤וֹת עַל־הַסֵּ֙פֶר֙ אֲשֶׁ֣ר קָֽרְא֔וּ לִפְנֵ֖י מֶ֥לֶךְ יְהוּדָֽה׃	Thus Yahweh hereby says, "Look, I am about to bring evil upon this place and upon its inhabitants, all the curses that are written on the document which they read in the presence of the king of Judah." (2 Chr 34:24)

The emphasis here on כל־האלות הכתובים על־הספר (as opposed to כל־דברי־הספר in 2 Kgs 22:16) highlights the increasing awareness of the scribes of the importance of the physical document rather than merely the text.

We cannot expect the biblical texts, with their diverse authorship, places, and times of composition, to equally demonstrate self-awareness of their own documentary character. As Carasik notes, even the book of Kings with its extensive documentary citations seems to have "no awareness of itself as a book. There is no introduction to the book, no colophon, and certainly no injunction that it be leaned or recited."[85] However, the biblical texts do occasionally betray this scribal orientation, as I hope to have shown above.[86] We must remember, as well, that those texts are documents and that their very existence, regardless of their contents, attests to the fact that there was a scribal culture that crafted material products. Most scholars would agree that the Hebrew Bible is a collection of documents, many of which show marked signs of significant revision, in many cases by multiple hands. This is the best evidence that the Hebrew Bible has scribal origins and audience, and was tradited by scribes for future generations of scribes.

85 Carasik 2006: 59.

86 For a discussion of scribalism and literacy as it pertains to the psalter in particular, see Dobbs-Allsopp 2015: 298–235. Drawing on Sanders (2006) Dobbs-Allsopp describes the kind of literacy on display in the Bible as "emergent" in order "to emphasize its nascent and fluid nature and to resist the idea that mere textuality immediately gives way to full fledged literate or monolithic conceptions" (318). See also Dobbs-Allsopp 2009.

Judean Scribes as an Epistemic Culture

What is now taken as a given in the study of Mesopotamian scribes of the first millennium, that the physicality of writing played a crucial, if not central, role in scribes' epistemic culture, needs to be demonstrated for the scribes who composed and tradited the Hebrew Bible. What was the distinctive interaction between epistemic subject, tools, process, and object that constructed knowledge—scribal knowledge in Judah?

In constructivist terminology, the knowledge subjects, of course, are the scribes who manufacture documents. Much has been written on their identity, and it seems that there were in ancient Judah scribal families and perhaps even tribes (Judg 5:14) and enclaves (Josh 15:15–16; Judg 1:11–12; 5:15–16) that were known for their skills with the pen and papyrus. Rollston has convincingly argued that there was some sort of formalized and/or standardized process of scribal training in the Iron II (and presumably carried into exile), regardless of whether there were formal schools.[87] In this process of inculcation and enculturation, scribes learned the tools, processes, and proper objects of knowledge.

The material culture of scribes is more or less well documented, both textually and archaeologically. Scribes would create documents (סְפָרִים, *passim*) by writing (כָּתַב, *passim*) using either an awl (צִפֹּרֶן, Jer 17:1) or a pen (עֵט, Jer 8:8; 17:1; Ps 45:2; Job 19:24) with ink (דְּיוֹ, 36:18), that was stored in a container (קֶסֶת, Ezek 9:11) expressly made for that purpose (קֶסֶת הַסֹּפֵר, Ezek 9:2–3). The choice of tool that would imprint the writing on the surface would depend on the surface that was prepared to receive the writing. That surface could be a scroll (מְגִלָּה, Ps 40:8; Jer 36:2, 4, 6, 14, 20,21, 23, 25, 27–29, 32; Ezek 2:9; 3:1–3; Zech 5:1–2) composed of individual panels (דְּלָתוֹת, Jer 36:23) presumably of papyrus or prepared animal skin, or a tablet (לוּחַ *passim*) that might be made of stone (e.g., Exod 24:12; but see also the Gezer Calendar), or a wax surface, and possibly even clay (though neither of these are explicitly attested in the biblical text). Alternatively, a scribe might imprint the writing on a non-transportable surface, such as a natural though prepared part of the landscape (e.g., bedrock of the Royal Steward Inscription or Siloam Inscription), a prepared monument (e.g., the Mesha Inscription), or plastered wall (e.g., the Deir ʿAlla text).

As part of their education and enculturation, scribes learned a number of skills and procedures. Presumably some of these items above were made by the scribe himself: the pen, for example, had to be manufactured and maintained, and scribes did so by utilizing a scribe's knife (תַּעַר הַסֹּפֵר, Jer 36:23), a

87 Rollston 2010: 129.

sub-cultural variant of a common tool (cf., as well, Exod 34:1, 4 in which Moses prepares stone tablets as a writing surface). Obviously, one of the most basic elements of their praxis would have been mastery of the alphabet, including its proper order, its writing, and its reading. This would involve conventions in marking verbalization (standard orthographies, word division, etc.). So, too, one had to learn the laying out of basic kinds of documents (making lines, margins, etc.). Under layout, we might include as well standard formulations appropriate for certain kinds of documents (letters, dockets, etc.). Scribes learned language as well. This would consist first of their own language, including technical/specialist vocabulary and its uses (legal, diplomatic, economic, etc.), as well as literary (obscure/rare terms; cf., poetic word pairs, neologisms). It might also include foreign languages useful for international correspondence (Aramaic, Egyptian, Akkadian).

Names and the Construction of Knowledge: ישׂראל as an Epistemic Object

As Radner has documented, the importance of remembering names in the wider ancient Near East was tremendous.[88] And, indeed, one of the primary objects of scribal process was the name. In constructivist terms, names served as epistemic objects to which knowledge subjects (that is, scribes) applied processes, and created knowledge to be further utilized for various purposes. As scribes, the processes applied to names were documentary: names of people were recorded in tax lists, rosters of personnel, and arranged, stacked, and rearranged in genealogies. Names of items were documented in inventories, sales contracts, etc. Names of places were listed in territorial tax appropriations, commercial and military itineraries. In short, names were manipulated documentarily to produce all sorts of useful knowledge that was applied to cultic, economic, governmental, and martial purposes.

But the process of scribes was not merely the manufacture of documents—documents and the writings they bore were also subjected to the process of visual inspection and interpretation. In both the manufacture of documents and in their visual inspection, scribes brought their material and epistemic cultures to bear—their pens and ink, their mastery of writing and language.[89] I recognize that this is a lot to say that scribes write things and read things. But

88 Radner 2005.

89 For a discussion of visual inspection *vis-à-vis* verbalization of documents, see Hyman 2006.

being conscious of the graphic and material nature of their processes, I will show, can bear fruit in the study of the narrative etymologies in the Hebrew Bible. To this end I will discuss the explanation or, I hope to show, *explanations* of the name "Israel" in Genesis 32 and 35.[90]

Gen 32:28–29

First the former. Gen 32:28–29 details how Jacob, as a result of his wrestling match with a divine stranger, obtains a new name:

וַיֹּ֥אמֶר אֵלָ֖יו מַה־שְּׁמֶ֑ךָ וַיֹּ֖אמֶר יַעֲקֹֽב׃
וַיֹּ֗אמֶר לֹ֤א יַעֲקֹב֙ יֵאָמֵ֥ר עוֹד֙ שִׁמְךָ֔ כִּ֖י אִם־
יִשְׂרָאֵ֑ל כִּֽי־שָׂרִ֧יתָ עִם־אֱלֹהִ֛ים וְעִם־אֲנָשִׁ֖ים
וַתּוּכָֽל׃ וַיִּשְׁאַ֣ל יַעֲקֹ֗ב וַיֹּ֙אמֶר֙ הַגִּֽידָה־נָּ֣א
שְׁמֶ֔ךָ וַיֹּ֕אמֶר לָ֥מָּה זֶּ֖ה תִּשְׁאַ֣ל לִשְׁמִ֑י וַיְבָ֥רֶךְ
אֹת֖וֹ שָֽׁם׃ וַיִּקְרָ֧א יַעֲקֹ֛ב שֵׁ֥ם הַמָּק֖וֹם פְּנִיאֵ֑ל
כִּֽי־רָאִ֤יתִי אֱלֹהִים֙ פָּנִ֣ים אֶל־פָּנִ֔ים וַתִּנָּצֵ֖ל
נַפְשִֽׁי׃

He said to him, "What is your name?" And he said, "Jacob." And he said, "Your name will no longer be called 'Jacob,' but rather 'Israel' for you have striven with a god (and with men), and triumphed." Jacob responded, and said, "Would you please tell me your name." He said, "What is this? You ask for my name?" And he blessed him right there. Jacob called the name of the place "Peni'el," for "I saw a god face to face, and my life was spared." (Gen 32:28–31)

The passage evinces a thriving onomastic complex, a sort of biome of conscious name-talk: not only does Jacob receive a new name as a result of the seemingly pointless tussle, but he inquires of the cryptic challenger's name, though he is rebuffed. Then Jacob, or rather, *Israel*—that is to say, Jacob (the writer still refers to him as Jacob, even after what has just happened!) names the place in honor of the weird encounter.

Here, the etymology of "Israel" offered by the scribe is rather straightforward; it follows one of the formal patterns (Form I) for etymologies identified by Fichtner in his classic form-critical study.[91] The hermeneutics applied by

90 Previous approaches to understanding the Hebrew Bible's many etymologies have generally labeled them either *Volksetymologien* or understood them as literary devices in the manner of punning and paronomasia. For the reasons I reject these approaches, see Cooley forthcoming.

91 Fichtner 1956.

the scribe to the name seem transparent; however, they are by no means a given. The interpretation of the moniker is *not* literal, in the sense the interpretation does not offer exclusively valid equivalencies for each of the name's components.[92] It should, rather, be qualified as a process of *synonymous* interpretation, the kind we see in Mesopotamian word lists like *malku-šarru*.[93] That is to say, the scribe chooses to identify the components of the name from a spectrum of *possible equivalents*, none of which are intrinsically prioritized. First, the scribe equates the theophoric element of the name אל with אלהים, actually, with עם אלהים. This is simple enough, but we should never take such equations as a given; he could have equated it with יהוה as is done in Gen 16:11. He could have even identified that אל with another deity, like *Baal*; though such an etymology would not have made the biblical author's editorial scalpel, one could easily imagine such an equation being offered in the northern kingdom of Israel by the priests of the temple of Baal at Samaria. Speculation aside, what is patently unclear in this passage is precisely *what* the identity of the אלהים is that the scribe is equating with אל. Traditionally, of course, it is equated with the god of Judah and Israel. But the writer does not say that explicitly, and the fact that the explanation of the theophoric element seems to include ועם־אנשים ("and with men") should give us pause from such an easy interpretation. As alluded to a number of times already, in his explanation the scribe also adds several syntactical particles (prepositions, conjunctions, etc.), reading the sentence ישׂר אל as if it were a piece of lyric stripped of the metrically laborious elements almost always found in prose, but so often frustratingly lacking in poetry.

The verbal component of the name, written ישׂר, is interpreted in Gen 32:29 with a rather rare root, שׂרה.[94] Based on the context within the story in Genesis 32, the root apparently means something like "to strive." The fact that this is a rare root is notable, and I will return to this point below. For the moment, however, I will briefly turn to Hos 12:3–5, which offers a number of word plays (rather than strictly etymologies) on the name ישׂראל, since like Gen 32:28–29, one of them utilizes the root שׂרה:

---

92 In the sense, however, that the interpretation of the name is derived from its written/graphic elements, one could, in concord with Uri Gabbay's definition, qualify the reading as "literal" (Gabbay 2009).

93 Hrůša 2010.

94 *HALOT*, 1354.

וְרִיב לַיהוָה עִם־יְהוּדָה וְלִפְקֹד עַל־יַעֲקֹב כִּדְרָכָיו כְּמַעֲלָלָיו יָשִׁיב לוֹ׃ בַּבֶּטֶן עָקַב אֶת־אָחִיו וּבְאוֹנוֹ שָׂרָה אֶת־אֱלֹהִים׃	Now Yahweh has a dispute with Judah, and punishes Jacob according to his ways, commensurate with his actions he repays him. In the womb he betrayed his brother, and with his strength he strove with God. (Hos 12:3–4)

Additionally, Hos 12:5 offers a further word play on the name that, according to the Massoretic vocalization, uses the root שׂור, perhaps a by-form of שׂרר, indicating "to rule":[95]

וַיָּשַׂר אֶל־מַלְאָךְ וַיֻּכָל בָּכָה וַיִּתְחַנֶּן־לוֹ בֵּית־אֵל יִמְצָאֶנּוּ וְשָׁם יְדַבֵּר עִמָּנוּ׃	He ruled over a messenger,[a] and he succeeded; he wept and implored him. At Bethel he found him and there he spoke with him.

a The awkward use of the preposition אל to govern the object מלאך appears to be deliberate in order to create the graphic (though not verbalized) word play, i.e., וישׂר אל = ישׂראל. Alternatively, see Andersen and Freedman 1980: 608.

As a verb the root used in Hos 12:5, שׂור, is not that common (in the *qal*: Judg 9:22; Isa 32:1; Hos 8:4; Prov 8:16; Est 1:32; 1 Chr 15:22 and cf., משׂרה in Isa 9:5–6). As a noun, in contrast, שׂר is *exceptionally* common (the verbal root itself is perhaps denominative). Thus, Hosea's second word play appears to correctly identify the verbal root in the name ישׂראל. From a scientific philological perspective the name appears to mean "may El rule," and conforms to normative Semitic naming practice.[96] In light of the Hosea passage, I return to

95 Sarna 2001: 405. Symmachus, the Vulgate, the Peshitta, and the Targums all follow this reading; see discussion in *HALOT* 313; see also Vermes 1975: 12–13. The reading here is contentious and the best arguments against it are twofold: (1) the LXX uses the same verb to translate וישׂר (ἐνίσχυσεν) in 12:5 as it does שׂרה in 12:4; and (2) the desire for the following verb in 12:5, ויכל to be in (con)sequence of וישׂר. As scholars have noted, however, the events of Jacob's life described in Hos 12:3–5 are not in order (e.g., the Bethel and Jabbok encounters appear in Hos 12:5 in the reverse order as they do in Genesis 28 and 32 respectively), and their equations with the description in Genesis are hardly clear, nor is it even certain as to who the subject of the verb ויכל is—Jacob or the messenger. For a representative discussion, see Andersen Freedman 1980: 600, 608–609.

96 Classically, see Noth 1928: 207–209; more recently, Rechenmacher 1997: 98.

the etymology in Gen 32:28–31. There, the scribe has chosen to interpret the verbal element in the name ישׂראל *not* with the common and transparent root indicating sovereignty, שׂור or שׂרר, but with the much more obscure root indicating strife, שׂרה.[97] That is to say, he has explained a common root with an obscure one.

Two more things are worth noting. First, the scribe who composed Gen 32:28–31 has interpreted the prefixing conjugation of the name ישׂראל not as a jussive, but rather as a preterite and has capitalized on the potential temporal overlap of that form with the perfect form. He does this in order to read the verb not as a wish expressed, but as alluding to a past event. Second, in similar discord with common name expectations, the syntax of the components of the name has been twisted by that scribe—the theophoric element should be the subject, not the object of the verb.

In constructivist terms, the scribe, the knowledge subject, has applied a process to a knowledge object, the name of the patriarch. The process employed the mental tools and resources of the scribal episteme: the application of synonymous words in interpretation (אל with אלהים), the use of rare or obscure words (שׂרה), a sophisticated consciousness of the verbal system (that a prefixing verbal form can be temporally antecedent), and the use of a poetic hermeneutic (the inference of absent grammatical and syntactical elements). The result of this expert process is knowledge that is in turn applied to the problem of Israel's history and identity within a narrative. Perhaps the knowledge actually was the basis of that narrative, perhaps it merely fleshed it out intellectually, giving it a scribally epistemic legitimacy. For the moment at least I am leaving that last question unresolved.

Gen 35:9–11

My discussion of the etymology of ישׂראל in Gen 32:28–31 above might have elicited some doubt, at least in terms of the level of sophistication with which I am characterizing the scribal author's hermeneutics. There is likely little doubt, however, that it is in fact an etymology, an attempt to explain the origin of the name. When it comes to Gen 35:9–11, however, the discussion begins by establishing it as an etymology. It is clear that the deity is renaming Jacob "Israel," as in Genesis 32. But most commentators mark that the renaming is lacking an explanation. Westermann, for example, considers the absence an

97 This is the opposite of what we would expect if Gen 32:28–31 were an example of folk etymology, which tend to explain obscure terms with more common ones (cf., Sarna, *Genesis* 2001: 405).

"embarrassment"—an overstatement, to say the least.[98] More commonly, commentators simply say that no explanation is required since one has already been given in Chapter 32:28–31, and here in 35:9, God is described as appearing to Jacob "again."[99] For example, Sarna sees Gen 32:28–31 and 35:9–11 in dialog with each other: in Gen 35:9–11, Jacob's new name is being *confirmed* by the deity, not given, since it was originally applied to Jacob not by Yahweh, but by the "celestial patron of Esau" east of the Jordan.[100] Let us examine the passage:

וַיֵּרָ֨א אֱלֹהִ֤ים אֶֽל־יַעֲקֹב֙ ע֔וֹד בְּבֹא֖וֹ מִפַּדַּ֣ן אֲרָ֑ם וַיְבָ֖רֶךְ אֹתֽוֹ׃ וַיֹּֽאמֶר־ל֥וֹ אֱלֹהִ֖ים שִׁמְךָ֣ יַעֲקֹ֑ב לֹֽא־יִקָּרֵא֩ שִׁמְךָ֨ ע֜וֹד יַעֲקֹ֗ב כִּ֤י אִם־יִשְׂרָאֵל֙ יִהְיֶ֣ה שְׁמֶ֔ךָ וַיִּקְרָ֥א אֶת־ שְׁמ֖וֹ יִשְׂרָאֵֽל׃ וַיֹּאמֶר֩ ל֨וֹ אֱלֹהִ֜ים אֲנִ֨י אֵ֤ל שַׁדַּי֙ פְּרֵ֣ה וּרְבֵ֔ה גּ֛וֹי וּקְהַ֥ל גּוֹיִ֖ם יִהְיֶ֣ה מִמֶּ֑ךָּ וּמְלָכִ֖ים מֵחֲלָצֶ֥יךָ יֵצֵֽאוּ׃	God appeared to Jacob again; when he came from Padan-Aram, he blessed him. And God said to him, "Your name is 'Jacob.' Your name will not be called 'Jacob' any more, rather, it will be 'Israel.'" And he called his name "Israel." And God said to him, "I am El-Shaddai: be fruitful and multiply; a nation and an assembly of nation will come from you; while kings will proceed from your loins." (Gen 35:9–11)

To summarize: the deity changes Jacob's name, and then says, seemingly out of the blue, "I am El-Shaddai."

In spite of the commentators' modest bewilderment and hermeneutical scrambling, from a form critical perspective this passage is very much a full etymology and conforms to a number of other recognized etymologies, in which the name is pronounced (קרא שם) and is then explained with a simple ויאמר statement or equivalent. This formula appears to be a kind of apocopated version of a more explicit form found in such passages as Gen 31:47–48, Josh 7:24–26; 2 Sam 5:20.[101] Another, similarly apocopated etymology can be found

98 Westermann 1995: 553.

99 Wenham 2000: 325.

100 Sarna 2001: 241–242.

101 For a discussion of implicit etymologizing, see Zakovitch 1980. Zakovitch argues that many of the Bible's explicit etymologies have actually been expanded to make them so by later redactors for whom the implicit etymology offered in the text they received was not explicit enough (or even missed the original etymology entirely).

in 1 Sam 7:12, where the prophet names a monument erected to commemorate the geographic limit of Yahweh's assistance:[102]

וַיִּקַּח שְׁמוּאֵל אֶבֶן אַחַת וַיָּשֶׂם בֵּין־ הַמִּצְפָּה וּבֵין הַשֵּׁן וַיִּקְרָא אֶת־שְׁמָהּ אֶבֶן הָעָזֶר וַיֹּאמַר עַד־הֵנָּה עֲזָרָנוּ יְהוָה׃	And Samuel took one stone and placed it between the Watchtower and the Tooth. And he called its name Eben Ha'ezer [lit., "The Stone of the Help"], and he said, "Up to here, Yahweh has assisted us."

Similarly, Abraham names Isaac, it seems, as a response to Sarah's joy (or embarrassment) by the deity in Gen 21:4, 6:

וַיִּקְרָא אַבְרָהָם אֶת־שֶׁם־בְּנוֹ הַנּוֹלַד־לוֹ אֲשֶׁר־יָלְדָה־לּוֹ שָׂרָה יִצְחָק׃ ...וַתֹּאמֶר שָׂרָה צְחֹק עָשָׂה לִי אֱלֹהִים כָּל־הַשֹּׁמֵעַ יִצְחַק־לִי׃	Abraham called his name (the one who was born to him, which Sarah had born for him) Isaac. ...And Sarah said, "Such a joy God has made me! Everyone who hears will laugh for me!"[a]

a I read Sarah's explanatory utterance in a pessimistic light, in contrast to e.g., Westermann 1995: 333–334; Sarna 2001: 146.

Here, the Priestly writer explains the situation surrounding the naming of the child, and though it is Sarah's utterance that provides the motivation, it is Abraham who actually does the naming. Isaac's name was already predicted by the god (but not explained by him!) in Gen 17:19.[103] Similar, but not formally identical is Gen 4:1, in which Eve names Cain:

102 Note, this verse is dismissed as an awkward insertion by Long, alleviating him from having to fit it into the rigid formalism he adopted from Fichtner (Long 1968: 36).

103 The priestly writer, with his positive spin on Sarah's laughter, is thus transforming the Yahwist's implied etymology of Isaac in Gen 18:9–15, which is much less generous. Notable in both these cases (P's stated etymology, as well as J's implied one) is that the scribal authors are reinterpreting the traditional name יצחק—as Sarna noted, a hyporisticon of יצחקאל, in which El (or even some other deity) is supposed to be the subject of the verb (Sarna 2001: 127, following Stamm 1950; but see also discussion in Westermann 1995: 269). This too is an act of onomastic exegesis.

וְהָאָדָם יָדַע אֶת־חַוָּה אִשְׁתּוֹ וַתַּהַר וַתֵּלֶד אֶת־קַיִן וַתֹּאמֶר קָנִיתִי אִישׁ אֶת־יְהוָה׃	Now, the man knew Eve, his woman, and she conceived and bore Cain. She said, "I have acquired/made(?) a man with Yahweh."[a]

a Here, I am unconcerned with what it means to "acquire/make" a man with Yahweh—I am simply just pointing to the form.

Though the actual act of naming is not noted, the motivation for the child's name is explained with a simple ויאמר (here, ותאמר) statement, as in 1 Sam 7:12 and Gen 21:6. Though the *causal* relationship between the act of naming and the situation surrounding the act is not explicit in Gen 21:6 and 4:1, the scribal author is in both situations still offering an etymology of the names in question, since he is positing an ontological relationship between the form of the name as it exists and its meaning *vis-à-vis* the circumstances of its bearer's origin. These are not the only examples of this form of etymology. Most commonly, the form is inverted with the ויאמר statement coming first: (Gen 30:7–8, 9–11, 12–13, 16–18, 19–20; 32:2–3; 38:28; Exod 15:15, 31; Josh 5:9; 1Kgs 9:12–13).[104] In short, the form of Gen 35:9–11 dictates that this is an etymology.

I am left to conclude that the scribal author, here the Priestly writer, is offering God's self-identifying statement in 35:9, אני אל שדי, as an explanation of the name ישראל. That is to say, the Priestly writer is deriving the eponymous name of the patriarch from the divine moniker אל שדי. What is the process by which the scribe does so? In constructivist terms, how does he, the knowledge subject apply his expert process to the epistemic object to create the knowledge that is further employed in the narrative?

First, the obvious: the אל of ישראל equals the אל in אל שדי. Here, unlike the etymology offered by the scribe responsible for Gen 32, there is no need for synonymous interpretation since the two words are graphically identical: אל is אל.

This means that somehow ישר must equal שדי. Making the leap between the two elements is not difficult, if we remove the blinders that privilege our scientific philology and *look* at the page, visually inspect the thing: it is not difficult to see the appearance of ישר and שדי is very similar. *Yod* is *yod*, *shin*

104 Cf., Gen 5:29; Judg 6:32; 1 Sam 4:21 (with לאמר); 1 Chr 4:9 (with לאמר כי); Exod 2:10 (with ותאמר כי), Exod 17:15–16 and Gen 26:22 (with ויאמר כי). Scanning the etymologies offered for the offspring of Jacob (Gen 29–30), shows considerable variation in form. In any case, explanatory conjunctions, such as כן, על, or על כן are clearly unnecessary.

and *sin* are indistinguishable without the diacritical dot, and *resh* and *dalet* are nearly identical in the Aramaic script that we utilize. But they are not only so in the Aramaic script—they are quite similar, as well, in the ancient Hebrew script. Furthermore, as Ada Yardeni demonstrates, scribes actually wrote, that is to say, *manufactured*, the two letters very similarly throughout their ancient history, whether in Hebrew or Aramaic scripts. In the former case, the letters have a large triangular head with either a short descending tail (for a *dalet*) or a long descending tail (for a *resh*).[105] In the Aramaic scripts utilized by Jewish scribes of the Second Temple Period, the *dalet* and *resh* are often identically constructed and are, indeed, frequently indistinguishable in appearance.[106] Our first year Hebrew students might confuse *dalet* and *resh* out of beer-fueled indolence. In sharp contrast, the scribe who penned Gen 35:9–11, employed his expertise in the technology of writing and the manufacture of documents to equate *dalet* and *resh* for scholarly purposes. (The letter order is mixed, of course, but this should not concern us since mixing letter orders is not unknown in the Bible's etymologies. For example, the name יעבץ in 1 Chr 4:9, using the root עבץ, is explained by means of the noun עֹצֶב, whose root is עצב.)

Returning to Gen 35:9–11, we see that the verbalization of the documented text actually *obscures* the origin of the name the author proposes. That is to say, the ancient scribe understands the word as a scribe: it can have an oral manifestation, no doubt, but for the scribe the written character of a word can also offer important ontological data. If similar sounds are enough to justify for the scribe the etymological relationship between a word and a name, then certainly similar appearances should be as well, particularly if the similar appearances are the result of an identical process of manufacture. The knowledge derived from the name ישׂראל was constructed by the Judean scribe by employing his scribal tools and scribal processes. New knowledge was thereby created, which, when applied to the issue of the relationship between Israel and his god revealed a deep, existential bond between the two who share a kind of grammatological genome.

Of course, new knowledge was universally suspect in the ancient Near East, and the Judean scribe did not think he was constructing knowledge. Rather, from an emic perspective he believed he was revealing the knowledge of another scribe, one in full command of expert skill and with the professional

105 Cf., what seems to be evidence of a scribe (or someone in scribal training?) graphically manipulating the two letters, perhaps in the place name of Arad, in an Iron Age ostracon from that site (Aharoni 1981: 112–113 [#99]).

106 E.g., Yardeni 2002: 165 (4QSam^{b}), 167 (4QJer^{a}), 169 (4QEx^{f}), 171 (1QIsa^{a}).

confidence to equate two nearly identical letters and rearrange the order of their documentation. According to the author of Gen 35:9–11, that expert scribe was the god of Israel. Like Marduk's graphic inversion of wedges that resulted in the reduction of Babylonia's punishment, noted above, El-Shaddai renamed Jacob and labeled him with a holy name derived from an epithet of the deity, one that was wholly appropriate for a holy people. By scribal means the eponymous ancestor, the people, and their god were permanently ontologically entangled with one another.

Conclusion

Those of us who are products of Western elementary education have inculcated in us the idea that writing is in essence and principal a means by which we graphically encode verbal speech—we are all hooked and flailing on the fishing line of phoneticism. Increasingly, however, those working with ancient Babylonian documents have come to recognize that, as van de Mieroop has recently remarked, "Cuneiform was not invented to represent spoken language."[107] Recognizing this fact has had, and will continue to have, a dramatic impact on how we understand Mesopotamian literature, to be sure. One quantum facet of this recognition is that the epistemology featured in Babylonian documents is indelibly permeated with its "writtenness," a result of their being the product of a craft process of expert professionals with distinctive skills and toolsets. Babylonian scribes had their own knowledge culture. While the historical depth and graphic sophistication was undoubtedly lesser in West Semitic alphabetic scribalism, we still need to consider the documentary character of the biblical text when attempting to describe the epistemology of Judean scribes. They too had skills and toolsets that distinguished them from their surrounding culture and which certainly had a cognitive impact on their worldview. I maintain that this is most clearly on display in the methods they used to interpret and describe the origin of the names of important people and places, such as that of the man Israel.

Bibliography

Aharoni, Yohanan

1981 *Arad Inscriptions*. Jerusalem: Israel Exploration Society.

107 van de Mieroop 2015: 79.

Andersen, Francis I. and Freedman, David Noel

1980 *Hosea: A New Translation with Introduction and Commentary*. AB 24. Garden City: Doubleday.

Avrahami, Yael

2012 *The Senses of Scripture: Sensory Perception in the Hebrew Bible*. LHB/OTS 545. New York: T&T Clark.

Beaulieu, Paul-Alain

1995 "An Excerpt from a Menology with Reverse Writing." *ASJ* 17: 1–13.

Bell, Daniel

1973 *The Coming of Post-Industrial Society: A Venture in Social Forecasting*. New York: Basic Books.

Ben Dov, Jonathan

2008 "Writing as Oracle and as Law: New Contexts for the Book-Find of King Josiah." *JBL* 127: 223–239.

Boman, Thorlief

1960 *Hebrew Thought Compared with Greek*, trans. Jules Moreau. Philadelphia: Westminster Press.

Bottéro, Jean

1974 "Symptomes, signes, ecritures en Mesopotamia ancienne." Pp. 70–193 in *Divination et rationalite*, ed. J.-P. Vernant. Paris: Éditions du Seuil.

1977 "Les noms de Marduk, l'ecriture et la 'logique' en Mesopotamie ancienne." Pp. 5–28 in *Essays on the Ancient Near East in Memory of Jacob Joel Finkelstein*. Mem. Conn. Acad. Arts Sci. 19, ed. Maria De Jong Ellis. Hamden, CT: Archon Books.

Bourdieu, Pierre

1977 *Outline of a Theory of Practice*, trans. Richard Nice. Cambridge: Cambridge University Press.

Brown, David

1999 *Mesopotamian Planetary Astronomy-Astrology*. CM 18. Groningen: Styx.

Carasik, Michael

2006 *Theologies of the Mind in Biblical Israel*. StBibLit 85. New York: Peter Lang.

Carr, David

2004 *Writing on the Tablet of the Heart: Origins of Scripture and Literature*. Oxford: Oxford University Press.

Cooley, Jeffrey L.

2011 "The Story of Saul's Election (1 Sam 9–10) in Light of Mantic Practice in Ancient Iraq." *JBL* 130: 247–261.

2013 *Poetic Astronomy in the Ancient Near East: The Reflexes of Celestial Science in Ancient Mesopotamian, Ugaritic, and Israelite Narrative*. HACL 5. Winona Lake, IN: Eisenbrauns.

f.c. *Names and Knowledge in Ancient Near Eastern Narrative: A Study in Etymology and Epistemology.* Cambridge: Cambridge University Press.

Crenshaw, James

1998 *Education in Ancient Israel: Across the Deadening Silence.* New York: Doubleday.

Dell, Katherine

2008 "Scribes, Sages and Seers in the First Temple." Pp. 125–144 in *Scribes, Sages, and Seers: The Sage in the Eastern Mediterranean World*, ed. Leo G. Perdue. Göttingen: Vandenhoeck & Ruprecht.

Dobbs-Allsopp, F.W.

2009 "Acrostic I. Ancient Near East and Hebrew Bible/Old Testament." Pp. 282–288 in *Encyclopedia of the Bible and Its Reception.* Vol. 1. *A-Anic*, ed. Dale Allison et al. Berlin: de Gruyter.

2015 *On Biblical Poetry.* Oxford: Oxford University Press.

Dowd, Ryan P.

2009 *The Wisdom of the Torah: Epistemology in Deuteronomy and the Wisdom* Literature. FRLANT 225. Göttingen: Vandenhoeck & Ruprecht.

Durkheim, Émile

1915 *The Elementary Forms of Religious Life*, trans. Joseph Ward Swain. London: George Allen & Unwin.

Fichtner, J.

1956 "Die etymologische Ätiologie in der Namengebung der geschichtlichen Bücher des Alten Testaments." *VT* 6: 372–396.

Fox, Michael

1987 "Qohelet's Epistemology." *HUCA* 58: 137–155.

2015 *Proverbs: An Eclectic Edition with Introduction and Textual Commentary.* Atlanta: Society of Biblical Literature.

Frahm, Eckart

2010 "Reading the Tablet, the Exta, the Body: The Hermeneutics of Cuneiform Signs in Babylonians and Assyrian Text Commentaries and Divinatory Texts." Pp. 93–142 in *Divination and Interpretation of Signs in the Ancient World.* OIS 6, ed. Amar Annus. Chicago: The Oriental Institute.

Frankfort, Henri. and Frankfort, H.A.

1946a "Myth and Reality." Pp. 3–27 in *The Intellectual Adventure of Ancient Man: An Essay on Speculative Thought in the Ancient Near East*, eds. Henri Frankfort et al. Chicago: University of Chicago Press.

1946b "The Emancipation of Thought from Myth." Pp. 363–388 in *The Intellectual Adventure of Ancient Man: An Essay on Speculative Thought in the Ancient Near East*, eds. Henri Frankfort et al. Chicago: University of Chicago Press.

Friedman, Richard Elliot

1987 *Who Wrote the Bible?* San Francisco: Harper & Row.

Gabbay, Uri

2009 "Deciphering Cuneiform Texts through Ancient and Modern Conceptions of Literal Meaning." Pp. 161–169 in *Le sens littéral des Écritures*, ed. Olivier-Thomas Venard. Paris: Cerf.

Galison, Peter Louis

2001 *Logic and Image: A Material Culture of Microphysics.* Chicago: University of Chicago Press.

Gericke, Jaco

2012 *The Hebrew Bible and Philosophy of Religion.* Atlanta: Society of Biblical Literature.

Grabbe, Lester L.

1995 *Priests, Prophets, Diviners, Sages: A Socio-Historical Study of Religious Specialists in Ancient Israel.* Valley Forge: Trinity Press International.

Hawking, Stephen and Mlodinow, Leonard

2011 *The Grand Design.* New York: Random House.

Hazony, Yoram

2012 *The Philosophy of Hebrew Scripture.* Cambridge: Cambridge University Press.

Healy, Mary and Parry, Robin Parry, eds.

2007 *The Bible and Epistemology: Biblical Soundings on the Knowledge of God.* Milton Keynes: Paternoster.

Hilgert, Markus

2009 "Von 'Listenwissenschaft' und 'epistemischen Dingen': Konzeptuelle Annäherungen an altorientalische Wissenspraktiken / Of 'Listenwissenschaft' and 'Epistemic Things': Conceptual Approaches to Ancient Mesopotamian Epistemic Practices." *J GEN PHILOS SCI* 40: 277–309.

Hrůša, Ivan

2010 *Die akkadische Synonymliste* malku = šarru*: Eine Textedition mit Übersetzung und Kommentar.* AOAT 50. Münster: Ugarit-Verlag.

Hyman, Malcom D.

2006 "Of Glyphs and Glottography." *L&C* 26: 231–249.

Irwin, William A.

1946 "The Hebrews." Pp. 223–360 in *The Intellectual Adventure of Ancient Man: An Essay on Speculative Thought in the Ancient Near East,* eds. Henri Frankfort et al. Chicago: University of Chicago Press.

Jacobsen, Thorkild

1946 "Mesopotamia." Pp. 125–219 in *The Intellectual Adventure of Ancient Man: An Essay on Speculative Thought in the Ancient Near East,* eds. Henri Frankfort et al. Chicago: University of Chicago Press.

Jastrow, Morris

1898 *The Religion of Babylonia and Assyria*. Boston: Ginn & Company.

Jeremias, Alfred

1913 *Handbuch der altorientalischen Geisteskultur*. Leipzig: Hinrichs.

Johnson, Dru

2013 *Biblical Knowing: A Scriptural Epistemology of Error*. Eugene: Cascade Books.

Knorr Cetina, Karin

1981 *The Manufacture of Knowledge: An Essay on the Constructivist and Contextual Nature of Science*. Oxford: Pergamon.

1999 *Epistemic Cultures: How the Sciences Make Knowledge*. Cambridge: Harvard University Press.

2001 "Objectual Practice." Pp. 175–188 in *The Practice Turn in Contemporary Theory*, eds. Karin Knorr Cetina et al. Abingdon: Routledge.

Koch-Westenholz, Ulla

1995 *Mesopotamian Astrology: An Introduction to Babylonian and Assyrian Celestial Divination*. CNI Publications 19; Copenhagen: Museum Tusculanum.

Kovacs, Brian

1974 "Is There a Class-Ethic in Proverbs?" Pp. 171–189 in *Essays in Old Testament Ethics*, eds. James L. John T. and Crenshaw Willis. New York: Ktav.

Kugler, Franz

1910 *Im Bannkreiss Babels: Panbabylonistische Konstruktionen und Religionsgeschichtliche Tatsachen*. Münster i.W.: Aschendorff.

Landsberger, Benno and von Soden, Wolfram

1965 *Die Eigenbegrifflichkeit der babylonischen Welt, Leistung und Grenze sumerischer und babylonischer Wissenschaft*. Darmstadt: Wissenschaftliche Buchgesellschaft.

Latour, Bruno

1987 *Science in Action: How to Follow Scientists and Engineers through Society*. Cambridge: Harvard University Press.

Latour, Bruno and Woolgar, Steve

1986 *Laboratory Life: The Construction of Scientific Facts*. Princeton: Princeton University Press.

Lenzi, Alan

2007 *Secrecy and the Gods: Secret Knowledge in Ancient Mesopotamia and Biblical Israel*. SAAS 19. Helsinki: Neo-Assyrian Text Corpus Project.

2014 "Revisiting Biblical Prophecy, Revealed Knowledge Pertaining to Ritual, and Secrecy in Light of Ancient Mesopotamian Prophetic Texts." Pp. 65–86 in *Divination, Politics and Ancient Near Eastern Empires*. ANEM/MACO 7, eds. Alan Jonathan and Lenzi Stökl. Atlanta: Society of Biblical Literature.

Lemaire, André

1981 *Les Écoles et la formation de la Bible dans l'ancien Israël*. OBO 39. Fribourg: Éditions Universitaires.

Lichtheim, Miriam

2006 *Ancient Egyptian Literature, Volume 2: The New Kingdom*. Berkeley: University of California Press.

Liverani, Mario

1973 "Memorandum on the Approach to Historiographic Texts." *Or* 42: 178–194.

Long, Burque.

1968 *The Problem of Etiological Narrative in the Old Testament*. BZAW 108. Berlin: Alfred Töpelmann.

Lynch, Michael

1993 *Scientific Practice and Ordinary Action: Ethnomethodology and Social Studies of Science*. Cambridge: Cambridge University Press.

Malul, Meir

2002 *Knowledge, Control and Sex: Studies in Biblical Thought, Culture and Worldview*. Tel-Aviv—Jaffa: Archaeological Center.

Mannheim, Karl

1936 *Ideology and Utopia: An Introduction to the Sociology of Knowledge*, trans. Louis Edward and Wirth Shils. London: Routledge.

Noth, Martin

1928 *Die israelitischen Personennamen im Rahmen der gemeinsemitischen Namengebung*. Stuttgart: W. Kohlhammer.

Niditch, Susan

1989 *Oral World and Written Word: Ancient Israelite Literature*. Louisville: Westminster John Knox Press.

Oppenheim, A. Leo

1977 *Ancient Mesopotamia: Portrait of a Dead Civilization*. Chicago: University of Chicago Press.

Puech, Emile

1988 "Les Écoles dans l'Israël préexilique: données épigraphiques." Pp. 189–203 in *Congress Volume: Jerusalem 1986*. VTSup 40. Leiden: Brill.

Radner, Karen

2005 *Die Macht des Namens: altorientalische Strategien zur Selbsterhaltung*. SANTAG 8. Wiesbaden: Harrassowitz.

Rechenmacher, Hans

1997 *Personennamen als theologische Aussagen: Die syntaktischen und semantischen Strukturen der satzhaften theophoren Personennamen in der hebräischen Bibel*. ATSAT 50. St. Ottilien: EOS Verlag.

Reiner, Erica

2002 *An Adventure of Great Dimension: The Launching of the Chicago Assyrian Dictionary*. TAPS 92/3. Philadelphia: American Philosophical Society.

Rendsburg, Gary

2001 "Hebrew Philological Notes (II)." *HS* 42: 187–195.

Rheinberger, Hans-Jörg

1997 *Towards a History of Epistemic Things: Synthesizing Proteins in the Test Tube*. Stanford: Stanford University Press.

Robson, Eleanor

2008 *Mathematics in Ancient Iraq: A Social History*. Cambridge: Cambridge University Press.

Rochberg, Francesca

1982 "Fate and Divination in Mesopotamia." Pp. 363–381 in *28. Rencontre Assyriologique International, Wien 1981*. AfOB 19, eds. H. H. and Hunger Hirsch. Horn: F. Berger.

1999 "Empiricism in Babylonian Omen Texts and the Classification of Mesopotamian Divination as Science." *JAOS* 119: 559–569.

2004 *The Heavenly Writing: Divination, Horoscopy, and Astronomy in Mesopotamian Culture*. Cambridge: Cambridge University Press.

2010 "Fate and Divination in Mesopotamia." Pp. 19–30 in *In the Path of the Moon: Babylonian Celestial Divination and Its Legacy*. SAMD 6, ed. Francesca Rochberg. Leiden: Brill.

2014 "The History of Science and Ancient Mesopotamia." *JANEH* 1: 37–60.

2016 "Where were the Laws of Nature Before there Was Nature?" Pp. 21–39 in *Laws of Heaven—Laws of Nature: Legal Interpretations of Cosmic Phenomena in the Ancient World / Himmelsgesetze—Naturgesetze: Rechtsförmige Interpretationen kosmischer Phänomene in der antiken Welt*. OBO 276, eds. Konrad Christoph and Schmid Uehlinger. Göttingen: Vandenhoeck & Ruprecht.

Rollston, Christopher A.

2010 *Writing and Literacy in the World of Ancient Israel*. Atlanta: Society of Biblical Literature.

Sanders, Seth L.

2009 *The Invention of Hebrew*. Urbana: University of Illinois Press.

Sarna, Nahum

2001 *JPS Torah Commentary: Genesis*. Philadelphia: Jewish Publication Society.

Schellenberg, Annette

2002 *Erkenntnis als Problem: Qohelet und die alttestamentliche Diskussion um das menschliche Erkennen*. OBO 188. Göttingen: Vandenhoeck & Ruprecht.

Schniedewind, William M.

2004 *How the Bible Became a Book: The Textualization of Ancient Israel*. Cambridge: Cambridge University Press.

2013 *A Social History of Hebrew: Its Origins through the Rabbinic Period*. New Haven: Yale University Press.

Selz, Gebhard J.

2011 "Remarks on the Empirical Foundation and Scholastic Traditions of Early Mesopotamian Acquisition of Knowledge." Pp. 49–70 in *The Empirical Dimension of Ancient Near Eastern Studies/Die empirische Dimension altorientalischer Forschungen*, ed. Gebhard J. Selz. WOO 6. Wien: Lit.

Shupak, Nili

2003 "Learning Methods in Ancient Israel." *VT* 53: 416–426.

Stamm, J.J.

1950 "Der Name Isaak." Pp. 33–28 in *Das Wort sie sollen lassen stahn: Festschrift für D. Albert Schädelin*. Bern: H. Lang & Cie.

Stavrakopoulou, Francesca

2013 "Materialist Reading: Materialism, Materiality, and Biblical Cults of Writing." Pp. 223–242 in *Biblical Interpretation and Method: Essays in Honour of John Barton*, eds. Katharine J. Paul M. and Dell Joyce. Oxford: Oxford University Press.

Starr, Ivan

1983 *The Rituals of the Diviner*. BiMes 12. Malibu: Undena Publications.

Strassmeier, J.N. and Epping, J.

1889 *Astronomisches aus Babylon, oder, das Wissen der Chaldäer über den gestirnten Himmel*. StML 11/41–44. Freiburg i.B.: Herder.

Tanselle, G. Thomas

1989 *A Rationale of Textual Criticism*. Philadelphia: University of Pennsylvania Press.

Trolle Larsen, Mogens

1987 "The Mesopotamian Lukewarm Mind: Reflections on Science, Divination and Literacy." Pp. 203–205 in *Language, Literature, and History: Philological and Historical Studies Presented to Erica Reiner*. AOS 67, ed. Francesca Rochberg-Halton. New Haven: American Oriental Society.

van de Mieroop, Marc

2015 *Philosophy before the Greeks: The Pursuit of Truth in Ancient Babylonia*. Princeton: Princeton University Press.

van der Toorn, Karel

2007 *Scribal Culture and the Making of the Hebrew Bible*. Cambridge: Harvard University Press.

Vayntrub, Jacqueline

2016 "The Book of Proverbs and the Idea of Ancient Israelite Education." *ZAW* 128: 96–114.

Veldhuis, Niek

1997 "Elementary Education at Nippur: The Lists of Trees and Wooden Objects." Ph.D. diss., Groningen.

2011 "Levels of Literacy." Pp. 68–89 in *The Oxford Handbook of Cuneiform Culture*, eds. Karen Eleanor and Radner Robson. Oxford: Oxford University Press.

2012 "Domesticizing Babylonian Scribal Culture in Assyria: Transformation by Preservation." Pp. 11–24 in *Theory and Practice of Knowledge Transfer: Studies in School Education in the Ancient Near East and Beyond. Papers Read at a Symposium in Leiden, 17–19 December 2008*, eds. W.S. W.H. and van Egmond van Soldt. Leiden: Nederlands Instituut voor het Nabije Oosten.

2014 "Intellectual History and Assyriology." *JANEH* 1: 21–36.

Vermes, Geza

1975 "The Impact of the Dead Sea Scrolls on Jewish Studies During the Last Twenty-Five Years." *JJS* 26: 1–14.

von Soden, Wolfram

1936 "Leistung und Grenze sumerischer und babylonischer Wissenschaft." *WG* 2: 411–464, 509–557.

1994 *The Ancient Orient: An Introduction to the Study of the Ancient Near East.* Grand Rapids: Eerdmans.

Wenham, Gordon

2000 *Genesis 16–50*. WBC 2. Nashville: Thomas Nelson.

Westermann, Claus

1995 *Genesis 12–36: A Continental Commentary*, trans. John J. Scullion, S.J. Minneapolis: Fortress.

Wiggerman, Frans

1993 "Mythological Foundations of Nature." Pp. 279–306 in *Natural Phenomena: Their Meaning, Depiction and Description in the Ancient Near East*, ed. D.J.W. Meijer. Amsterdam: Der Koninklijke Nederlandse Akademie Van Wetenschappen Verhandelingen.

Wilson, John

1969 "The Instruction of Amen-em-opet." Pp. 421–425 in *Ancient Near Eastern Texts Relating to the Old Testament*, ed. James B. Pritchard. Princeton: Princeton University Press.

Winckler, Hugo

1904 *Die Weltanschauung des alten Orients.* JEOL 1/1. Leipzig: Pfeiffer.

Winitzer, Abraham

2011 "Writing and Mesopotamian Divination: The Case of Alternative Interpretation." *JCS* 63: 77–94.

Wright, G. Ernest

1952 *God Who Acts: Biblical Theology as Recital.* London: SCM Press.

Yardeni, Ada

2002 *The Book of Hebrew Script: History, Paleography, Script Styles, Calligraphy & Design.* London: British Library.

Young, Ian M.

1998 "Israelite Literacy: Interpreting the Evidence." *VT* 48: 239–253, 408–422.

Zakovitch, Yair

1980 "Explicit and Implicit Name-Derivations." *HAR* 4: 167–181.

CHAPTER 12

Dating *EAE*. When was the Astrological Series *Enūma Anu Ellil* Created?

Nils P. Heeßel

Since Ernst Weidner's groundbreaking reconstruction of the astrological omen series *Enūma Anu Ellil*, its creation has been dated to the second half of the second or the beginning of the first millennium BCE. As Weidner (1941–44: 176) put it:

> Wann man das ältere Material durch umfangreiche Ergänzungen aus jüngerer Zeit aufgefüllt und das Ganze in die Form der grossen astrologischen Serie Enûma Anu Enlil gebracht hat, ist eine schwer zu beantwortende Frage. Ein sehr hohes Alter, wie man früher wohl glaubte, kommt der Serie als solcher gewiss nicht zu. Solange unsere Kenntnisse Stückwerk sind, bleibt eine sichere Entscheidung unmöglich, jedenfalls spricht aber nach unserem heutigen Wissen die Wahrscheinlichkeit dafür, dass die Serie Enûma Anu Enlil ein Werk babylonischer Kompilatoren aus der zweiten Hälfte des 2. oder dem Anfang des 1. Jahrtausends ist.

While this was no doubt a big step forward considering the fact that before Weidner the series *Enūma Anu Ellil* (henceforth: *EAE*) was even dated to the beginning of the second millennium BCE, the "question, which is difficult to answer"—when exactly within this long period from the second half of the second to the beginning of the first millennium BCE the astrological omens were compiled into the series named *EAE*—has been discussed several times since.

Over time, a *communis opinio* evolved among scholars that regarded the Kassite period in the second half of the second millennium as the date of compilation of *EAE*. However, this paper tries to show through careful re-examination of the available evidence that the general conviction of a second millennium date for *EAE* rests on highly shaky evidence and should at best be regarded as an educated guess. The argument is offered here as a sign of admiration to Francesca Rochberg, who spent her lifetime exploring the astrological tradition of Mesopotamia and explaining the astrological series *Enūma Anu Ellil*.

 | DOI 10.1163/9789004363380_013

"Standardized" Texts vs. Texts of the Series *EAE*

Before reviewing the alleged evidence for a second millennium origin of *EAE*, we must first clarify what we are actually talking about when speaking of "the time of compilation of the astrological series *EAE*." Do we mean the date, when the texts reached a kind of fixed, "standardized" form, as we find it in the texts of the first millennium libraries? Or do we mean the format of a series that is characterized by catch-lines to the incipit of the next tablet and a tablet numbering and series title in the colophon?

Recent research in the textual development of omen texts[1] has shown that these texts achieved a fixed form quite early in Kassite times. While the Old Babylonian omen texts display a great variety, the Kassite texts appear in a fixed, standardized form; even direct duplicates are not infrequent. These texts remain relatively stable and are handed down with few, yet distinctive changes until the end of cuneiform culture. Such changes in these omen texts between the Kassite period and the first millennium BCE copies include omitting and adding omens, adding a second apodosis, replacing protases or apodoses and replacing or deleting unintelligible or obsolete words. However, while such changes affect the text, its general outline remains basically the same, making it clearly recognizable even over many centuries.

Such standardized texts, whether from Kassite Babylonia or from Middle Assyrian Assur, could be included into different text series. The forming of textual series already began in the Old Babylonian period but gained real momentum in the second half of the second millennium BCE. For the extispicy omens it can be shown that there existed at least four different extispicy series: in addition to an old Babylonian series, of which little is known, a middle Babylonian series, a middle Assyrian series and the first millennium series called *bārûtu* in late times can be adduced. The latter three all used the same kind of standardized texts to form a series; the standardized texts are therefore used like text modules in different series.

In view of all this, it is clear that a simple resemblance between texts, e.g. between a Middle Babylonian text and a tablet of the series *EAE* is not indicative of the existence of the series *EAE* in middle Babylonian times. In order to determine the time of compilation of the astrological series *EAE* we are looking for the first texts that can be securely connected with the series named *Enūma Anu Ellil*. In this respect, the only clear evidence is a text bearing a tablet numbering and naming the series *EAE* in its colophon. However, a catch-line and an identical text without the kind of changes between non-serialized earlier texts

1 Jeyes 2000; Heeßel 2011 and 2017; Rutz 2013: 321–323; Koch 2015: 59–66.

and serialized first millennium texts as observed in omen texts may also suffice for an unambiguous identification of a tablet of the series *EAE*.

The *communis opinio*: Discussions of the Date of Redaction of *EAE* in Scholarship since Weidner's Time

It was the honoree herself who presented the first thorough discussion of the textual history of *EAE* since Weidner's time. In the second chapter on the textual history of the lunar eclipse omens within her edition of *EAE* tablets 15–22 (Rochberg-Halton 1988:18–25), she made many valuable observations concerning the date of compilation and transmission of *EAE* material. Yet she refrained from giving an exact date for the creation of the series *EAE*, rather stressing that the lunar eclipse omens form "the only group of celestial omens that have what might be described as a continuous manuscript tradition, i.e., from the Old Babylonian through the Middle Babylonian and Middle Assyrian periods to the codified version known in the Nineveh library and other Neo-Assyrian and Neo-Babylonian scribal centers" (Rochberg-Halton 1988: 7). In other previous publications, however, Rochberg set the date of the compilation of the *EAE* series in the time of the Kassite dynasty.[2] While discussing Old Babylonian celestial divination texts, Rochberg (1999: 417) wrote:

> ... Weidner's statement that *Enūma Anu Enlil* was likely to be a composition of the end of the second or beginning of the first millennium can be revised. Certainly however, if Weidner meant the composition of the standard 70-tablet series, this recension was indeed a product of the Middle Assyrian/Middle Babylonian period, as the non-standard character of the Old-Babylonian texts confirms.

Ulla Koch-Westenholz (1995: 42) formulated her opinion carefully when she wrote that "some form of the series *EAE* existed under that name already in the 13th century, as is shown not only by the famous fragment from Hattusas but also by VAT 9740+11670, a Middle Assyrian tablet which has traces of *EAE*'s title in the colophon."[3] Hermann Hunger and David Pingree (1999:12) noted,

2 Rochberg-Halton 1987: 327: "The process by which these texts reached their final form is nowhere explained or even mentioned in our sources, but it is assumed to be the work of Kassite period transcribers and editors." See also Rochberg-Halton 1984: 127.

3 This proposed identification of the famous scholar Esagil-kīn-apli as the "redactor of *EAE*" is based on a misinterpretation of the "Exorcist's Manual" (main text: KAR 44). The fact that

"the elaboration of the Enūma Anu Enlil series had already begun in the 2nd half of the 2nd millennium, as is the case with other Babylonian 'canonical' texts." And David Brown (2000: 254) summed up his extensive review of the available evidence in 2000: "To the best of my knowledge, then, it appears that EAE drew heavily on OB material, some of which was already known by the title (proto-*EAE*), and was put into its standard form during the Kassite period." This consensus among specialists in astrological and astronomical texts about a date of redaction of EAE in the second half of the second millennium is echoed in general overviews as well as specialized studies, too many to cite them all here in detail.[4]

In recent times, Andrew George discussed the routes of transmission of the series on the occasion of publishing a new manuscript for EAE tablet 20.[5] He showed for this tablet of EAE that the assumed two reconstructed recensions are in fact not so strictly separated as the previously known material suggested, with many texts sharing traits of both recension A and B at the same time. George attributed this to the long-known fact that there was no single form of the series EAE, which was handed down unchanged over time, as at least two different versions of EAE tablet 56 can be detected in Assurbanipal's library in Nineveh and in late Babylonian Uruk (Weidner, 1941–44: 181). While this is a phenomenon deserving more study in the future as many parts of EAE are now becoming available in modern editions,[6] it does not imply per se that earlier related textual material must belong to the series named *Enūma Anu Ellil*.

The Assumed Evidence for a Second Millennium Date for the Redaction of EAE

Different evidence has been put forth to substantiate the claim of a second millennium date for EAE:

- two Old-Babylonian catalogues listing the title of EAE
- a Hittite omen fragment from Boğazköy, bearing the title of the series

EAE is listed in the second list by Esagil-kīn-apli as belonging to his curriculum of *āšipūtu* texts does not imply his involvement in its redaction. Cf. also al-Rawi/George 2006: 54.

4 Pongratz-Leisten 1999: 24; Maul 2003: 52; Jakob 2003: 527; Brown 2007: 461; Hunger 2012: 96; Koch 2015: 162.

5 Al-Rawi/George 2006: 50.

6 Gehlken 2012; van Soldt 1995; Verderame 2003.

- a text from Emar giving the title of *EAE*
- two, sometimes three middle Assyrian texts (VAT 9740+, Ass. 10145, BM 121034), one of which bears a colophon with tablet number and the fragmentary title of the series *EAE*.

This evidence can be summarized into two basic arguments. On the one hand, a second millennium date for the title of the series *EAE*, is attested in Old-Babylonian catalogues, in a Hittite text and in a text from Emar, and on the other hand three middle-Assyrian texts actually belonging to the series *EAE*. We will review the alleged evidence one by one.

David Brown drew attention to two citations of the title of *EAE* in Old Babylonian literary catalogues.[7] AO 5393 (de Genouillac 1930: pl. 67) l. 55 (iv 4) contains u_4-an-né, and the literary catalogue U. 17900H (Kramer 1961: 172) ll. 49–50 notes the phrase with its Akkadian equivalent: u_4-an-né, *i-nu* AN *ù* ᵈ*en-líl*.[8] Brown, following Kramer 1961: 176, interprets this as the title of a "proto-*EAE*" and argues, "these titles show beyond doubt that a form of *EAE* existed in the OB period, though this fact is not widely stated on the secondary literature." (Brown 2000: 248f.). However, Leo Oppenheim (1977: 374, n.66) already doubted whether these titles in literary catalogues refer to the astrological series. In fact, his doubts are well grounded, as we shall see presently.

The title of the series *enūma Anu Ellil* "When Anu and Ellil" consists of the first three words of the beginning of a quite famous mythological account of the creation of the "signs of heaven and earth" by the three great gods Anu, Ellil and Ea. This short account was put into Sumerian and Akkadian at the beginning of the first tablet of *EAE*. The fact that this mythological narrative was connected with astrological omens is already attested in two astrological texts, one from Emar and one from Ḫattuša. The Emar text is cited by Brown (2000: 254) as providing evidence for the use of the title of the series *EAE* in the second millennium BCE in Emar.[9] However, in the Emar text Msk 74235a+ (Arnaud 1987: no. 652.) the mythological account is not at the beginning of the text but at its end. This cannot be taken as the title of *EAE*, even if the phrase x *mim-ma an-nu-ú* "all these (omens belong to?)" appears before the mythological account. The use of *EAE* as a title occurs in a Hittite version of this account, in the Boğazköy text KUB 34 no. 12 (Riemschneider 2004: 119f.) line 1, which

7 Brown 2000: 248, 254 and also Brown 2004: 120.

8 For this text see Gadd/Kramer 1963: no. 123, pl. 136.

9 Brown 2000:254. Brown refers to this account in the Emar text as being "in Sumerian"; however, the text is clearly Akkadian with some logographic writings.

reads: [*ma-a*]-*an* ^{d}*a-nu-uš* ^{d}[*en-líl-aš* …]. In the following lines lunar omens are listed, but the text is too damaged for translation.[10]

The first line of the mythological account at the beginning of *EAE* gave the series its name, but this does not necessarily mean that earlier attestations of this line or the whole narrative must refer to the series *EAE*. On the contrary, it is much more likely that this mythological account existed independently before the creation of *EAE* and that the Old Babylonian attestations in the two literary catalogues refer to this literary narrative rather than to the series *EAE*. The first clear connections between this mythological narrative and astrological omens, then, is the Hittite version KUB 34 no. 12, which is also corroborated by the Emar text Msk 74235a+. While the Hittite fragment offers little more than the title and a few signs of omens just enough to surmise that these omens concern the moon, the Emar text can be identified as an earlier version, a "forerunner" of lunar omens, which later formed tablets 21 and 22 of the series *EAE*.[11] But the text lacks any evidence that it is a tablet of the series *EAE*, as no catch-line, tablet number or series title is provided and the text differs substantially from *EAE* 21–22.

The occurrence of the "title" of *EAE* in second millennium texts, therefore, cannot be taken as evidence for an early date for the series *EAE*. But the second part of the argument involving several middle Assyrian texts may still prove the point. The argument rests mainly on the Assur text VAT 9740 + 11670, which Weidner first drew attention to as being a manuscript of the 20th tablet of *EAE*.[12] Weidner refrained from assigning a date to the tablet, but Francesca Rochberg-Halton dated the tablet in her edition to the middle-Assyrian period.[13] As VAT 9740 not only offers the text of *EAE* 20, but has also the catch-line to *EAE* 21 and a partly preserved colophon that can be restored with confidence as mentioning the series *EAE*, it is quite certain that VAT 9740+ is indeed a manuscript of tablet 20 of the series *EAE*.

15' [DIŠ *ina* ^{iti}BÁRA U$_4$ 14-KA]M AN.MI GAR-*ma ina* IM.U$_{18}$.LU SAR-⸢*ma*⸣ *ina* [I]M.S[I.SÁ]

16' *iz-ku ina* EN.NUN.AN.⸢ÚSAN⸣ SAR-*ma ina* EN.NU[N.MURUB$_4$.BA *iz-ku*]

17' [DIŠ DUB 20$^{?}$-KÁM UD AN ^{d} EN].LÍL *šà-ṭir ba-a-ri*

10 Riemschneider, DBH 12, 119f. provides an edition of this text.

11 Note that A 2407, for which see at present the copy of F. Köcher in Weidner 1954–56, pl. II and the edition Rochberg-Halton 1988: 217f. ("source Y"), is also a forerunner to *EAE* 20.

12 Weidner 1954–56: 80f.

13 Rochberg-Halton 1988: 19 and 174.

Subsequent scholarship followed Rochberg-Halton's Middle Assyrian date for this tablet, and the tablet was repeatedly cited as direct evidence for a Middle Assyrian origin of the series *EAE*.[14] However, al-Rawi and George 2006: 36, on the occasion of collating VAT 9740+, corrected that date and stated that this "is a Neo-Assyrian tablet, not a Middle Assyrian one." As VAT 9740+ displays a clear late Neo-Assyrian ductus without any Middle Assyrian sign forms, al-Rawi and George's dating of the tablet is correct.[15]

With this new date for VAT 9740+, the strongest argument for a second millennium origin of the series *EAE* is refuted. The other two middle-Assyrian texts cited as manuscripts of *EAE* are Assur 10145 and BM 121034.[16] Both texts, however, lack any evidence of belonging to *EAE*; the incipits, catch-lines and colophons of both Assur 10145[17] and BM 121034[18] are missing. While Assur 10145 is to a large extent parallel to *EAE* 20 and should be considered as a "forerunner" to that text, the text of BM 121034 goes clearly back to Old Babylonian material, which is similar but only partly parallel to *EAE* 15. These three Middle Assyrian texts—VAT 9740+, Assur 10145, and BM 121034—, therefore, cannot be used as evidence for a second millennium date of *EAE*.

Conclusion

The alleged evidence for a second millennium date for the series *EAE* does not stand up to scrutiny. The citation of the "title of *EAE*" in second millennium texts cannot be regarded as valid proof, and Middle Assyrian astrological omen texts are not clearly marked as belonging to *EAE*; if the series is mentioned in texts from Assur, it is in first millennium texts only.

From this review of the evidence, therefore, it is obvious that the first clear evidence for the series *EAE* comes from the first millennium BCE. But which are the first attestations of *EAE* in the first millennium? As there are neither dated tablets of this series from Assur nor texts that might allow a more precise dating within the Neo-Assyrian period, we must look to Kalḫu/Nimrūd for

14 Brown 2000; Jakob 2003: 527 fn. 107.

15 A cuneiform copy and edition of the tablet VAT 9740 + 9743 + 11670 + 11787 + 11837 (+) 9881 (+) 9945 will be published in the author's forthcoming *Divinatorische Texte III: Astrologische Omina* as part of the series *Keilschrifttexte aus Assur literarischen Inhalts* (*Wissenschaftliche Veröffentlichungen der Deutschen Orient-Gesellschaft*).

16 Brown 2000; Jakob 2003: 527 fn. 107.

17 A copy of Ass. 10145 (= A 2407) by F. Köcher was published in Weidner, AfO 17 (1954–56) pl. II, an edition of the text by F. Rochberg-Halton 1988: 217f. ('source Z').

18 Copy and edition by F. Rochberg-Halton 1988: 273–279 ('Appendix 1').

the first instances of *EAE*. Most of the quite numerous astrological omen texts from the library of the Nabû temple in Kalḫu are quite fragmentarily preserved; two texts, however, might shed light on our question: CTN IV/8 is probably a manuscript of *EAE* 31[19] and was written, according to the colophon, in 787 BCE by the personal scholar (*ummânu*) of king Adad-nērārī III. Unfortunately, due to a break it is unclear whether the colophon is complete and whether or not the series *EAE* was mentioned. Another fragment, CTN IV/26, contains in the colophon] $^{d+}$EN.LÍL [, but it is uncertain whether this is part of the title of *EAE* or of the catch-line. Neither CTN IV/8 nor CTN IV/26 is clear proof for an early 8th century origin of the series *EAE*. The currently first undisputed evidence for the existence of the series *EAE* is, therefore, to be found within the texts of the famous scholar Nabû-zuqup-kēnu, who copied the first preserved manuscripts of *Enūma Anu Ellil* in the year 716 BCE.[20]

Therefore, the question raised by Weidner in 1944 about the date of compilation of *EAE* has not been satisfactorily answered yet. The *communis opinio* about a redaction in Kassite times does not stand up to scrutiny; as the review of the evidence shows, there is no clear proof for the existence of *EAE* prior to 716 BCE. Considering the fact that there are no manuscripts nor other evidence for *EAE* attested in Ḫattuša and Emar or in middle Assyrian Assur, the date of creation of *EAE* can be set anytime between 1200–716 BCE.

Bibliography

Arnaud, Daniel

1987 *Recherches au pays d'aštata. Emar VI/4. Textes de la bibliotheque : transcriptions et traductions*, Paris: Ed. Recherche sur les Civilisations.

Brown, David

2000 *Mesopotamian Planetary Astronomy-Astrology*. CM18. Groningen: Styx.

19 And certainly not a manuscript of *EAE* 20 as this author claimed in Heeßel 2010: 167, fn. 20.

20 Nabû-zuqup-kēnu copied the *EAE* tablets K. 2686+, K. 3129, K. 5283 and 91-5-9, 97 in 716 BCE, see Lieberman 1990: 209 and Millard 1994: 123.

Should the tablet K. 5281, a manuscript of *EAE* 6, indeed be proven to be copied in 718 BCE, as claimed by Lieberman 1990: 204, n. 223 it would be even two years older. However, as the name of the eponym is almost completely broken other restaurations and other dates seem possible than Lieberman's reading *Zēru-ibni*. Millard 1994: 102 thus dates this tablet to the eponym *Mutakkil-Aššur* who was eponym in 706 BCE. Collation shows that the sign in question is MU rather than NUMUN, which speaks in favour of Millard's reading.

2004 Review article of Beate Pongratz-Leisten, 1999. *Herrschaftswissen in Mesopotamien. Formen der Kommunikation zwischen Gott und König im 2. Und 1. Jahrtausend v. Chr.* [= SAAS 10.] Helsinki: State Archives of Assyria Project. *ZA* 94: 112–121.

2007 "Mesopotamian Astral Science." Pp. 460–472 in *The Babylonian World*, ed. Gwendolyn Leick. New York: Routledge.

Gadd, Cyril J. and Kramer, Samuel Noah

1963 *Literary and Religious Texts.* First Part, UET 6/I. London: British Musuem Press.

Gehlken, Erlend

2012 *Weather Omens of* Enūma Anu Enlil. *Thunderstorms, Wind and Rain* (*Tablet 44–49*). CM 43, Leiden: Brill.

de Genouillac, Henri

1930 *Textes religieux sumériens du Louvre* I. TCL 15. Paris: Geuthner.

Heeßel, Nils P.

2010 "Neues von Esagil-kīn-apli. Die ältere Version der physiognomischen Omenserie *alamdimmû*" Pp. 139–187 in *Assur-Forschungen*. ed. S.M. Maul and N.P. Heeßel. Wiesbaden: Harrassowitz.

2011 "'Sieben Tafeln aus sieben Städten'. Überlegungen zum Prozess der Serialisierung von Texten in Babylonien in der zweiten Hälfte des zweiten Jahrtausends v. Chr." Pp. 171–195 in *Babylon. Wissenskultur in Orient und Okzident*, ed. Eva Cancik-Kirschbaum et al. Topoi. BSAW 1. Berlin: de Gruyter.

2017 "Zur Standardisierung und Serialisierung von Texten während der Kassitenzeit am Beispiel der Opferschau-Omina." Pp. 219–228 in *Karduniaš. Babylonia under the Kassites*, ed. Alexa Bartelmus and Katja Sternitzke. Berlin: de Gruyter.

Hunger, Hermann

2012 "Die Wissenschaft der babylonischen Astronomen." Pp. 95–103 in *Wissenskultur im Alten Orient. Weltanschauung, Wissenschaften, Techniken, Technologien*, Münster 2002, ed. Hans Neumann. Colloquien der Deutschen Orient-Gesellschaft 4. Wiesbaden: Harrassowitz.

Hunger, Hermann and Pingree, David

1999 *Astral Sciences in Mesopotamia.* HdO 1. Abteilung. Der Nahe und Mittlere Osten. 44. Band. Leiden: Brill.

Jakob, Stefan

2003 *Mittelassyrische Verwaltung und Sozialstruktur*. CM 29. Leiden: Brill.

Jeyes, Ulla

2000 "A Compendium of Gall-Bladder Omens Extant in Middle Babylonian, Nineveh and Seleucid Versions." Pp. 345–373 in *Wisdom, Gods and Literature.* Fs. W.G. Lambert, ed. Andrew R. George and Irving L. Finkel. Winona Lake, IN: Eisenbrauns.

Koch-Westenholz, Ulla

1995 *Mesopotamian Astrology. An Introduction to Babylonian and Assyrian Celestial Divination*. CNI Publications 19. Copenhagen: Museum Tusculanum.

Koch, Ulla

2015 *Mesopotamian Divination Texts: Conversing with the Gods. Sources from the First Millennium BCE*. GMTR 7. Münster: Ugarit-Verlag.

Kramer, Samuel Noah

1961 "New Literary Catalogue from Ur." *RA* 55: 169–176.

Lieberman, Stephen J.

1990. "Canonical and Official Cuneiform Texts: Toward an Understanding of Assurbanipal's Personal Tablet Collection." Pp. 305–336 in *Lingering over Words: Studies in Ancient Near Eastern Literature in Honor of William L. Moran*, ed. Tzvi Abusch et al. Atlanta: Scholars Press.

Maul, Stefan M.

2003 "Omina und Orakel. A." Pp. 45–88 in *Reallexikon der Assyriologie und Vorderasiatschen Archäologie*, Band. 10. 1./2. Lieferung, ed. Dietz Otto Edzard. Berlin: de Gruyter.

Millard, Alan

1994. *The Eponyms of the Assyrian Empire 910–612 BC*. SAAS 2. Helsinki: Helsinki University Press.

Oppenheim, A. Leo

1977 *Ancient Mesopotamia. Portrait of a Dead Civilization*. Revised Edition. Chicago: University of Chicago Press.

Pongratz-Leisten, Beate

1999 *Herrschaftswissen in Mesopotamien. Formen der Kommunikation zwischen Gott und König im 2. Und 1. Jahrtausend v. Chr.* SAAS 10. Helsinki: State Archives of Assyria Project.

Al-Rawi, Farouk N. H and George, Andrew R.

2006 "Tablets from the Sippar Library XIII. Enūma Anu Ellil XX." *Iraq* 68: 23–57.

Riemschneider, Kaspar K.

2004 *Die akkadischen und hethitischen Omentexte aus Boğazköy*. Dresdner Beiträge zur Hethitologie 12. Dresden: TU Dresden.

Rochberg-Halton, Francesca

1984 "Canonicity in Cuneiform Texts." *JCS* 36: 127–144.

1987 "The Assumed 29th *aḫû* Tablet of *Enūma Anu Enlil*." Pp. 327–350 in *Language, Literature and History: Philological and Historical Studies Presented to Erica Reiner*, ed. Francesca Rochberg-Halton. AOS 67. New Haven: The American Oriental Society.

1988 *Aspects of Babylonian Celestial Divination. The Lunar Eclipse Tablets of Enūma Anu Enlil*. AfO Beiheft 22. Horn: Ferdinand Berger & Söhne.

1999 "Continuity and Change in Omen Literature." Pp. 415–425 in *Munuscula Mesopotamica*, ed. B. Böck, E. Cancik-Kirschbaum, and Th. Richter. AOAT 267. Münster: Ugarit-Verlag.

Rutz, Matthew

2013 *Bodies of Knowledge in Ancient Mesopotamia. The Diviners of Late Bronze Age Emar and Their Tablet Collection*. AMD 9. Leiden: Brill.

van Soldt, Wilfred H.

1995 *Solar Omens of Enuma Anu Enlil: Tablets 23 (24)–29 (30)*. PIHANS 73, Leiden: Nederlands Instituut voor het Nabije Oosten.

Verderame, Lorenzo

2003 *Le Tavole I–VI della serie astrologica Enūma Anu Enlil*. NISABA 2, Messina: Di.Sc.A.M.

Weidner, Ernst

1941–44 "Die astrologische Serie Enûma Anu Enlil." *AfO* 14: 172–195, 308–318.

1954–56 "Die astrologische Serie Enûma Anu Enlil (Fortsetzung)." *AfO* 17: 71–89.

CHAPTER 13

Bricoleurs in Babylonia: The Scribes of *Enūma Anu Enlil*

Eduardo A. Escobar and Laurie E. Pearce

ina amāt Anu u Antu lišlim

Introduction | Social Constructivism and the Scribes of *Enūma Anu Enlil*

Our paper considers a socio-intellectual biography of a *ṭupšar Enūma Anu Enlil* named Šamaš-ēṭir—including members of his clan and the peers with whom he associated—during the height of astronomical knowledge production in Uruk, ca. 60–150 SE. Using data gathered from scholarly colophons and the corpus of legal texts[1] and with the support of network visualizations, we argue that the title *ṭupšar Enūma Anu Enlil* was of both social and epistemological significance. We propose, moreover, that to call oneself a *ṭupšar Enūma Anu Enlil* (henceforth, *ṭ EAE*)—which Erica Reiner translated as "expert in celestial matters"[2]—carried with it historical awareness deeply rooted in the culture of astronomical knowledge production in Hellenistic Babylonia. Based on our study of Šamaš-ēṭir's biography and social networks, we posit the existence of historically and socially engaged astronomers, *bricoleurs* who mediated the realms of intellectual and legal activity, and whose connections to the men of economic and social privilege served to promote their own standing in the world of Babylonian science. We find these tactics of legitimation well-represented within the intellectual biographies of two scholars from the Ekur-zākir clan: Šamaš-ēṭir and Nidintu-Anu.

The late appearance and use of the title *ṭ EAE* has troubled scholars, for it seems—to put it simply—out of place, at odds with a period of Babylonian scholarship that has been characterized by advancements in mathematical

1 The legal texts from Hellenistic Uruk are available in the digital project Hellenistic Babylonia: Text, Images and Names (HBTIN: http://oracc.org/hbtin), directed by Laurie Pearce.

2 Reiner 1995: 63.

 | DOI 10.1163/9789004363380_014

astronomy; indeed, the title has been deemed anachronistic. In her 2000 "Scribes and Scholars: The *ṭupšar Enūma Anu Enlil*," Francesca Rochberg herself stated, "[i]t is difficult to imagine what relevance the content of EAE omen apodoses could have had for Babylonians in Hellenistic Babylonian society, particularly inasmuch as no evidence that the omens were consulted survives from this period."[3] Similarly, its appearance is infrequent, as scribes did not employ the title with any predictable regularity, "[w]ith the appearance of mathematical astronomical texts in the Seleucid period, the use of the term *ṭ* EAE occurs primarily in the colophons of ephemerides, but was not attached to every scribe who wrote or possessed astronomical texts."[4] Moreover, it appeared in combination with scribal titles in the colophons to texts that had no connection to the world of astral science.

In revisiting this mysterious title, we have expanded the traditional toolsets and approaches utilized in cuneiform studies, adapting social constructivist approaches from the History of Science and employing visualizations at home in Social Network Analysis (SNA) with the aim of shedding light on two questions: first, what makes a *ṭupšar Enūma Anu Enlil*? and second, what strategies did scribes so titled employ in the legitimation of their social and intellectual output?

Rochberg explored the historicity of the title *ṭ* EAE and its relation to *ṭupšarrūtu* as a category of knowledge. She narrowed in on two key aspects of late Babylonian astronomy that required further investigation: its socio-historical setting and its epistemological goals.[5] Rochberg's interest in the social dimension of astronomical knowledge was parallel to methodological discussions regarding the social construction of scientific knowledge that most immediately impacted studies of the Scientific Revolution.[6] The claim that scientific knowledge was socially constructed brought new life to venerable figures in the history of science, including, notably, Galileo. For Mario Biagioli, Galileo was, in addition to an astronomer, an *emblem maker* and *bricoleur*:

> To say that Galileo was simply lucky with his patronage strategies—as to say that he was just an exceptional scientist—is to ignore the more general socio-historical processes that made possible his unusual career and framed his strategies for the legitimation of Copernicanism and mathematical physics. Rather, I would say that Galileo was a great *bricoleur*.

3 Rochberg 2000: 372.

4 Rochberg 2000: 365–366.

5 Rochberg 2000.

6 Shapin 1996.

> Many of the ingredients of his career, from telescopes to courts, were already there. The bricolage was not.[7]

That the social strategies of Galileo could affect the production and reception of his astronomical research is immediately suggestive for the study of ancient astronomy. The concept of a Babylonian bricolage has not yet been employed to reshape our still limited understanding of the social life of Babylonian astronomy, nor have Assyriologists fully considered the impact of social ingredients and strategies of legitimation in the environs of the Rēš temple of Uruk, the setting from which our richest source materials for the ṭ EAE derive. Admittedly, our ingredients for constructing the careers of Babylonian astronomers fall short in comparison to the quantity and scope of documentary sources available to Early Modernists. The gaps in our knowledge are substantial. Indeed, little is known about the social dimension of Babylonian astronomy due to the incomplete nature of our sources, our lack of explicit biographical evidence, and the dearth of archaeological evidence for astronomical instrumentation. It is particularly important, therefore, to frame our questions with precision and historiographical awareness, to collaborate in bringing source materials together from different subspecialties of cuneiform studies, and to pay heed to *the scaffolding of our thoughts*, to echo the title of this volume.

Any study of ancient astronomy brings with it methodological challenges, beginning with the problem of importing modernist demarcations between science, religion, and magic, or establishing, tacitly or explicitly, binary oppositions between astronomy/astrology, rationality/irrationality,[8] or internalism/externalism.[9] There is no simple way past these categories, as Geoffrey Lloyd

7 Biagioli 1990: 258.

8 Recently, Rochberg (2015) examined how rationality has affected the study of Babylonian intellectual history. Although we follow Biagioli in utilizing the concept of bricolage as a representative characteristic of social constructivism in science, the term had a different function in the structural anthropology of Claude Lévi-Strauss. In *The Savage Mind* (1966: 16–17), he utilized the term to characterize the mythic mind: 'There still exists among ourselves an activity which on the technical plane gives us quite a good understanding of what a science we prefer to call 'prior' rather than 'primitive,' could have been on the plane of speculation. This is what is commonly called 'bricolage' in French … [t]he characteristic feature of mythical thought is that it expresses itself by means of a heterogeneous repertoire which, even if extensive, is nevertheless limited. It has to use this repertoire, however, whatever the task in hand because it has nothing else at its disposal. Mythical thought is therefore a kind of intellectual 'bricolage.'

9 Shapin 1992.

stated: "[w]e cannot, on pain of distortion, impose our own conceptual framework. Yet we have to."[10]

In the present work, we delineate a boundary between scholarly and legal activities; this distinction is productive and pragmatic for the purposes of the project, which seeks to visualize the degree of the scholars' social location as reflected in their participation in legal activities. A visualization of these two separate spheres of socio-intellectual life—to be dealt with more fully in our discussion of networks—can be seen in Figure 13.1.

This hard boundary between scholarly and legal worlds is admittedly artificial. We are not claiming that these were two unconnected worlds, or that the ancient scholars themselves disassociated themselves from the realm of legal texts in the production of astronomical knowledge. On the contrary, the realities of ancient knowledge production were decidedly messier, as the career arcs of those scholar-scribes responsible for the recording of numerous contracts demonstrate. In this way, the concept of a Babylonian bricolage provides a productive framework, as it aids in characterizing the production of ancient

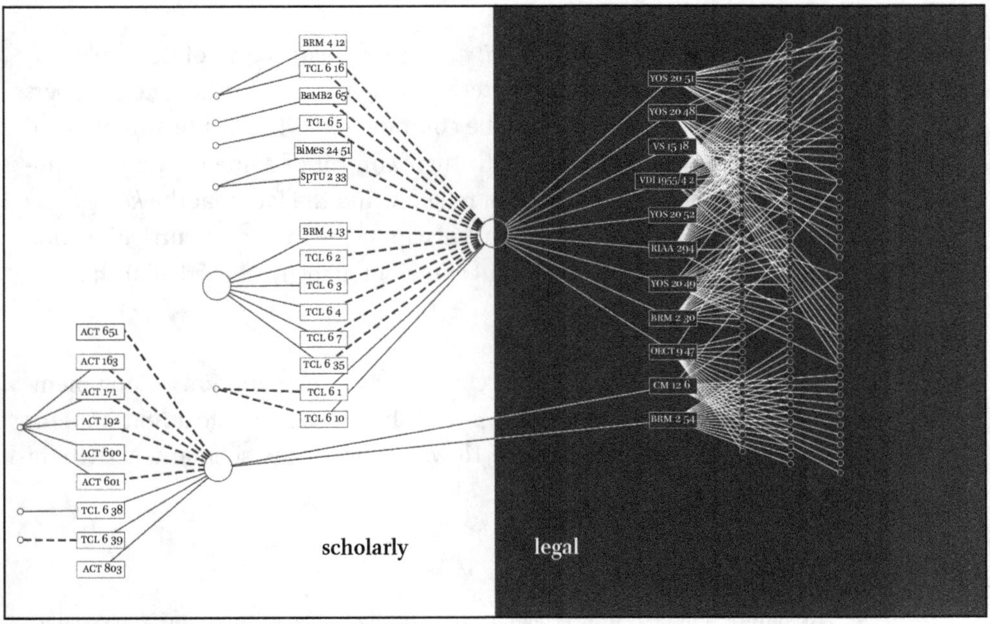

FIGURE 13.1 *Late Babylonian scholarly and legal texts often occupy two different worlds in modern scholarship. This visualization demonstrates that scribes were connected to both worlds.*

10 Lloyd 2004: 2.

knowledge as multi-faceted, and, in addition, it accurately describes the modern historiographical process of piecing together disparate and fragmentary documents and disciplinary specializations, towards the identification of *bricoleurs* in Babylonia. In the following sections, we posit that Rēš temple astronomers could construct their intellectual identities by appealing to history as a tool of knowledge legitimation and that network visualization can bring us closer to mapping and understanding the ways in which astronomer-scribes mediated the realms of intellectual and economic life.

Astronomer *Bricoleurs* | History as a Tool for Socio-Intellectual Legitimation

Certain patterns emerge in consideration of the production of scholarly knowledge at the Rēš temple and its environs, principal among them the idea that knowledge production is tied directly to familial training, especially between father and son. The record of the Sîn-lēqe-unninni family demonstrates that

> … they possessed a strong *esprit de corps* and conceived of themselves as an aristocracy of the mind, fosterers of an age-old tradition, and keepers of mysteries impenetrable to the common man. This acute awareness of their eminence, combined with their control of some of the most important branches of the official religion, and the fact that the *kalûtu* was officially recognized as a separate discipline, with its own curriculum and its own literary corpus, no doubt contributed to give a particular shape to their intellectual quest.[11]

The consistency and insularity that characterizes collaboration among members of the Sîn-lēqe-unninni family establishes a social context and impetus for the production of knowledge.[12] However, when the scholarly activity of

11 Beaulieu 2000: 16.

12 Ossendrijver (2011a, b [especially p. 215]) expands on Neugebauer's (1955/1983: 14–16) comments and promotes the notion of "collaboration" to embrace not only the fact that a vast majority of colophons to scholarly texts identify both tablet owner (*ṭuppi* PN) and scribe (*qāt* PN) and that familial connections often exist between them, but also, and more importantly, that the owner-scribe relationship expresses that connection between

members of other prestigious families is considered in light of the collaborative model, a different picture emerges. It is on the outliers to this pattern of familial scribal training that we focus our attention.

Even more than descendants of Ah'utu and Hunzû, members of the Ekur-zākir family contributed to the outpouring of scientific knowledge in Hellenistic Uruk. But their deviations from the well-documented collaboration between members of agnatic lines in the Sîn-lēqe-unninni clan is readily apparent, and the contacts between one Ekur-zākir scholar, Nidintu-Anu / Anu-bēlšunu, and other scholar-scribes demonstrate the extent to which that is so. Nidintu-Anu owned only two texts produced by his son Mannu-iqâp,[13] and none by his son Anu-ah-iddin;[14] this particular genealogical line of scholar-scribes ends with them. Nidintu-Anu's other collaborations included members of the Ah'utu and Hunzû clans. His first attestations as tablet owner occur in 97 and 98 SE, in texts written by the scribe Anu-uballiṭ / Nidintu-Anu // Hunzû.[15] While these encounters point to a career scattered with contacts across the social fabric of the Uruk community, in the early stages of his career Nidintu-Anu established his scholarly bona fides through his connections as scribe to the socially prominent Anu-balāssu-iqbi // Ah'utu, *rabû rēš āli ša Uruk*, "chief mayor and temple delegate."[16] Against the social and genealogical homogeneity of the Sîn-lēqe-unninni clan,[17] Nidintu-Anu engages with scholars and the urban elite—actors in the legal texts with which he is connected—across family lines, demonstrating the existence of a social location for *bricoleurs* who supported the production of scholarly knowledge. Nidintu-Anu interacts with members of the Ah'utu and Hunzû families,[18] although his collaborations in scholarly texts occur primarily with Anu-ah-ušabši / Ina-qibīt-Anu / Anu-uballiṭ // Ekur-zākir, as shown in Figure 13.2, which demonstrates the strength and insularity of connections between Sîn-lēqe-unninni scribes.

mentor-student and thus reflects a conduit for the transmission of intellectual content and methodologies.

13 BiMes 24 51; SpTu 2 33.

14 See Ossendrijver 2011b Table 1, rows 19, 21.

15 TCL 6 16, notably a copy of EAE 56; and BRM 4 12, a copy of tablet 55 of the *barûtu*.

16 TCL 6 1 and 10, written in 90 and 91 SE, respectively, are the earliest known tablets produced by Nidintu-Anu. In these two texts, he bears only the title *mašmaš Anu u Antu*. The inventory of Nidintu-Anu's scholarly output is presented in Ossendrijver 2011b, Table 1.

17 Beaulieu 2000; Stevens (2013: 217) notes that of the seventy tablets with fully or partially preserved colophons which record activity of members of the Ekur-zākir and Sîn-lēqe-unninni families, only ten demonstrate inter-family collaboration.

18 See Ossendrijver 2011b Table 1, and Figure 13.1.

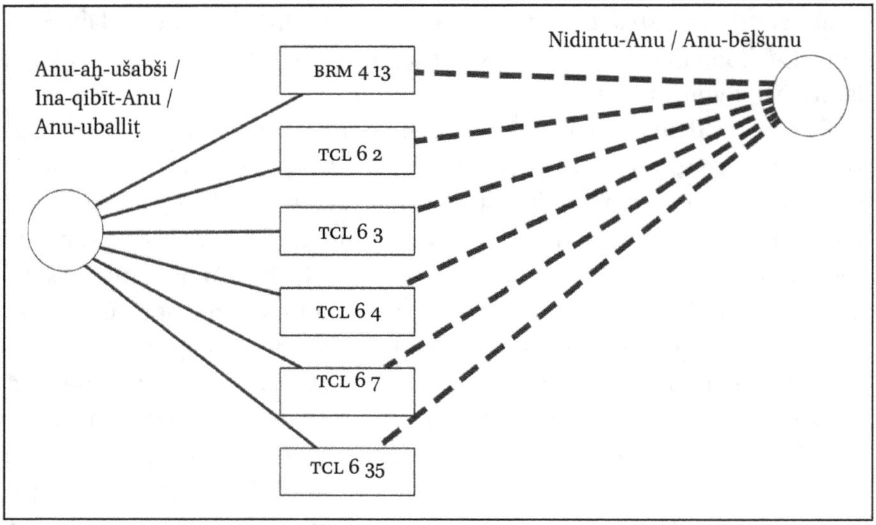

FIGURE 13.2 *Collaboration between unrelated Ekur-zākir scribes: tablet scribe Anu-aḫ-ušabši/ Ina-qibīt-Anu and owner Nidintu-Anu/Anu-bēlšunu.*

But attention should be drawn to the fact that Anu-ah-ušabši, along with Šamaš-ēṭir / Ina-qibīt-Anu / Anu-ah-ušabši / Šibqat-Anu, "floated," without clear scholarly interaction with members of their own family line. They are two of Uruk's three scholar-scribes who sport the three professional titles: *mašmaššu, aḫû rabû ša Rēš, ṭupšar Enūma Anu Enlil,* "exorcist, big brother of the Rēš temple, scribe of *Enūma Anu Enlil*"; the third, also named Anu-ah-ušabši, is the son of the Kidin-Anu who figures in Šamaš-ēṭir's construction of his scribal identity, as we discuss below. It is in that exceptional dislocation from genealogical scribal lines that we explore these scribes' careers for evidence and/or explanation of bricolage in the production of Babylonian scientific knowledge and the identity of a Babylonian *bricoleur* (Figure 13.3).

Šamaš-ēṭir / Ina-qibīt-Anu / Šibqat-Anu // Ekur-zākir's status as a scholar-scribe is evident more from his productivity and scribal titulature than in well-documented and traceable familial connections to the scribal community. Colophons in which his name appears demonstrate noteworthy features, including a tripartite professional titulature, and most importantly for the present argument, a unique, undated, colophon recording the scribal activity at the beginning of the Seleucid Era (ca. 20 SE) of Kidin-Anu, the father of another fully titled scribe. This evidence provides a rare glimpse at factors that may contribute to the socio-intellectual identity of a *ṭ EAE* and his activity as a *bricoleur*.

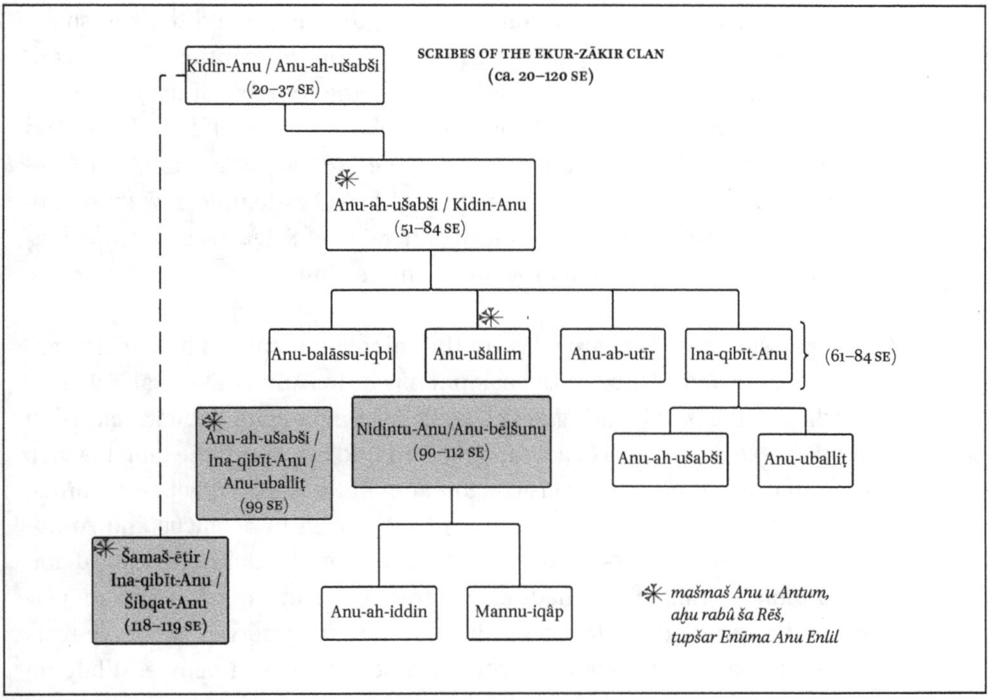

FIGURE 13.3 *Family tree of Ekur-zākir scribes between years ca. 20–120 SE, indicating kinship relations (solid lines). Certain actors (grey boxes) show no scholarly activity along agnatic lines. Šamaš-ēṭir claims connection (dotted line) to Kidin-Anu (20–37 SE), a progenitor of an active scribal line that continues in the era of Šamaš-ēṭir almost a century later. Tripartite title bearers indicated by:* ✻

One of the most curious colophons inscribed on texts circulating among the Uruk astronomers appears on TCL 6 38, a ritual for daily offerings in the Rēš temple, a text copied by Šamaš-ēṭir.[19] The colophon speaks both to his social location and his intellectual program:

> [43](Written by) the hand of Šamaš-ēṭir, son of Ina-qibīt-Anu, son of Šibqat-Anu.
> [44–46]Wax-table (containing) the rites for the Anu-worship, the pure *šuluḫḫu*-purification rites, the ritual regulations of kingship, including(?) the divine *šuluḫḫu*-purification rites of the Rēš-temple, the Ešgal-temple, the Eanna-temple and the (other) temples of Tiranna, the ritual activities

19 Boiy 2010 begins his alignment of the various branches of the family to which Šamaš-ēṭir belongs with a description of the exceptional colophon of TCL 6 38 (=TU 38).

> of the exorcists, the lamentation priests, the singers and the craftsmen, all of them, who are subservient to the ..., not to mention all that concerns the apprentices, in accordance with the contents of the tablets, [47–50]which Nabopolassar, king of the Sealand, carried off from Uruk. And at that time(?) Kidin-Anu from Uruk, the exorcist of Anu and Antu, descendant of Ekur-zakir, the high priest of the Rēš-temple, saw these tablets in the land of Elam and during the reign of Seleucus and Antiochus, the kings, he copied and brought (them) to Uruk.[20]

The expression *qāt*(*i*) *ša Šamaš-ēṭir* in the colophon identifies him as the tablet's scribe. Also identified in the colophon is Kidin-Anu // Ekur-zākir who—according to the usual pairing of titles and roles among individuals named in colophons (Ossendrijver's collaborators)—should have been the tablet owner. Rather than a contemporaneous collaborator, Kidin-Anu's activity is recorded almost a century earlier (ca. 20–37 SE) in the reign of Seleucus and Antiochus.[21] The colophon reports that this predecessor of Šamaš-ēṭir viewed and copied tablets in Elam that had been carried there in the reign of Nabopolassar and returned them to Uruk, where, in turn, they served as source texts from which Šamaš-ēṭir would eventually produce his own oeuvre. While no Kidin-Anu is documented as a scribe in the Uruk scholarly network, Boiy's reconstruction of the family trees of Uruk scribes establishes that this Kidin-Anu was the family member from which two scholarly branches emanate,[22] including the one to which Šamaš-ēṭir belonged. Another son of Kidin-Anu, Anu-ah-ušabši / Kidin-Anu // Ekur-zākir, who bears the same three-part scribal title as Šamaš-ēṭir, was active in the second half of the first century of the Seleucid era (51–84 SE [= 261–228 BCE]; see Table 13.2). The undated colophon of TCL 6 38 provides a heroic narrative that situates Kidin-Anu's professional activity at the dawn of the Seleucid Era, when the intellectual power of the Ekur-zākir clan was in its infancy. Kidin-Anu's role as a witness to prebend sales[23] reflects the social

20 Translation is that of Linssen 2004: 179.

21 The colophon references Seleucus and Antiochus. Assuming that the kings are presented according to the usual pattern employed in referencing co-regencies—the older king first, the co-regent second—the rulers referenced should be Seleucus I Nicator (311–281 BCE) and Antiochus I Soter (281–262 BCE, appointed co-regent in 292 BCE). The period from 292–281 BCE antedates any of Anu-ah-ušabši's scribal activity by a minimum of twenty years; that span of time, a common designation of the duration of a generation, would be the appropriate period in which to assign the Kidin-Anu known from other colophons as Anu-ah-ušabši's father.

22 Boiy 2010: 177.

23 Kidin-Anu is attested as a witness in legal documents from the first forty years or so of the Seleucid era: VS 15 51 (--.i.15 SE); BRM 2 4 (20.iv.29 SE); YOS 20 17 (26.xi.35 SE); BRM

circles in which he traveled and for which he may have been an interlocutor. By attributing to his seemingly ahistorical collaborator a role in the perpetuation of scientific knowledge at the start of the Seleucid era, Šamaš-ēṭir promulgated fiction in the service of ideological ends.

That Šamaš-ēṭir creates this pious fiction in the colophon of TCL 6 38 is striking in view of the unusual (but not unattested) absence of references to family connections between him and his collaborators, a pattern so prevalent among scholar-scribes that Diodorus Siculus makes note of it.[24] His adroit negotiation of social location is further evident from his role as tablet owner; Šamaš-ēṭir exclusively owns tablets written by members of the Sîn-lēqe-unninni clan.[25] With no apparent evidence that a father, grandfather, or uncle whose standing in the community of scholars provided him with training and social capital to legitimate his own authority,[26] Šamaš-ēṭir, whose legacy does not include scholarly roots or offshoots, achieves these goals by strengthening his social ties to members of other clans and to prominent individuals and scribes associated with the business world. In this, he reflects the historical figure he invokes in the colophon: Nabopolassar, who achieved success in establishing "legitimacy within one ideological discourse,"[27] transforming the traditionally pejorative claim to have been "a son of a nobody" into a personal and positive usage supporting his construction of social and cultural identity.[28]

---

2 10 (2.vi.37 SE); YOS 20 20 (19.xii.41 SE). The iconography of his seals (Wallenfels 1994: no. 125A, 180) represents *apkallu* figures, suggesting a family connection with the intellectual heritage of cuneiform literature.

24 "For among the Chaldaeans the scientific study of these subjects is passed down in the family, and son takes it over from father, being relieved of all other services in the state. Since, therefore, they have their parents for teachers, they not only are taught everything ungrudgingly but also at the same time they give heed to the precepts of their teachers with a more unwavering trust. Furthermore, since they are bred in these teachings from childhood up, they attain a great skill in them, both because of the ease with which youth is taught and because of the great amount of time which is devoted to this study." Diodorus Siculus. Library of History II 29.4, trans. Oldfather 1933: 447.

25 See Table 13.1.

26 Stevens (2013: 226) identifies Kidin-Anu as a recent ancestor of Šamaš-ēṭir and locates the production of TCL 6 38 in the earlier stages of Šamaš-ēṭir's career. We wholeheartedly agree with her suggestion that Šamaš-ēṭir was constructing his scribal identity.

27 Beaulieu 2003: 2*.

28 Rocío Da Riva (2008: 4–5 with note 20) summarizes recent research that establishes Nabopolassar's origins as son of a governor of Uruk, a quisling in cahoots with the Assyrian overlords of Uruk. Thus, the administrative records contradict the "son of nobody" biography Nabopolassar promulgates in the royal inscription C32 I 8: *māru lā mammānam anākū-ma*. Of the attestations of that expression, it is only Nabopolassar who uses it with

Nabopolassar's programmatic claims for historical legitimacy resonate in this colophon on a Hellenistic scholarly text. So, too, does the antiquarianism of the Uruk list of kings and scholars, which traces the professional legitimacy of the scholars back to the antediluvian sage.[29] A literary genealogy, it promoted the restoration of the Anu cult at Seleucid Uruk[30] and drew attention to royal patronage of the scribal arts and scholarly knowledge by Nikarchos, a fictionalized royal manifestation of Anu-uballiṭ-Nikarchos of the administrative record, who, in fact, had participated in the restoration of the Anu temple.[31] Finally, it emphasized the social location of scribal authority, that of the Sîn-lēqe-unninni clan in particular.[32]

Just as the Uruk king list mediates the transition of scholarly knowledge tradition from the Neo-Babylonian down through its Achaemenid and Seleucid period heirs,[33] TCL 6 38 may do so for the production of knowledge in the Ekur-zākir clan, connecting the scholar-scribe Šamaš-ēṭir to the earliest generation of this innovative era of the production of scientific knowledge—namely to Kidin-Anu, the father of the first scholar scribes known to bear the three titles "exorcist, big brother of the Rēš temple, and *ṭ* EAE."[34] Of particular interest in the TCL 6 38 colophon is not just the reference to antiquity but the chronological, geographical, and cultural connection between astronomical

reference to himself, a clear marker of the tendentiousness that permeates many aspects of his legacy.

29 W.20030/7 (= BaghMB 2 89). Among the various studies of this text, Lenzi's 2008 article is particularly relevant to the topic at hand. The seal of Anu-ah-ušabši / Kidin-Anu // Ekur-zākir, published as AUWE 19: 183, depicts an *apkallu*, the so-called "fish-man sage." Wallenfels (1994: 40) notes that all but one example of *apkallu* iconography appeared on Uruk seal impressions belonging to descendants or relatives of Kidin-Anu // Ekur-zākir. Thus, the social privileging of members of the Sîn-lēqe-unninni clan through the literary genealogy did not preclude members of the Ekur-zākir clan from invoking an association with the venerable *apkallu* through iconographic means.

30 Beaulieu 1993, 2003.

31 As recorded in YOS 1 52. See discussions in Beaulieu 1992: 68; 1993: 47; Doty 1988; Lenzi 2008: 138, 164.

32 The text mentions Anu-bēlšunu / Nidintu-Anu / Anu-uballiṭ // Sîn-lēqe-unninni, grandfather of the Anu-belšunu who wrote and owned the text. Anu-bēlšunu *grandpère* was the only member of the Sîn-lēqe-unninni clan to appear in CM 12 6, and thus he is connected to Šamaš-ēṭir. Pearce and Doty (2000) trace the career of both scribes.

33 As Stevens (2013: 216) notes, the Ekur-zākir family was active and productive in the intellectual life of fourth century Uruk, as evidenced by the library of Iqīšâ, the contents of which date from ca. 325–300 BCE. This correlates well with Kidin-Anu's activity in the first decades of the Seleucid Era.

34 As depicted in Figure 13.3.

knowledge and ancient Babylonia; this association, seemingly well attested in later periods in the history of science, is considered in the conclusions below.

The nature of Šamaš-ēṭir's participation in this process extends beyond his status as a Rēš temple astronomer and an interlocutor with the Sîn-lēqe-unninni clan. Although Table 13.1 identifies connections between Šamaš-ēṭir and single individuals in the network of tablet owners and scholars, it cannot adequately illustrate his relationships to the many parties—principals and witnesses—who populate the legal texts he scribed. For these, we will turn to SNA as an aide to the recovery and reporting of the other component of Šamaš-ēṭir's professional life.

Network Visualizations and the Construction of Scientific Biographies

The idea that scientific knowledge production can be understood in terms of dynamic networks between human and non-human actors (in the present case, cuneiform texts) was fundamental to the work of Bruno Latour since the early 1980s.[35] Although this paper will not delve into the ontological consequences of assigning agency to cuneiform texts, we nevertheless acknowledge that the production of bipartite networks that set cuneiform texts and scholars on similar axes is rooted in Science and Technology Studies and related literature that holds untapped potential for ancient studies.[36] What this all means for the present purposes is that there is a theory-laden realm to unpack from beneath the lines, circles, and boxes that constitute our visualizations. Here, however, we utilize network visualizations for the pragmatic aim of clarifying the nature of social ties and revealing relationships that may previously have gone unnoticed. In order to make these connections evident for the reader, it will be necessary to outline a roadmap: What are networks and how should we read them?

Networks are composed of nodes (actors) and edges (connections between actors). In our bipartite networks, nodes are represented using two conventions: circles, which signify individual scribes, and rectangles, which represent the individual texts with which those scribes interacted. Dotted lines represent

35 Latour himself provides an overview of the literature of actor network theory in his introduction to *Reassembling the Social* (2005: 10–11).

36 In *Biographies of Scientific Objects*, Daston et al. (2000), productively explored the ontological status of objects and their relations to the production of scientific knowledge. Relatedly, anthropologically rooted studies have argued for the importance of materiality in the study of cuneiform cultures, for example Taylor 2011.

owners of tablets, whereas solid-line edges connecting nodes represent all other relationships (roles) of actors in a text. Note in the visualizations below, the length of these lines is determined aesthetically for the sake of highlighting particular points of interest.[37] Collectively, these components generate visual maps that can aid the historian in the composition of biographies of scholars and astronomers and the production of knowledge itself. The contribution of this approach is apparent in our exploration of the network visualization of our astronomer Šamaš-ēṭir and his contemporaries.

The egocentric network depicted in Figure 13.4 focuses on Šamaš-ēṭir and illustrates his connections to scholarly and legal activity, including the text and

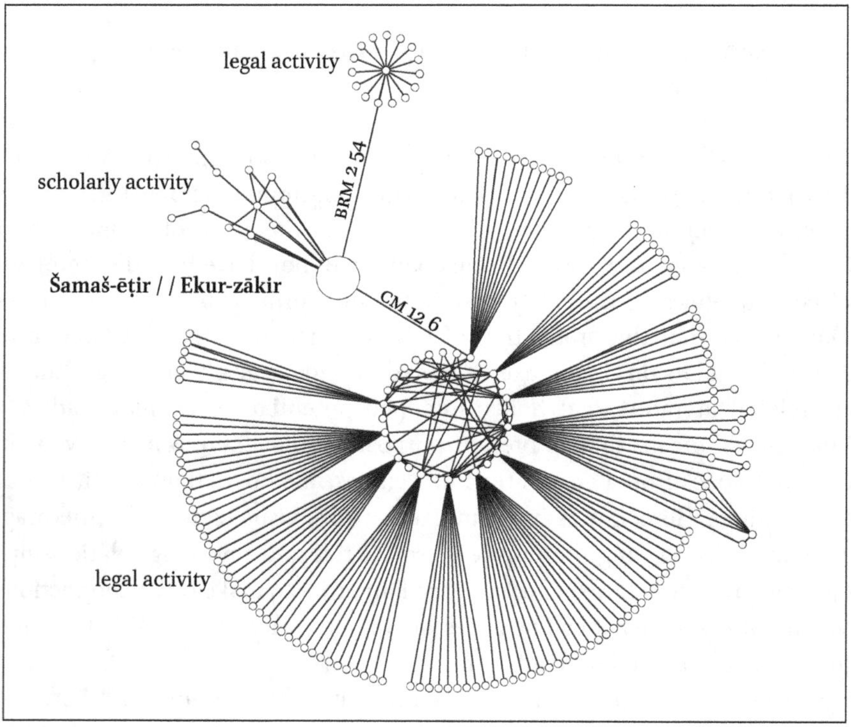

FIGURE 13.4 *Egocentric network demonstrating Šamaš-ēṭir's degrees of connectedness to legal and scholarly activity. In many ways, he proves to be an independent actor in the Uruk network of scribes.*

37 For example, we have not weighted our edges according to the computed strength of social connections, and, furthermore, the length of these edges is not representative of spatial distance or strength of social ties. The present visualizations can be characterized as being both data-driven (in that we have employed tabular data to generate the visualizations) and infographic (in that we have manipulated the visualizations, including the size of nodes, in order to communicate points of data key to our narrative).

individuals with whom he interacted.[38] He is connected to a distinctive group of scholarly texts and actors.[39] Through one of the two legal texts that he wrote (CM 12 6,[40] the other being BRM 2 54[41]), he is indirectly connected to a larger community of actors in the legal world. The nature and consequences of these connections are explored further in Figure 13.5.

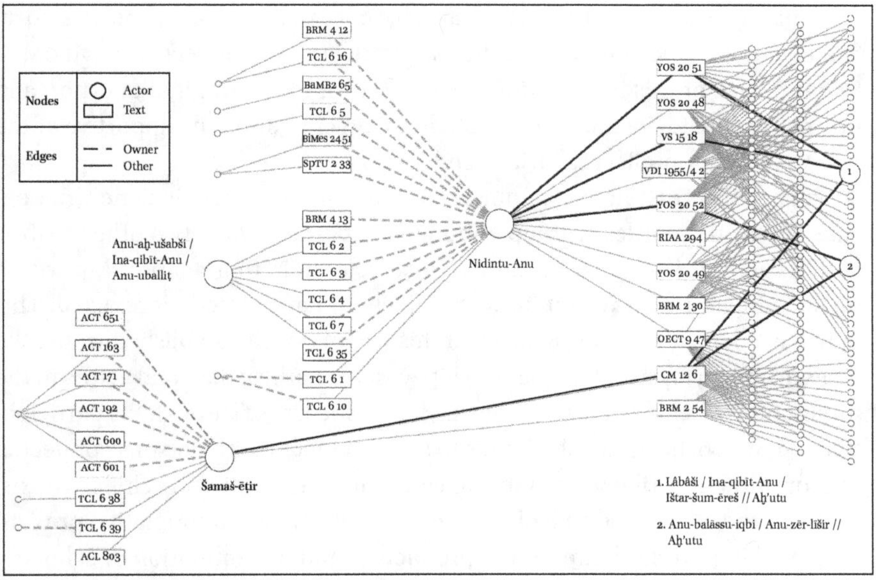

FIGURE 13.5 *This visualization illustrates the two degrees that separate Šamaš-ēṭir and Nidintu-Anu. Although Šamaš-ēṭir is not well-connected to the world of legal activities, his role as scribe of* CM *12 6 locates him in a social environment inhabited by Nidintu-Anu, who was prolific in both legal and scholarly activity. Heavy black lines emphasize the actors that co-occur in* CM *12 6 and connect the two scribes.*

---

38 In this basic visualization no distinction is made between node types (text/actor).

39 TCL 6 38, 39; ACT 163, 171, 192, 600, 601, 651, 803. This group of texts can be found in the lower left corner of Figure 13.5.

40 Šamaš-ēṭir and a scribe from the Sîn-lēqe-unninni clan, Anu-bēlšunu / Nidintu-Anu / Anu-uballiṭ //Sîn-lēqe-unninni, co-occur in CM 12 6. This Anu-bēlšunu is the elder of two scribes of that name. His career, first as a legal, then as a scholarly scribe, as Pearce and Doty (2000) discuss, further demonstrates interaction across social and professional contexts. The remainder of Šamaš-ēṭir's associations with members of the Sîn-lēqe-unninni family occur as owner of scholarly texts written by Anu-ab-utīr / Anu-bēlšunu (ACT 163, 171, 192, 600, 601). Note the paucity of members of the Sîn-lēqe-unninni clan in the legal corpus overall, where they occur in less than 10% of the name instances, which further underscores the social location of that family in the world of scholarship.

41 BRM 2 54 is a duplicate of BiMes 24 28. For simplicity's sake, we have not included reference to the duplicate throughout this study.

Two other individuals who appear in CM 12 6 also appear in texts written by the productive and socially prominent scholar-scribe Nidintu-Anu / Anu-bēlšunu // Ekur-zākir, whose career is firmly established in the realms both of scholarly and legal texts (as Figure 13.1 also demonstrates). These two individuals are Anu-balāssu-iqbi / Anu-zēr-līšir // Ah'utu, who appears as a witness in both CM 12 6 and YOS 20 52, and Lâbâši-Anu / Ina-qibīt-Anu / Ištar-šum-ēreš // Ah'utu, who serves in that same capacity in CM 12 6, VS 15 18, and YOS 20 51. Šamaš-ēṭir and Nidintu-Anu are thus separated by two degrees through CM 12 6. In the face of this information, we can return to our original question and utilize the data to postulate the existence of an environment supportive of the notion of socio-intellectual bricolage.[42]

Šamaš-ēṭir is distinctive among the scholar-scribes in Hellenistic Uruk because of his three titles, a group of honorifics borne by only two other scribes, and because of his claim, in the TCL 6 38 colophon, to a relationship with a historical figure with whom he was unlikely to have directly interacted. This may have been a motivating factor in his use of a well-established trope, the *pious fiction*, to lay claim to relationship with the earliest days and actors in the production of scholarly knowledge in Hellenistic Uruk. Through his participation in CM 12 6, he established links to and interacted with persons connected to a prominent scribe with a strong career in both legal and scholarly texts. Through the actors tied to Nidintu-Anu, Šamaš-ēṭir stood in relative proximity to this important figure for the production and dissemination of scholarly knowledge.

Network visualization can more holistically represent the interplay between the scholarly and legal worlds of late Babylonian scholarship. We have found that particularly within the Ekur-zākir clan, social ties occur across clans and social spheres and indeed, beyond the internalist environs of the Rēš temple and its documents. Furthermore, the contributions of actors such as Nidintu-Anu, which were not immediately conspicuous using traditional Assyriological methods, became central to navigating the connections between scholarly and legal scribal activity (as in Figure 13.1). Finally, it is hoped that this section and the network visualizations employed therein have demonstrated for the reader the ways in which a diagrammatic and visual approach to history can aid in both the narration of intellectual biographies as well as the generation of new research questions that were previously obfuscated in muddy prosopography or unsynthesized data.

42 In her 2014 investigation of means of transmission of knowledge of the genre of Babylonian Chronicles to the biblical author(s) of 1–2 Chronicles, Caroline Waerzeggers postulates that although many degrees separate them, members of the Judean exile may have been connected to members of the social classes who had access to the production and storage of the Babylonian Chronicles texts in sixth-century BCE Babylonia.

Conclusions

A great deal of research remains to be done to more fully answer the question that initiated our collaboration, namely, what makes a scribe a *ṭupšar Enūma Anu Enlil*? In our micro-biography of Šamaš-ēṭir (see section **Astronomer *Bricoleurs* | History as a Tool for Socio-Intellectual Legitimation**) we suggest his scholarly identity was tethered to a historical memory of Babylonian erudition. Yet the unusual colophon of TCL 6 38 in which Šamaš-ēṭir secures his intellectual lineage to a great ancestor named Kidin-Anu—an intellectual hero who returned astronomical knowledge to its rightful territory after retrieving it from Elam—may not be unusual at all. The idea that the intellectual genealogy of science is rooted in Babylonia, finds echoes in later periods of the history of science.

Šamaš-ēṭir, our *bricoleur*, shares a family resemblance to a later scholar named Abū Sahl al-Faql b. Nawbakht, who, around the turn of the 9th century CE, recounted an etiological narrative regarding the Babylonian origins of scientific knowledge in Baghdad.[43] According to Abū Sahl, astronomical knowledge and "indications of future things were already known and described in the books of the Babylonians. It was from those books that the Egyptians had learned their craft, and the Indians have also employed it in their own country."[44] Abū Sahl explains that moral corruption led to the loss of knowledge. And that which remained of Babylonian scholarship was eventually destroyed and captured by Alexander and the Greeks, who took the knowledge as their own and "copied what was gathered in the libraries and offices of the city of Istakhr...into Greek and Coptic."[45] But historical memory survived invasion and destruction, and it would come to pass that Persian scholars would eventually come to recover and translate back into Persian the great books of Greeks:

> He [Ardashīr I] copied all that had fallen into their hands, and pursued the very few remaining sciences that had survived in Iraq. He gathered together all that had been dispersed, and collected all that which was scattered. His son Shapur, who succeeded him, persisted in this policy, until all those books were copied into Persian: among them the books of Hermes the Babylonian who ruled over Egypt, Dorotheus the Syrian, Phaedrus the Greek from the city of Athens which was famous for its

43 The story is received through the Fihrist of al-Nadim, who recounts various such etiologies for the origins of science, for which, see Saliba 2007: 29–31.

44 Saliba 2007: 31.

45 Saliba 2007: 32.

> learning, Ptolemy the Alexandrian, Farmāsp the Indian. They then explained those books and taught them to the people, just as they have learned from those books, which were originally from Babylon.[46]

As George Saliba explains, Abū Sahl's legendary tale of the origins of the sciences may well have been recounted as a social tactic for elevating his Persian origins.[47] Furthermore, and of particular interest to the study of pre-modern intellectual history, is the way in which Abū Sahl characterized scientific knowledge, not as an emblem of progress but as an ancient reminder of loss. Seen in this light, the goals of the celestial scholar were ultimately historical; the recovery of knowledge from the distant past would always lead, ultimately, to Babylonian knowledge, *ṭupšarrūtu*. Given the similarity in attitudes towards knowledge, it is remarkable that there existed between Abū Sahl and Šamaš-ēṭir a cultural and chronological gulf of more than a thousand years.

The intellectual identities of late Babylonian scholars and, in particular, the *ṭupšar Enūma Anu Enlil* were at least partially constructed by a historical and epistemological interest in their own discipline of celestial expertise. Further analysis will tell if these motivations can be discerned in the content of the scholarly texts (for example, in the relationship between historiography and retrodiction). It appears that this historical interest was at least socially motivated. Strikingly, scribes bearing the most extensive and prominent collection of professional titles, including *ṭ EAE*, exhibit the fewest extensive familial connections. These outlier scribes (or "floaters" as we have called them), are in fact the most compelling actors for the intellectual history of science, as they provide a glimpse, however fragmentary, into the legitimation practices of ancient astronomers. Our study of Šamaš-ēṭir and his peers may shed light on other such figures who have yet to receive the same level of biographical research.

In 2000, Rochberg surmised (already anticipating her 2016 work on the role of Nature in the production of scientific knowledge in Babylonia) that "perhaps the relationship between the scholars and the temple need not be understood in a utilitarian way, at least with respect to the cult ... [c]elestial divination constituted a body of knowledge conceived of as 'divine,' in the sense that the gods both produced the signs in nature and the scribes attributed 'authorship' of *EAE* to the god Ea."[48] We conclude, similarly, that the utilitarian interpretation of the title *ṭupšar Enūma Anu Enlil* is methodologically limited. Where the title becomes most informative for the intellectual history of science, given the present state of the evidence, is in bringing to light epistemic

---

46 Saliba 2007: 31–32.

47 Saliba 2007: 33.

48 Rochberg 2000:372.

attitudes and the role of authority in knowledge-making rather than solely the "utilitarian" matter of recurring temple functions or only reconstructing training patterns of familial clans from colophon evidence. We conclude that the key to understanding the title depends on defining how scribes at or in proximity to the Rēš temple at Uruk constructed *ṭupšarrūtu* "knowledge." In this study, which exposes the potential of collaborations between historians of science and of social and economic history, we have not aimed to be comprehensive either in plumbing the full gamut of intersections evident in the documentation of each discipline or in the explication of additional patterns suggesting the social construction of scholarly knowledge.

Unlike members of the prominent Sîn-lēqe-unninni clan, who "possessed a strong *esprit de corps* and conceived of themselves as an aristocracy of the mind, fosterers of an age-old tradition, and keepers of mysteries impenetrable to the common man,"[49] important scholar-scribes of the Ekur-zākir clan, including notably Šamaš-ēṭir, lacked the genealogical social capital of the Sîn-lēqe-unninni line, a fact reflected in his location seemingly outside the prebendary system, serving not as a participant but only as a scribe to the sale of shares in the *gerseqqûtu*-prebend. Šamaš-ēṭir employed a historiographic trope to legitimate his scholarly bona fides among the constellation of the intellectual elite. Šamaš-ēṭir's biography is representative of a socio-intellectual bricolage that extends well beyond the traditional internalist narrative of a hermetic temple astronomer. In his appeal to the historical origins of knowledge in his use of tripartite titulature, and in his collaborations in scholarly and legal contexts with members of other elite clans, Šamaš-ēṭir may be our best current indicator of what makes a *ṭupšar Enūma Anu Enlil*.

Acknowledgment

Taking the collaborative model of the Uruk scribes to heart, we two Assyriologists gratefully acknowledge the contribution of UC Berkeley undergraduate Emily Wang, who processed the tabular data for SNA and the generation of the visualizations. Her continuing interest in the use of SNA in historical study is an outgrowth of her work with Eduardo Escobar in Summer 2016 under the auspices of the Student Mentoring and Research Teams (SMART) program at UC Berkeley. Thus, to the SMART program, our appreciation is also due.

49 Beaulieu 2000: 16.

TABLE 13.1 *Texts mentioning Šamaš-ēṭir / Ina-qibīt-Anu / Šibqat-Anu // Ekur-zākir*

Text	Date SE	Collaborator	Content	Role	Title
BiMes 24 28 || BRM 2 54	5.iv.150?	not applicable	Prebend sale	scribe	*ṭupšarru*
CM 12 6 || VS 15 32	21.x.119	not applicable	Real estate sale	scribe	*ṭupšarru*
ACT 171	Lost	Anu-ab-utīr / Anu-bēlšunu / Nidintu-Anu //Sîn-lēqe-unninni	Auxiliary table, Moon B	owner	*mašmaš Anu u Antu Urukū*
ACT 163	Lost	Anu-ab-utīr / Anu-bēlšunu / Nidintu-Anu //Sîn-lēqe-unninni	Auxiliary table, Moon B	owner	*mašmaš Anu u Antu aḫu rabû ša Rēš ṭupšar* EAE[a]
ACT 192	13.iii.118	Anu-ab-utīr / Anu-bēlšunu / Nidintu-Anu //Sîn-lēqe-unninni	Synodic table, daily Moon B	[owner?][b]	lost
ACT 600	12.vii.118	Anu-ab-utīr / Anu-bēlšunu / Nidintu-Anu //Sîn-lēqe-unninni	Synodic table, Jupiter A	owner	<*kalû*> *Anu u Antu Urukū*
ACT 601	–.–.118	Anu-ab-utīr / Anu-bēlšunu / Nidintu-Anu //Sîn-lēqe-unninni	Synodic table, Jupiter A	owner	*mašmaš Anu u Antu Urukū*
ACT 651	lost	lost	Daily positions, Jupiter A	owner	lost
ACT 803	lost	lost	Procedural Text, Mars A	scribe	lost
TCL 6 38	n.d.	Kidin-Anu // Ekur-zākir	Ritual	scribe	none
TCL 6 39–40	n.d.	Anu-uballiṭ / Nidintu-Anu //Hunzû	Ritual	scribe	*mašmaš Anu u Antu*

a The heavily reconstructed title follows Neugebauer 1955/1983: 17.

b The tablet is damaged; we follow the reconstruction of Ossendrijver 2011a, Table 1.

TABLE 13.2 *Texts mentioning Anu-ah-ušabši / Kidin-Anu // Ekur-zākir*

Texts	Date SE	Collaborator	Content	Role	Title
HANE/M 8 177–179	28.viii.51	not applicable	Prebend sale	Prebend holder	*ṭupšarru*
BRM 2 17	22.iii.67	not applicable	Assignment of workers	witness	*ṭupšarru*
BRM 2 18	20.vii.68	not applicable	Quitclaim	witness	*ṭupšarru*
BRM 2 21	25.vi.75	not applicable	Real estate sale	neighbor to sold property	*ṭupšarru*
BiMes 24 42	—/—/—	not applicable	Quitclaim	beneficiary	*ṭupšarru*
OECT 9 12 || 13	—/—/—	not applicable	Dowry gift, slave girl	witness	*ṭupšarru*
BRM 4 7	25?.iv.61	Anu-balāssu-iqbi / Anu-aḫ-ušabši / Kidin-Anu	Bilingual ritual	owner	*mašmaš Anu u Antum, aḫu rabû ša Rēš Urukū*
BRM 4 8	26.vi.61	Anu-ušallim / Anu-aḫ-ušabši / Kidin-Anu	Bilingual hymn	owner	*mašmaš Anu u Antum , aḫu rabû ša Rēš Urukū*
UCP 9–9	-.-.61	Anu-balāssu-iqbi / Anu-aḫ-ušabši / Kidin-Anu	*Enūma Anu Enlil*	owner	*mašmaš Anu u Antum, aḫu rabû ša Rēš, ṭupšar Enūma Anu Enlil*
UVB 15 37	-.-.61	Anu-balāssu-iqbi / Anu-aḫ-ušabši / Kidin-Anu	Hymn/prayer	owner	*mašmaš Anu u Antum aḫu rabû ša Rēš ṭupšar Enūma Anu Enlil*
K.3753	21.iv.65		Calendar text	owner	*mašmaš Anu u Antu aḫu rabû ša Rēš ṭupšar Enūma Anu Enlil*
TCL 6 19	12.v.84	Ina-qibīt-Anu / Anu-aḫ-ušabši / Kidin-Anu	Omen	owner	*mašmaš Anu u Antum*

Bibliography

Beaulieu, Paul-Alain

1992 "Antiquarian Theology in Seleucid Uruk." *ASJ* 14: 47–75.

1993 "The Historical Background of the Uruk Prophecy." Pp. 41–52 in *The Tablet and the Scroll : Near Eastern Studies in Honor of William W. Hallo*, eds. Mark E. Cohen, Daniel C. Snell, and David B. Weisberg. Bethesda: CDL Press.

2000 "The Descendants of Sîn-lēqe-unninni." Pp. 1–16 in *Assyriologica et Semitica: Festschrift für Joachim Oelsner anläßich seines 65. Geburtstages am 18. Februar 1997*, eds. Joachim Marzahn and Hans Neumann. AOAT 252. Münster: Ugarit-Verlag.

2003 "Nabopolassar and the Antiquity of Babylon." Pp. 1*–9* in *Hayim and Miriam Tadmor Volume*, eds. Israel Eph'al, Joseph Aviram, and Hillel Geva. Eretz-Israel 27. Jerusalem: Israel Exploration Society.

Biagioli, Mario

1990 "Galileo the Emblem Maker." *Isis* 81: 230–258.

Boiy, Tom

2010 "TU 38, Its Scribe and His Family." *RA 104*: 169–178.

Da Riva, Rocío

2008 *The Neo-Babylonian Royal Inscriptions*. GMTR 4. Münster: Ugarit-Verlag.

Daston, Lorraine, ed.

2000 *Biographies of Scientific Objects*. Chicago: University of Chicago Press.

Diodorus

1933 *Diodorus of Sicily* I. trans. C.H. Oldfather London: W. Heinemann.

Doty, L.T.

1988 "Nikarchos and Kephalon." Pp. 95–118 in *A Scientific Humanist. Studies in Memory of Abraham Sachs*, eds. Erle Leichty, Maria deJong Ellis, and Pamela Gerardi. Philadelphia: The University Museum.

Latour, Bruno

2005 *Reassembling the Social: An Introduction to Actor-Network Theory*. Oxford: Oxford University Press.

Lenzi, Alan.

2008 "The Uruk List of Kings and Sages and Late Mesopotamian Scholarship." *JANER* 8: 137–169.

Lévi-Strauss, Claude

1966 *The Savage Mind*. Chicago: University of Chicago Press.

Linssen, Marc

2004 *The Cults of Uruk and Babylon: The Temple Ritual Texts as Evidence for Hellenistic Cult Practice*. CM 25. Leiden: Brill/Styx.

Lloyd, G.E.R.

2004 *Ancient Worlds, Modern Reflections: Philosophical Perspectives on Greek and Chinese Science and Culture*. Oxford: Oxford University Press.

Neugebauer, Otto

1955/1983 *Astronomical Cuneiform Texts*, vols I–III. Berlin: Springer.

Ossendrijver, Mathieu

2011a "Exzellente Netzwerke: Die Astronomen von Uruk." Pp. 631–644 in *The Empirical Dimension of Ancient Near Eastern Studies = Die empirische Dimension altorientalischer Forschungen*, eds. Gebhard Selz. Wiener Offene Orientalistik 6. Berlin: Lit Verlag.

2011b "Science in Action: Networks in Babylonian Astronomy." Pp. 213–221 in *Babylon : Wissenskultur in Orient und Okzident*, eds. Eva Christiane Cancik-Kirschbaum, Margarete van Ess, and Joachim Marzahn. Topoi 1. Berlin: de Gruyter.

Pearce, Laurie E. and L.T. Doty

2000 "The Activities of Anu-belšunu, Seleucid Scribe." Pp. 331–342 in *Assyriologica et Semitica: Festschrift für Joachim Oelsner anläßich seines 65. Geburtstages am 18. Februar 1997*, eds. Joachim Marzahn and Hans Neumann. AOAT 252. Münster: Ugarit-Verlag.

Reiner, Erica

1995 *Astral Magic in Babylonia*. TAPS 85. Philadelphia: American Philosophical Society.

Rochberg, Francesca

2000 "Scribes and Scholars: The *ṭupšar Enūma Anu Enlil*." Pp. 359–375 in *Assyriologica et Semitica, Festschrift für Joachim Oelsner anläßich seines 65. Geburtstages am 18. Februar 1997*, eds. Joachim Marzahn and Hans Neumann. AOAT 252. Münster: Ugarit-Verlag.

2015 "The Babylonians and the Rational: Reasoning in Cuneiform Scribal Scholarship." Pp. 209–246 in *In the Wake of the Compendia: Infrastructural Contexts and the Licensing of Empiricism in Ancient and Medieval Mesopotamia*, ed. J. Cale Johnson. Berlin: de Gruyter.

2016 *Before Nature: Cuneiform Knowledge and the History of Science*. Chicago: University of Chicago Press.

Saliba, George

2007 *Islamic Science and the Making of the European Renaissance*. Cambridge, MA: The MIT Press.

Shapin, Steven

1992 "Discipline as Bounding: The History and Sociology of Science as Seen through the Externalism-Internalism Debate." *Hist. Sci.* 30: 333–369.

1996 *The Scientific Revolution*. Chicago: The University of Chicago Press.

Stevens, Kathryn

2013 "Secrets in the Library: Protected Knowledge and Professional Identity in Late Babylonian Uruk." *Iraq* 75: 211–253.

Taylor, Jonathan

2011 "Tablets as Artefacts, Scribes as Artisans." Pp. 5–31 in *The Oxford Handbook of Cuneiform Culture*, eds. Karen Radner and Eleanor Robson. Oxford: Oxford University Press.

van Dijk, J. and W.R. Mayer

1980 Texte aus dem Rēš-Heiligtum in Uruk-Warka. BaghMB 2. Berlin: Gebr. Mann.

Waerzeggers, Caroline

2014 "Locating Contact in the Babylonian Exile: Some Reflections on Tracing Judean-Babylonian Encounters in Cuneiform Texts." Pp. 131–146 in *Encounters by the Rivers of Babylon: Scholarly Conversations Between Jews, Iranians, and Babylonians in Antiquity*, eds. Uri Gabbay and Shai Secunda. TSAJ 160. Tübingen: Mohr Siebeck.

Wallenfels, Ronald

1994 *Uruk : Hellenistic Seal Impressions in the Yale Babylonian Collection*. AUWE 19. Mainz am Rhein: P. von Zabern.

CHAPTER 14

The Date of the Accession of Nabonidus to the Throne of Babylon: A Reappraisal of the Evidence

Grant Frame

The date of the accession of Nabonidus to the throne of Babylon in 556 BCE, or more accurately the earliest attestations of him being accorded the title "king of Babylon," has been a matter of some discussion in the past. In particular, it has been believed that some texts continued to be dated according to the reign of Lābâši-Marduk after Nabonidus had replaced him as ruler (e.g., Dougherty 1929: 74; Parker and Dubberstein 1956: 13; von Voigtlander 1963: 150–151 and 159; Beljawski 1971: 208–209; Röllig 1983: 409; Beaulieu 1989: 86–88; da Riva 2008: 16).[1] Beaulieu (1989: 88) says: "[Nabonidus] was probably recognized as king as early as May 25 in central Babylonia (Babylon and Nippur), but outlying regions would have recognized Lābâši-Marduk until the end of June." This study will re-examine the question of the overlap in dating and exactly when Nabonidus was first called "king" in a cuneiform text.

Dr. Francesca (Chessie) Rochberg and I first met in the Fall of 1974, when we both began doctoral studies in Assyriology at the Oriental Institute of the University of Chicago, she arriving there from the University of Pennsylvania and me from the University of Toronto. We were frequently in the same Akkadian, Sumerian and Mesopotamian history classes, and it was a blessing to have such a kind and friendly person there with me. On numerous occasions before class, we students would meet and discuss problematic passages in the Akkadian or Sumerian text to be read that day. At times in class, the professor would acknowledge the difficulty of a passage or word or even individual sign in a text and state that one really needed to look at the original tablet, and not just the copy, since that might help clarify matters. This brief study is the result of following that sound advice and looking at a tablet rather than just relying

* My thanks must be expressed to J.A. Brinkman and J. Jeffers for looking over the manuscript of this article. Any errors are, of course, the responsibility of the author. Abbreviations follow those of the Chicago Assyrian Dictionary.

1 The documents are not dated *arki* Lābâši-Marduk, as had been done on similar earlier occasions, after the deaths of Aššur-nādin-šumi, Nergal-ušēzib and Kandalānu (see Brinkman/Kennedy 1983: 16 fn. 2, fn.5 and G.1, and p. 49 L.160 and L.163; but possibly note p. 13 Bn.1).

 | DOI 10.1163/9789004363380_015

on the published copy. It is offered here as a small token of my esteem and friendship.

The latest four texts that have been identified as attesting to the reign of Lābâši-Marduk are listed in Parker and Dubberstein 1956: 13 and Goetze 1944: 44. The dates and provenances given for them there are:[2]

22-II-accession year (June 1, 556), Babylon—Sack 1976: 287–288; YOS 16 forthcoming (YBC 4012)
9-III-accession year (June 17, 556), Sippar?—Strassmaier in *Actes du 8e Congrès International* no. 15 (BM 60224, AH 82-9-18,189)
11-III-accession year (June 19, 556), Uruk—YOS 16 forthcoming (YBC 3817)
12-III-accession year (June 20, 556), Sippar?—Evetts, *Lab.* 1 (BM 92721, 82-7-14,224 [lines 39–40])

Copies and/or editions of three of these have been published, and the entry for each of them deserves a brief comment. Although YBC 4012 (June 1, 556) was stated by Goetze to come from Babylon, the actual place of composition is not given in the text. The text deals with silver taken to Babylon at the end of the previous year (the third year of Neriglissar) by the *qīpu* of Eanna. The text is certainly part of the Eanna archive and was composed in the presence of Nabû-aḫḫē-bulliṭ, the temple scribe (of Eanna), which might indicate that it was drawn up at Uruk.[3]

Strassmaier, *Actes du 8e Congrès International* no. 15 (BM 60224, June 17, 556) is missing its beginning and appears to deal with a bundle (of scrap metal) given to a smith to be turned into ingots. It does not state where the transaction took place. The tablet, however, comes from the British Museum registration series AH 82-9-18, which comprised twelve cases and two parcels, all but one case said to have come from Sippar (Abu Habbah); "the other was from Babylon, with a few Dailem pieces" (Reade 1986: xxxiii).

Evetts, *Lab.* 1 (BM 92721, June 20, 556) records the issuing of bundles of reeds for various purposes over a period of time, beginning in the first month of

2 The dates are given day-month-year, with the month being cited in Roman numerals. With regard to the various statements in texts about the length of the reign of Lābâši-Marduk—9 months in Berossos and 3 months in the Uruk king list—see Beaulieu 1989: 86–86. Economic texts are dated according to his accession year in months Nisannu, Ayyaru and Simanu (Goetze 1944: 44; Parker and Dubberstein 1956: 13; and see below).

3 P.-A. Beaulieu kindly provided information on this text and YBC 3817 to me. Copies of the two texts will appear in the forthcoming YOS 16 by K. Nemet-Nejat. For Nabû-aḫḫē-bulliṭ as temple scribe of Eanna, see Kümmel 1979: 144.

Neriglissar's third year (557), but with the deliveries not being recorded in strict chronological order. Three or four deliveries were made in the reign of Neriglissar, and seven or more were dated to the accession year of Lābâši-Marduk, the latest being on 12-III, but some of the others took place on 12(+1)-II, 22-II, 2-III, 4-III, and 9-III (lines 7, 21, 14, 16 and 32 respectively). The text itself, however, may have been composed after the time of the last delivery (12-III) since the last section of the text, the section that includes that date, reads:

39 [...SIMU]G?.⸢MEŠ ITI.SIG$_4$⸣ U$_4$.12.KÁM
40 [MU.SAG.NAM.LUGAL.L]A ^{m}*la*-[*ba*]-⸢*a-ši*⸣-^{d}ŠÚ LUGAL E.KI
41 [x x x x x-*l*]*u*$_4$ *šá* LÚ.SIMUG.⸢MEŠ⸣ ITI.SIG$_4$
42 [U$_4$.x.KÁM MU.SAG].NAM.LUGAL.LA ^{m}*la-ba-a-ši*-^{d}ŠÚ LUGAL [E.KI]
43 PAP ⸢4 LIM 4⸣ [M]E 70 *gu-zu-ul-l*[*u*$_4$]
44 *te-lit*

Line 39 is preceded by a line ruling.[4] The text does not state where it was composed, but it comes from the registration series 82-7-14. Although the tablets in this series in theory come "entirely from Abu Habbah," J.E. Reade (1986: xxxiii) has noted that it does contain intrusive elements. This text is used by Bongenaar in his study of the Ebabbar temple at Sippar (Bongenaar 1997: 397) and is not discussed in Sack's study of the reign of Neriglissar (Sack 1994).

The four earliest texts that have been identified as attesting to the reign of Nabonidus are listed in Parker and Dubberstein 1956: 13. The dates and provenances given for them there are:

15-II-accession year (May 25, 556), NA.ŠU.ŠA.KU.NA—BE 8/1 39 (CBS 6101)
1-III-accession year (June 9, 556), Sippar?—VAS 6 65 (VAT 5031)
18-III-accession year (June 26, 556), Sippar—Strassmaier, *Nbn.* 1 (BM 75214, AH 83-1-18, 544)
23-III-accession year (July 1, 556), Uruk—YOS 6 1 (YBC 3986)

Based on this material, it has been suggested that the base of Nabonidus' support may have been at Nippur, where the earliest text, BE 8/1 39 (May 25, 556), was found (Beaulieu 1989: 87), and that texts from Sippar and Uruk continued to date events by the reign of Lābâši-Marduk after Nabonidus had replaced him as ruler. It has been suggested that Sippar and Uruk did so because they were not initially in support of the change in ruler. In particular, it has been

4 The text was kindly collated by J. Fincke, who noted that Evetts "copied what he thought or 'knew' was on the tablet, but not what is actually there."

thought that Sippar may have been reluctant to abandon its allegiance to Lābâši-Marduk because Lābâši-Marduk's family had a connection to Sippar, where his father (Neriglissar) may have had responsibilities over the Ebabbar and possibly even been a high official there before he became king.[5] While it would have taken several days for the news of a change in ruler at Babylon to reach Uruk, it would not have taken almost a month. No reason has been suggested for why Uruk would not have immediately recognized Nabonidus as king, but it has been suggested that when Nabonidus visited southern Babylonia at the beginning of his first regnal year, he held court at Larsa and dealt with matters involving the Eanna there, rather than at the more important city of Uruk, where the Eanna temple was located, because of tension between him and the establishment at Uruk due to the latter's "reluctance to recognize him as king the year before " (Beaulieu 1989: 88).

An examination of the evidence indicates that Nabonidus was not recognized as king as early as previously thought and that there is no clear proof that any overlap in dating according to the reign of Lābâši-Marduk occurred. Each of the four texts dated by the reign of Nabonidus mentioned above will be looked at in the order given by Parker and Dubberstein.

(1) BE 8/1 39 (CBS 6101)—The tablet is a debt note, stating that five *kur* of dates are to be paid on the seventeenth day of Tašrītu (VII). Collation of the tablet shows that Clay's copy of the date upon which the text was composed is not accurate. The text was not composed on the fifteenth day of Ayyāru (ITI.GU$_4$), but rather on the fifteenth day of Ulūlu (ITI.KIN), thus on September 20, not May 25. The text has:

(ITI.KIN). The debt was to be paid just over a month after the tablet was composed. The tablet actually states that it was drawn up at URU.*na-šu*? *šá ku-na-a* (see Zadok 1978: 324),[6] which may have been located in the region of Nippur because the tablet was found there.

(2) VAS 6 65 (VAT 5031)—Kugler had argued that VAS 6 65 shows that Nabonidus was already king in the month of Ayyaru since he was on the throne on the first day of Simanu (Kugler 1924). VAS 6 65 reads:

5 See Wiseman 1956: 38. Beaulieu (1989: 87 n. 17) notes that the undated Neo-Babylonian economic text CT 55 673 mentions a Neriglissar who was *qīpu* of Ebabbar; see also MacGinnis 1994: 31–32 no. 33.

6 Note, however, that this place is not listed in Zadok 1985. Collation shows that the reading of the third sign as ŠU is not certain.

1 [x x (x)] x.ḪI.A *šá* MU.SAG.NAM.LUGAL.E ^{md}AG-I LUGAL TIN.TIR.⌜KI⌝
2 [TA U$_4$.1.KÁ]M *šá* ITI.SIG$_4$ EN *qí-it šá* ITI.ŠE 10 ITI *u$_4$-mu* 3 ÉRIM.M[EŠ]
3 [x (x) *dul*?]-*lu* GIŠ.IG ESIR.ḪI.A 9(ILLIMU$_4$) ME ÉRIM.MEŠ *ina* UGU *pa-ḫe-e* GIŠ.MÁ.⌜MEŠ⌝
4 TA U$_4$.3.KÁM *šá* ITI.KIN EN U$_4$.28.⌜KÁM⌝ 26 *u$_4$-mu* ÉRIM.MEŠ *gab-bi*
5 *ina* UGU É *ziq-qur-rat* 8 LIM 5 ME 80 ÉRIM.MEŠ *dul-lu i-te-ép-šu*

Although it has been stated that this text indicates that Nabonidus was ruler on the first day of Simanu, this is not certain. Even if [TA U$_4$.1.KÁ]M is to be restored at the beginning of line 2—and this is a completely reasonable restoration—the text would state that work had been carried out in the accession year of Nabonidus from the first day of Simanu until the end of Addaru. While the text *may* indicate Nabonidus was on the throne on the first day of Simanu, all we can be sure of is that he was on the throne at the end of Addaru. The work may have begun before his accession to the throne, during the reign of Lābâši-Marduk, but the scribe saw no need to state that fact since Nabonidus had replaced him and the first months of the year could reasonably be thought to be part of Nabonidus' accession year.

Kugler (1924) argues that the ziggurat mentioned in line 5 must be that of Sippar, in large part because he feels it is unlikely to be ones at Harran, Larsa or Ur; however, Beaulieu (1989: 30–31 and 34) argues that work on that structure probably took place in the last years of the king's reign, after he had returned from Tema. Thus, the place at which this document was drawn up must remain uncertain.

(3) Strassmaier, *Nbn.* 1 (BM 75214)—The beginning of the tablet, which includes the date, is only partially preserved. The text lists various sums of silver either paid to particular individuals or remaining to be paid to them. Almost a century ago, Weissbach questioned the reading of the (only partially preserved) month name in the date, suggesting that it might be ŠE (Addaru) rather than SIG$_4$ (Simanu). He states "das Monatsideogramm ist nicht vollständig erhalten und könnte unter Umständen XII und III sein, falls STR.s Wiedergabe des Zeichenrestes nicht ganz genau sein sollte" (Weissbach 1925: 360). However, collation of the tablet supports the reading of the month as Simanu: [ITI.SI]G$_4$ U$_4$.18.KÁM MU.SAG.N[AM.LUGAL.L]A ^{d}AG-I LUGAL TIN.TIR.KI (line 4). The traces of the month are:

Thus, Nabonidus was clearly king by the eighteenth day of Simanu in his accession year.

This text comes from the British Museum registry collection AH 83-1-18,544, where the AH stands for Abu Habbah, the modern name for ancient Sippar. J.E. Reade (1986: xxxiv) has noted that while much of the material with the registry number AH 83-1-18 came from Sippar, "there is no reason to suppose that all of it did." However, the fact that the Ebabbar temple is mentioned in line 3 (*a-na é-babbar-ra id-*[*din*?]) of the introductory section of the text (lines 1–4) and that two of the people appearing in the text have names than include the theophoric name Šamaš (Šamaš-šarru-uṣur and Šamaš-zēru-ibni, lines 7 and 21 respectively) would suggest that this text may well have come from Sippar.

(4) YOS 6 1 (YBC 3986)—The date is not found at the end of the tablet, but rather at the beginning of the text. The first five lines read:

1 ITI.SIG$_4$ U$_4$.23.KÁM
2 8 MA.NA SÍG.ḪI.A [md]AG-*ú-ṣur*
3 *it-ta*[!](text: GA)-*ši*
4 MU.SAG.NAM.LUGAL.LA
5 [md]AG-I LUGAL E.KI

The text carries on with a list of amounts and the names of individuals. As noted by Goetze (1944: 44), "It seems doubtful whether III/23 can be acknowledged as the date of the Uruk tablet ... It is true that the tablet begins with the date III/23 in line 1, but the *šattu rēš šarrūti* is mentioned not before line 3, a full line separating the two principal parts of the alleged date. No other Uruk tablet of Nabū-na'id is earlier than the sixth month of the accession year." I agree with him that we cannot be certain that this text was composed on 23-III of Nabonidus' accession year, although it may very well have been. It must be noted that this text does not actually state that it was composed at Uruk, although it does seem to fit in the Eanna archive. Thus, we cannot be certain that this text was composed on 23-III of Nabonidus' accession year or be absolutely certain it was composed at Uruk.

In conclusion, an overlapping in the dating of documents according to the reign of a recently deceased king and that of the new king would not be surprising at even the most peaceful transition of power due to the time necessary to get news of the transition in authority in Babylon to outlying regions of Babylonia. Such an overlapping would also not be surprising when the transition in power occurred as the result of the violent overthrow or murder of the previous ruler and/or when there was some uncertainty about who exactly would become the new king and/or when there was some uncertainty if the new ruler would not also be deposed in the immediate future. As is well known, Berossos states that

Lābâši-Marduk (Laborosoardokhos), although a child, was murdered by his friends because of his evil ways; the conspirators then came to an agreement that one of their number, Nabonidus (Nabonnedos), become the new ruler (Verbrugghe/Wickersham 2001: 60; Beaulieu 1989: 88–90).[7] Thus, an overlap in the dating of texts by Lābâši-Marduk and Nabonidus would not be unexpected. However, the earliest clear attestation of Nabonidus with the title king is not on May 25 (15-II; BE 8/1 39) at a site possibly close to Nippur in central Babylonia, as has previously been assumed, but on June 26 (18-III; Strassmaier, *Nbn.* 1), likely at Sippar, in northern Babylonia. VAS 6 65 cannot be taken as proof that he had ascended the throne—or been acknowledged as ruler—by June 9 (the first day of Simanu), although this certainly remains possible. Lābâši-Marduk is last mentioned in a date formula from Sippar on June 20 (12-III), although that the actual date the tablet was composed may have later than that, but still in Simanu (Evetts, *Lab.* 1). While it is possible that there was an overlap in dating documents according to the reigns of Lābâši-Marduk and Nabonidus, there is no unequivocal proof of such an overlap. Since the earliest text(s) dated by the reign of Nabonidus likely come from Sippar, there is no reason to assume that Sippar continued to date texts by the reign of Lābâši-Marduk instead of the new king Nabonidus after the latter seized the throne because the family of Lābâši-Marduk had a connection to that city or to assume that Nabonidus' base of support came from central Babylonia (i.e., the area around Nippur). Nor is there any evidence that texts at Uruk were dated to the reign of Lābâši-Marduk after Nabonidus assumed the kingship and thus no reason to assume that Nabonidus went to Larsa rather than Uruk early the next year due to antipathy toward the latter city for not having acknowledged his rule immediately.

Bibliography

Beaulieu, P.-A.

1989 *The Reign of Nabonidus, King of Babylon 556–539 B.C.* YNER 10. New Haven: Yale University Press.

Beljawski, V.A.

1971 "Der politische Kampf in Babylon in den Jahren 562–556 v. Chr." Pp. 197–215 in *In Memoriam Eckhard Unger: Beiträge zu Geschichte, Kultur und Religion des alten Orients*, ed. M. Lurker. Baden-Baden: Valentin Koerner.

7 Nabonidus referred to Lābâši-Marduk as "his (Neriglissar's) young son who had not (yet) learned proper behavior," DUMU-*šu ṣa-aḫ-r*[*i*] / *la a-ḫi-iz ri-id-di* (Schaudig 2001: 517 3.3[a] iv 38′–39′).

Bongenaar, A.C.V.M.
1997 *The Neo-Babylonian Ebabbar Temple at Sippar: Its Administration and Prosopography*. PIHANS 80. Leiden: Nederlands Instituut voor het Nabije Oosten.

Brinkman, J.A. and Kennedy, D.A.
1983 "Documentary Evidence for the Economic Base of Early Neo-Babylonian Society: A Survey of Dated Babylonian Economic Texts, 721–626 B.C." *JCS* 35: 1–90.

Da Riva, R.
2008 *The Neo-Babylonian Royal Inscriptions: An Introduction*. GMTR 4. Münster: Ugarit-Verlag.

Dougherty, R.P.
1929 *Nabonidus and Belshazzar: A Study of the Closing Events of the Neo-Babylonian Empire*. YOR 15. New Haven: Yale University Press.

Goetze, A.
1944 "Additions to Parker and Dubberstein's Babylonian Chronology." *JNES* 3: 43–46.

Kümmel, H.M.
1979 *Familie, Beruf und Amt im spätbabylonischen Uruk: prosopographische Untersuchungen zur Berufsgruppen des 6. Jahrhunderts v. Chr. in Uruk*. ADOG 20. Berln: Mann.

Kugler, F.X.
1924 "Nachweis, dass Nabū-na'id bereits im Airu 556 v. Chr. Herr der Stadt Sippar war." Pp. 405–408 in *Sternkunde und Sterndienst in Babel: assyriologische, astronomonische und astralmythologische Untersuchungen*, volume 2/2/2. Münster: Aschendorffsche Verlagsbuchhandlung.

MacGinnis, J.
1994 "Neriglissar *qīpu* of Ebabbar" *NABU* 1994/2: 31–32 no. 33.

Parker, R.A. and Dubberstein, W.H.
1956 *Babylonian Chronology 626 B.C.–A.D. 75*. Brown University Studies 19. Providence: Brown University Press.

Reade, J.E.
1986 "Rassam's Babylonian Collection: The Excavations and the Archives." Pp. xiii–xxxvi in *Catalogue of Babylonian Tablets in the British Museum, Volume VI: Tablets from Sippar 1*, ed. E. Leichty. London: British Museum Publications.

Röllig, W.
1983 "Lābāši-Marduk." *Reallexikon der Assyriologie und Vorderasiatischen Archäologie* 6/5–6: 409.

Sack, R.H.

1976 "Some Remarks on Sin-Iddina and Zērija, *qīpu* and *šatammu* of Eanna in Erech ... 562–56 B.C." *ZA* 66: 280–291.

1994 *Neriglissar—King of Babylon*. AOAT 236. Kevelaer: Butzon & Bercker; Neukirchen-Vluyn: Neukirchener Verlag.

Schaudig, H.

2001 *Die Inschriften Nabonids von Babylon und Kyros' des Grossen samt den in ihrem Umfeld entstandenen Tendenzschriften: Textausgabe und Grammatik*. AOAT 256. Münster: Ugarit-Verlag.

Verbrugghe, G.P. and Wickersham, J.M.

2001 *Berossos and Manetho, Introduced and Translated: Native Traditions in Ancient Mesopotamia and Egypt*. Ann Arbor: University of Michigan Press.

von Voigtlander, E.N.

1963 *A Survey of Neo-Babylonian History*. PhD dissertation. University of Michigan.

Weissbach, F.H.

1925 "Zur neubabylonischen Chronologie." *StOr* 1: 359–369.

Wiseman, D.J.

1956 *Chronicles of Chaldaean Kings (626–556 B.C.) in the British Museum*. London: British Museum Publications.

Zadok, R.

1978 "The Nippur Region during the Late Assyrian, Chaldean and Achaemenian Periods Chiefly According to Written Sources." *IOS* 8: 266–332.

1985 *Geographical Names According to New- and Late-Babylonian Texts*. RGTC 8. Wiesbaden: Reichert.

CHAPTER 15

Intercalary Months in Achaemenid Elamite Administrative Documents from Persepolis

Matthew W. Stolper

Francesca Rochberg began a Festschrift contribution on periodicities and period-relations by citing the example of the calendric relationship 19 sidereal years = 235 lunar months. She remarked that the nineteen-year cycle of intercalation was standardized in the last quarter of sixth century BCE (that is, approximately during the reign of Darius I) and fixed by the second year of Xerxes (Rochberg 2014: 211, 215). Part of the evidence for this consensus (e.g., Britton and Walker 1996: 46, Hunger and Pingree 1999: 200; Britton 2007: 121f.; Hunger 2009: 132) has come from correlating Babylonian sources with Achaemenid Elamite administrative texts from Persepolis.

Synchrony between Old Persian and Babylonian months became evident when the versions of the inscriptions of Darius I at Bisotun (DB) were deciphered. Where the Old Persian and Elamite versions dated the killing of Gaumata to the 10th day of the month Bāgiyādi (Old Persian: *Bāgayādaiš māhyā* 10 *raucabiš ϑakatā āha* [i 55f.]; Elamite: 10 *nan* ^{AN}[ITI^{MEŠ} ^{AN}]*Bagiyatišna pirka* [i 42f.]), for example, the Babylonian version dated it to the 10th day of Tašrītu (UD.10.⸢KÁM⸣ *ša* ITI DU$_6$ [23]), and similarly wherever day number

* The Persepolis Fortification Archive Project at the Oriental Institute, from which these observations arise, has received timely support from the Andrew W. Mellon Foundation, the Farhang Foundation, the Getty Foundation, the Iran Heritage Foundation, the National Endowment for the Humanities, the National Geographic Society Committee for Research and Exploration, the PARSA Community Foundation, the Roshan Cultural Heritage Institute, the University of Chicago Women's Board, and other donors and organizations. Thanks are due to Annalisa Azzoni, Gian Pietro Basello, Michael Jursa, Wouter F.M. Henkelman and Seunghee Yie for comments, information, references, and suggestions. Responsibility for errors of substance and judgment remains with the author.

Draft editions and images of many of the unpublished Persepolis Fortification texts and tablets cited here by the sigla Fort. and PF-NN (many of the editions still preliminary, uncollated and uncorrected) are displayed on line by the Persepolis Fortification Archive Project at the Online Cultural and Historical Research Environment (OCHRE, see https://oi.uchicago.edu/research/ochre/projects.html). High-quality images of many of these tablets are also available online from InscriptiFact (see http://www.inscriptifact.com/).

 | DOI 10.1163/9789004363380_016

and month name were preserved (Schmitt 2003: 16; de Blois 2006: 43; tabulated in Basello 2000: 81 table 14).

As evidence from Achaemenid Elamite administrative texts excavated at Persepolis began to emerge, Arno Poebel (1938) used it to establish the sequence of the Old Persian months, filling in the gaps left by the main Bisotun text. He also observed the mention of an intercalary sixth month of year 19 in the text eventually published as PF 0894 (1938: 136ff.). Poebel took the mentions of intercalation as his first evidence for the absolute position of the sixth month (1938: 138 n. 22). He also invoked the intercalated VIb of year 19 to support his erroneous attribution of the Fortification Archive to the reign of Artaxerxes I (1939, 301f.).

When Richard Hallock established that the Archive comes from the reign of Darius I, he added new references to intercalations in years 16, 19, 22 and 24 in as yet unpublished Persepolis Fortification texts. He could assert that the preponderance of Persepolitan evidence outweighed a single questionable contradictory Babylonian text (1942: 231). When George Cameron published Elamite administrative texts from the Persepolis Treasury Archive, it was evident that intercalary months mentioned there from regnal years of Darius I, Xerxes, and Artaxerxes I also corresponded exactly to the same months in the same regnal years known from Babylonian texts (1948: 34). With Hallock's publication (1969) of more than 2,000 Elamite texts from the Fortification Archive, the Aramaic epigraphs on Elamite texts also demonstrated continued use of Babylonian month names at Persepolis and their correspondence with Elamite and transcribed Iranian month names.[1] The Elamite documents further corroborated the correspondence of Achaemenid Persian, Achaemenid Elamite and Babylonian intercalary months for the middle regnal years of Darius I.

As early as 1942, Parker and Dubberstein cited Persepolis texts quoted by Poebel and communicated by Cameron and Hallock among evidence for

---

1 *ʾb* (month V, e.g., PF 0968, PF 1816, PF 1819), *ʾlwl* (month VI, e.g., PF 0855, PF 0858, PF 0968), *tšry* (month VII, e.g., PF 1717, PF 1820), *mrḥšwn* (month VIII, e.g., PF 1717), *šbṭ* (month XI, e.g., PF 0867), *ʾdr* (month XII, e.g., PF 0867); Basello 2000: 127; 2002: 15, Schmitt 2003: 16 . 5. Also, in epigraphs on Elamite Fortification documents available after Hallock 1969: *n(y)sn* (month I, e.g., PF-NN 1228, Fort. 1740-001), *kslw* (month VI, PFa 08). Months II, IV and X (*ʾyr, tmwz, ṭbt*) are not yet attested in epigraphs on Elamite tablets, but *ṭbt* occurs in a monolingual Aramaic Fortification text (PFAT 021), and transcribed as TEB⌜H⌝T in the Fortification text in Greek script (Fort. 01771: Stolper and Tavernier 2007: 3f., 24 with literature). No intercalary months are attested in Aramaic epigraphs or monolingual Aramaic documents. An Achaemenid Elamite administrative document of unknown provenience refers to month I of regnal year 15 (probably of Darius I) by the transcribed Babylonian-Aramaic name, ^{AN}*nu-šá-an-na*! (YBC 16813:07, see Jones and Stolper 1986: 249ff.).

intercalary months in the Babylonian calendar (1942: 7; 1956: 7f.). As recently as 1997, Walker remarked on the correspondence of intercalary months in Babylonia and Persepolis, illustrating the complementarity of the sources with a table that shows denser evidence from Persepolis than from Babylonia to document some intercalations (1997: 23f.; Basello 2000: 131ff.).

The Persepolitan evidence for intercalations in the years documented by the Fortification Archive—regnal years 13–28 of Darius I, but with occasional references to administrative transactions as early as year 4—has become still denser as about 4,200 additional Elamite Fortification texts have become available, and the larger sample adds to the Achaemenid Elamite terminology and usage with which intercalations are marked.

Usage

Poebel discussed one of the Elamite words used to mark intercalary months (*meša(na)* 1938: 137f.). Hallock highlighted the specific chronological usage of other words with wider uses in his glossary (1969 s.vv. *appukana*, *-memana*, *mešana*), nuances also taken up in the dictionary of Hinz and Koch 1987, and noted by Basello 2002: 15, Schmitt 2003: 17, and de Blois 2006: 44. Much of what follows merely supplements Basello's compendious survey (2000: 127ff., 179ff.).

Explicit Identification of Intercalary Months

mešana (vars. *mešanna, mešan, mešanu,* and *miš*⌈*na*⌉ [Fort. 1659-101:06])

Poebel 1938: 137f., taking the lemma as *meša*, "second, other, following, later, further, additional etc."; Hallock 1969: 730 "'intercalary' ... lit. pps. 'additional,'" similarly Basello 2000: 127; Hinz and Koch 1987: 916f. "*nachträglich ... nachträglich hinzugefügt* im Sinne von *Schalt*-Monat ... nachträglich (*eingeschoben*)...". Hallock adduced the element *meša* in adverbial expressions in royal inscriptions (*mešameraka* "apart from" [Hallock: "in addition to"], corresponding to Old Persian *apataram* (*hačā*) "apart from, elsewhere," Akkadian *elat* "except, in addition"; *mešamerašae* "later," corresponding to Old Persian *aparam* "later," Akkadian *ana arkātu ūmu*), also attested once in independent adverbial use (supplementary rations for personnel *appa meša kinnuka* "who came later (or: in addition)" PF-NN 2393:06).

Contrasted with *appukana*, "first, former, earlier": PF 1943:03, PF-NN 2493: [02], 10, 19, Fort. 1952-101:08′, [17′] (referring to months VI and VIb of Darius I year 19). Contrasted with *irpi* "previous, former": PT 10 r.3 (referring to months VI and VIb of Xerxes year 2); PT 27:13 (referring to months XII and XIIb of Xerxes year 12).

All attestations modify transcribed Iranian month names (Karbašiyaš [VI], Miyakannaš [XII]); none modify Elamite month names (Belilit [VI], Aššetukpi [XII]).[2]

Identifies these months:

VIb Darius I 19	(PF 0660, PF 0894, PF 1943, PF-NN 2493 [6 times], Fort. 1952-101, Fort. 2132-104, Fort. 2282-101);
XIIb Darius I 22	(PF 0764, PF-NN 1558, Fort. 2321-102, Fort. 1659-101);
XIIb Darius I 24	(PF 0246, PF 0297, PF 0765, PF 0998, PF 0999, PF 1257, PF 1844, PF-NN 1062, PF-NN 1172, PF-NN 1984, PF-NN 1752, PF-NN 2187);
XIIb Darius I 27	(PF 0864, PF-NN 1789);
VIb Darius I 30	(PT 1963-19);
XIIb Darius I 32	(PT 2, see Cameron 1965: 187);
VIb Xerxes 2	(PT 10, PT 11);
XIIb Xerxes 12	(PT 27);
XIIb Artaxerxes I 5	(PT 79);

XIIb of year without preserved number (PF 0973).

(-)*memana*

Hallock 1969: 729 "'intercalary(?).' ...cf. -*memana* used after numerals"; Hinz and Koch 1987: 913 concur. If not a repeated error for *mešana*, backformed from 2-(*um*)*memana*, "second" (with possessive-attribute ending -*na*), as Hallock hints.

Identifies month VIb (with ditto-ed transcribed Iranian name) of Darius I year 19 (Fort 1901A-101); month XIIb (with transcribed Iranian name) of Darius I year 22 (PF 0869, PF 0870, PF 0966, PF 1718:07; PF-NN 1486, restored in PF-NN 2470).

2-(*um*)*me*(*ma*)*na*

"Second" (with the same markers of ordinals used *passim* with year-numbers, numbers of tablets in a series and elsewhere). Also in the phrase *appa*

2 Here and with respect to other markers of intercalation, this is a limit only on attestation, not on usage, since Elamite month names occur far less often than transcribed Iranian names, and almost exclusively in texts from the northwestern parts of the administrative area covered by the Persepolis archive, in the direction of Susa (the Fahliyān region, Henkelman 2008: 44, 88, 118).

2-memanna lit. "which (is) the second," (Stolper 2004: 76). Contrasted with *appukana*, "first, former, earlier" (Fort. 1555-101:15′, Fort. 2025-101:28′, 32′ vs. 34′). All attestations modify transcribed Iranian month-names.

Identifies these months:

XIIb Darius I 16	(PF 0881, Fort. 1937-101, Fort. 1555-101, Fort. 2025-101);
VIb Darius I 19	(PF 1790);
XIIb Darius I 22	(PF 1814, PF-NN 1870, Fort. 0863-110, Fort. 0875-102);

anomalous month XIIb Darius I 23 (Fort. 2340-103, in a context that also cites year 24, below).

2-edana

Hallock 1969: 684 "'second, for two'"; Hinz and Koch 1987: 391 "*also, damit.*" Probably an inanimate counterpart of the grammatical marker found in 2 (or: 3)-*be*-(*ud*)-*da*-*na*, "both, all three" (Hinz and Koch 1987: 177).

Identifies transcribed Iranian month XII and XIIb[3] of Darius I year 27 (PF 0879:09, cf. Basello 2000: 129; 2002 15 n. 18).

KI+MIN, "ditto"

In PFS 0873 and PF 0875, both referring to month XIIb of Darius I year 22 (Basello 2002: 15 n. 18); cf. ^{AN}KI+MIN (= ^{AN}*karbašiyaš*) *mešana* Fort. 1952-101:[17′], 28′, referring to month VIb of Darius I year 19, ^{AN}⸢KI+MIN⸣ (= [^{AN}] ⸢*miya*⸣*kan*[*naš*]*na*) *2-memanna* Fort. 0863-110:14f., referring to month XIIb of Darius I year 22; ^{AN}KI+MIN (= ^{AN}*karbašiš*) *appuka*(*na*) Fort. 1901A-101:09′, 11′, ^{AN}KI+MIN KI+MIN-*na* (= ^{AN}*karbašiš appukana*) ibid. 12′, all referring to pre-intercalary month VI of Darius I year 19.

Repeated Month Name

mikannaš kutte mikanna "month XII and month XII(b)" PF-NN 0221:15f. in a list of 5 months, Darius I year 22 implied.

Explicit Identification of Pre-intercalary Months

appuka(*na*)

Hallock 1969: 669 *appuka*: "'formerly,'" *appukana* "'first (tablet…),' 'former' (law…), 'earlier' (sixth month, in contrast to *mešana*, 'later' or 'intercalary'…), lit. 'of formerly.'" Hinz and Koch 1987: 78 *appuka*: "*vordem, früher*"; *appukana* "*erste,*

3 That is, understanding *Mikannaš 2-edana* as "both months XII," reducing the emendation required by the arithmetic of the text to the name of month XI, <Samiyamaš>.

wörtlich *frühere.*" The short form, without possessive-attributive *-na*, mostly in adverbial use in administrative texts, corresponds in royal inscriptions to Old Persian *paruvam* "previously," Akkadian *ina panātu-* "before" (citations in Hinz and Koch 1987: 78[4]). In calendric context *appukana* is usual; occasional uses of *appuka* (Fort. 1901A-101:05′, 06′, 08′, 09′ vs. *appukana* ibid. 11′, KI+MIN-*na ibid.* 12′; Fort. 1952-101:28′ vs. *appukana* ibid. 05′, 17′) are probably not erroneous, considering both the number of attestations and the counterpart variants *irba, irpi* ~ *iripna* (below). All attestations are with transcribed Iranian month names.

Contrasted with *mešana* "later" and 2-*memana* "second" (above). Used alone, identifies these months:

pre-intercalary XII Darius I 16	(Fort. 2025-101:34′);
pre-intercalary VI Darius I 19	(Fort. 1901A-101:05′ etc., Fort. 1952-101:05′ etc.);
pre-intercalary XII Darius I 22	(PF-NN 1460, Fort. 1229-107, Fort. 1736-102).

irpi (*irpina, irba*)

Hallock 1969: 704 "'former, formerly.' Adjectival phrase *irpi* + *-na* serves as sg. adj." A marginal note in Hallock's copy of Hallock 1969 adds "P[re]A[chaemenid] E[lamite] *ur*(*i*)*pu*(-)." Hinz and Koch 1987: 778 "*früher, vorher,* synon. ap-pu-qa ... mE dürfte das Stichwort urp gelautet haben, vgl. mE ur-pu-ub-ba, ur-pu-up-pi." The short form, without possessive-attributive—*na*, corresponds in a difficult passage of DB to Old Persian *frataram* (iii 26; elsewhere without temporal nuance, Schmitt 2014: 177), Akkadian *maḫrû*. Hallock recognized a cognate adverbial form *ir-ba-en* "previously" in PFa 30:8 (Hallock 1978: 130, with Hinz and Koch 1987: 767), *ir-pi-in* in PF-NN 0812 (Hallock marginalia, reading with Henkelman [personal communication] against Hinz and Koch 1987: 778), corroborated by *ir-pan-na* (Fort. 1975-101:22′, 24′, 26′, 28′, 30′) and *ir-ba* (Fort. 1912-103:53′, 54′), all in similar context, referring to travel completed "previously," and by *mušimme sap ir-ba-en*! (text: HU) *lika* "an account was delivered as before" Fort. 1248-102:05. The phrase *tuppi ir-pa-*⌈*na*⌉ Fort. 2027-104:05′, in broken context at end of text, replaces the frequent *tuppi hi appukana*, "this is the first tablet" (recording a journal for a given commodity, place and period).[5]

4 In DB Elamite i 55f., the phrase *sap appa anka appuka* probably does not correspond to the counterpart Old Persian *pasāva yaϑā* "after" (i 72, so Hallock 1969: 669; Hinz and Koch 1987: 78) but to the counterpart Akkadian *libbû ša panātūa* (29, cf. Schmitt 1980: 113).

5 Hallock 1978: 116 suggested that the account PFa 32 is the second tablet implied by the phrase that tags the journal PF 1943 as the first tablet (*tuppi hi appukana*, line 41). In other journals,

Contrasted with *mešana* "later, etc." (PT 27); contrasted with (-)*memana* (Fort. 1901A-101).

Identifies these months:

pre-intercalary VI Darius I 19	(Fort. 1901A-101:51′ and probably 17′ and 18′);
pre-intercalary VI Xerxes 2	(PT 10);
pre-intercalary XII Xerxes 12	(PT 27),
all with transcribed Iranian month names.	

Implicit Indications of Intercalary Months

Passages in some texts indicate intervals from month N through month N_2, a total of x months, where the total implies that an intercalary month is counted, for example ^{AN}ITI^{MEŠ} 10-*na* … ^{AN}ITI^{MEŠ} ^{AN}*karbašiyanamamar kuš* ^{AN}ITI^{MEŠ} ^{AN}*šurumarnama* "for 10 months … (they consumed the rations) from month VI through month II" (i.e., 9 months plus an intercalary VIb or XIIb of an unspecified year) PF 0975 (Hallock 1969: 281 n. b to PF 0975). Other examples include PF-NN 2493:28 (7 months from month III through month VIII of year 19[6]); PF-NN 0728:13–15 (10 months, from month VI of Darius I year 22 through month II of year 23). This list is not exhaustive; more examples are likely to appear when unpublished texts, especially journal fragments, are reviewed and collated. Comparable passages indicate intervals that do <u>not</u> include known intercalary months (below).

PF-NN 2589 refers to 13 months of year 24[7] (Henkelman 2008: 286 n. 648), without naming the intercalary month of that year (texts of the same form, Hallock's Category K1, commonly record allocations for a full calendar year, below). In PF 1736, the total amount entered for a full year implies a thirteen-month year 24 (Hallock 1969: 475 note a to PF 1736).

---

the tablet number runs as high as 14 (*tuppi hi* 14-*me-na* Fort. 1310-101:02″). Contrast [*tup*]*pi* 2-⌜*umme*⌝*manna tar*⌜*mak*?⌝, probably "second and final tablet" Fort. 1916B-101:30, see Azzoni and Stolper 2015: 56f.

6 The intercalation is explicit in earlier entries in the same journal (02, 10, 19) and it is elided in another entry (25, below).

7 The year-number 24 is clear in the Elamite cuneiform text (collated by Henkelman [personal communication]), but the number is 23 in the Aramaic docket (personal communication, Annalisa Azzoni). A comparable anomaly (or error) appears in Fort. 2340-103, which refers first, in a broken context, to year 24, then to month XIIb of year 23 (collated).

beptika

The only pertinent term used in connection with an Elamite month name as well as transcribed Iranian month names, *beptika*, prompts confusion and ingenuity. It is attested until now only in memoranda recording outlays of rations, written at stations to the northwest of Persepolis (Fahliyān region, Henkelman 2008: 286 n. 648). It has not appeared in serial lists of such transactions compiled at Persepolis.

Qualifying Hanamaka, Iranian month X, Darius I year 19: PF 1069, PF 1070, PF 1073; qualifying Aššetukpi (and variants), Elamite month XII: PF 1046, PF 1049, PF 1057, PF-NN 1469, all Darius I year 22, and Fort. 2346-103, Darius I year 20+⸢2?⸣; qualifying Hadukannuya, Iranian month I, Darius I year 23: PF 1053, PF-NN 1223. That *beptika* qualifies an intercalary month is assured by PF 1046, which counts the interval from Elamite month X through Elamite month XII *beptika* as 4 months.[8]

Hallock 1950: 242 n. 22; 1969: 75, 677 "'intercalary,' (twelfth month), 'shifted(?)' (first or tenth month following an intercalation)," drawing attention (1950: 242 n. 22) to cognate verbal forms of *bepti-* in DB, meaning "'to revolt' ('to shift [allegiance]'?")."

On Hallock's view that *beptika* may refer not only to an intercalary month but also to a later month displaced by a recent intercalation Hartner (1985: 747) pronounced "It goes without saying that this is utterly improbable." He proceeded from the suppositions that *beptika* always and only means "intercalary" and that the 19-year intercalary cycle was not established in Persia until the 24th year of Darius. He inferred that different Elamite and Persian notions about key astronomical events on which calendars were fixed produced intercalations that would be unacceptable in Babylonia—Xb in year 19 and Ib in year 23. He confessed that the outcome of two different practices of intercalation in the same environment was difficult to accept, but concluded that no other explanation of the "apparent contradictions in the surviving texts" was reasonable. He excluded the possibility of better evidence from additional texts of the kind adduced here. Among things that "went without saying," Hartner did not consider the other Persepolitan evidence for intercalary months in years 19 and 22, or the use of cognates with *beptika* either in royal inscriptions[9] or in administrative texts, or who might have promulgated a heterodox local

8 Hallock took this indication of intercalation as the primary evidence for the absolute positions of the Elamite months (1969: 75).

9 But the volume editor, Ilya Gershevitch, inserted a supporting note (*apud* Hartner 1985: 748 n. 1) commenting on the uses of *bepti-* in the meaning "rebel," which suggested to him a common sense of "pretension." Intercalary months, he suggested, were thought of as "pretended,

intercalation, or how regional administrative and recording systems might have dealt with two or even three simultaneous modes of intercalation.

As others have commented, this argument and the conclusion are insupportable. Hinz and Koch (1987: 181f.), Basello (2002: 15 n. 18), Schmitt (2003: 17 n. 9), de Blois (2006: 44 n. 26) and Henkelman (2008: 286 n. 648) all found Hartner's apparent ignorance of *mešana* or other explicit markers of intercalation and of the well-documented intercalations of years 19 and 22 to be fatal to his thesis and its implications. Koch (*apud* Hinz and Koch 1987: 181f., translating *beptika* as "*in Verzug geraten, verzögert ...* in Verbindung mit Schaltmonaten," concurring with Hallock) suggested that the passages in which *beptika* followed months other than XII were lapses of inattentive scribes. Henkelman admitted the possibility that the region where all the texts that use *beptika* were produced was for a time out of step with the generally regular intercalation practice.[10] De Blois reasserted Hallock's interpretation that *beptika* "does not mean 'intercalated' but rather, more broadly, 'displaced by intercalation'" (2006: 44). He elaborated this view in two ways, procedural and philological.

As to procedure, de Blois explained that "scribes at Persepolis," needing to make sure that outlays over periods of several months would add up correctly, were anxious to insert reminders of whether the year in question had twelve or thirteen months. Intercalary and post-intercalary months were "hostile" (as de Blois glosses *beptika*) because they threatened accuracy. As an example of the possibility of error, he adduced PT 79, where the interval from month VII through month XIIb (*mešana*) of Artaxerxes I year 5 is incorrectly given as 6 months rather than 7.

The texts that use *beptika*, however, were memoranda of single transactions. They are not the work of scribes at Persepolis. The late-stage register-type documents that were compiled at Persepolis (journals and accounts, in

---

soi-disant," and he compared Middle Persian references to intercalary days as "stolen" ("quite likely in continuation of Achaemenid usage").

10 But ignorance of regular intercalation cannot have been general in the region. In the letter-order PF 1790, the director of the Persepolis institution authorizes outlays at four locales in the region (Hunar, Liduma, Hidali, and Atek) during 19 months, including 13 months of year 19, specifically naming the intercalated VIb of that year.

A pre-Achaemenid "Babylonian" intercalation is attested in the Fahliyān region of western Fars. The Neo-Babylonian legal text BM 79013 is dated at Hādalu (Hidali) in Babylonian XIIb of the accession period of one of the Neo-Elamite rulers named Tammaritu, (Leichty 1983: 153f.; Brinkman and Kennedy 1983: 61; cf. Henkelman 2003: 255 n. 6), either the intercalation of 654 BCE (-653 in Britton 2007: 121 fig. 5) or the intercalation of 646 BCE (-645). The form and language of the document, the curse on violators, and the names of parties are all Babylonian. But who decreed the intercalation?

Hallock's terminology), where accurate yearly or multi-year totals were a concern, indicate intercalation in the other ways indicated above, or else disregard it (below).[11]

As to philology, de Blois's appendix (2006: 49–51) discussed the stems *bepti-* and *peti-* (*piti-*)[12] in connection both with the Old Persian counterpart *hamiçiya-* and with the Achaemenid and pre-Achaemenid Akkadian counterparts *nakāru, nakru*. He observed not only that in the versions of DB and DNb, forms of Elamite *pepti-* correspond to forms of Akkadian *nakāru* and its cognates as well as to phrases with Old Persian *hamiçiya-* but also that in the oldest known continuous Elamite text the animate noun *pitir* almost certainly means "enemy" (König 1965 No. 2) and also that in the single Middle Elamite quasi-bilingual *pitir*, "enemy" corresponds to Akkadian *nakru* (Steve 1967 Nos. 31–32). On the basis of the correspondence with Akkadian *nakāru*, he revived Herzfeld's proposal to derive *hamiçiya-* not from the consensus **miça-* "contract" (literature in Schmitt 2014: 189) but from the root represented by Old Persian *miϑa-* "wrong, falsehood" (Herzfeld 1938: 184ff., 259ff.). He concluded that months called *peptika*, "which the underlying Persian dictation thus presumably called *hamiçiya-*," were understood as "hostile" or "opposing" (2006: 51).[13]

Attested Achaemenid Elamite forms of *peti-* (*piti-*), however, are not confined to royal inscriptions that refer to people as hostile and to administrative texts that characterize months as vexatious. The intransitive-passive form *pitika* is a frequent marker of a debit item in livestock accounts; Hallock proposed "transferred" (1969: 745) and Hinz and Koch (1987: 225) preferred "*verloren gegangen*." Most germane among other contexts are passages in letters that discuss the change of the seal of Parnaka, the chief administrator of the Persepolis institution, where *pitika* is collocated with *appuka*, "previous(ly)," the word elsewhere

11 PT 79 does not really illustrate arithmetic confusion arising from intercalation. It accounts for the seven-month interval from VII through XIIb in two parts, potentially overlapping, one of 6 months, the other of 4 months. The omission of a month does not account for the arithmetic error in the text's total (corrected by Hallock 1960: 100).

12 For de Blois these are graphic variants, not morphologically or lexically distinctive.

13 The presumed "underlying Persian dictation" apparently assumes Gershevitch's conclusion that the Elamite of the administrative tablets always and everywhere represented contents that were spoken and read out as Old Persian or a colloquial Iranian dialect (Gershevitch 1971, 1979, 1983). This implausible view of things (see especially Henkelman 2011: 586–595, 614–622) is especially inappropriate to the texts that use *beptika*, written in the Fahliyān region, since other texts from the same region are rich in telltales of Elamite morphology and syntax that contrast with apparent calques–"elamographs" in Gershevitch's terminology—in texts from other places (Henkelman 2008: 87f.).

used to identify pre-intercalary months: *halmi appa appuka unini halmi hube pitika* "the seal that previously (was) mine, that seal is *pitika*" PF 2067:10–12, PF 2068:12–14. The same collocation recurs in *halmi appukana pitik* "the previous sealed (document authorizing receipt of travel rations) is *pitik*" PF-NN 1465:09. The same nuance, though without contrasting *appuka*, is in *halmi mušinna pitika inni tingiš* "the sealed (document) of account is *pitika*, they did not send it" Fort. 1203-101: 01" and Fort. 1761-101:29 (Fisher and Stolper 2015: 9).[14] In none of these contexts is a nuance like "hostile" tenable. In all of them "replaced," "displaced," or "lost" is required.

As a glance at the Akkadian dictionaries suffices to recall, uses of Akkadian *nakāru*, the long-standing counterpart of Elamite *peti-*, are also not confined to royal inscriptions referring to hostility. Primary and derived stems of the verb also indicate (among other nuances) alteration, removal, discarding, and change of position (but not, evidently, temporal displacement).

It is plausible, then, to suppose that *peti-* and *nakāru* had partly overlapping ranges of meaning: in the royal inscriptions, having to do with hostility and resistance, but also in practical texts, having to do with change and displacement. In that case, Old Persian *hamiçiya-* reflects part of the range of Elamite *peti-* (*bepti-*) and Akkadian *nakāru* in a high rhetorical register (with the narrow ranges of concern typical of royal incriptions—construction, legitimation, and *res gestae*). It is irrelevant to the meanings of *peti-* and *bepti-* in lower registers (with the wider ranges of concern that administrative and legal documents serve).

The core of this suggestion, that *peti-* corresponds to Akkadian *nakāru* and carries the meaning of "change" as well as "be hostile," was already put forward by Cameron (1942: 216, 1948: 96f., citing the texts that would be published as PF 2067 and PF 2068 and translating *pitika* as "lost"). Probably because Cameron made this observation to help explain words that he read incorrectly,[15] it has been ignored.

I do not suggest that administrative uses of Elamite *bepti-*, *piti-* "translate" Akkadian *nakāru*, but I also do not accept that they render an underlying Old Persian *hamiçiya-*. Elamite *peti-* and Akkadian *nakāru* have wide ranges of meaning and usage, only part of which corresponds to the narrow range of Old Persian *hamiçiya-*. Like de Blois I concur with Hallock's interpretation,

---

14 In Fort. 1761-101:29, read *hal-mi* (not: ŠE.BAR^{MEŠ}) (collated).

15 **gisa betiš* for *gisapattiš* (Iranian **gaiθapatiš*, "overseer of livestock," Tavernier 2007: 422), **betikamaš* for *battikamaš* "message, report" (Iranian **patigāma-* Tavernier 2007: 410). It seems likely that a similar idea lay behind Hallock's suggested translations of *pitika* and *beptika*.

that *beptika* refers to months as "displaced," just as other administrative uses of *pitika* refer to livestock, seals, and documents as "displaced," "replaced," or "lost." Unlike de Blois, I doubt that any of the administrative uses of *beptika* or *pitika* convey a nuance of "hostility." On this line of surmise, *beptika* "displaced," that is, "disordered" is an infrequent replacement for *mešana* "later," to identify an intercalary month, roughly in the same way *irba* (*irpi*) is an infrequent replacement for *appukana* to identify a pre-intercalary month.

Summary of Usage

XII/XIIb Darius I 16	2-(*um*)*me*(*ma*)*na*;
	appuka(*na*)
VI/VIb Darius I 19	*mešana*, *-memana*, 2-(*um*)*me*(*ma*)*na*, KI+MIN;
	appuka(*na*), *irba*, KI+MIN;
	implicit
XII/XIIb Darius I 22	*mešana*, *-memana*, 2-(*um*)*me*(*ma*)*na*, KI+MIN, MN+MN, *beptika*;
	appuka(*na*);
	implicit
XIIb Darius I 24	*mešana*
	implicit
XII/XIIb Darius I 27	*mešana*, 2-*edana*
VIb Darius I 30	*mešana*
XIIb Darius I 32	*mešana*
VI/VIb Xerxes 2	*mešana*;
	irpi
XII/XIIb Xerxes 12	*mešana*;
	irpi
XIIb Artaxerxes I 5	*mešana*

Omission of Intercalary Months

What remains unclear is why a few clerks in one part of the Persepolis administrative region needed to highlight post-intercalary months as "displaced," a fact of no use or interest to the central accounting offices at Persepolis itself,[16] particularly considering that others wrote documents that pass over intercalary months in silence.

PF 0743 and Fort. 1808-102, for example, record outlays for divine offerings for 12 months of Darius I year 22, called a whole year (*bel kappatinna*), without including intercalated XIIb (Hallock 1969: 224 n. b). PF-NN 2589 offers a counter-example, a terse document of the same kind recording outlays for the 13 months of year 24 (above, n. 7). Most documents of this kind record issues for 12-month periods, whether or not the year ended with an intercalation (e.g., PF 745, PF 746, PF 749, PF 758, noted by Hallock 1969: 224ff.), or even included an intercalation (PF-NN 2087, year 19). These 12-month outlays seem to anticipate a separate issue for the intercalary month (Henkelman 2008: 210).

The expectation of separate memoranda for issues in intercalary months can also account for other documents that refer to 12-month intervals in years with intercalated months[17] or shorter intervals ending with a pre-intercalary VI or XII.[18] With less comfort, the same expectation might account for memoranda that refer to intervals that omit intercalary VIb[19] or intervals that omit month XIIb.[20] The scarcity of documents that fulfill this expectation could be a sampling artifact. For all examples, counterpart texts of the same formal and functional types are available that <u>do</u> include the intercalary months, that is, that do not anticipate separate memoranda for issues in intercalary months.

16 I am also at a loss to explain 1 ^{AN}ITI^{MEŠ} *hatuma bel 27-na* ^{AN}ITI^{MEŠ} *hanamakašna irpina* "(rations received) during 1 month of year 27, former? month X" PF-NN 2509:05–09, unless *irpina* is not adjectival, modifying the month name, but adverbial, describing the following ration amounts as "formerly" on hand "for" the recipient (PN-*na* ... *nimak*). It cannot imply a "displaced" month X, since the intercalation of year 27 was the third month following, XIIb.

17 Of lexical interest is ^{AN}ITI^{MEŠ} *ziklina kuš* ^{AN}ITI^{MEŠ} *aššetupikna da-la-ru* [PAP] 12 ^{AN}ITI^{MEŠ} *hatu*[*ma be*]*l 22-mena* "the interval ? from (Elamite) month I through (Elamite) month XII, during [a total of] 12 months, year 22" Fort. 0247-101:07–12. Other examples include PF 0776, PF 0967, PF 1060 (year 22), PF-NN 0836, PF-NN 0963, PF-NN 1053, etc. (year 24) PF 0794 (year 27).

18 Examples include PF 1746, PF-NN 0082 (year 19), PF 0910, PF 0997, PF 1001, PF 1050, PF 1051, PF 1172 etc. (year 22), PF 0862, PF-NN 0438, PF-NN 0563, PF-NN 1388 etc. (year 24).

19 Examples include PF 0299 (months VI, VII, VIII, IX, total 4 months, year 19), PF 1159 (month I through month XII, total 12 months, year 19), PF 1705 (months IV, V, VI, total 3 months, and 15 days of month VII, year 19), PF-NN 1832 (VI through XII, total 7 months, year 19).

20 Examples include PF-NN 0223 (6 months, from month IX [year 22 implicit] through month II, year 23), PF-NN 1255 (month XII of year 22, month I of year 23, total 2 months), PF-NN 1553 (months XI, XII [year 22 implicit], I, II, III, IV, V,VI, VII, VIII, IX, X, year 23).

Especially suggestive is PF-NN 1999, a letter-order from Parnaka. The date on which the order is delivered is indicated as day 24, month VII of Darius I year 22. The unusual dating may arise from the fact that the order authorizes both past and future payments for a 12-month period, namely, months IV, V, VI, VII, VIII, IX, X, XI, and XII, summed up as 9 months of year 22, as well as months I, II, and III, summed up as 3 months of year 23. It strongly suggests that as of mid-year the general director of the Persepolis administrative institution was not aware that there was to be an intercalated XIIb—because it had not yet been announced? because the 19-year intercalary cycle was not yet fixed, predictable practice? It also suggests that at least of some of the memoranda mentioning intervals that omit intercalary months were drawn up before the intercalations were promulgated.[21]

But how are we to interpret entries in journals that omit intercalary months? These are serial lists of transactions at district centers, compiled at Persepolis to document inputs, outlays, and balances for administrative periods of half a year, a whole year or several years. These are the documents that ought to show some concern for things adding up. And they were not drawn up in ignorance of intercalations, because they were compiled after the accounting periods were past.

Hallock noted such omissions in PF 1943:12ff., 22f., 32f., 35f. (1969: 529 nn. j, l, v, and y), a journal recording grain transactions in Darius I year 19,[22] that also has an entry that <u>does</u> note the intercalated VIb (02–04). The same omission occurs in PF 1942:15ff., also in a journal compiling grain transactions in year 19. PF 1957, compiled in year 23 but listing grain transactions in year 22 (35f.), enters outlays for a period from month V (year 22 implicit) through month II (year 23 implicit), counted as 10 months, hence omitting XIIb of Darius I year 22 (ll. 12–13; the grossly erroneous amount of the entry is booked as part of the total outlay of year 22 [Hallock 1969: 565 nn. k and s]).[23]

Especially suggestive in this connection is PF-NN 2493, listing wine transactions in year 19. It mentions the pre-intercalary VI and/or intercalary VIb in six

21 PF 1790 (above n. 10), another letter-order from Parnaka, undated, offers a counter-example. It authorizes outlays of sheep/goats during 19 months, including month VIb of year 19. Comparison with PF-NN 1999 allows us to suppose that at least part of the authorization was retroactive, that the order was given late in the 19-month period, after the intercalation was announced or even after it had elapsed. Other damaged letter-orders authorize payments for intercalated VIb of year [19] (Fort. 2132-104, perhaps not a letter-order, but a receipt of Hallock's Category H) and for pre-intercalary month XII of year 22 (Fort. 1229-107).

22 With a counterpart balanced account, PFa 32, see Hallock 1978: 116.

23 In PF 1092, the apparent omission of an intercalated XIIb of year 22 is certainly an illusory product of a scribal error, as Hallock's note explains (1969: 317 n. d).

entries (above). It also includes an entry for outlays in a period of 8 months (24ff.), from month III through month VI, and from month VII through month XII. The immediately following entry (27f.) covers an interval of 7 months, from month III through month VIII, including implicit VIb. The conspicuous omission of intercalary VIb in line 26 must therefore be deliberate.

The individual entries in these journals were based on memoranda of outlays that were collected at district centers and sent to Persepolis to be compiled (Hallock 1969: 55f.; Jones and Stolper 2008: 30f.; Henkelman 2008: 136ff.; Azzoni and Stolper 2015: 5ff., 43ff., etc.). If the source texts included an intercalated month, the journal entry included it. If the sources did not include an intercalated month, the journal entry omitted it. But if, in the ordinary run of things, separate source texts for outlays in intercalary months were expected, the journal did not express the expectation or treat the absence of the information as an error, as other sources of error were sometimes noted (Stolper 2017).

These oversights suggest that at and around Persepolis in the reign of Darius I, intercalations sometimes affected administrative practice and recording, and sometimes they did not. At the time and place of the primary transactions, when commodities were deposited, transferred or disbursed, known intercalations were documented, but other intercalations were not always anticipated. Down the information stream, when periodic accounts were compiled, the focus was on summing up documents (rather than realities), even if the documents did not always account for the whole year, and even then the sums were sometimes slipshod (Hallock 1969: 55).

I do not suggest that in intercalary months some workers went without rations, livestock without fodder, or gods without offerings because some administrators did not have a grip on the calendar. I suggest only that managers of stockpiles held some supplies in reserve for unscheduled payments of the kind implied by some letter-orders and of the kind required by unanticipated, unannounced intercalations.

A modern observer who works closely with administrative archives becomes aware that the "normal run of things" painstakingly reconstructed from minute documentation was never the operation of clockwork. One of the reasons for which the records were made and preserved was to police performance, which entailed error checking, but opportunities for non-systematic error abounded. Erratically recording outlays of rations, fodder, and offerings and then including these vagaries in final accounts was within the realistic range of error—perhaps the acceptable range of error—of the Persepolis administrative institutions. Even if some local administrators thought of royal expedients meant to bring the sun, moon and months together as annoying "displacements," nevertheless intercalations and their consequences provoke more perplexity in modern observers than they did in most ancient clerks.

Conclusion

For practical purposes, a single calendar and a single regime of intercalation prevailed throughout the Persepolis administrative area in the reign of Darius I and his successors. Months called by Iranian, Elamite and Aramaic-Babylonian names were the same months. The months and intercalations were synchronous with those of Babylonia.

Among the few Persepolis documents from the earliest documented years of the Fortification Archive the expected XIIb in Darius I year 13 is not attested. Minimum numbers of attestations (explicit or implied) of subsequent intercalations in Fortification texts are as follows (updating Walker 1997: 23; Basello 2000: 133 table 26):[24]

XIIb Darius I 16:	8 attestations in 4 texts.
VIb Darius I 19:	31 attestations in 14 texts.
XIIb Darius I 22:	28 attestations in 28 texts.
XIIb Darius I 24:	13 attestations in 13 texts.
XIIb Darius I 27:	3 attestations in 3 texts.

Attestations in Persepolis Treasury texts are unchanged:[25]

VIb Darius I 30:	1 attestation in 1 text.
XIIb Darius I 32:	1 attestation in 1 text.
VIb Xerxes 2:	2 attestation in 2 texts.[a]
XIIb Xerxes 12:	1 attestation in 1 text.[b]
XIIb in Artaxerxes I year 5:	1 attestation in 1 text.

a Walker 1997: 23 enters 3 attestations, but I find only PT 10 (referring to pre-intercalary VI and intercalated VIb), and PT 11.

b Walker 1997: 23 enters 2 attestations, but I find only PT 27 (referring to pre-intercalary XII and intercalated XIIb).

24 Texts qualifying months as *beptika* are included. Mentions of pre-intercalary months are included. Entries in journals, each representing one or more source documents, are counted separately. Numbers are minimums because further implicit attestations are expected (above).

25 Cameron 1948: 34 (cf. 14f.) with Hallock *apud* Cameron 1965: 182, cited by Walker 1997: 25 n. 14.

In pre-Achaemenid Assyria and Babylonia, notices of intercalations were framed as pronouncements of the king (ABL 401, 1258 [see Parpola LAS 2: 285, SAA 13 4 and 5], ABL 956 [Esarhaddon, see Parpola LAS 190, SAA 10 253], YOS 3 115 [Nabonidus, see Parpola LAS 2: 504]) or communicated between temple administrators (YOS 3 15 and 196 [Cyrus or Cambyses, see Parpola LAS 2: 504]). The existence of such notices reflects the circumstance that the nineteen-year intercalary cycle was not yet predictable in practice (Parker and Dubberstein 1956: 1; Hunger 1980: 298). Comparable royal or official notices of intercalations to be observed have not appeared in the Persepolis archives. Who decreed intercalations—or, perhaps, on whose calculations or advice Darius decreed them—and how they were communicated throughout the administrative networks is not indicated.

The Persepolitan evidence is consistent with the consensus that the nineteen-year cycle was not established in civilian use until about 500 BCE and perhaps not until the early reign of Xerxes. The combination of dense documentation of intercalary months with occasional implications that intercalations were overlooked and especially the letter-order PF-NN 1999 suggest that intercalations were not predictable in the most basic administrative operations before 500 BCE. If the great organizer Darius ever said to his scholars "Give me a calendar," as Walker ventured to imagine (1997: 24), it was not until late in his reign. But why (to continue this fantasy) would he do so? What frustrated clerk or underappreciated scholar could persuade Darius or Xerxes that the predictability of intercalation was more important than the monarch's prerogative of decreeing it?

Abbreviations

DB, DNb	multilingual Achaemenid royal inscriptions cited according the the sigla assigned by Kent 1953, extended by Mayrhofer 1978 and Schmitt 2009
Fort.	(as in Fort. 1659-101) Elamite Persepolis Fortification tablet or text recorded by the Persepolis Fortification Archive Project; the first four digits indicate the box from which the tablet came (Hallock 1969: 1, Jones and Stolper 2008: 37ff.) and the last three digits are an arbitrary identification number
MN	month name
PF	Elamite Persepolis Fortification text published in Hallock 1969
PFa	Elamite Persepolis Fortification text published in Hallock 1978

PF-NN	Elamite Persepolis Fortification text cited from draft edition by Richard T. Hallock, collated and corrected by Wouter F.M. Henkelman
PN	personal name
PT	Persepolis Treasury text published in Cameron 1948
PT 1963	Persepolis Treasury text published in Cameron 1965

Bibliography

Azzoni, Annalisa and Matthew W. Stolper

2015 "From the Persepolis Fortification Archive Project, 5: The Aramaic Epigraph *ns(y)ḥ* on Elamite Persepolis Fortification Documents." *ARTA* 2015.004. (http://www.achemenet.com/document/ARTA_2015.004-Azzoni-Stolper.pdf).

Basello, Gian Pietro

2000 Problemi Calendariali nelle Fonti Elamiche di Età Achemenide. Laurea Thesis in Iranistica, Università degli Studi di Bologna.

2002 "Elam and Babylonia: the Evidence of the Calendars." Pp. 13–36 in *Ideologies and Intercultural Phenomena*, ed. A. Panaino and G. Pettinato. MS 3. Milan: Università di Bologna / IsIAO.

de Blois, François

2006 "Lunisolar Calendars of Ancient Iran." Pp. 39–52 in *Proceedings of the 5th Conference of the Societas Iranologica Europaea, I: Ancient & Middle Iranian Studies*, ed. A. Panaino and A. Piras. Milan: Mimesis.

Brinkman, J.A. and D.A. Kennedy

1983 "Documentary Evidence for the Economic Base of Early Neo-Babylonian Society: a Survey of Dated Babylonian Economic Texts, 721–626 BC." *JCS* 35: 1–90.

Britton, John P.

2007 "Calendars, Intercalations and Year-Lengths in Mesopotamian Astronomy." Pp. 115–132 in *Calendars and Years: Astronomy and Time in the Ancient Near East*, ed. J.M. Steele. Oxford: Oxbow.

Britton, John and Christopher Walker

1996 "Astronomy and Astrology in Mesopotamia." Pp. 42–67 in *Astronomy before the Telescope*, ed. C. Walker. London: British Museum.

Fisher, Michael T. and Matthew W. Stolper

2015 "Achaemenid Elamite Administrative Tablets, 3: Fragments from Old Kandahar, Afghanistan." *ARTA* 2015.001. (http://www.achemenet.com/pdf/arta/ARTA_2015.001-Fisher-Stolper.pdf).

Cameron, George G.

1942 "Darius' Daughter and the Persepolis Inscriptions." *JNES* 1: 214–218.

1948 *Persepolis Treasury Tablets*. OIP 65. Chicago: University of Chicago Press.

1965 "New Tablets from the Persepolis Treasury." *JNES* 24: 167–192.

Gershevitch, Ilya

1971 "Editor's Preface." Pp. 1–9 in R.T. Hallock, The Evidence of the Persepolis Tablets (advanced printing of Chapter 11 of *The Cambridge History of Iran, 2: The Median and Achaemenian Periods* [1985], ed. I. Gershevitch). Cambridge: Cambridge University Press.

1979 "The Alloglottography of Old Persian." *TPS* 1979: 114–190.

1983 "Extrapolation of Old Persian from Elamite." Pp. 51–56 in *Kultur und Geschichte der Achämenidenzeit und ihr Fortleben*, ed. H. Koch and D.N. Mackenzie. AMI 10. Berlin: Dietrich Reimer.

Hallock, Richard T.

1942 "Darius I, the King of the Persepolis Tablets." *JNES* 1: 230–232.

1950 "New Light from Persepolis." *JNES* 9: 237–252.

1960 "A New Look at the Persepolis Treasury Tablets." *JNES* 19: 90–100.

1969 *Persepolis Fortification Tablets*. OIP 92. Chicago: University of Chicago Press.

1978 "Selected Fortification Texts." *CDAFI* 8: 109–136.

Hartner, Willy

1985 "The Old Iranian Calendars." Pp. 714–792 in *The Cambridge History of Iran, 2: The Median and Achaemenian Periods*, ed. I. Gershevitch. Cambridge: Cambridge University Press.

Henkelman, Wouter F.M.

2003 "Defining Neo-Elamite History." *BiOr* 60: 251–263.

2008 *The Other Gods Who Are: Studies in Elamite-Iranian Acculturation based on the Persepolis Fortification Texts*. AchHist 14. Leiden: Nederlands Instituut voor het Nabije Oosten.

2011 "Cyrus the Persian and Darius the Elamite: a Case of Mistaken Identity." Pp. 577–634 in *Herodot und das Persische Weltreich/Herodotus and the Persian Empire*, ed. R. Rollinger, B. Truschnegg and R. Bichler. CLeO 3. Wiesbaden: Harrassowitz.

Herzfeld, Ernst

1938 *Altpersische Inschriften*. AMI 1. Berlin: Dietrich Reimer.

Hinz, Walther and Heidemarie Koch

1987 *Elamisches Wörterbuch*. AMI 17. Berlin: Dietrich Reimer.

Hunger, Hermann

1980 "Kalender." *RlA* 5 3/4: 297–303.

2009 "Schaltmonat." *RlA* 12/1–2: 130–132.

Hunger, Hermann and David Pingree

1999 *Astral Sciences in Mesopotamia*. HdOr 44. Leiden: Brill.

Jones, Charles E. and Matthew W. Stolper

1986 "Two Late Elamite Tablets at Yale." Pp. 243–254 in *Fragmenta Historiae Elamicae, Mélanges offerts à M.J. Steve*, ed. L. De Meyer, H. Gasche and F. Vallat. Paris: Recherche sur les Civilisations.

2008 "How Many Persepolis Fortification Tablets Are There?" Pp. 27–50 in *L'archive des Fortifications de Persépolis: État des questions et perspectives de recherches*, ed. P. Briant, W.F.M. Henkelman and M. Stolper. Persika 12. Paris: de Boccard.

Kent, Roland G.

1953 *Old Persian: Grammar, Texts, Lexicon*. 2nd rev. ed. AOS 33. New Haven: American Oriental Society.

König, Friedrich Wilhelm

1965 *Die elamischen Königsinschriften*. AfOB 16 Graz: Ernst Weidner.

Leichty, Erle

1983 "Bel-epuš and Tammaritu." *AnSt* 33: 153–155.

Mayrhofer, Manfred

1978 *Supplement zur Sammlung der altpersischen Inschriften*. SÖAW, phil.-hist. Kl. 338. Vienna: Österreichische Akademie der Wissenschaften.

Parker, Richard T. and Waldo H. Dubberstein

1942 *Babylonian Chronology 626 B.C.–A.D. 45*. SAOC 24. Chicago: University of Chicago Press.

1956 *Babylonian Chronology 626 B.C.–A.D. 75*. BUS 19. Providence, RI: Brown University Press.

Poebel, Arno

1938 "The Names and the Order of the Old Persian and Elamite Months during the Achaemenian Period." *AJSL* 55: 130–141.

1939 "The King of the Persepolis Tablets: the Nineteenth Year of Artaxerxes I." *AJSL* 56: 301–304.

Rochberg, Francesca

2014 "Periodicities and Period Relations in Babylonian Celestial Sciences." Pp. 211–218 in *Extraction and Control: Studies in Honor of Matthew W. Stolper*, ed. M. Kozuh, W.F.M. Henkelman, C.E. Jones and C. Woods. SAOC 68. Chicago: Oriental Institute.

Schmitt, Rüdiger

1980 "Zur babylonischen Version der Bīsutūn-Inschrift." *AfO* 27: 106–126.

2003 *Meno-Logium Bagistano-Persepolitanum, Studien zu den altpersischen Monatsnamen und ihren elamischen Wiedergaben*. SÖAW, phil-hist. Kl. 705. Vienna: Österreichische Akademie der Wissenschaften.

2009 *Die altpersischen Inschriften der Achaimeniden. Editio minor mit deutscher Übersetzung*. Wiesbaden: Reichert.

2014 *Wörterbuch der altpersischen Königsinschriften*. Wiesbaden: Reichert.

Steve, M.-J.

1967 *Tchoga Zanbil (Dur-Untash), III: Textes élamites et accadiens de Tchoga Zanbil*. MDAI 41. Paris: Paul Geuthner.

Stolper, Matthew W.

2004 "Elamite." Pp. 60–94 in *The Cambridge Encyclopedia of the World's Ancient Languages*, ed. Roger D. Woodard. Cambridge: Cambridge University Press.

2017 "Investigating Irregularities at Persepolis." Pp. 741–823 in Die Verwaltung im *Achämenidenreich–Imperiale Muster und Strukturen/ The Administration of the Achaemenid Empire—Tracing the Imperial Signature*, ed. B. Jacobs, W.F.M. Henkelman and M.W. Stolper. CLeO17. Wiesbaden: Harrassowitz.

Stolper, Matthew W. and Jan Tavernier

2004 "From the Persepolis Fortification Archive Project, 1: An Old Persian Administrative Tablet from the Persepolis Fortification." *ARTA* 2007.001. (http://www.achemenet.com/pdf/arta/2007.001-Stolper-Tavernier.pdf).

Tavernier, Jan

2007 *Iranica in the Achaemenid Period (ca. 550–330 B.C.)*. OLA 158. Leuven: Peeters/Departement Oosterse Studies.

Walker, C.B.F.

1997 "Achaemenid Chronology and the Babylonian Sources." Pp. 17–25 in *Mesopotamia and Iran in the Persian Period: Conquest and Imperialism 539–331 B.C.*, ed. J. Curtis. London: British Museum.

CHAPTER 16

The Importance of Experts: Agents in the Transfer of Astral Knowledge between Hellenistic Mesopotamia and the Greek-speaking World

Zoë Misiewicz

In the second century BCE, the Greek citizens of Larisa in Thessaly erected a monument honoring the *Chaldaios astronomos* Antipatros for his contributions to their polis. The term "Chaldaean" was eventually to become synonymous with "astrologer" among Greek and Roman authors, and this inscription provides some of the most concrete evidence for a Chaldaean astronomer/astrologer making a contribution in the west. More often than not, solid epigraphical evidence of this sort is unfortunately lacking when we seek to understand the relationship between Mesopotamian astronomers/astrologers[1] and the Greco-Roman world; the existence of some connection is evident in various texts (and Francesca Rochberg has made numerous valuable contributions to the analysis of this evidence), but the individuals involved in the sharing of knowledge tend to be either unknown or known only from relatively unreliable second- or third-hand sources. Nevertheless, by combining the epigraphic evidence with the narrative accounts, we can begin to construct a plausible picture of some of the agents responsible for the cross-cultural sharing of astronomical/astrological ideas in the Hellenistic period. This picture can be clarified by contrast with a situation that seems to have been less effective for the sharing of knowledge.

Both epigraphical evidence and common narrative accounts suggest that experts who established themselves for the long term in new locales played a significant role in the spread of Mesopotamian astrological/astronomical knowledge to the Greek-speaking world, while stories of Alexander's experiences in Babylon suggest that non-experts, even those in close contact with

1 One point that Rochberg (e.g., 2004: 11) has made most forcefully is that the distinction between Mesopotamian astronomers and astrologers is a modern one. In Mesopotamia, at least in the early first millennium BCE, these professionals were known as *ṭupšarrū Enūma Anu Enlil*, scribes of (the celestial omen series) *Enūma Anu Enlil*, and their responsibilities included both observing celestial phenomena in the sky and interpreting their significance for affairs on Earth.

 | DOI 10.1163/9789004363380_017

experts, were not always effective in passing on knowledge of the practices that they witnessed. The term "experts" is deliberately a rather vague one, meant to indicate those scholars working within the Mesopotamian tradition who had a particular knowledge of the astral sciences. In the Neo-Assyrian period, they were known as *ṭupšarrū Enūma Anu Enlil*, and the extensive sources that survive from that time present us with a detailed picture of their activities; surviving documentation for later periods is unfortunately less complete, though the title at least persists. However, the goal of this contribution is not to delineate the precise titles and responsibilities of these scholars within Hellenistic Mesopotamia and the surrounding regions, but rather to consider some of the ways in which they came into contact with Greek-speakers and spread some forms of their knowledge to the Greco-Roman world. That Mesopotamian astral knowledge of various sorts did appear in various Greek texts is at this point beyond dispute, thanks in part to the contributions of Francesca Rochberg,[2] but we are left with the vexing question of how exactly it got there. This is not a question that can receive a definitive answer on the basis of the sources currently available to us, but it is to be hoped nevertheless that a consideration of a few specific cases will prove valuable.

The first point I will make here is perhaps an obvious one, that close but relatively casual contact between Greek-speakers and Mesopotamian experts is not in itself sufficient for the effective transfer of scholarly knowledge, as will be demonstrated by a consideration of narratives concerning Alexander and his entourage among the Babylonians. To counterbalance this negative argument, I will then present evidence for another approach to the sharing of knowledge: the migration of Near Eastern[3] experts to Greek-speaking regions, where they could settle for the long term and gain renown for their knowledge, potentially passing on their expertise to students over many years. The successful migration of a Near Eastern *astronomos* to a Greek-speaking region is demonstrated particularly by the case of the Antipatros, which is

2 For example, Rochberg-Halton (1984); Rochberg-Halton (1987).

3 I use this term cautiously, aware of Van Dongen's (2014) warning that it can give a misleading appearance of unity to disparate regions while overemphasizing the differences between these regions and Greece. Alternative terms present other problems. The material under consideration here is the cuneiform astrological/astronomical tradition, which was centered in Mesopotamia (specifically Babylon and Uruk in the Hellenistic period), while the people involved were often described as "Chaldaeans," including one who was a native of northern Syria. This is a discussion of interactions between groups, so although their different names may suggest a separation between them, the contacts between them should be foremost in mind.

documented by solid epigraphic evidence. The plausibility of a Near Eastern immigrant establishing a school of astronomical knowledge in his new home is supported by the stories concerning Berossos, which are problematic in their details but nevertheless provide evidence that Greek and Roman authors saw this as something that might well be done.

Alexander among the Babylonians

Mesopotamian textual sources offer considerable evidence for the acceptance of foreign conquerors by the cuneiform scribal elite. Clancier (2011: 761) has discussed the situation in Hellenistic Uruk in particular, emphasizing that "[t]he old urban nobility tried to please the political powers". Traditional Mesopotamian scribal culture had already survived the Persian conquest of 539 BCE, finding a home in the temples even when Akkadian was no longer the language of administration,[4] and this continuity may have depended on the flexibility of the scholars in adapting to new circumstances. The quick acceptance of foreign conquerors seems to have applied in the case of Alexander as well; after his conquest of the Persian empire, the writers of the astronomical diaries (a cuneiform text genre that included economic and political statements along with astronomy) were quick to begin referring to their new ruler by the traditional title "king of the world" (LUGAL $\check{S}U_2$), as seen in a diary for 331 BCE.[5]

The Greek tradition too presents "Chaldaean" experts as eager to appear helpful to their new rulers, although their motives were sometimes questioned.[6] The interactions of these experts with Alexander in particular are recounted in several sources, most notably Arrian's *Anabasis* (2nd century CE) and the *Library of History* of Diodorus Siculus (1st century BCE). These chief sources are histories written well after the fact, which means that their reliability is questionable, but in part this actually strengthens the point that I am making here: the interaction of Mesopotamian scholars with non-specialist Greeks

4 Clancier 2011: 758.

5 van der Spek 2003: 297, 298.

6 Smelik (1978: 108) has interpreted the stories about Alexander as evidence for Alexander's "misunderstanding and distrust towards the Babylonian priesthood." Diodorus Siculus (17.112) presents Alexander as initially heeding the warnings of the Chaldaeans, until a Greek philosopher persuaded him to disregard them, while Arrian (7.17) portrays Alexander as suspicious from the outset. As the divergences between the two accounts suggest, they are not reliable sources for this level of detail.

did not result in the effective communication of scholarly ideas. As a case in point, we will consider one particularly striking example: the reported interactions of Alexander and his entourage with the Babylonian scholars immediately preceding his arrival in Babylon and during his stay in that city. It must be emphasized yet again that these stories are second-hand (or worse), and are therefore to be taken with a very large grain of salt. Nevertheless, they may offer some valuable insights into the relationship between non-specialist foreigners and Mesopotamian scholars, with implications for the difficulty of knowledge exchange in such situations.

Both Arrian (7.16–17) and Diodorus Siculus (17.112) relate that the Chaldaeans warned Alexander of danger upon his approach to Babylon, with Diodorus attributing their knowledge to astrological interpretation. This narrative detail is perhaps not surprising; the Babylonians had a reputation for making predictions, and the accumulation of negative omens contributed to the story being told about Alexander in these histories, which culminated in his death in Babylon. The initial warning of the Chaldaeans might readily be seen as nothing more than a literary device. However, another related narrative has attracted more interest among modern historians, since a full understanding of the situation seems to rely on a greater knowledge of Babylonian practices than the Greek authors themselves possessed. Briefly, Alexander is said once to have arisen from his throne in Babylon, whereupon "some obscure fellow, a criminal on parole according to some"[7] came and sat upon it. The eunuchs allowed him to remain there, which was apparently a source of confusion for Alexander. To the Greek authors, this situation seems to have represented merely an interesting anecdote. Arrian imbues it with a sense of general foreboding, saying that it "strengthened the seers' impression that nothing good lay in store for Alexander."[8] However, modern scholars[9] have perceived this situation as a more deliberate performance of a standard Mesopotamian ritual, that of the substitute king. This was a ritual carried out in order to prevent calamity (i.e., death) from befalling the king upon the observation of a negative celestial omen, particularly a lunar eclipse. The king temporarily vacated his throne, which was occupied instead by a less important person, such as a criminal, who would suffer the negative consequences of the omen in the king's place. This substitute was anointed, dressed royally, and perhaps ultimately killed,

7 τῶν τινα οὖν ἠμελημένων ἀνθρώπων, οἱ δὲ καὶ τῶν ἐν φυλακῇ ἀδεσμῳ ὄντα λέγουσιν Arrian *Anabasis* 7.24, trans. Pamela Mensch. Diodorus (17.116) recounts a similar version of this story.

8 ᾗ δὴ καὶ μᾶλλον ἐπ' οὐδενὶ ἀγαθῷ ξυμβῆναι αὐτῷ οἱ μάντεις ἐξηγοῦντο: Arrian *Anabasis* 7.24, trans. Pamela Mensch.

9 Such as Briant 2002: 863; Nawotka 2010: 373; Ossendrijver 2015: 224–225; Smelik 1987: 102, with references to many other scholars.

while the king himself pretended to be a regular person; Neo-Asssyrian letters addressed to the king while this ritual was underway refer to him as "the farmer, my lord."[10]

Perceiving this ritual behind the mysterious circumstances described by Arrian and Diodorus would make sense of the Babylonians' seemingly strange behavior. If this interpretation is correct, and there was some attempt by the Babylonian scholars to carry out the substitute king ritual on behalf of Alexander, this brings us quickly to our next point about the circulation of knowledge: the meaning behind these actions seems to have been rapidly lost in the telling, if it was ever understood to begin with.[11] This has consequences for the role of laypeople, even those who were directly exposed to Mesopotamian divinatory practices, in passing on that knowledge to others. In brief, the presence of soldiers, courtiers, and other members of Alexander's entourage among the Babylonian scholarly elite, even while a ritual was being performed to ward off the consequences of a negative celestial omen, does not seem to have led to any particular understanding of Babylonian astronomical/astrological practices; at least by the time the story was recounted to another person, the whole purpose of the action seems to have been lost.

This highlights one of the differences between technical scholarly knowledge, such as that required for the interpretation of omens and the accomplishment of the relevant rituals, and other types of knowledge that might be exchanged more casually. Studies of knowledge exchange between the ancient Near East and Greece have often focused on myths and other narratives, where striking parallels have been observed. In considering how such narratives were shared, scholars such as López-Ruiz (2010: 36) and others have emphasized domestic contexts such as "[s]torytelling around the loom," while Haubold (2013: 23) lists various "mechanisms of transmission: domestic contexts, especially in bilingual populations; war and trade; supra-regional cults and religious festivals; and travelling experts." My goal here is to look at one particular type of knowledge, the astronomical/astrological knowledge that played a crucial role in Mesopotamian society, and consider which of these mechanisms might be more or less effective as a means of passing on this knowledge to others.

10 Parpola (1993) offers numerous examples, including the first two texts in the volume.

11 This argument is of course dependent on the assumptions that some event involving a stranger on Alexander's throne did in fact take place and that this stranger was meant to be a substitute king. These assumptions are not trivial. However, they are supported by two arguments: first, that the report of the strange incident serves no clear literary purpose, so it is less likely to have been inserted simply for the sake of the narrative; and second, that the ritual of the substitute king provides a logical explanation for behavior that otherwise appears inexplicable.

As seen in the case of Alexander among the Babylonians, conquest in itself was not always sufficient; conquerors with no particular interest in the details of the Babylonian astral sciences and their associated rituals might be present while one of these rituals was performed and yet fail to appreciate its significance or at least to explain its significance to others who were not present. Such interactions with Babylonian experts might well have fuelled the common stories about the Chaldaeans' great knowledge of the stars and their significance, but there remains a gap between impressionistic accounts of that sort and the more precise technical details discussed previously by Rochberg (Rochberg-Halton 1987).

Again, this is not a surprising conclusion, but it bears stating explicitly. Ultimately, everyone involved in this situation was driven by their own motivations; in the case of the historians, this meant presenting a certain narrative that contained an appropriate amount of drama and foreshadowing. Their goal was not to present an accurate description of Babylonian ritual practice or the astrological interpretations that lay behind it; nor, reliant as they were on other sources, would this have been a realistic possibility even had they desired it. When searching for the agents responsible for the dissemination of complex astrological formulations, it will prove more profitable to consider those with a particular interest in the subject, or at least a particular interest in exploiting the subject for their own ends. This brings us to the discussion of Berossos.

Berossos

Any discussion of Berossos is fraught with difficulties, particularly when he is considered in relation to astrology/astronomy, since his connection to these disciplines has been rejected outright by many.[12] I will survey this issue briefly, but my argument here is ultimately not dependent on resolving the question of what the historical Berossos did or did not do. Instead, I will emphasize the much simpler point that later Greek and Roman writers constructed narratives about his behavior that must have seemed plausible to their readers.

12 Most damning are the comments of G. De Breucker (2012: 27), the most recent editor of the Berossos fragments. He found that the astronomical-astrological fragments attributed to Berossos are Hellenistic in character and therefore not Babylonian, which led him to conclude that they were wrongly attributed to Berossos. De Breucker was by no means the first to reject these fragments, but his deep familiarity with the Berossos corpus gives his arguments extra weight.

Regardless of whether Berossos himself established a school of astronomical/ astrological learning on Cos, as he was reputed to have done, this was seen as reasonable behavior for a Mesopotamian-trained expert of astral science. The widespread idea of Berossos the "Chaldaean" astrological expert settling in Cos, combined with the more reliable epigraphic evidence of Antipatros that will be discussed in what follows, suggests that we must take seriously the idea of the immigrant scholar as a source of astral knowledge.[13]

Berossos is known primarily as the author of the *Babyloniaca*, a work that purports to tell the history of Babylonia to a Greek-speaking audience. This work has not survived intact, but fragments and testimonia were collected by Jacoby in his monumental work *Die Fragmente der griechischen Historiker*, which has now been updated as *Brill's New Jacoby*. Jacoby attributed the astronomical/astrological material to a "Pseudo-Berossos," and the authenticity of this material remains somewhat disputed (cf. Steele 2013). The Roman writer Vitruvius (fl. 1st c. BCE) reports that Berossos settled on the island of Cos and taught astrology there, but in light of Berossos' dubious connection to astrology in general, this account has been dismissed by some as purely fictive.

However, it may be possible to develop a more nuanced understanding of Berossos' relationship to astronomy/astrology, which can clarify the question of who might be considered a Chaldaean expert among the Greeks and Romans. This understanding depends upon several key points, which have been discussed in detail by Steele (2013): (1) the astronomy/astrology in the (pseudo-)Berossos fragments does not represent the highest level of Mesopotamian astral knowledge of that time; (2) these fragments do show some connection to broader Mesopotamian cosmological ideas, such as those found in *Enūma Eliš*; (3) the fragments show evidence of Greek influence; (4) Berossos was writing for a Greek-speaking audience; and (5) not all Babylonians, even those associated with the temple, would automatically have been familiar with the most advanced forms of astronomical/astrological knowledge.

These arguments, taken together, lead to another possible interpretation of Berossos' relationship with the astral sciences: that he did in fact profess to have some knowledge of these topics and used to his advantage the Chaldaeans' high reputation in the field, but was not conversant with the most complex technical astronomy that certain other Babylonian scholars were producing at the time. This is in keeping with Haubold's (2013: 16) view of how Berossos crafted an identity for himself in Book I of the Babyloniaca: "Berossos

---

13 I use the term "immigrant" rather than a phrase like Burkert's (1983) "itinerant" because the model that appears here is one of scholars establishing themselves for the long term in a single new locale, rather than moving continuously from place to place.

exploits longstanding convergences between Greek and Mesopotamian cosmogonic traditions… to present himself as a Chaldaean philosopher-priest and keeper of a universal archive of human knowledge," in part by "tap[ping] into orientalizing Greek fictions about Mesopotamia." To a Greek audience, the archetypal Chaldaean was an expert in the astral sciences whose knowledge was deserving of great respect, and Berossos could benefit from being perceived in this way even if his own astronomical/astrological knowledge was not so extensive.

This characterization of Berossos may resolve certain problems, but it also introduces new complexities regarding the role of experts in the circulation of knowledge. In particular, it raises the possibility that some of the experts involved may have been somewhat lacking in expertise. Mladen Popović (2014) has used the term "levels of learning" to account for the fact that the Jewish scholars of Qumran apparently had access to some forms of Babylonian astronomy but not to the most advanced developments. He explains this in terms of social networks, positing that "the Jewish scholarly elite may not have been of the same standing as the Babylonian scholarly elite." Moreover, even within the Mesopotamian temple milieu, not all scholars were equally involved in all fields of knowledge. As an example, we can consider the case of Nidinti-Anu, an exorcist at the Reš temple in Uruk who lived several generations after Berossos. He was a scribe or "owner" of several tablets, including one from the astrological series *Enūma Anu Enlil*, so that he can be considered in connection with the broader community of astronomers/astrologers.[14] The most likely implication of his "ownership" of tablets in the temple library of Uruk is that he held a temple position.[15] Yet Nidinti-Anu probably still did not "belong to the 'hard-core' astronomers—those who were active in the field of mathematical astronomy, and presumably observational astronomy."[16]

It should come as no surprise that the interaction of ideas across cultural boundaries is not a simple process, since it ultimately depends on individuals with individual aims and individual backgrounds. Nevertheless, what we have seen so far is that concepts can be easily misconstrued by lay observers who are focused on other goals, so that simply being in the company of experts does not in itself suffice for the transfer of knowledge; that ancient readers accepted the idea of an expert moving to a new land and establishing a school there

14 Ossendrijver 2011: 220.

15 Ossendrijver 2011: 220.

16 Ossendrijver 2011: 220.

as a likely means of spreading knowledge; and that even partial knowledge of an esteemed and exotic field was enough to reap the rewards of an expert persona.

Berossos' *Babyloniaca* presents evidence for an educated Mesopotamian making a deliberate effort to convey Mesopotamian knowledge of some sort to the Greek-speaking world, and Vitruvius' remark that Berossos established a school on Cos suggests that behavior of this sort was not considered unusual: regardless of whether Berossos himself moved to Cos to teach astrology, it was something that a Babylonian would conceivably do. The case of Antipatros, which will be discussed below, further strengthens this argument by presenting indisputable evidence that at least one "Chaldaean" expert in the astral sciences did in fact establish himself successfully in a Greek-speaking locale. However, the case of Berossos' purported astronomical/astrological knowledge should also serve as a warning that the Mesopotamian expert who makes an effort to share his knowledge with foreigners for his own benefit may not be quite the expert that he claims to be.

Antipatros

Not much more is known about Antipatros than the little mentioned above, derived from an honorary inscription produced in Larisa in the second century BCE. The body of the inscription reads as follows:

> Ἐπειδὴ Ἀντίπα-
> τρος Ἀντ̣ιπάτρου Ἱεροπολίτης τῆς Σελευκί-
> δος, πεπολιτογραφημένος [δὲ] ἐν Ὁμολίῳ
> ὑπάρχων Χαλδαῖος ἀστρονόμος, ἐνδη-
> μῶν ἐν τῇ πόλει ἡμ̣ῶν ἀπὸ χρόνων, ἔν τε
> τοῖς κατὰ τὴν ἀναστροφὴν πέφηνεν ἄξιος
> τῆς ἡμετέρας πόλεως καὶ τῆς ἰδίας πατρί-
> δος, ἔτι δὲ καὶ τοῦ μαθήμα̣[τος ο]ὗ μεταχειρί-
> ζεται [ἐπ]ιδίδωσιν [ἑαυτὸν---ε]νως τοῖς
> ἐντυγχ̣άνου̣σ̣ι̣[ν----------][17]

17 Gallis 1980: 250–251. To the best of my knowledge, this important inscription has not previously appeared in English translation. I have reproduced Gallis' Greek text without changes, except for the direction of the penultimate bracket.

> Since Antipatros, son of Antipatros, a Hieropolitan of Seleucis, who has been admitted to citizenship in Homolion, a Chaldaean *astronomos*, who has lived in our city for some time, during which, by his behavior, he has appeared worthy of our city and of his own fatherland, and also of the learning which he practices, devotes himself....

Syrian Hierapolis, the hometown of Antipatros, was a city with a long history. Located west of the Euphrates and about a hundred kilometres northeast of Aleppo, it was known as Nappigu in the early first millennium BCE before being renamed Līta-Aššur upon its conquest by the Neo-Assyrian ruler Shalmaneser III (r. 858–824).[18] The city then came to be known as Bambyke, an appellation that persisted into the middle of the first millennium CE,[19] even after it had reportedly been formally re-founded as Hierapolis by Seleukos I Nikator (r. 305–281 BCE).[20] It was also known as Manbog in Aramaic and Syriac.[21] In his analysis of Hellenistic Syria, Millar (2006: 22) highlights this city as the "one place outside of Phoenicia and Judea" where "a real continuity is traceable"; in other words, it was a place where earlier traditions lived on.

The description of Antipatros as "Chaldaean" deserves some consideration. In Greek and Latin texts, the term eventually came to refer to any astrologer, but it originally referred to a geographic or ethnic origin. Strabo of Amasia, writing in Greek in the late first century BCE or early first century CE, found it necessary to elaborate on the various uses of the word (*Geography* 16.1.6):

> Ἀφώριστο δ' ἐν τῇ Βαβυλῶνι[22] κατοικία τοῖς ἐπιχωρίοις φιλοσόφοις τοῖς Χαλδαίοις προσαγορευομένοις, οἳ περὶ ἀστρονομίαν εἰσὶ τὸ πλέον· προσποιοῦνται δέ τινες καὶ γενεθλιαλογεῖν (οὓς οὐκ ἀποδέχονται οἱ ἕτεροι). ἔστι δὲ καὶ φῦλόν τι τὸ τῶν Χαλδαίων καὶ χώρα τῆς Βαβυλωνίας ὑπ' ἐκείνων οἰκουμένη.... ἔστι δὲ καὶ τῶν Χαλδαίων τῶν ἀστρονομικῶν γένη πλείω· καὶ γὰρ Ὀρχηνοί τινες προσαγορεύονται καὶ Βορσιππηνοὶ καὶ ἄλλοι πλείους, ὡς ἂν κατὰ αἱρέσεις ἄλλα καὶ ἄλλα νέμοντες περὶ τῶν αὐτῶν δόγματα.

> And in Babylon is marked off a settlement for the native philosophers called Chaldaeans, who are for the most part concerned with *astronomia*. And some of them, whom the others do not accept, also profess to cast

18 Grayson 1996: 19.

19 Sinclair and Gatier 2015.

20 Cohen 2006: 173.

21 Millar 2006: 22.

22 For this passage I have followed the Greek text of Radt (2005), which reads Βαβυλῶνι rather than Βαβυλωνίᾳ.

> nativities. And there is also a certain tribe of the Chaldaeans and a region of Babylonia inhabited by them.... And there are also several clans of Chaldaean astronomers/astrologers (ἀστρομικῶν): and they are called variously Orcheni or Borsippeni or other names, as if they were divided into several sects that respect different doctrines concerning these matters.

In the Antipatros inscription, the use of the phrase "Chaldaean *astronomos*" rather than "Chaldaean" alone suggests that the term was meant in a geographic or ethnic sense and not simply as a synonym for "astrologer." One possibility is that the inhabitants of Hieropolis in general were considered Chaldaeans at this time; Millar (2006: 23) mentions this possibility and also proposes that Antipatros in particular might have studied astronomy in Babylonia.

Bowersock (1983) has equated the Antipatros of this inscription with the Antipater mentioned in Vitruvius' *De Architectura* (9.6.2). This is the same passage mentioned above in reference to Berossos, where the Roman architect, writing in the first century BCE, praises the Chaldaeans as foremost in astrology and goes on to mention several Chaldaean astrologers by name:

> primusque Berosus in insula et civitate Coo consedit, ibique aperuit disciplinam, postea †studens† Antipater, iterumque Achinapolus, qui etiam non e nascentia sed ex conceptione genethlialogiae rationes explicatas reliquit.[23]

> First Berosus settled on the island and in the city of Cos and introduced the subject there, then Antipater, and again Achinapolus, who handed down genethlialogical reckonings derived not only from birth but from conception.

The equation of these two Chaldaean astrologers, the Antipatros of the Larisa inscription and the Antipater of Vitruvius, has been accepted by several other scholars.[24] Nevertheless, we should be hesitant in using Vitruvius' statement as a source for any further information about the life of Antipatros, both because the reliability of this passage has been seriously questioned in the case of Berossos and because the key phrase concerning Antipatros is in need of some emendation. After examining the problems with the word *studens* and also with one proposed emendation, Bowersock (1983) ultimately suggests that the word may simply be a late gloss, and this is certainly the safest interpretation; while it is tempting to follow the translation of Verbrugghe and Wickersham (2000: 35)

23 I follow Bowersock's (1983) text of this problematic passage.

24 E.g., Millar 2006: 23 and Dickie 2003: 98.

and see Antipatros as a student of Berossos on Cos, this would place too much weight on a single problematic word.

Fortunately, in some ways, the details of the situation are unimportant. At the broadest level, Vitruvius' statement can be taken as evidence for the success of a "Chaldaean" astrologer in the Greek-speaking west. While even this basic claim can be disputed in the case of Berossos, the Larisa inscription provides reliable documentation for the presence of at least one such Chaldaean astrologer/astronomer in Thessaly. The inscription itself shows that he achieved some measure of success, gaining citizenship and being commemorated in an inscription, while Vitruvius' remark seems to demonstrate that his fame was both widespread and lasting—regardless of whether Vitruvius knew anything more than his name. Amongst all the uncertainties of historical narratives and dubious attributions, the Antipatros inscription offers a secure mooring post: an individual by the name of Antipatros, described as a Chaldaean and known for his astrological/astronomical accomplishments, arrived in Thessaly from Syria and made his home there.

To sum up, we have examined three different cases that seem to involve the interaction of Greek-speakers with Mesopotamian astrologers/astronomers. In the story related by Arrian and Diodorus, Alexander had arrived in Babylon accompanied by an inauspicious celestial omen or prophecy that may have required the performance of a standard apotropaic ritual. To the historians at some remove from these events, the significance of this ritual and its association with a negative omen were completely lost, suggesting some of the limitations inherent in the sharing of knowledge with non-specialist observers. The case of Berossos is problematic for a different reason, since there is considerable uncertainty surrounding his relationship with astronomy/astrology, but the staying power of a remark like that of Vitruvius, that Berossos brought Chaldaean astronomical/astrological knowledge to Cos, suggests that this was perceived by his readers as a relatively typical behavior. Finally, the case of Antipatros in Thessaly provides solid evidence that at least one "Chaldaean" astronomer/astrologer from Syria did in fact relocate to Greece, where he was apparently welcomed by his new neighbors and achieved great success.

The idea of the foreign expert established in Greece is by no means a new one, nor is it restricted to a certain time period or area of knowledge. Burkert (1992) has devoted significant attention to the migrant craftsmen of the early first millennium BCE, while Bryce (1999) has argued that the Anatolians living in Mycenaean Greece in the second millennium BCE must have included cuneiform scribes with the necessary training to send letters in the Hittite language to the Hittite court. On a similar note, Brosius (2014) stresses the importance of the trained professional scribe even after the development of the

Greek alphabetic script, rejecting the common notion that the alphabet could be easily learned by anyone and therefore eliminated the need for the literate expert. Although some of these examples deal with fields other than astronomy/astrology, they are nevertheless indicative of a line of thought similar to the one expressed here: that it is worth considering the role of the learned expert in the transmission of knowledge.

Throughout all of this, one other important means of knowledge transfer has been overlooked. I have focused on Near Eastern experts travelling west, but of course travel was not unidirectional. While Greek-speaking laypeople who found themselves among Babylonian scholars may not have appreciated the knowledge with which they were surrounded, the situation would presumably have been quite different for those who travelled to Babylon or Uruk with a specific interest in astronomical/astrological learning. Toomer (1988: 359) has hypothesized that Hipparchus attained knowledge of Babylonian astronomical techniques and data by travelling personally to Babylon and working closely with the Babylonian experts there, and there is some evidence for this type of knowledge sharing in the related divinatory field of extispicy: Arrian (*Anabasis* 7.18.1–5) recounts a story concerning a Greek interpreter of liver omens, the brother of one of Alexander's Companions, who was living in Babylon and apparently refined his divinatory practice under the influence of the Babylonians (Flower 2008: 131). All of these examples together represent only a tiny fraction of the evidence for the circulation of astronomical, astrological, or divinatory knowledge around the ancient Mediterranean; as the work of Francesca Rochberg has shown, this is an extraordinarily rich topic deserving of further investigation.

Bibliography

Bowersock, G.W.

1983 "Antipater Chaldaeus." *The Classical Quarterly*, New Series 33.2: 491.

Briant, P.

2002 *From Cyrus to Alexander: A History of the Persian Empire*. trans. P.T. Daniels. Winona Lake, IN: Eisenbrauns.

Brosius, M.

2014 "Some Remarks on the Channels of the Transmission of Knowledge in the Ancient Mediterranean World." Pp. 285–300 in *Kulturkontakte in Antiken Welten: vom Denkmodell zum Fallbeispiel: Proceedings des internationalen Kolloquiums aus Anlass des 60. Geburtstages von Christoph Ulf, Innsbruck, 26. bis 30. Januar 2009*, ed. R. Rollinger and K. Schnegg. Colloquia antiqua 10. Leuven: Peeters.

Bryce, T.

1999 "Anatolian Scribes in Mycenaean Greece." *Hist.* 48: 257–264.

Burkert, W.

1983 "Itinerant Diviners and Magicians." Pp. 115–120 in *The Greek Renaissance of the Eighth Century B.C.: Tradition and Innovation*, ed. R. Hägg. Stockholm: Paul Åströms Förlag.

1992 *The Orientalizing Revolution*. Cambridge: Harvard University Press.

Clancier, P.

2001 "Cuneiform Culture's Last Guardians: The Old Urban Notability of Hellenistic Uruk." Pp. 752–773 in *The Oxford Handbook of Cuneiform Culture*, ed. K. Radner and E. Robson. Oxford: Oxford University Press.

Cohen, G.

2006 *The Hellenistic Settlements in Syria, the Red Sea Basin, and North Africa*. Berkeley: University of California Press.

De Breucker, G.

2012 *De Babyloniaca van Berossos van Babylon. Inleiding, editie en commentaar*. Ph.D. diss. Groningen.

2013 "Berossos: His Life and Work." Pp. 15–28 in *The World of Berossos: Proceedings of the 4th International Colloquium on "The Ancient Near East between Classical and Ancient Oriental Traditions," Hatfield College, Durham 7th–9th July 2010*, ed. J. Haubold, G.B. Lanfranchi, R. Rollinger, J.M. Steele. CleO 5 Wiesbaden: Harrassowitz.

Dickie, M.

2003 *Magic and Magicians in the Greco-Roman World*. New York: Routledge.

Flower, M.

2008 *The Seer in Ancient Greece*. Berkeley: University of California Press.

Gallis, K.

1980 "New Inscriptions from Larissa." *AAA* 13: 246–263.

Grayson, A.K.

1996 *Assyrian Rulers of the Early First Millennium BC II (858–745 BC)*. RIMA 3. Toronto: University of Toronto Press.

Haubold, J.; Lanfranchi, G.B.; Rollinger, R.; and J.M. Steele, eds.

2013 *The World of Berossos: Proceedings of the 4th International Colloquium on "The Ancient Near East between Classical and Ancient Oriental Traditions," Hatfield College, Durham 7th–9th July 2010*. CleO 5 . Wiesbaden: Harrassowitz.

Haubold, J.

2013 *Greece and Mesopotamia: Dialogues in Literature*. New York: Cambridge University Press.

López-Ruiz, C.

2010 *When the Gods Were Born: Greek Cosmogonies and the Near East*. Cambridge, MA: Harvard University Press.

Millar, F.

2006 *The Greek World, the Jews, and the East.* Chapel Hill: University of North Carolina Press.

Nawotka, K.

2010 *Alexander the Great.* Newcastle upon Tyne: Cambridge Scholars Publishing.

Oppenheim, A.L.

1969 "Divination and Celestial Observation in the Last Assyrian Empire." *Centaurus* 14: 97–135.

Ossendrijver, M.

2011 "Science in Action: Networks in Babylonian Astronomy." Pp. 213–221 in *Babylon. Wissenkultur in Orient und Okzident / Science Culture Between Orient and Occident,* ed. E. Cancik-Kirschbaum, M. van Ess, J. Marzahn. Topoi – BSAW 1. Boston: de Gruyter.

2015 "The Story of the Magi in the Light of Alexander the Great's Encounters with Chaldaeans." Pp. 217–231 in *The Star of Bethlehem and the Magi: Interdisciplinary Perspectives from Experts on the Ancient Near East, the Greco-Roman World, and Modern Astronomy,* ed. G.H. van Kooten and Peter Barthel. Leiden: Brill.

Parpola, S.

1970–83 *Letters from Assyrian Scholars to the Kings Esarhaddon and Assurbanipal,* 2 vols. AOAT 5. Kevelaer: Verlag Butzon & Bercker.

1993 *Letters from Assyrian and Babylonian Scholars.* SAA 10. Reprinted 2014. Winona Lake, IN: Eisenbrauns.

Popović, M.

2014 "Networks of Scholars: The Transmission of Astronomical and Astrological Learning Between Babylonians, Greeks and Jews." Pp. 153–193 in *Ancient Jewish Sciences and the History of Knowledge in Second Temple Literature,* ed. S. Sanders and J. Ben-Dov. New York: New York University Press.

Radt, S.

2005 *Strabons Geographika Band 4: Buch XIV–XVII: Text und Übersetzung.* Göttingen: Vandenhoeck & Ruprecht.

Rochberg(-Halton), F.

1987 "Elements of the Babylonian Contribution to Hellenistic Astrology." *JAOS* 108: 51–62.1984 "New Evidence for the History of Astrology." *JNES* 43: 115–140.

2004 *The Heavenly Writing: Divination, Horoscopy, and Astronomy in Mesopotamian Culture.* Cambridge; New York: Cambridge University Press.

2010 *In the Path of the Moon: Babylonian Celestial Divination and Its Legacy.* SAMD 6. Leiden: Brill.

Sinclair, T. and Gatier, P.

2015 "Places: 658480 (Hierapolis/Bambyke). Pleiades." <http://pleiades.stoa.org/places/658480> [Accessed: January 25, 2016 3:06 pm].

Smelik, K.A.D.

1987 "The 'Omina Mortis' in the Histories of Alexander the Great." *Talanta* 10/11: 92–111.

Steele, J.M.

2013 "The 'Astronomical Fragments' of Berossos in Context." Pp. 99–113 in *The World of Berossos: Proceedings of the 4th International Colloquium on "The Ancient Near East between Classical and Ancient Oriental Traditions," Hatfield College, Durham 7th–9th July 2010*, ed. J. Haubold, G.B. Lanfranchi, R. Rollinger, J.M. Steele. CleO 5 Wiesbaden: Harrassowitz.

Toomer, G.J.

1988 "Hipparchus and Babylonian Astronomy." Pp. 353–362 in *A Scientific Humanist: Studies in Memory of Abraham Sachs*, ed. E. Leichty, M. de Jong Ellis, P. Gerardi. OPSNKF 9. Philadelphia: University Museum.

van der Spek, R.J.

2003 "Darius III, Alexander the Great and Babylonian Scholarship." Pp. 289–346 in *A Persian Perspective: Essays in Memory of Heleen Sancisi-Weerdenburg*, ed. W. Henkelman and A. Kuhrt. Leiden: Nederlands Instituut voor het Nabije Oosten.

Van Dongen, E.

2014 "The Concept of 'the Near East': A Reconsideration." Pp. 253–267 in *Kulturkontakte in Antiken Welten: vom Denkmodell zum Fallbeispiel: Proceedings des internationalen Kolloquiums aus Anlass des 60. Geburtstages von Christoph Ulf, Innsbruck, 26. bis 30. Januar 2009*, ed. R. Rollinger and K. Schnegg. ColAnt 10. Leuven: Peeters.

Verbrugghe, G.P. and J.M. Wickersham

2000 *Berossos and Manetho, Introduced and Translated*. Ann Arbor: University of Michigan Press.

Westenholz, A.

2007 "The Graeco-Babyloniaca Once Again." *ZA* 97: 262–313.

CHAPTER 17

Diodorus on the Chaldeans

Alexander Jones and John Steele

The passage on the Chaldeans (Χαλδαῖοι) in Book 2 of the mid first century BCE historian Diodorus's *Historical Library* (*Bibliotheca Historica*), though it is the most extensive and detailed account in the extant Greco-Roman literature of the Babylonian scholarly milieu in which astronomy and astrology flourished during the second half of the first millennium BCE, tends to receive less attention than comparable passages in Strabo's *Geography* 16.1.6 and Pliny the Elder's *Natural History* 6.121–123. The present article has the modest but, we hope, useful goal of appraising Diodorus's information in the light of our more direct knowledge of the Babylonian astral sciences and their practitioners.

Of the original forty books (βιβλία) of the *Historical Library*—fewer than half of which survive—Books 1 through 6 ostensibly covered history up to the Trojan War, with 1–3 devoted to the major "barbarian" peoples, and 4–6 to the Greeks. Following Book 1, which, after preliminaries, is wholly about Egypt, Diodorus turns in Book 2 to Asia, treated nation by nation: the Assyrians, Chaldeans, Medes, Indians, Scythians, Amazons, Hyperboreans, Arabs, and the inhabitants of an island in the remote southern Ocean as described by one Iambulus. The order is partly determined by chronology but chiefly by geography, proceeding from the nearer peoples to the more remote and in some cases positively fanciful.[1]

What and how much Diodorus has to say about these peoples varies considerably. The first twenty-eight of the sixty sections into which Book 2 is subdivided in modern editions present a narrative history of Assyria, the great part of which concerns its first king, Ninus, and his queen Semiramis, followed by a brief account of their son Ninyas and a precipitate rush through thirty generations to the reign of Sardanapalus, the last king of Assyria. The comparatively brief passage (2.32–34) on the Medes is also narrative, and there are a few narrative passages elsewhere. On the other hand, the predominant mode in Book 2 for peoples other than the Assyrians and Medes is ethnographic and descriptive geography, in relation to which the nominal chronological limit of the Trojan War is generally disregarded. But even the narrative history does not respect this bound (which Diodorus sets at 779 years before the end of the

1 Iambulus's work may in fact have been a novel that Diodorus mistook for an authentic narrative of travel.

 | DOI 10.1163/9789004363380_018

Peloponnesian War, i.e., 1184 BCE), since the rather distorted account of the fall of the Assyrian Empire to the Medes refers to events of the late seventh century BCE. In other words, Diodorus readily conflates non-Hellenic with pre-Hellenic.

Diodorus is generally regarded, with reason, as a largely derivative historian who depended, for long stretches of his work, on a few principal authors—sometimes acknowledged, sometimes not—supplemented by occasional material from other sources and intermittent personal interventions, especially in the non-narrative parts.[2] His Assyrian and Median narratives were evidently based in great part on the *Persika* of Ctesias (ca. 400 BCE), who is cited by name eleven times. Though lodged between the Assyrians and the Medes, however, Diodorus's account of the Chaldeans (2.29–31), which is wholly ethnographic, probably owes little or nothing to Ctesias, and may in fact be a personal and eclectic amalgam assembled by Diodorus from multiple sources.[3] The transitional passages at its beginning and, especially, at its end make its character as a digression from the continuity of the Assyrian-Median historical narrative obvious.

A peculiarity of the Chaldean passage, in comparison with other ethnographic passages in Book 2, is that it is not about a people in the broad sense but about a specific scholarly community in Babylon; otherwise Babylonia receives no separate treatment in the *Historical Library*, but is subsumed into the Assyrian narrative and into narrative passages in the later books. Diodorus perhaps intended the passage to provide background to anecdotes elsewhere in his work in which Chaldeans were prominent. These include the prophecy of the Chaldeans to Alexander the Great that he would die if he entered Babylon (17.112), and their foretelling to Antigonus that Seleucus would become ruler of Asia and that Antigonus would die in battle against him (19.55, repeated 21.1); both these incidents are in fact glancingly alluded to in 2.31. But already in 2.24 Diodorus had introduced Belesys, general of the Babylonians and confidant of Arbaces, the general of the Medes, as "the most noteworthy of the priests whom the Babylonians call Chaldeans," an expert in astral science (ἀστρολογία) and divination (μαντική), and this might have been the immediate provocation for introducing a digression on the Chaldeans just a few paragraphs later.

---

2 For a sustained plea for seeing Diodorus's work as reflecting some degree of authorial control beyond mechanical digesting of sources, see Sacks 1990.

3 In this we agree with Boncquet 1987: 170–173, who also provides a useful review of previous (and generally inconclusive) speculations about Diodorus's source or sources for the Chaldean passage. However, we are inclined to be more sceptical than Boncquet of the hypothesis that the imprint of the Stoic Posidonius can be detected in Diodorus.

Curiously, in Book 2 he makes no reference to the fact that in his treatment of Egypt he has twice (1.28 and 1.81) written that the Egyptians claimed to have taught the Chaldeans their astral wisdom; nor do those passages give any hint that a fuller description of the Chaldeans is to come. Nevertheless Diodorus's presentation of the Chaldeans in 2.29–31 appears to combine genuine information about Babylonian scholars and scholarship with less reliable or positively false information that seems to reflect a source that ascribed aspects of Egyptian or Greco-Egyptian astral science to the Babylonians.

For the text of 2.29–31 we have consulted the critical editions of Vogel (1888) and Eck (2003), omitting *apparatus criticus* but alerting the reader to the relatively few major textual uncertainties in our notes. We have tended to present a comparatively conservative text, retaining some problematic and controversial readings from the manuscript tradition and reserving discussion of possible corrections for the notes. The notes are keyed to the text and translation by chapter and section number.

Text and Translation

(29) [1]ἡμῖν δ' οὐκ ἀνάρμοστον εἶναι δοκεῖ περὶ τῶν ἐν Βαβυλῶνι καλουμένων Χαλδαίων καὶ τῆς ἀρχαιότητος αὐτῶν βραχέα διελθεῖν, ἵνα μηδὲν παραλείπωμεν τῶν ἀξίων μνήμης. [2]Χαλδαῖοι τοίνυν, τῶν ἀρχαιοτάτων ὄντες Βαβυλωνίων, τῇ μὲν διαιρέσει τῆς πολιτείας παραπλησίαν ἔχουσι τάξιν τοῖς κατ' Αἴγυπτον ἱερεῦσι· πρὸς γὰρ τῇ θεραπείᾳ τῶν θεῶν τεταγμένοι πάντα τὸν τοῦ ζῆν χρόνον φιλοσοφοῦσι, μεγίστην δόξαν ἔχοντες ἐν ἀστρολογίᾳ. ἀντέχονται δ' ἐπὶ πολὺ καὶ μαντικῆς, ποιούμενοι προρρήσεις περὶ τῶν μελλόντων, καὶ τῶν μὲν καθαρμοῖς, τῶν δὲ θυσίαις, τῶν δ' ἄλλαις τισὶν ἐπῳδαῖς ἀποτροπὰς κακῶν καὶ τελειώσεις ἀγαθῶν πειρῶνται πορίζειν. [3]ἐμπειρίαν δ' ἔχουσι καὶ τῆς διὰ τῶν οἰωνῶν μαντικῆς, ἐνυπνίων τε καὶ τεράτων ἐξηγήσεις ἀποφαίνονται. οὐκ ἀσόφως δὲ ποιοῦνται καὶ τὰ περὶ τὴν ἱεροσκοπίαν,

(29) [1]It appears to us not unfitting to discourse briefly about the people in Babylon called Chaldeans and their antiquity, so that we leave out none of the things worthy of mention. [2]Well then, the Chaldeans, being among the most ancient Babylonians, have a rank in the subdivision of the state comparable to the priests in Egypt; for having been assigned to the cultivation of the gods, they practice philosophy for all the time of their life, having a very great repute in astral science (*astrologia*). They also adhere to a great extent to divination (*mantike*), making forecasts about the things that are going to happen, and they undertake to supply turnings-away (*apotropai*) of evil things and fulfilments of good things, some (of the evils and goods) by purifications, some by sacrifices,

ἄκρως ἐπιτυγχάνειν νομίζοντες. τὴν δὲ τούτων μάθησιν ἁπάντων οὐχ ὁμοίαν ποιοῦνται τοῖς τὰ τοιαῦτ' ἐπιτηδεύουσι τῶν Ἑλλήνων.

some by various other incantations (*epodai*). [3]They have experience in divination by birds, and pronounce interpretations of dreams and monstrosities. And they also perform matters of extispicy in a manner not lacking in wisdom, considering it to be supremely successful. The education (*mathesis*) that they carry out for all these things is not like those of the Greeks who practice such things.

[4]παρὰ μὲν γὰρ τοῖς Χαλδαίοις ἐκ γένους ἡ τούτων φιλοσοφία παραδέδοται, καὶ παῖς παρὰ πατρὸς διαδέχεται, τῶν ἄλλων λειτουργιῶν πασῶν ἀπολελυμένος. διὸ καὶ γονεῖς ἔχοντες διδασκάλους, ἅμα μὲν ἀφθόνως ἅπαντα μανθάνουσιν, ἅμα δὲ τοῖς παραγγελλομένοις προσέχουσι πιστεύοντες βεβαιότερον. ἔπειτ' εὐθὺς ἐκ παίδων συντρεφόμενοι τοῖς μαθήμασι, μεγάλην ἕξιν περιποιοῦνται διά τε τὸ τῆς ἡλικίας εὐδίδακτον καὶ διὰ τὸ πλῆθος τοῦ προσκαρτερουμένου χρόνου. [5]παρὰ δὲ τοῖς Ἕλλησιν ὁ πολὺς ἀπαράσκευος προσιὼν ὀψέ ποτε τῆς φιλοσοφίας ἅπτεται, καὶ μέχρι τινὸς φιλοπονήσας, ἀπῆλθε περισπασθεὶς ὑπὸ βιωτικῆς χρείας. ὀλίγοι δὲ παντελῶς ἐπὶ φιλοσοφίαν ἀποδύντες ἐργολαβίας ἕνεκεν παραμένουσιν ἐν τῷ μαθήματι, καινοτομοῦντες ἀεὶ περὶ τῶν μεγίστων δογμάτων καὶ τοῖς πρὸ αὐτῶν οὐκ ἀκολουθοῦντες.

[4]For among the Chaldeans the philosophy of these things is transmitted by descent (*ek genous*), and son succeeds father, having been freed from all other duties (*leitourgiai*). Hence having teachers who are also parents, at the same time they learn all things ungrudgingly, and they obey the instructions, because they trust more confidently. And moreover being nourished in mathematics/learnings (*mathemata*) straightway from childhood, they profess a great competence because of the ease of learning belonging to that age, and because of the quantity of time spent in diligent employment. [5]Among the Greeks, on the other hand, many is the unprepared person who takes up philosophy late in the day, and after working diligently up to a point, he gives up, being tugged away by the needs of life. A few, however, stripping themselves entirely for philosophy for the sake of gain, remain in the learned pursuit (*mathema*), always coming up with newfangled ideas about the most important teachings and not following their predecessors.

[6]τοιγαροῦν οἱ μὲν βάρβαροι διαμένοντες ἐπὶ τῶν αὐτῶν ἀεὶ βεβαίως ἕκαστα λαμβάνουσιν, οἱ δ' Ἕλληνες τοῦ κατὰ τὴν ἐργολαβίαν κέρδους στοχαζόμενοι καινὰς αἱρέσεις κτίζουσι· καὶ περὶ τῶν μεγίστων θεωρημάτων ἀλλήλοις ἀντιδοξοῦντες διχονοεῖν ποιοῦσι τοὺς μανθάνοντας καὶ τὰς ψυχὰς αὐτῶν πλανᾶσθαι, τὸν πάντα βίον ἐν αἰώρᾳ γενομένας καὶ μηδὲν ὅλως πιστεῦσαι δυναμένας βεβαίως. τὰς γοῦν ἐπιφανεστάτας αἱρέσεις τῶν φιλοσόφων εἴ τις ἀκριβῶς ἐξετάζοι, πλεῖστον ὅσον εὑρήσει διαφερούσας ἀλλήλων καὶ περὶ τῶν μεγίστων δοξῶν ἐναντία δοξαζούσας.

[6]For that very reason the barbarians, because they always stick to the same things, take hold of all things securely, whereas the Greeks, because they make a random shot for the reward of lucre, found novel sects; and by holding contrary opinions to one another about the most important objects of contemplation (*theoremata*), they cause the students to doubt (*dichonoein*) and (cause) their souls to wander, seesawing (*en aiora genomenas*) their whole lives, and able to trust nothing whatsoever securely. At any rate, if one were to make an accurate scrutiny of the most conspicuous sects of the philosophers, one would find them differing ever so much from each other and entertaining opposite positions on the most important opinions.

(30) [1]οἱ δ'οὖν Χαλδαῖοι τὴν μὲν τοῦ κόσμου φύσιν ἀΐδιόν φασιν εἶναι καὶ μήτε ἐξ ἀρχῆς γένεσιν ἐσχηκέναι μήθ' ὕστερον φθορὰν ἐπιδέξεσθαι· τήν τε τῶν ὅλων τάξιν τε καὶ διακόσμησιν θείᾳ τινὶ προνοίᾳ γεγονέναι, καὶ νῦν ἕκαστα τῶν ἐν οὐρανῷ γενομένων οὐχ ὡς ἔτυχεν οὐδ' αὐτομάτως, ἀλλ' ὡρισμένῃ τινὶ καὶ βεβαίως κεκυρωμένῃ θεῶν κρίσει συντελεῖσθαι. [2]τῶν δ' ἄστρων πολυχρονίους παρατηρήσεις πεποιημένοι, καὶ τὰς ἑκάστου κινήσεις τε καὶ δυνάμεις ἀκριβέστατα πάντων ἀνθρώπων ἐπεγνωκότες, πολλὰ τῶν μελλόντων συμβαίνειν προλέγουσι τοῖς ἀνθρώποις. [3]μεγίστην τέ φασιν εἶναι θεωρίαν καὶ δύναμιν περὶ τοὺς πέντε ἀστέρας τοὺς πλάνητας καλουμένους, οὓς ἐκεῖνοι κοινῇ μὲν ἑρμηνεῖς

(30) [1]The Chaldeans say that the nature of the cosmos is eternal and neither had a coming-to-be from the beginning nor is going to obtain a passing-away later; that the arrangement and order of everything took place through a certain divine providence, and that now each of the things happening in the heavens is brought about not at random or spontaneously but by some defined and securely ratified (*kekyromene*) judgment of the gods. [2]Having made observations of the stars over many years, and found out the motions and powers of each of them most accurately of all people, they foretell to people many of the things that are going to happen. [3]They say that the

ὀνομάζουσιν· ἰδιᾳ̂ δὲ τὸν ὑπὸ τῶν Ἑλλήνων Κρόνον ὀνομαζόμενον, ἐπιφανέστατον δὲ καὶ πλεῖστα καὶ μέγιστα προσημαίνοντα, καλοῦσιν Ἡλίου, τοὺς δ᾽ ἄλλους τέτταρας ὁμοίως τοῖς παρ᾽ ἡμῖν ἀστρολόγοις ὀνομάζουσιν, Ἄρεως, Ἀφροδίτης, Ἑρμοῦ, Διός.

greatest contemplation (*theoria*) and power pertains to the five stars called "wanderers" (*planetes*), which they collectively name "interpreters"; particularly, the one named Kronos by the Greeks, which is most conspicuous (*epifanestaton*) and signifying (*prosemainon*) the most and greatest things, they call "of the Sun," while they name the other four in the same way as the astronomers (*astrologoi*) among us, (as the stars) "of Ares," "of Aphrodite," "of Hermes," "of Zeus."

4διὰ τοῦτο δ᾽ αὐτοὺς ἑρμηνεῖς καλοῦσιν, ὅτι, τῶν ἄλλων ἀστέρων ἀπλανῶν ὄντων καὶ τεταγμένῃ πορείᾳ μίαν περιφορὰν ἐχόντων, οὗτοι μόνοι πορείαν ἰδίαν ποιούμενοι τὰ μέλλοντα γίνεσθαι δεικνύουσιν, ἑρμηνεύοντες τοῖς ἀνθρώποις τὴν τῶν θεῶν ἔννοιαν. τὰ μὲν γὰρ διὰ τῆς ἀνατολῆς, τὰ δὲ διὰ τῆς δύσεως, τινὰ δὲ διὰ τῆς χρόας προσημαίνειν φασὶν αὐτοὺς τοῖς προσέχειν ἀκριβῶς βουληθεῖσιν. 5ποτὲ μὲν γὰρ πνευμάτων μεγέθη δηλοῦν αὐτούς, ποτὲ δὲ ὄμβρων ἢ καυμάτων ὑπερβολάς, ἔστι δὲ ὅτε κομητῶν ἀστέρων ἐπιτολάς, ἔτι δὲ ἡλίου τε καὶ σελήνης ἐκλείψεις, καὶ σεισμούς, καὶ τὸ σύνολον πάσας τὰς ἐκ τοῦ περιέχοντος γεννωμένας περιστάσεις ὠφελίμους τε καὶ βλαβερὰς οὐ μόνον ἔθνεσιν ἢ τόποις, ἀλλὰ καὶ βασιλεῦσι καὶ τοῖς τυχοῦσιν ἰδιώταις. 6ὑπὸ δὲ τὴν τούτων φορὰν λέγουσι τετάχθαι τριάκοντα ἀστέρας, οὓς προσαγορεύουσι βουλαίους θεούς· τούτων δὲ τοὺς μὲν ἡμίσεις τοὺς ὑπὲρ γῆν τόπους ἐφορᾶν, τοὺς δ᾽ ἡμίσεις ⟨τοὺς⟩ ὑπὸ τὴν γῆν, τὰ κατ᾽ ἀνθρώπους ἐπισκοποῦντας ἅμα καὶ τὰ κατὰ τὸν

4They call them interpreters, because, whereas the other stars are nonwandering and having a single revolution with an orderly (*tetagmene*) journey, these alone, making a journey each of its own, indicate the things that are going to happen, interpreting for people the intention of the gods. They say that they signify, to those who have chosen to pay accurate attention, some things by their rising, some by their setting, and some by their color. 5For (they say that) they sometimes show the magnitudes of the winds, sometimes the excesses of rainstorms or heat waves, and sometimes the risings of comets, and also eclipses of Sun and Moon, and earthquakes, and in a word all the circumstances arising from the environment, being beneficial and harmful not only to peoples and places but also to kings and random individuals. 6They say that below the motion of these are placed thirty stars, which they designate as council gods; (they say that) half of

οὐρανὸν συμβαίνοντα. διὰ δ' ἡμερῶν δέκα πέμπεσθαι τῶν μὲν ἄνω πρὸς τοὺς κάτω καθάπερ ἄγγελον ἕνα τῶν ἀστέρων, τῶν δ' ὑπὸ γῆν πρὸς τοὺς ἄνω πάλιν ὁμοίως ἕνα, καὶ ταύτην ἔχειν οὗτους φορὰν ὡρισμένην καὶ περιόδῳ κεκυρωμένην αἰωνίῳ.

these oversee the places above the Earth, and half (oversee the places) below the Earth, watching over both people and the things occurring in the heavens; (they say that) at intervals of ten days, one of the stars of those above is sent to the ones below like a messenger, and again similarly one of those below the Earth to the ones above, and (that) they have this motion defined and ratified for an eternal period.

[7]τῶν θεῶν δὲ τούτων κυρίους εἶναι φασι δώδεκα τὸν ἀριθμόν, ὧν ἑκάστῳ μῆνα καὶ τῶν δώδεκα λεγομένων ζῳδίων ἕν προσνέμουσι. διὰ δὲ τούτων φασι ποιεῖσθαι τὴν πορείαν τόν τε ἥλιον καὶ τὴν σελήνην καὶ πέντε τοὺς πλάνητας ἀστέρας, τοῦ μὲν ἡλίου τὸν ἴδιον κύκλον ἐν ἐνιαυτῷ τελοῦντος, τῆς δὲ σελήνης ἐν μηνὶ τὴν ἰδίαν περίοδον διαπορευομένης.

[7]They say that there are lords of these gods, twelve in number, to each of which they allot one month and one of the twelve so-called zodiacal signs (*zodia*). They say that the Sun and the Moon and the five wandering stars make their journey through these, with the Sun completing its own circle in a year, and the Moon making the complete traversal of its own circuit (*periodos*) in a month.

(31) [1]τῶν δὲ πλανήτων ἕκαστον ἴδιον ἔχειν δρόμον καὶ διηλλαγμένως καὶ ποικίλως χρῆσθαι τοῖς τάχεσι καὶ τῇ τῶν χρόνων διαιρέσει. πλεῖστα δὲ πρὸς τὰς γενέσεις τῶν ἀνθρώπων συμβάλλεσθαι τούτους τοὺς ἀστέρας ἀγαθά τε καὶ κακά· διὰ δὲ τῆς τούτων φύσεώς τε καὶ θεωρίας μάλιστα γινώσκειν τὰ συμβαίνοντα τοῖς ἀνθρώποις. [2]πεποιῆσθαι δέ φασι προρρήσεις ἄλλοις τε βασιλεῦσιν οὐκ ὀλίγοις καὶ τῷ καταπολεμήσαντι Δαρεῖον Ἀλεξάνδρῳ καὶ τοῖς μετὰ ταῦτα βασιλεύσασιν Ἀντιγόνῳ τε καὶ Σελεύκῳ τῷ Νικάτορι, ἐν ἅπασι δὲ τοῖς ῥηθεῖσιν εὐστοχηκέναι δοκοῦσιν· ὑπὲρ ὧν ἡμεῖς τὰ κατὰ μέρος ἐν οἰκειοτέροις

(31) [1](They say that) each of the planets has its own course and is subject to variations and complexities in its speeds and its division of the time intervals. (They say that) these stars contribute numerous goods and evils for the nativities of people; (and they say that) by means of the nature and sight of them, above all, they (*scil.* the Chaldeans) know the things that happen to people. [2]They say that they have made forecasts to not a few kings, and in particular to Darius when he was about to make war against Alexander, and to Antigonus and Seleucus Nicator

ἀναγράψομεν καιροῖς. [3]προλέγουσι δὲ καὶ τοῖς ἰδιώταις τὰ μέλλοντα συμβαίνειν οὕτως εὐστόχως ὥστε τοὺς πειραθέντας θαυμάζειν τὸ γενόμενον καὶ μεῖζον ἢ κατ᾽ ἄνθρωπον ἡγεῖσθαι.

who reigned after these things, and they appeared to have hit the mark in all their utterances; we will write the particulars about these things on more appropriate occasions. [3]They foretell also to private persons the things that are going to happen, in such a well-aimed manner that those who have experienced it are astonished at the event and believe that it is superhuman.

[4]μετὰ δὲ τὸν ζῳδιακὸν κύκλον εἴκοσι καὶ τέτταρας ἀφορίζουσιν ἀστέρας, ὧν τοὺς μὲν ἡμίσεις ἐν τοῖς βορείοις μέρεσι, τοὺς δ᾽ ἡμίσεις ἐν τοῖς νοτίοις τετάχθαι φασί, καὶ τούτων τοὺς μὲν ὁρωμένους τῶν ζώντων εἶναι καταριθμοῦσι, τοὺς δ᾽ ἀφανεῖς τοῖς τετελευτηκόσι προσωρίσθαι νομίζουσιν, οὓς δικαστὰς τῶν ὅλων προσαγορεύουσιν. [5]ὑπὸ πάντα δὲ τὰ προειρημένα τὴν σελήνην φέρεσθαι λέγουσιν, ἔγγιστα μὲν τῆς γῆς οὖσαν διὰ τὴν βαρύτητα, διαπορευομένην δ᾽ ἐν ἐλαχίστῳ χρόνῳ τὸν ἑαυτῆς δρόμον, οὐ διὰ τὴν ὀξύτητα τῆς φορᾶς, ἀλλὰ διὰ τὴν βραχύτητα τοῦ κύκλου. [6]ὅτι δὲ τὸ φῶς ἀλλότριον ἔχει καὶ διότι τὰς ἐκλείψεις ποιεῖται διὰ τὸ σκίασμα τῆς γῆς παραπλήσια λέγουσι τοῖς Ἕλλησιν. περὶ δὲ τῆς κατὰ τὸν ἥλιον ἐκλείψεως ἀσθενεστάτας ἀποδείξεις φέροντες οὐ τολμῶσι προλέγειν οὐδ᾽ ἀκριβῶς ὑπὲρ ταύτης περιγράφειν τοὺς χρόνους. [7]περὶ δὲ τῆς γῆς ἰδιωτάτας ἀποφάσεις ποιοῦνται, λέγοντες ὑπάρχειν αὐτὴν σκαφοειδῆ καὶ κοίλην, καὶ πολλὰς καὶ πιθανὰς ἀποδείξεις εὐποροῦσι περί τε ταύτης καὶ περὶ τῶν ἄλλων τῶν κατὰ τὸν κόσμον· ὑπὲρ ὧν τὰ κατὰ μέρος διεξιέναι τῆς ὑποκειμένης ἱστορίας ἀλλότριον

[4]After the zodiacal circle, they mark off twenty-four stars, half of which, they say, are situated in the northern regions, half in the southern, and they count those of them that are seen as belonging to the living, but they consider the invisible ones to be determined for the dead, and they call these judges of all. [5]They say that the moon travels below all the aforesaid, being closest to the Earth because of its weight, and making the complete traversal of its own course in the least time, not because of the swiftness of its motion but because of the brevity of the circle. [6]They say similar things to the Greeks to the effect that its light is not its own, and that it makes eclipses through the shadowing of the earth. They adduce very weak demonstrations concerning the eclipse with respect to the Sun and do not dare to forecast or determine the times for this (eclipse) accurately. [7]They make very peculiar pronouncements about the Earth, saying that it is bowl-shaped and concave, and they abound in many plausible demonstrations about this

εἶναι νομίζομεν. [8]τοῦτο μέντοι γε διαβεβαιώσαιτ' ἄν τις προσηκόντως, ὅτι Χαλδαῖοι μεγίστην ἕξιν ἐν ἀστρολογίᾳ τῶν ἁπάντων ἀνθρώπων ἔχουσι καὶ διότι πλείστην ἐπιμέλειαν ἐποιήσαντο ταύτης τῆς θεωρίας.

and about the other matters with respect to the cosmos; but we consider it to be outside the scope of the present history to go into each detail about these things. [8]But one would be assured of this, that the Chaldeans have the greatest competence in astronomy/astrology of all people, and that they took the greatest care of this science.

[9]περὶ δὲ τοῦ πλήθους τῶν ἐτῶν ἐν οἷς φασι τὴν θεωρίαν τῶν κατὰ τὸν κόσμον πεποιῆσθαι τὸ σύστημα τῶν Χαλδαίων, οὐκ ἄν τις ῥᾳδίως πιστεύσειεν· ἐτῶν γὰρ ἑπτὰ καὶ τετταράκοντα μυριάδας καὶ τρεῖς ἐπὶ ταύταις χιλιάδας εἰς τὴν Ἀλεξάνδρου διάβασιν γεγονέναι καταριθμοῦσιν, ἀφ' ὅτου τὸ παλαιὸν ἤρξαντο τῶν ἄστρων τὰς παρατηρήσεις ποιεῖσθαι. [10]καὶ περὶ μὲν Χαλδαίων ἀρκεσθησόμεθα τοῖς ῥηθεῖσιν, ἵνα μὴ μακρότερον ἀποπλανώμεθα τῆς οἰκείας ἱστορίας· περὶ δὲ τῆς Ἀσσυρίων βασιλείας ὡς ὑπὸ Μήδων κατελύθη προειρηκότες ἐπάνιμεν ὅθεν ἐξέβημεν.

[9]Concerning the number of years in which, they say, the college (*systema*) of the Chaldeans has made the science of matters concerning the cosmos, one would not easily believe; for they count 473,000 years to the crossing over of Alexander, from when they began of old the observations of the stars. [10]Let us be satisfied with what has been said about the Chaldeans, so that we should not any longer digress from the proper history; having previously spoken about the kingdom of the Assyrians and how it was destroyed by the Medes, we will proceed next from the place where we digressed.

Notes

29.2 *rank … comparable to the priests in Egypt.* In 1.73 Diodorus writes that the priests in Egypt hold one of the three parts of Egypt, are exempt from taxes, and come second only to the king in reputation and resources.
29.2 *philosophy.* Similarly Strabo, 16.1.6, speaks of the Chaldeans as philosophers.
29.2. *astral science.* The word *astrologia* had essentially the same scope as *astronomia*, originally covering pursuits that we would characterize as astronomy, but also applicable to astrology from the time of astrology's diffusion in the

Greco-Roman world in the late Hellenistic period.[4] For Diodorus, it definitely signifies astronomy in some passages, for example 12.36 à propos of Meton of Athens, and this is likely to be true in general. In the present context astrology seems to be subsumed into divination.

29.2–3. *divination*. The practice of divination from a wide variety of natural and invoked phenomena is well attested in Babylonia. The most frequent form of divination was through the interpretation of observed events by consulting collections of conditional omen statements. Many texts contain collections of particular types of omens. These texts were usually arranged into multi-tablet series devoted to a particular type of divination. For example, the series *Enūma Anu Enlil* contains celestial omens (i.e. omens from the appearance and behavior of the moon, sun, stars, planets and certain metrological phenomena), *Šumma ālu* contains terrestrial omens (i.e. omens from daily life in a city and the behavior of animals), *Šumma izbu* contains omens from monstrous births, and the series *bārûtu* contains omens from the liver of a sheep (i.e. extispicy). All of the divinatory practices mentioned by Diodorus—by the behaviours of birds, by the interpretation of dreams, by the birth of malformed animals, and by extispicy—are known to be Babylonian practices. For the Babylonians, divination was a process by which the gods sent messages to mankind. If these messages were unfavorable, for example predicting the death of the king, the failure of crops, defeat in a battle, or general hardship in the lands, rituals known as *namburbû*—literally "(ritual for) undoing of it"[5]— could be performed to make the ill-effect pass by. These rituals customarily involved the purification of the affected individual, ritual actions such as the presentation of food or other offerings to the gods, destruction of the portentous object (if appropriate), the performance of incantations, and, occasionally, sacrifice.[6]

29.3–4. *education*. This passage is a fairly accurate reflection of the training of the scribes who wrote scholarly texts in Babylonia during the last few centuries BCE (and possibly earlier). From Hellenistic Babylon we have a document from the Esagila temple which states that the son of a scholar has shown himself capable of making astronomical observations and carrying out the other duties of an astronomer, and will therefore take over the role held at an earlier time by his father.[7] From Uruk we have considerable evidence from colophons on astronomical and other scholarly tablets that there was a small network of

4 Hübner 1990.

5 Caplice 1974: 8.

6 Caplice 1974: 9–13.

7 Rochberg 2000.

scribal families in which sons followed fathers into the scholarly professions. The sons were trained both within their family and in other families in something like a reciprocal arrangement for apprentices.[8]

29.5–6. *Among the Greeks*. Diodorus's disparagement of Greek philosophical education as inferior to that of the Chaldeans and other non-Greeks ("barbarians") is hard to parallel, though it accords with other passages (9.9, 10.7) in which he is severely critical of philosophers as hypocritical and lazy.[9]

29.5. *many is the unprepared person ... late in the day*. A variant manuscript reading would translate as "the man who approaches many (teachers? or subjects?) without preparation takes up philosophy late in the day."

29.6. *they cause the students to doubt*. One may compare Galen, *De libris propriis* 39–41, describing the intellectual crisis he experienced after studying under a succession of philosophers of various sects and contradictory teachings, a crisis that he resolved by turning to mathematics as a model of reasoning.

29.6. *seesawing*. αἰώρᾳ (literally, a swing or a swinging motion) is Henri Étienne's persuasive emendation (1559) of text garbled throughout the manuscript tradition.

30.1. *the nature of the cosmos is eternal*. Diodorus's remark that the cosmos is "eternal" and that its "arrangement and order of everything happened through a certain divine providence" seem to reflect the story of the creation of the universe found in the Babylonian creation epic *Enūma Eliš*, in which the Earth and the celestial realms were created by Marduk from the body of Tiāmat. Marduk creates the stars, the Moon and the Sun to regulate the months and the years. In doing so, Marduk gives the celestial bodies and order and regularity:

> He fashioned heavenly stations for the great gods, and set up constellations, the patterns of the stars. He appointed the year, marked off divisions, and set up three stars each for the twelve months. After he had organized the year, he established the heavenly station of Nēberu to fix the stars' intervals. That none should transgress or be slothful he fixed the heavenly stations of Enlil and Ea with it. Gates he opened on both sides, and put strong bolts at the left and the right. He placed the heights (of heaven) in her (Tiāmat's) belly, he created Nannar, entrusting to him the night. He appointed his as the jewel of the night to fix the dates, and month by month without ceasing he elevated him with a crown.[10]

8 Ossendrijver 2011a, 2011b; forthcoming.

9 Sacks 1990: 189.

10 *Enūma Eliš* V 1–14, translated by Lambert 2013: 99.

30.1 *is going to obtain* (ἐπιδέξεσθαι). An alternate manuscript reading would translate as "has obtained" (ἐπιδέξασθαι). Eck (2003: 156) defends this reading, on the grounds that the notion of a cosmos continuing its existence into the future without limit is an Aristotelian one unparalleled in Mesopotamia; with the past tense, the statement would be taken as a denial that a great deluge had occurred in the past.

30.2. *Having made observations*. Diodorus now turns to the Chaldeans' reputation for celestial divination. Collections of omens which can be used to interpret observed celestial phenomena are preserved from as early as the Old Babylonian period (first half of the second millennium BCE).[11] The standardized terminology for astronomical phenomena found in these texts implies a pre-existing tradition of basic astronomical observation. By the mid-first millennium BCE, this tradition had developed into a practice of regular observation and the systematic recording of those observations in texts known today as Astronomical Diaries.[12] Diodorus refers to observations of the "stars" made "over many years." Later, at 31.9, Diodorus says that observations began 473,000 years before Alexander the Great's crossing the Hellespont (334 BCE)—see the discussion below.

The reference to the "stars" is probably intended to be taken as a broad term indicating astronomical phenomena generally, and involving the sun, moon, and planets as well as the fixed stars; the Greek word *aster* is generally applicable to fixed stars, planets, the sun and moon, as well as constellations. The Akkadian word *kakkabu* too has the general sense of an astronomical body and can refer to a star, a constellation, a planet, or even a comet. Diodorus says that from these observations the Chaldeans were able to deduce the "motions" (*kineseis*) of the stars, by which he probably means simply astronomical phenomena, and their "powers," which presumably means the astrological influences. This latter notion suggests influence from Greek philosophy in Diodorus's presentation, since in Babylonia celestial phenomena are understood as signs from the gods rather than having direct influence on human affairs.

30.3. *the five stars called "wanderers."* Diodorus continues by saying that the Chaldeans paid particular attention to the five planets, which he says that they named the "interpreters." This name bears no relation to the Akkadian term for the planets as a group, *bibbu*, literally meaning "wild sheep." Diodorus lists the planets by their Greek theophoric name in the following order: Saturn, Mars, Venus, Mercury and Jupiter. This planetary order is not the standard Babylonian order of the planets which is commonly found in cuneiform texts of the

11 Rochberg 2006.

12 Sachs and Hunger 1988, 1989, 1996.

late first millennium BCE (Jupiter, Venus, Mercury, Saturn, Mars), nor does it have an obvious Greek origin.

30.3. *Kronos*. Diodorus picks out Saturn (the "star of Kronos") for special attention, saying that it is most "conspicuous" and "signifying the most and greatest things," and that the Chaldeans called it "of the Sun." It is not clear why Saturn should be the most "conspicuous" of the planets since it is neither the brightest nor the slowest moving. Furthermore, there is no indication within Babylonian astrology that Saturn is more significant than other planets. The relationship between Saturn and the Sun is found in cuneiform texts, however. For example, in MUL.APIN, Saturn is at one place referred to as "Saturn, also called the Scales (or) star of the Sun."[13] The association of planets with other heavenly bodies is a common feature of Babylonian omen texts.[14]

30.3. *they name the other four in the same way as the astronomers among us.* This is a curious statement if meant literally, since of course the Babylonians neither spoke Greek nor would have associated the planets with Greek divinities. Perhaps Diodorus merely means that they identified the planets by divinities who were considered to be equivalent, or at least to have correspondence, to the divinities of the conventional Greek theophoric names "star of Hermes" (Mercury), "star of Aphrodite" (Venus), "star of Ares" (Mars), and "star of Zeus" (Jupiter). Various names for the planets were used in scholarly cuneiform texts. In the late period, the most commonly used names were descriptive, such as MÚL.BABBAR "white star" (Jupiter) and GU$_4$.UD "jumping" (Mercury), but theophoric names such as Inana/Ištar (Venus) and Ninurta (Mercury), some of which could be used for more than one planet, are also attested, especially in astrological and non-astronomical scholarly texts.[15] Planet names could be written with either the star determinative MÚL or the divine determinative DINGIR. Collectively, the planets were normally known as the "wild sheep," but they could on occasion be referred to as, for example, the "six gods" (referring to the five planets plus the sun) in MUL.APIN II i 7.

30.4. *intention* (ἔννοιαν). This is Dindorf's (1828) emendation of εὔνοιαν ("good will"); the transmitted reading seems inappropriate—though defended by Eck (2003)—since many omen texts indicated malevolent intentions on the part of the gods.

13 MUL.APIN II i 64 (Hunger and Pingree 1989: 86). For other references to Saturn as the Sun, see van der Sluijs and James 2013 and Brown 2000: 57.

14 See, for example, Brown 2000: 53–81 and Reiner 2004.

15 Brown 2000: 54–81; Rochberg 2011.

30.4. *They say that they signify*. Diodorus asserts that the “rising,” presumably the first appearance, “setting,” presumably last appearance, and color of the planets predict a variety of natural phenomena on Earth including the strength of winds, rain, heat waves, and earthquakes, as well as celestial phenomena including comets and eclipses of the sun and moon, which affect countries, kings and everyday people. Traditional Babylonian celestial divination of the kind epitomized by the omen series *Enūma Anu Enlil* indeed pays attention to the dates, location, and appearance of a planet at its first and last appearance, as well as its apparent color. However, there are also many omens which are concerned with other phenomena of the planets, most notably their position among the constellations. A notable feature of Babylonian celestial divination which is correctly referred to by Diodorus is the prediction of a celestial phenomenon as the apodosis of another celestial omen, with the implicit understanding that this predicted phenomenon can then be interpreted as an omen in its own right. In particular, a number of references to eclipses in omen apodosis are found in *Enūma Anu Enlil*. Weather and other natural phenomena are also commonly predicted in celestial omens as well as in Late Babylonian astrological texts such as TCL VI 19 and 20.[16] It is interesting, however, that Diodorus does not mention the prediction of human affairs directly. References to good and bad harvests, wars, the death of the king, rising and falling values in the market, etc., are very common in both *Enūma Anu Enlil* and late astrological texts. It is also interesting that Diodorus does not mention nativities here, reserving mention of them for the next section.

30.6–7. Following the discussion of general celestial divination, Diodorus turns to the constellations and the zodiac. This topic is resumed, after a digression on personal astrology, in 31.4.

30.6. He begins by saying that “below” (ὑπό, following the transmitted text) the motion of the planets are placed thirty stars. The notion of heavenly bodies being above (i.e., farther away) or below (i.e., closer) each other is not known from Babylonian texts; the particular case here, of the constellations being below the planets, is very strange even in a Greek cosmological context, and most editors therefore accept Letronne’s 1839 emendation, ὑπέρ (“above”).[17] Another conceivable interpretation would take *hypo* in the sense “subordinate to,” but this too would not correspond to any known Babylonian or Greek doctrine.

30.6. *thirty stars*. The reference is puzzling and has led to considerable discussion among modern scholars, both for the number and for the original identity

16 Hunger 1976.

17 Letronne 1839: 584, reprinted 1883: 499.

of these "stars."[18] Most scholars have assumed that the number thirty is an error for thirty-six. This was first suggested by the biblical scholar Wilhelm Gesenius, adducing as a reference to the same entity the later passage 31.4 that speaks of twenty-four "stars" in addition to the twelve zodiacal constellations; Gesenius supposed that Diodorus was speaking of the Egyptian decans, erroneously ascribing them to the Chaldeans.[19] Most subsequent editors have adopted Gesenius's correction. However, starting with Weidner, recent scholars have tended to identify in the passage a reference to the so-called "Three Stars Each" texts (often misleadingly referred to as "Astrolabes" in modern scholarship).[20] In reality, taking 30.6–7 + 31.4 all together, Diodorus is scrambling together three distinct systems: the Egyptian decans, the Babylonian Three Stars Each, and the zodiac which was both Babylonian and, by adoption, Egyptian and Greco-Egyptian.

Section 30.6, considered in isolation, makes sense understood as a description of the Egyptian decans in their original function (traceable back to the First Intermediate Period, late third millennium BCE) as a belt of thirty-six constellations whose successive risings subdivided the night into hours.[21] Although the intervals between first appearances of actual constellations can hardly have been so strictly regular, Egyptian "star-clock" schemes assigned one decan to each ten-day interval in the twelve 30-day months of the Egyptian calendar year. Diodorus's mention of these ten-day intervals is not only a compelling reason for identifying the stars in question as the decans, but also for emending the "thirty" to "thirty-six." The decans were regarded as divine beings having deep associations with the netherworld and with funerary rituals and beliefs.[22] The comment that "they have this motion defined and ratified for an eternal period" is in agreement with the statement in the Babylonian creation epic *Enūma Eliš* (quoted above) that Marduk set up three stars for each month, and this could be seen as a direct reference to the Three Stars Each texts, but the expression is not sufficiently distinctive to rule out the possibility

18 See, for example, Weidner 1915; van der Waerden 1949; Lambert 1987; Oelsner and Horowitz 1997–98.

19 Gesenius 1821: 333–334.

20 Weidner 1915: 63. A notable exception is Oelsner & Horowitz (1997–98), who assume that thirty stars is correct and refers to texts which contain lists of ten stars for each of the three celestial paths of Ea, Anu and Enlil. However, internal evidence from Diodorus' text discussed in what follows argues in favor of correcting the text to thirty-six.

21 Kákosy 1982: 163. Further insights into the decans in later Egyptian and Greco-Egyptian astrology can be expected from J.F. Quack's forthcoming monograph on the topic.

22 Kákosy 1982: 182–184.

that this too is a reflection of Egyptian beliefs about the decans and their cyclic appearances.

In 30.7, twelve superior divinities associated with the months and with the signs of the zodiac are specified as lords of the foregoing thirty-<six> stars. This too is entirely intelligible in terms of Egyptian astrology, now of the late period from the third century BCE onward, closely following the importation into Egypt of the Babylonian zodiac.[23] Without being deprived of their divine status or their nominal role as "hour-watchers," the decans were mapped on the zodiac, three decans to a sign, so that each decan became equated with one of the three 10° segments of a zodiacal sign.

Section 31.4 marks itself as a continuation of 30.7 through the shared reference to the zodiac. However, considered on its own, it appears quite clearly now to be describing the schemes of the Three Stars Each texts. These texts assign three stars, one from each of the "paths" of Ea (northern sky), Anu (middle sky), and Enlil (southern sky), to each month of the schematic 360-day year. This makes a total of thirty-six, not constituting a continuous belt of successively rising stars like the decans, but rather twelve sets of three stars loosely aligned in the north-south direction. Diodorus's account is confused, however, in two crucial respects. Firstly, the stars of Anu were never equated with the signs of the zodiac; all that they have in common is being twelve in number. Secondly, the statements that they are "judges" and that those that are visible (i.e. above the horizon) belong to the living while those that are invisible (below the horizon) belong to the dead is apparently a resumption of the functions of the decans. The expression is sufficiently distinct from that of 30.6 so that it is not certain whether Diodorus means the thirty-<six> stars of 30.6 to be the same as the twelve-plus-twenty-four stars of 31.4.

30.6 *eternal period* (περιόδῳ ... αἰωνίῳ). Another possible translation is "a period of an aeon," that is, a vast but not infinite interval of cyclic regeneration.

30.7. *there are lords of these gods, twelve in number*. The notion of zodiacal constellations through which the Sun, Moon and planets pass is attested already in the early Babylonian astronomical compendium MUL.APIN (I iv 31–II i 8) and by the end of the fifth century BCE it had developed into the concept of the division of the path of the Moon (and other heavenly bodies) into twelve signs of the zodiac of equal length.[24]

31.1. *each of the planets has its own course*. In his discussion of the Sun, Moon and planets, Diodorus notes that each of the planets is "subject to variations

23 Greenbaum & Ross 2010: 156–158.

24 Steele 2007.

and complexities in its speeds and its division of the time intervals." Subdivisions of the synodic arcs of the planets are already presented in MUL.APIN (II i 44–67), but it is only in the mathematical astronomy of the Achaemenid and Seleucid periods that we find a discussion the variation in length and duration of the synodic phases and the speeds of the planets.

31.1. *the nativities of people*. Diodorus turns now to the use of the zodiac in the prediction of the lives of individuals from their birth dates. Horoscopic astrology appeared in Babylonia towards the end of the fifth century BCE, more or less contemporaneously with the Babylonian development of the concept of the zodiac. The practice of horoscopic astrology is attested both by preserved horoscopes which contain astronomical data (principally the positions of the sun, moon and planets at the moment of birth),[25] and by texts which contain omens for the life or an individual or astrological associations which could be used for interpreting horoscopes.[26] Diodorus does not go into detail but refers to the stories he gives elsewhere concerning Alexander (17.112) and Antigonus and Seleucus Nicator (19.55)—though the prophecies of the Chaldeans in those stories are actually more in keeping with the practice of astral omen interpretation referring to kings and kingdoms, best known to us from the late Neo-Assyrian period, than with the later personal astrology.

31.5. *the moon travels.* Diodorus next outlines a cosmological system in which the Moon is closest to the Earth "on account of its weight," that it shines by light "not its own" (i.e., reflected sunlight), and that it causes solar eclipses by casting a shadow on the Earth's surface. There is no evidence that this is a Babylonian cosmological system. Similarly, a couple of lines later, Diodorus says that the Earth is bowl-shaped, which is again not a Babylonian cosmological concept.

31.8–9. *the Chaldeans have the greatest competence in astronomy/astrology of all people.* Diodorus's suming up accounts for the Chaldeans supreme competence in astral science on the grounds that they have been observing the stars for 473,000 years. This number may have its origin in one of the various tradition of king lists that are known from Babylonia. For example, one version of the Sumerian king list (Ash. WB 62) allots a period of 456,000 years to the six kings who ruled before the deluge, although other versions give periods roughly half this long.[27] Berossos assigns a period of 432,000 years to the same antediluvian kings. As for the antiquity of Babylonian astronomical observations, though classical sources transmit a variety of (to us) absurdly large time intervals, it is curious that several cluster near Diodorus's figure: 490,000 years

25 Rochberg 1998.

26 For example, TCL VI 14 and LBAT Nos. 1588–1593.

27 See most recently Friberg 2007: 240.

according to Berossos and Critodemus (as reported by Pliny the Elder, *Natural History* 7.193), 480,000 according to Julius Africanus (as reported by Syncellus, ed. Dindorf p. 31), and 470,000 according to Posidonius (as reported by Cicero, *De divinatione* 1.36 and 2.97).

Bibliography

Boncquet, J.

1987 *Diodorus Siculus (II, 1–34) over Mesopotamië: een historische kommentaar.* Verhandelingen van de Koninklijke Academie voor Wetenschappen, Letteren en schone Kunsten van België, Klasse der Letteren, Jaargang 49, Nr 122. Brussels: Koninklijke Academie voor Wetenschappen, Letteren en schone Kunsten van België.

Brown, D.

2000 *Mesopotamian Planetary Astronomy-Astrology*. Groningen: Styx.

Caplice, R.I.

1974 *The Akkadian Namburbu Texts: An Introduction.* Malibu: Undena Publications.

Eck, B.

2003 *Diodore de Sicile: Bibliothèque historique Tome II Livre II.* Paris: Les Belles Lettres.

Friberg, J.

2007 *A Remarkable Collection of Babylonian Mathematical Texts*. New York: Springer.

Gesenius, W.

1821 *Philologisch-kritischer und historischer Commentar über den Jesaia. Zweyter Theil.* Leipzig: Friedr. Christ. Wilh. Vogel.

Hübner, W.

1990 *Die Begriffe "Astrologie" und "Astronomie" in der Antike: Wortgeschichte und Wissenschaftssystematik. Mit einer Hypothese zum Terminus "Quadrivium."* Stuttgart: Franz Steiner.

Hunger, H.

1976 "Astrologische Wettervorhersagen." *ZA* 66: 234–260.

Lambert, W.G.

1987 "Babylonian Astrological Omens and their Stars." *JAOS* 107: 93–96.

2013 *Babylonian Creation Myths.* Winona Lake, IN: Eisenbrauns.

Letronne, A.-J.

1839 Review of L. Ideler, Ueber die Ursprung des Thierkreises. *Journal des Savants* 1839: 480–495, 527–539, 577–592, and 651–668.

1883 "Sur l'origine du zodiaque grec et sur plusieurs points de l'uranographie et de la chronologie des Chaldéens." Pp. 458–530 in *Oeuvres choisies de A.-J. Letronne membre de l'institut*, ser. 2, vol. 1. Paris: Ernest Leroux.

Oelsner, J., and Horowitz, W.

1997–98 "The 30-Star-Catalogue HS 1897 and the Late Parallel BM 55502." *AfO* 44–45: 176–185.

Ossendrijver, M.

2011a "Exzellente Netwerke: Die Astronomen von Uruk." Pp. 631–644 in *The Empirical Dimension of Ancient Near Eastern Studies*, ed. G. Selz and K. Wagensonner. Vienna: LIT.

2011b "Science in Action: Networks in Babylonian Astronomy." Pp. 213–221 in *Babylon: Wissenskultur in Orient und Okzident*, ed. E. Cancik-Kirschbaum, M. van Ess and J. Marzahn. Berlin: de Gruyter.

(fc.) "Astronomy and the Babylonian Scribal Families." In *Handbook of Hellenistic Astronomy*, ed. A.C. Bowen and F. Rochberg. Leiden: Brill.

Reiner, E.

2004. "Constellation into Planet." Pp. 3–15 in *Studies in the History of the Exact Sciences in Honour of David Pingree*, ed. C. Burnett, J.P. Hogendijk, K. Plofker and M. Yano. Leiden: Brill.

Rochberg, F.

1998 *Babylonian Horoscopes*. Philadelphia: American Philosophical Society.

2000 "Scribes and Scholars: The ṭupšar Enūma Anu Enlil." Pp. 359–376 in *Assyriologica et Semitica: Festschrift für Joachim Oelsner*, ed. J. Marzahn and H. Neumann, Münster: Ugarit-Verlag, 2000.

2006 "Old Babylonian Celestial Divination.," Pp. 336–348 in *If a Man Builds a Joyful House: Assyriological Studies in Honor of Erle Verdun Leichty*, ed. A.K. Guinan et al. Leiden: Brill.

2011 "The Heavens and the Gods in Ancient Mesopotamia: The View from a Polytheistic Cosmology." Pp. 113–132 in *Reconsidering Revolutionary Monotheism*, ed. B. Pongratz-Leisten. Winona Lake, IN: Eisenbrauns.

Sachs, A. and Hunger, H.

1988 *Astronomical Diaries and Related Texts from Babylonia. Volume I: Astronomical Diaries from 652 B.C. to 262 B.C.* Vienna: Österreichishe Akademie der Wissenschaften.

1989 *Astronomical Diaries and Related Texts from Babylonia. Volume II: Astronomical Diaries from 261 B.C. to 165 B.C.* Vienna: Österreichishe Akademie der Wissenschaften.

1996 *Astronomical Diaries and Related Texts from Babylonia. Volume III: Astronomical Diaries from 164 B.C. to 61 B.C.* Vienna: Österreichishe Akademie der Wissenschaften.

Sacks, K.S.
1990 *Diodorus Siculus and the First Century.* Princeton: Princeton University Press.

Steele, J.M.
2007 "Celestial Measurement in Babylonian Astronomy." *Annals of Science* 64: 293–325.

Van der Sluijs, M.A., and James, P.
2013 "Saturn as the 'Sun of Night' in Ancient Near Eastern Texts." *AuOr* 31/2: 279–321.

Van der Waerden, B.L.
1949 "Babylonian Astronomy. II. The Thirty-Six Stars." *JNES* 8: 6–26.

Vogel, F.
1888 *Diodori Bibliotheca Historica.* Vol. 1. Leipzig: Teubner.

Weidner, E.
1915 *Handbuch der babylonischen Astronomie.* Leipzig: J.C. Hinrichs.

CHAPTER 18

Observing Eclipses: An Optical Question in *Problemata* 15.11

Alan C. Bowen

Introduction

In book 15 of the *Problemata*, the eleventh *problema* opens with the question:

> Why is it that during eclipses of the Sun, if one observes [them] through a sieve or through leaves (for example, [the leaves] of a plane tree or another broad-leafed tree) or by joining the fingers of one hand to the other, the beams become lunes on the ground?

This is intriguing: not only is there a paucity of texts from Antiquity or the Middle Ages that even mention the mechanics of observing eclipses, none other that I am aware of addresses puzzles that arise as one actually looks at them. But our curiosity will yield to bewilderment when this question is followed immediately by another,

> Is it because the [Sun's] light is rounded, that is, a cone, just as [it is] if it shines through a small opening with suitable angles?

a question followed in turn directly by this explanatory suggestion:

> A reason [for this] is that there are two cones, the one from the Sun to the opening and the one which is from there to the ground; and they share the same vertex. Thus, with [the cones] disposed in this way, when the upper [cone] is cut off from above by a circle, there will be a lune of light opposite on the ground, since the rays come from the arc of the lune and since the [openings] in fingers and sieves are, as it were, small openings. This is why [the arc of the lune] is clearer than through large openings.

What are we to make of this? After all, why ever would one suppose that the Sun is a point source of light and, thus, that solar eclipses are instantaneously total (see Figure 18.1, p. 354)? And what does any of this talk of solar cones have

 | DOI 10.1163/9789004363380_019

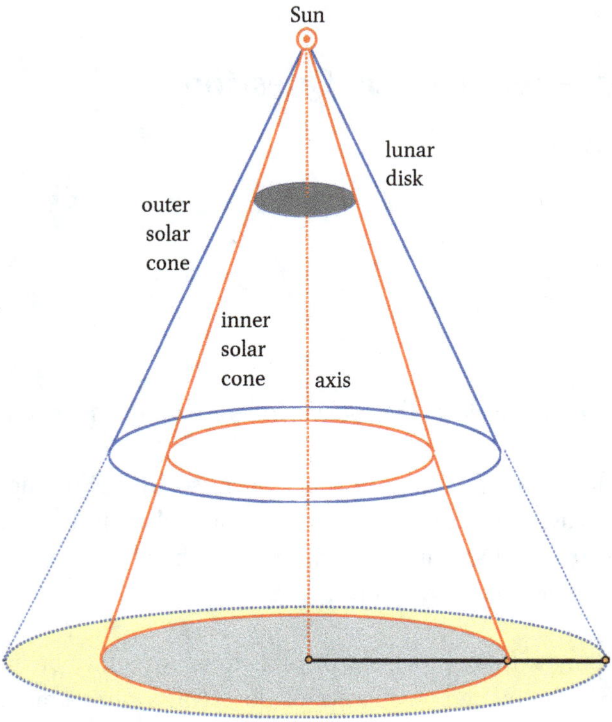

FIGURE 18.1. *The solar cones during a solar eclipse* [Prob. *15.11*]. *In this configuration, all solar eclipses are total and occur instantaneously. That is, any observer inside the inner solar cone will see a total eclipse; whereas any observer outside it will not detect any eclipse at all.*

to do with our seeing lunes on the ground? Surely, by itself, the suggestion can account at best for the inner curve of a lune (see Figure 18.2, p. 355).

My aim on this occasion is to develop a reading of 15.11 that advances to the stage where the interlocutor or reader realizes that the answer to the second question is "No" and begins to see a somewhat different framework for answering the first or opening question.

To accomplish this, we must first deal as best we can with the question of how to read the *Problemata*. But *caveat lector*. Given any ancient text, there is, of course, no compelling argument that there is one and only one way to read it properly. Indeed, scholarly experience surely gives greater credence to the view that there are many ways to read or interpret such a text and that, for the most part, their success or failure is a matter of their assumptions and the concerns that they address. Still, in so far as it is our common task to make sense of such texts as communications from their authors to their readers,

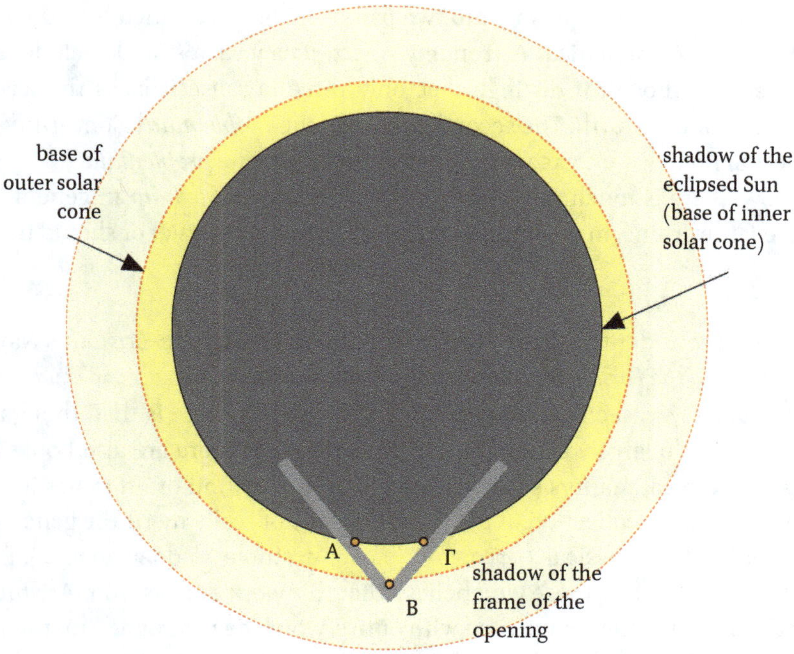

FIGURE 18.2. *The ground illuminated by the Sun during a solar eclipse* [Prob. *15.11*]. *The area illuminated within the shadow of the material forming the opening,* ΑΒΓ, *is not a lune.*

it is true that any reading that abandons this effort must count as a failure of interpretation. This is, however, my criticism of a dominant trend in current scholarship devoted to the *Problemata*. To supplant this trend, then, we must begin by considering the *Problemata* as a historical document.

The *Problemata*

The earliest and most authoritative witness to the *Problemata* is the 10th-century codex Parisinus graecus 2036.[1] In this codex, the *Problemata* comes without any explanatory front matter: after a folio page which gives the title and then lists the contents, the codex simply presents a collection of over several thousand questions divided into 900 sections or chapters that are organized into 38 books.[2] Each section or chapter is called a *problema*.

1 Par. gr. 2036 contains two works: the *Problemata* and Longinus' *De sublimitate*.

2 The list of contents on f. 1v of the codex includes only 37 books: it omits book 37, "All the *Problemata* Concerning the Body as a Whole," which, curiously enough, is not numbered (see f. 175r).

The codex divides book 15 into two parts. The first is entitled "All [the *Problemata*] in Mathematical Astronomy" («ὅσα μαθηματικά») and includes four *problemata*, though it distinguishes only three because it lacks the leading question of the fourth. The second part, "All [the *Problemata*] Concerning the Celestial Bodies" («ὅσα περὶ τὰ οὐράνια»), includes nine *problemata*.[3]

Now, there is much more to be said about the *problemata* in general but, I give fair warning, in what follows most of it is controversial or should be.

The Title

To begin, consider the title of the work given in this codex: *Aristotle's* Natural Problems *in the Form of a Compilation* ('Ἀριστοτέλους φυσικὰ προβλήματα κατ' εἶδος συναγωγῆς, see Par. gr. 2036, f. 1r). The difficulty here is that, though the *problemata* do largely concern natural phenomena, there are also books that focus on ethical matters such as justice and moderation or on topics in law.[4] Thus, the entire collection of *problemata* does not really merit the general rubric 'natural' or 'physical' («φυσικά») (cf., for example, Cadden 2006: 4). Especially telling against those who believe that this work derives from Aristotle is that book 15 is largely concerned with optics, which he regards as a mathematical science, which, though it is closer to natural science than geometry, say, is still not a natural science (φυσική) *per se* (see *Phys.* 2.2).

The Author

Next, there is, in fact, no good reason to hold that the text is by Aristotle anyway. Granted, there is clear evidence in Aristotle's extant works that he wrote a *Problemata*, which has not survived. Indeed, on the strength of passages in Par. gr. 2036 that are parallel to passages in Aristotle's surviving works, some scholars have maintained that Aristotle, though hardly the author of the collection in its entirety, either wrote some of its books or directed their compilation. But, as others have noted, this line of argument is not compelling.

After all, if passage B is later than passage A and parallel to it, it does not follow that A is the source (direct or indirect) of B. To make the case, one must supply independent argument, for example, that no one else in A's time or before wrote *problemata*.

---

3 On these titles, see Bowen 2015. Note that the *problemata* in book 15 of the codex are not numbered continuously as they are today in our editions and translations; rather, numeration starts anew in each part. Though the edition of the Greek text in Appendix A (see p. 369) preserves this numeration of parts, I will follow modern practice in the translation in Appendix B (see p. 373) and commentary by writing of *Prob.* 1–4 in part 1 and of *Prob.* 5–13 in part 2.

4 See also the listing in Table 18.1 on p. 358.

But the real problem with this focus on sources is that, even if one could indeed establish that A is either a proximate or ultimate source of B, it still would not follow that the *meaning* or *sense* of B in its context is captured by A. Thus, for instance, if passage A is expository and occurs in some Aristotelian treatise, and passage B is found in a *problema*, passage B will not have the commitment of passage A. That is, where passage A may assert a truth in the course of an argument, passage B will remain agnostic (see pp. 360–361) and have perhaps a very different purpose.

In sum, the classical tools traditionally brought to bear on the question of the authorship of the *Problemata* fail to secure knowledge either of the text's dependencies and origins or of its meaning and purpose.

In any case, the question of who wrote the *Problemata* is complicated. In fact, as I hope to make clear, it is far better to maintain that the text has no single author but a sequence of authors, none of whose names has been preserved save Aristotle's.

Now, whether Aristotle's name is still worth preserving is another question. Those who have historically connected the compilation to Aristotle seem to have been motivated in part by the desire to secure cachet for their labors on this work. Thus, in medieval times, as Cadden notes (2006: 8–9), editors and translators maintained Aristotle's authorship because exclusion of the *Problemata* from the canon of Aristotle's works would have marginalized the *Problemata* even further than it was by virtue of its peculiar nature and contents. Still, I see no pressing need to follow them in this nowadays.

The Text

But were it not enough to undermine the title that it ascribes to this work, the codex itself also gives reason to doubt the historical validity or worth of our modern editions of book 15 at the least and, incidentally, challenges our reflexive notions of textual authorship and meaning.

In this codex, in another hand dated by Louis to the 12th century,[5] there is on f. 1r–v to the left of the listing of book numbers and their titles, a lacunose column of numbers indicating book numbers in some other compilation (see Table 18.1, p. 358).

5 See Louis 1986: 199; 1991–1994, 1.xlii. The ink in this hand is blacker than the reddish black ink of the scribe of the codex. Unlike the scribe, this redactor, who is also responsible for the rightmost column of numbers puts a diaeresis over «ι»; and, in his abbreviation of «περί», «επ », the horizontal stroke in «π» stops at the vertical strokes [see f. 1v, entry for book 36].

TABLE 18.1 *Listing for books 14–18 in Par. gr. 2036, f. 1r*

ϊγ	ιδ́	ὅϲα περὶ τὰ κράϲειϲ	ϊϛ
ϊδ	ιέ	ὅϲα μαθηματικὰ ·γ· καὶ περὶ τὰ ουρ(ά)νια	θ̈
	ιϛ́	ὅϲα περὶ τὰ ἄψυχα	ϊγ[a]
	ιζ́	ὅϲα περὶ ἔμψυχα	γ̈
ϊε	ιή	ὅϲα περὶ φιλολογίαν	ϊ

13	14	All the [*Problemata*] Concerning Mixtures	16
14	15	All the [*Problemata*] in Mathematical Astronomy [a] and [All the *Problemata*] Concerning Celestial Bodies	9
	16	All the [*Problemata*] Concerning Inanimate Things	13
	17	All the [*Problemata*] Concerning Animate Things	3
15	18	All the [*Problemata*] Concerning the Love of Words	10

[a] The number of *problemata* in book 16 originally written here has been corrected and is illegible. But the same correction occurs in the book itself in the numeration of the last *problema*, where it is clear that the numeral has been corrected from «ιβ» ('12') to «ιγ» ('13') [f. 78r].

Certainly, given the subsequent textual history of this work, it should come as no surprise that there may have been different arrangements of this compilation prior to Par. gr. 2036. In any event, if this is indeed the significance of the leftmost column on f. 1, that is, if book 15 in the codex corresponds to book 14 in a compilation now lost, then our modern editions of book 15 may well be illusory in so far as they pretend to present a text on the basis of Par. gr. 2036 that is earlier than this codex.

The problem may be cast as one of the degree to which book 15 in the codex is dependent on earlier versions. Lourus Filius (1999) has edited an Arabic version of the *Problemata*, which he argues is a translation of a Greek version that is no longer extant. In this Arabic version, which Filius attributes to Ḥunain ibn Ishāq (AD 809–873), the *problemata* found in book 15 of the codex are split into books 16 and 17. The Arabic book 16 has but three *problemata*: 16.1–2 correspond respectively to 15.1 and 15.3 of the codex (without the lines in 15.3 that now belong to our 15.4). Moreover, the articulation of 16.2 is substantially different and 16.3 has no counterpart in the codex. As for the Arabic book 17, it contains only two *problemata* 17.1–2, which correspond to 15.1–2 in Part 2 of the codex (our 15.5–6). But, again, there are significant divergences. Thus, not only is it clear that that the Greek version translated into Arabic does not figure in Par. gr. 2036, it would seem that even prior to Par. gr. 2036 a rearrangement of books could entail substantial variation in the selection and articulation of the *problemata* constituting them.

So, did the scribe of book 15 of the codex simply copy an earlier exemplar in Greek or did he in the course of copying this exemplar consult other manuscripts to produce the version that we see?[6] In my view, there is no good way to settle this question. So, given my preference for surer footing, I maintain that the way forward is to finesse such questions by editing Par. gr. 2036 by itself and to base all analysis on that text alone. It is, after all, the earliest and our surest witness to the Greek text of the *Problemata*. Accordingly, in what follows, I have included an edition of the two *problemata* in book 15 relevant to our discussion that is faithful to what one reads in this codex, barring some minor corrections needed to make sense of particular lines.

A Living Text

The *Problemata* was, apparently, not a text such as Aristotle's *De caelo* or Plato's *Timaeus*, texts that were copied and translated through the ages by those primarily intent on accessing their meaning. It is instead a living document that was not only read and copied but also transformed and adapted as it made its way through history. It has no single author or determinate co-authors because it was not just read, it was *used* by many who remain unnamed and who adapted it by selection, supplementation, rearrangement—sometimes in the course of translation—to different contexts and purposes at different times and in different places. Par. gr. 2036 preserves but one of those transformations. How this particular transformation compares to what there was earlier and how long ago it was made are unknowable. At the most we may assert only that this manuscript preserves a text that was read and used in the 10th century and later.

It follows from this that it may well be misleading and unhelpful to say that the *Problemata* is a Peripatetic or Stoic work. Such conjecture not only privileges unduly the early history of the *Problemata*, a history that is largely unknown, it also pretends that such a single designation can do justice to this collection and its active, transformative use over a period of roughly a millennium.

So much for the *Problemata* in general. Next, let us move to further matters no less controversial and consider how to read its numerous sections or chapters.

6 It would seem that, at the least, he is responsible for the listing of contents on f. 1.

The Typical problema

Each *problema* in book 15, as elsewhere, is not a single, isolated question. Rather, it is a structured set of questions opening with a leading question that begins with «διὰ τί» meaning "Why." This leading question may be followed by an affirmative sentence or two. But, in book 15 at least, it is more typically followed directly by one or more secondary questions. Each of these secondary questions opens with «ἢ ὅτι», though there is one *problema*, 15.12, in which the secondary question begins with «ἢ διότι», which has the same meaning as «ἢ ὅτι». If there is but one secondary question, "Is it because" will often suffice. But, given that «ἤ» often does not indicate an alternative to the primary question but substitutes a more specific question anticipating *an* answer to the leading question (cf. Smyth 1971: §2860), "Rather/Precisely, it is because" may sometimes be preferred. When there are several secondary questions, however, «ἤ» may signify a real disjunction as in "Is it because…? Or is it because … or because…?." These secondary questions may each followed by their own affirmative sentence or sentences.

The challenge, of course, is to interpret this structure. As I see it, the critical points to keep in mind are, first, that the *problemata* all seek factual information concerning material and efficient causation—the "Why?" questions do not broach teleology. Second is that the secondary questions invite no more than the answers "Yes" or "No." Third, and here is an equally critical point, these secondary questions lack the particles and phrases inviting either affirmation or negation.[7] In other words, these secondary questions inquire without indicating any expectation of what the appropriate answer might be. This has the important consequence that the sentences typically following them should not be understood as the premises of an argument concluding in the proposition reached by reformulating the secondary question. Instead, these sentences should be treated as considerations that clarify the sense or meaning of the secondary question, all the while still leaving its answer open.

In sum, the *problemata* are not, I propose, curiously disguised statements of doctrine.

This is a subtle point. All questions have contexts that give them their meaning. A questioner's suggesting a context does not preclude its rejection, especially when these contexts are offered in clarifying a question posed without bias to a particular answer. The upshot is that we may now distinguish between reading each *problema* and the *Problemata* itself as an inconsistent, poorly organized body of doctrine obfuscated by a use of interrogative sentence-forms

7 The particles «οὐ», «ἆρ' οὐ», «οὐκοῦν», e.g., and the phrase «ἄλλο τι» invite affirmation; «μή», «ἆρα μή», or «μὴ οὖν» (= «μῶν») invite negation: see Smyth 1971: §§2650–2655.

that may be ignored and reading it as a collection of open, articulated queries intended to provoke real thought.[8] This means that the inconsistency that many detect in the *Problemata* may be an epiphenomenon of getting the interlocutor or reader to think and not a fault after all.

The Solar and Visual Cones

With this understanding of the basic structure of any given *problema* in Par. gr. 2036, let us now return to 15.11.[9] As I have already pointed out, the questioner's suggestion that the answer might come by reflection on the solar cones can account at best for the inner curve of the shadow cast on the ground (see Figure 18.2, p. 355). To explain why this shadow is a lune (see Figure 18.3, p. 362), one needs an answer to *problema* 15.6:[10]

> Why does the Sun make not figures that are rectilinear but circles as it passes through the quadrilaterals, for example, in [wicker] bulwarks?

But this leading question is immediately retracted as a misfire and replaced with a secondary question:

> Rather, is it because the emission of visual rays is a cone and the base of a cone is a circle, so that the rays of the Sun appear circular at whatever object they strike?

This substitute question is then amplified:

> Certainly, the shape due to the Sun must be contained by straight lines, since the rays are straight [lines]. Indeed, whenever straight [lines] strike a straight [line], they produce a rectilinear [figure]; and this happens in the case of the [Sun's] rays too, since, where they shine through, they strike the line of the wicker-work (which is straight), and [since] these [rays] are straight, with the result that the emission [of the Sun's rays] will be to a straight line.

8 The question of the larger context in which this was to occur raises that of the readership of this treatise and why it was disseminated. Suffice to say for now, that the *Problemata* was not part of the standard curriculum at medieval universities.

9 See Appendix A, p. 372 for the text and Appendix B, p. 376–377 for a translation.

10 See Appendix A, p. 370–371 for the text and Appendix B, p. 373–375 for a translation.

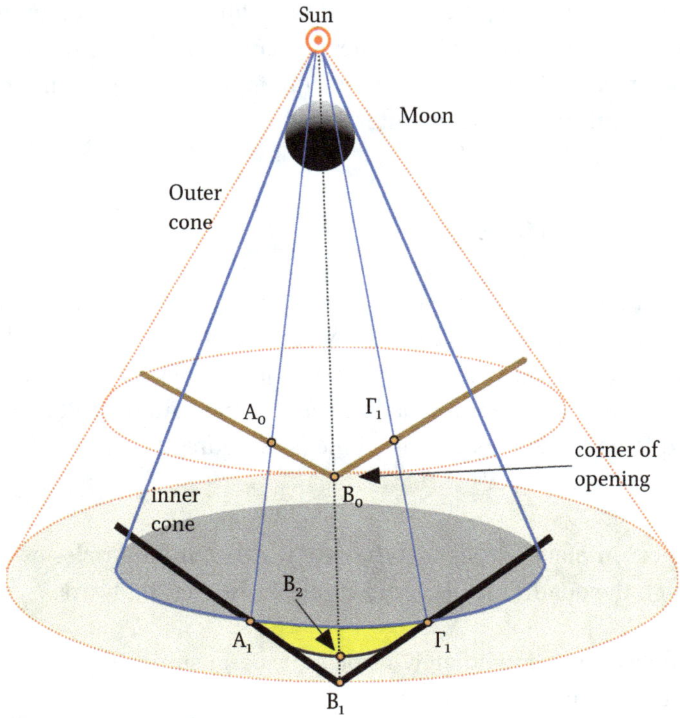

FIGURE 18.3. *The Sun's rays during an eclipse as they pass through a small regular-angled opening to the ground to produce an area of light seen as a lune* [Prob. *15.11*]. *The outer cone of solar rays has the inner edge of the corner of the small opening on the circumference of its base. The inner cone is defined by the Moon, which effects the area eclipsed. The challenge is to explain why the area* $A_1B_1\Gamma_1$ *is seen as the lune* $A_1B_2\Gamma_1$.

The upshot is that the challenge to explain the shape of the area illuminated on the ground by the Sun as it passes through wicker-work may require distinguishing between what is illuminated and what is seen and that this may require in turn distinguishing the nature of the cone in which light propagates from the Sun and the nature of the cone of visual flux (see Figure 18.4, p. 363, Figure 18.5, p. 364).

Just how the visual cone might contribute to an explanation is stated next:

> But, because [the visual rays] departing from the eyes to the ends of the straight lines are weak, what is in the angles [formed by the Sun's rays on the ground] is not seen. Rather, as much of the straight line as is *within* the [visual] cone effects a line [that we see], whereas the rest [of the

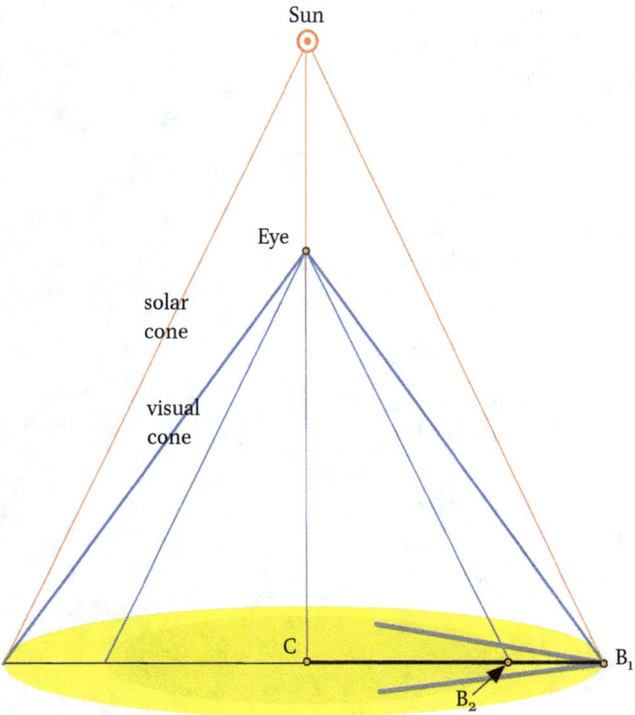

FIGURE 18.4. *The solar and visual cones* [Prob. *15.6*].

> straight line] does not; instead, the visual rays reporting what they see escape notice.

The operative contention here is twofold: first is that the rays towards the sides of the visual cone are weaker than those towards the cone's axis; and second is that vision fails, that is, we do not see an object, when the visual rays emitted towards it are not noticed, that is, when the rays fail to make a report. This contention is then supported and amplified by remarks about other occasions when visual perception fails:

> Certainly, we do not see many things to which our visual ray penetrates, for example, things in the dark. Similar to this too is the fact that a square appears polygonal and, if one stands at a greater distance, as a circle. That is to say, since the emission of our visual rays is a cone, when the shape is set at a point far off, the [visual rays] departing from our eyes to the angles [of the shape] do not see because they are weak and few [in number] as the distance becomes greater; whereas the visual rays striking at the center [of the shape] persist because they are crowded together and

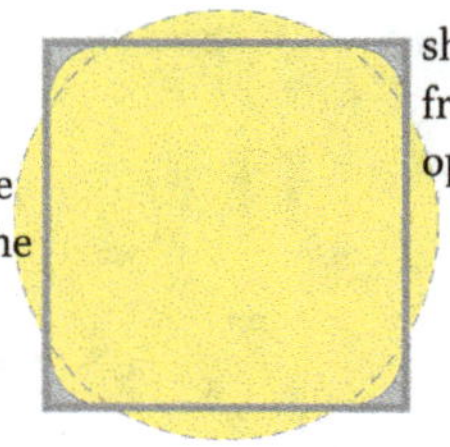

The corner angles of the opening may be less or more than 90°and its sides may vary in length.

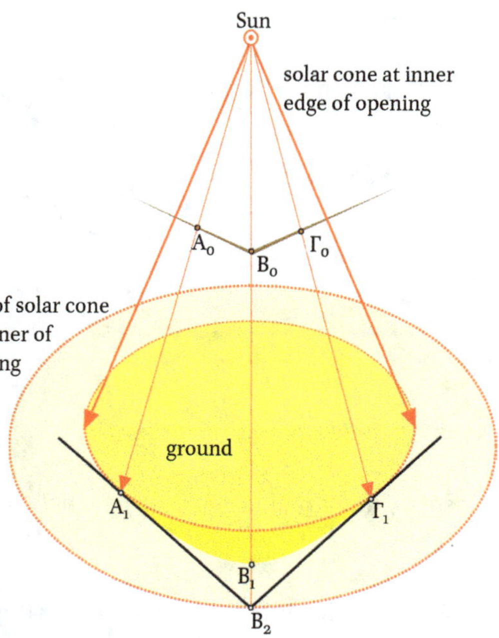

The challenge is to explain why $A_1B_1\Gamma_1$ appears as $A_1B_2\Gamma_1$.

FIGURE 18.5. *The Sun's rays passing through a small regular opening to the ground to form a meeting of straight lines seen as curved* [Prob. *15.6*].

> strong. Thus, when the shape is near, [the visual rays] can also see what is in the angles but, when the same [shape] is far off, they cannot.

These remarks are meant to aid the interlocutor in reflecting on why we see only a circular arc of light on the ground at the corner of the shadow of the opening. Of obvious importance is the suggestion that not only are the rays to the sides of the visual cone weaker than those towards its axis, they are also fewer in number as the distance to the object increases. Less obvious, but equally important, is the hint that any explanation of what we see must take due account of the role of light in our vision of objects.

But herein lies a further problem for the interlocutor or reader: Is there a coherent way in this instance to maintain that that there is an illuminated area on the ground that we cannot see? In other words, is the geometry and nature of the solar and visual cones indicated here really capable of explaining what is seen?

Certainly, they are not coherent as presented in *problema* 15.6, a fact evident in 15.11 as well. After all, given especially that the openings are small, the solar cone will illuminate an area on the ground that is interior to the shadow of a given opening as well as an area that is exterior. Moreover, this exterior area will be visible. Accordingly, to explain what is seen on the ground when the Sun's light passes through this opening, it is a mistake to limit the base of the visual cone to the interior of the shadow of the opening. In other words, whatever is seen at the corner of the shadow of this opening—be it a curved area of light, a lune of light, or whatever—it will not be due to the nature of the rays near the sides of the visual cone. In fact, what would seem to be required is some account of perceptual judgment when the visual cone includes shaded areas or darkness at its base (see Figure 18.6, p. 366).

The Need for Experience

There are numerous occasions in book 15 when the text or, better, the questioner, points to something outside the text, something that his interlocutor sees or experiences and his reader must assume if the text is to make sense.[11] Such passages indicate a context outside the text, a real or imaginary context that is to be experienced. So far as 15.11 is concerned, this would include "looking" at the patterns of light cast on the ground by the Sun through small openings during an eclipse or considering a diagram of this. In either case, it would appear quite readily that the setup in 15.11 is incorrect so far as solar eclipses are concerned.[12]

My suspicion is that the experience has ultimately to be real. At the least, this would seem a *logical* consequence of the leading questions of 15.6 and

11 See 15.100/911b12, p. 25 n. 5. Note: "15.100" means "book 15, line 100," counting from the first line of book 15 in the codex. "911b12" is page 911, second column, line 12 in Bekker 1831.

12 Indeed, it would be far better to hold that they occur when the Moon comes into a cone with its base at the Sun and its apex at a relatively short distance under the Earth's surface. Perhaps, one aim of 15.6 and 11 is to get the interlocutor to reconsider the underlying geometry of the solar cone by which objects cast shadows.

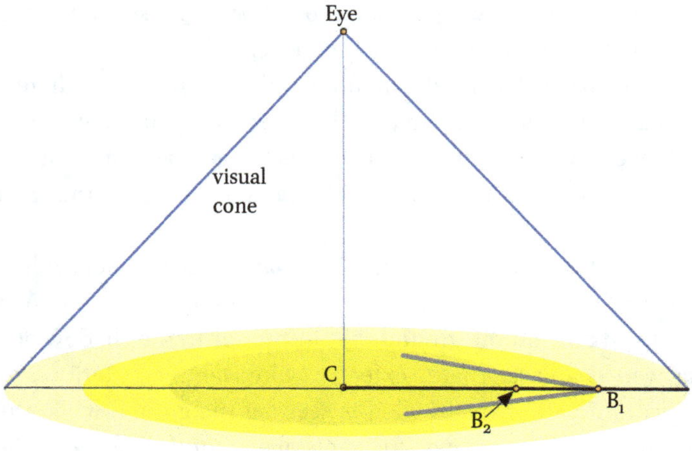

FIGURE 18.6. *The ground illuminated during a solar eclipse* [Prob. *15.6*]. *The challenge here, as in 15.11, may be to explain why the illuminated area at the very shadow of the corner appears curved as a consequence of how we see light at the border of a shadow* [*cf. Ptolemy,* Optica].

15.11, since they each introduce as facts claims that are plainly not part of ordinary experience and since imagination is in these instances of no help in deciding what actually is the case. Moreover, the leading question in both 15.6 and 15.11 makes one wonder why it is that we can supposedly see the corner of the shadow of the frame of the opening but not the illuminated ground *at* that corner. In short, the claims that we see a curved line or a lune toward the corner of the opening and the geometry of the visual cone suggested to account for this would appear absurd. Thus, in responding to such *problemata*, any interlocutor/ reader will be obliged to go and look in order to determine the status of the fact assumed by the why-is-it question of these two *problemata*. Of course, whether this is actually intended is another matter, though it would certainly make for better pedagogy if it were.

Interpreting Fuzziness

Now, should the interlocutor/reader undertake to verify the fact assumed by the leading question of 15.11, what would he or she see? Fuzziness. That is, one would see that the inner edge of the opening is not clearly defined. There are several reasons for this. First is that the Sun is not a point source of light, that is, it is not the apex of cone of light, as the questioner suggests. Moreover, given that the Sun has magnitude, its rays do not travel in parallel, as was supposed, for example, in the calculations of the Earth's circumference attributed to

Eratosthenes (see Cleomedes, *Cael.* 1.7). In short, the shadow of what defines the opening will have a penumbra that is lighter towards the opening. Second, and of increasing importance as the width of the material constituting the opening decreases, is that the shadow of its inner edge also loses definition because of the diffraction of light by the edge itself.[13]

But whether the fuzziness of the shadow cast by what defines the opening gives the impression of curvature directly to all or most observers or whether seeing it as curved requires effort much like that in seeing shapes in clouds remains, for me at least, an open question. Indeed, I suggest, the ultimate aim of *Prob.* 15.6 and 15.11 may well be to encourage reflection on how best to analyze certain optical phenomena which seem resistant to purely geometrical analysis.[14] The questioner introduces a theory of perception (see Appendix C, p. 378) but it too seems inadequate. So will some other theory be successful or does the answer lie in other aspects of human psychology? That is the issue, which, I suggest, the interlocutor/reader is ultimately expected to address.

Conclusion

To the modern reader who takes from books their arguments or ideas for reflection, the *Problemata* is a puzzle. What is one to do with such an overwhelming list of articulated questions, many of which would seem to be impenetrable? After all, what really are we to do with such questions as "Why is having sex useful against diseases caused by phlegm?" (*Prob.* 1.50a). Our world has plainly changed to the point that such questions articulated as they are make little real sense and can only amuse. And so the best that we do today is to treat them as lifeless artifacts and to contextualize them, typically by determining their sources. This is not the case for book 15. Here we can still detect the vitality and intellectual challenge of the *Problemata* as a whole. Its questions as articulated, though they may seem odd, even wrong, they are still alive because they are not so absurdly at odds with either our experience or our understanding of what is the case. That is, addressing them does not entail a radical break in the kind of discourse. To the contrary, these questions challenge us to engage

---

13 Diffraction, a phenomenon not defined and studied until the 17th century, is the process by which light or, indeed, anything that propagates in waves, generates interference patterns when it passes by the edges of objects or through narrow openings. Relevant to the present discussion is the fact that, by virtue of diffraction, the edges of shadows are fuzzy, not sharp. For images of the effect of a diffraction at the edge of a penny, see Figure 18.7, p. 368.

14 Among these, I would include the Moon Illusion—the fact that the Moon seems larger to us at the horizon than when it is overhead (see Ross and Plug 2002).

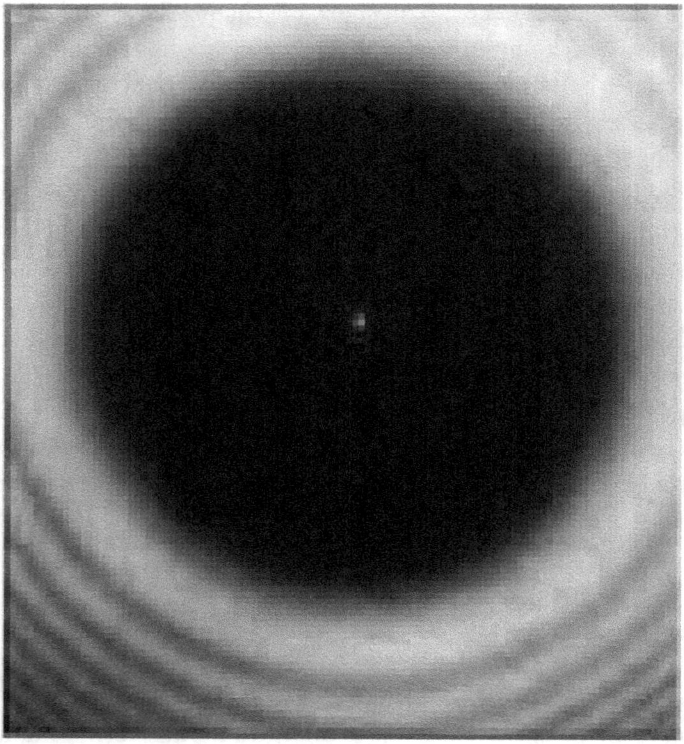

FIGURE 18.7. *Diffraction of light by a penny placed midway between the light source and a screen.*

them by applying our wits to our experience. So, during the next solar eclipse, go outdoors and look. Under the conditions of the leading question of 15.11, what do you see on the ground? A lune of light? Why?

Acknowledgments

This chapter reworks material that I had the great pleasure of presenting to the members of Project Melo: On Euclidean Optics at the University of Lisbon on 16 Sep 2014. I am most grateful to them, especially to Bernardo Mota and Henrique Leitaõ, for their comments and most gracious hospitality. The argument of this chapter has also benefitted by presentation at the conference "Ptolemy's Philosophy and Its Greek, Arabic and Hebrew Reception" at Georgetown University, Washington, DC on 3–4 May 2016. I also thank Fabio Acerbi for his careful reading of a draft of this chapter. I can only hope that my dear friend, the distinguished honorand of this volume, will find *Prob.* 15.11 as tantalizing as I do.

Appendix A

Introduction

The edition of *Prob.* 15.6 and 15.11 given here does not aim to provide a critical text for which Par. gr. 2036 is the best witness but to present these *problemata* as they appear in this codex, with only those emendations that are needed to make sense of them. The apparatus records these emendations and the warrant for them along with suggestions by later editors that are not accepted in the text but still merit consideration.

The line numbers to the left of the Greek text start at line 1 of book 15 as it appears in the codex. I add folio numbers as a convenience. To the right of the Greek text are the page, column, and line numbers in Bekker 1831.

Sigla

Y^a	Parisinus graecus 2036	
Y^{a2}	12th-century redactor of Par. gr. 2036	
Bart.[1]	Bartholomeus de Massana	see Bartholomeus 1482
Bart.[2]	Bartholomeus de Massana	see Bartholomeus 1501
Gaza	Theodorus Gazae	see Theodorus 1501
Eras.	Desiderius Erasmus	see Erasmus 1550
Syl.	Friedrich Sylburg	see Sylburg 1585
Buss.	Ulco Cats Bussemaker	see Bussemaker 1857
Bekk.	Immanuel Bekker	see Bekker 1831
Ruelle	Charles-Émile Ruelle	see Ruelle 1922
Fors.	Edward Seymour Forster	see Forster 1927

Parisinus graecus 2036.87–127 [*Prob.* 15.6]

(f. 90v)

B διὰ τί

Ὁ ἥλιος διὰ τῶν τετραϖλεύρων διέχων οὐκ εὐθύ
γραμμα ϖοιεῖ τὰ σχήματα ἀλλὰ κύκλους, οἷ
ον ἐν τοῖς ῥιψίν; ἢ ὅτι ἡ τῶν ὄψεων ἔκϖτω 91
σις κῶνός ἐστιν, τοῦ δὲ κώνου κύκλος ἡ βάσις,
ὥστε πρὸς ὃ ἂν ϖροσϖίϖτωσιν αἱ τοῦ ἡλίου ἀκτῖ
νες κυκλοτερεῖς φαίνονται; ἀναγκαῖον μὲν γάρ
ἐστι καὶ τὸ ὑϖὸ τοῦ ἡλίου σχῆμα ὑϖ' εὐθειῶν ϖε
ριέχεσθαι, εἴϖερ αἱ ἀκτῖνες εὐθεῖαι· ὅταν γὰρ εὐ
θεῖαι ϖρὸς εὐθεῖαν ϖροσϖίϖτω|σιν, εὐθύγραμ b1
μον ϖοιοῦσιν. ἐϖὶ δὲ τῶν ἀκτίνων συμβαίνει
τοῦτο· ϖρὸς εὐθεῖαν γὰρ ϖροσϖίϖτουσι τὴν
τοῦ ῥιϖὸς γραμμήν, ἢ δι' οὗ διαλάμϖουσιν,
καὶ αὗται εὐθεῖαί εἰσιν, ὥστε ϖρὸς εὐθεῖαν ἔ
f. 91r σται ἡ ἔκϖτωσις. ἀλλὰ διὰ τὸ ἀσθενεῖς εἶναι τὰς
ἀϖοσχιζομένας ἀϖὸ τῶν ὄψεων ϖρὸς τὰ ἄκρα
τῶν εὐθειῶν, | οὐχ ὁρᾶται τὰ ἐν ταῖς γωνίαις, b1
ἀλλ' ὅσον μὲν τῆς εὐθείας ἐνυϖάρχει ἐν τῷ κώ
νῳ, ϖοιεῖ αὐτήν, τὸ δὲ λοιϖὸν οὐ ϖοιεῖ ἀλλὰ λαν
θάνουσιν αἱ ὄψεις ἐϖιϖίϖτουσαι· ϖολλὰ γὰρ οὐχ ὁ
ρῶμεν ἐφ' ἃ διϊκνεῖται ἡ ὄψις, οἷον τὰ ἐν τῷ σκό
τει. ὅμοιον δὲ τούτῳ καὶ τὸ τὸ τετράγωνον ϖολυ
γωνοειδὲς φαίνεσθαι, ἐὰν | δὲ ϖλέον ἀφιστῇ, κύ b2
κλον. ὄντος γὰρ κώνου (τῆς) τῶν ὄψεων ἐκϖτώσεως,
ἀφισταμένου τοῦ σχήματος εἰς τὸ ϖόρρω, αἱ μὲν
εἰς τὰς γωνίας ἀϖοσχιζόμεναι τῶν ὄψεων δι
ὰ τὸ ἀσθενεῖς εἶναι καὶ ὀλίγας οὐχ ὁρῶσι, ϖλέον(ος)
τοῦ ἀϖοστήματος γινομένου, αἱ δὲ εἰς τὸ μέσο(ν)
ϖροσϖίϖτουσαι ἀθρόαι καὶ ἰσχυραὶ | οὖσαι διαμέ b2

90 ῥίψεσιν Y^{a}, *in proiectionibus* Bart.^{1,2}: *in cratibus* Gaza, ἐν τοῖς ῥιψίν agnosc. Sylb. et scrips. Buss., vide v. 100 98–99 *lineam quae est motus* Bart.^{1}, *lineam quae est foraminis* Bart.^{2}, vide ad v. 91 99 *aut per quam translucent* Bart.^{1,2}, *aut ad quodlibet per quod liceat translucere* Gaza, ἧ δι' οὗ λάμϖουσι Eras., ἧι διαλάμϖουσι Ruelle 100 αὑταὶ Bekk. 101 ἀσθενὲς Y^{a}, corr. Y^{a2} | τοὺς Y^{a} 108 τούτωι Y^{a2} 109 ἀφιστῆι Y^{a2} 113 ὀλίγαι Y^{a}, corr. Sylb. | ϖλέον Y^{a}, corr. Y^{a2}

νουσιν. ἐγγὺς μὲν οὖν ὄντος τοῦ σχήματος δύνα
ται καὶ τὰ ἐν ταῖς γωνίαις ὁρᾶν, πόρρω δὲ αὐ
τοῦ γινομένου ἀδυνατοῦσιν. διὸ καὶ ἡ ϖεριφε
ρῆς ἀϖαγομένη εὐθεῖα φαίνεται καὶ ἡ σελήνη
ὑϖὸ εὐθειῶν περιέχεσθαι δοκεῖ τῇ ὀγδόῃ, ἐὰν
μὴ κατὰ τὸ πλάτος ἀλλὰ κατὰ τὴν περιέ
χουσαν γραμ|μὴν αἱ ὄψεις ϖροσϖίϖτωσιν. ἐγγὺς b30
μὲν γὰρ οὔσης τῆς ϖεριφερείας δύνανται
διακρίνειν αἱ ὄψεις ὅσῳ ἐγγύτερόν ἐστι θάτε
ρον θατέρου μέρους τῆς ϖεριφερείας, ϖόρ
f. 91v ρω δὲ γινομένης οὐ διαισθάνεται ἀλλὰ δοκεῖ
αὐτὴ ἐξ ἴσου εἶναι. διὸ καὶ εὐθεῖα φαίνεται.

116–117 *postest est quae sunt in angulis videre* Bart.[1,2], *anguli quoque aspici possunt* Gaza, δύνανται Eras. 117–118 πόρρω δὲ αὐτοῦ γινομένου Sylb., πλείονος δ' αὐτοῦ Fors.: ϖλεῖον δὲ αὐτῶν γενομένου Y[a] 126–127 *videtur ipsa* Bart.[1,2], *cernere sibi videtur* Gaza, δοκεῖ αὐτῇ om. εἶναι Eras.. δοκεῖ αὐτῆ…εἶναι Bekk.

Parisinus graecus 2036.189–207 [*Prob.* 15.11]

(f. 92v)
z διὰ τί ἐν ταῖς τ(οῦ)
Ἡλίου ἐκλείψεσιν, ἐάν τις θεωρῇ διὰ κοσκίνου ἢ φυλλω(ν),
οἷον πλατάνου ἢ ἄλλου πλατυφύλλου, ἢ τοὺς δακτύ
λους τῆς ἑτέρας χειρὸς ἐπὶ τὴν ἑτέραν ἐπιζεύξας,
μηνίσκοι αἱ αὐγαὶ ἐπὶ τῆς γῆς γίνονται; ἢ ὅτι | ὥσ 912b
περ δι᾽ ὀπῆς ἐὰν λάμπῃ εὐγωνίου τὸ φῶς, στρογγύλο(ν)
καὶ κῶνος γίνεται; αἴτιον δὲ ὅτι δύο γίνονται κῶνοι
ὅ τε ἀπὸ τοῦ ἡλίου πρὸς τὴν ὀπὴν καὶ ὃ ἐντεῦ
θεν πρὸς τὴν γήν, καὶ συγκόρυφοι. ὅταν οὖν ἐχόν
των οὕτως ἄνωθεν κύκλῳ ἀποτέμνεται, ἔσται μη
νίσκος ἐξ ἐναντίας ἐπὶ τῆς γῆς τοῦ | φωτός· ἀ b20
πὸ τοῦ μηνίσκου γὰρ τῆς περιφερείας γίνονται
αἱ ἀκτῖνες, αἱ δὲ ἐν τοῖς δακτύλοις καὶ κοσκίνοις
οἷον ὀπαὶ γίνονται. διὸ ἐπιδηλότερον γίνεται ἢ
διὰ μεγάλων ὀπῶν. ἀπὸ δὲ τῆς σελήνης οὐ γίνον
f. 93r ται, οὔτε ἐκλειπούσης οὔτε ἐν αὐξήσει οὔσης ἢ
φθίσει, διὰ τὸ μὴ ἀκριβεῖς τὰς ἀπὸ τῶν | ἄκρων b25
αὐγὰς εἶναι, ἀλλὰ τῷ μέσῳ φαίνειν, ὁ δὲ μηνίσκος
μικρὸν τὸ μέσον ἔχει.

198 οὕτως ‹τοῦ› ἄνωθεν κύκλῳ ἀποτέμνεται, ‹τι› ἔσται Fors. 201 μικραὶ δὲ pro αἱ δὲ coni. Fors. 201–203 *qui autem sunt in digitis et cribris non foramina fiunt, propter quod manifestius fiunt quibus per magna foramina* Bart.[1], *qui autem sunt in digitis et cribris velut foramina fiunt, propter quod manifestius fiunt quibus per magna foramina* Bart.[2], *digitis tamen et cribris pro foraminum exiguitate res certius explicatur quam cum per foramina ampliora transmittitur* Gaza 206–207 *eclipsatum autem paruum medium habet* Bart.[1,2]

Appendix B

Prob. 15.6 Translation

Why does the Sun make not figures that are rectilinear but circles as it passes through the quadrilaterals, for example, in [wicker] | bulwarks?

Rather, is it because the emission of visual rays is a cone and the base of a cone is a circle,[1] so that the rays of the Sun appear circular at whatever object they strike?[2] Certainly, the shape due to the Sun[3] must be contained by straight lines, since the rays are straight [lines]. Indeed, whenever straight [lines] | strike a straight [line], they produce a rectilinear [figure]; and this happens in the case of the [Sun's] rays too,[4] since, where they shine through, they strike the line of the wicker-work (which is straight), and [since] these [rays] are straight, with the result that the emission [of the Sun's rays] will be to a straight line.[5]

1 15.90–91 [911b5–6]: cf. *Prob.* 3.9 872a35–37.

2 15.90–93 [911b5–7]: the question involves consideration of two processes—the Sun's shining through the wicker quadrilaterals and the eye's apprehension of what appears on the ground.
15.93 [911b7] κυκλοτερεῖϲ: if the arcs at the corners of the quadrilateral opening are to be circular, then the axis of the visual cone is at right angles to the ground, that is, the observer is looking downwards towards his or her feet (see Figure 18.4, p. 363); otherwise the curve on the ground will be elliptical. Moreover, though the opening may have angles that are obtuse and acute as well as sides of unequal length, there are clear limits to such variation: see Figure 18.5, p. 364.

3 15.94 [911b8] τὸ ὑϖὸ τοῦ ἡλίου ϲχῆμα: *scil.* the shape illuminated on the ground.

4 15.98 [911b11] τοῦτο. The contention [15.97–101] is that the Sun's rays striking wicker bulwarks is an instance of its rays striking straight lines and producing a rectilinear figure [15.15.93–97].

5 15.100 [911b12] αὗται εὐθεῖαι (these [rays] are straight). The force of the demonstrative «αὗται», the reading in the codex, is that the questioner is pointing to a diagram, a possibility obviated by Bekker's emendation «αὐταί». I say "to a diagram" because, as the questioner maintains next, we do not actually see these straight lines on the ground when the Sun's light passes through wicker-works or the like.
15.100 [911b12–13] ϖρὸϲ εὐθεῖαν (to a straight line): not "in a straight line," as Louis 1991–1994, 61 and Mayhew 2011, 465 would have it, since the sentence concerns the shape of the shadow cast: cf. 911b9–10.

But, because [the visual rays] departing from the eyes[6] to the ends of the straight lines are weak, what is in the angles [formed by the Sun's rays on the ground] is | not seen. Rather, as much of the straight line as is *within* the [visual] cone effects a straight line [that we see], whereas the rest [of the straight line] does not; instead, the visual rays reporting what they see escape notice.[7]

Certainly, we do not see many things to which our visual ray penetrates,[8] for example, things in the dark. Similar to this too is the fact that a square appears polygonal and, if one stands at | a greater distance, as a circle. That is to say, since the emission of our visual rays is a cone, when the shape is set at a point far off, the [visual rays] departing from our eyes to the angles [of the shape] do not see because they are weak and few [in number] as the distance becomes

6 15.102 [911b13–14] τὰϲ ἀποσχιζομέναϲ ἀπὸ τῶν ὄψεων: Mayhew 2011, 463–464 follows Forster in deleting «ἀπό» and translates "the visual rays cut off from the visual rays towards…." But not only is it left unclear just what these rays are cut from or by—a problem also in Hett's version [1953, 335], which retains «ἀπό»—the participle is repeated at 15.112 [911b22] where "cut off" makes little sense. It is better to construe «ἀποσχιζομέναϲ» as a participle in the middle voice: cf. Louis 1991–1994, 2.61 *qui partent des yeux*.

7 151.101–103 [911b13–17]: the thesis is that when we look at a straight line, the visual rays projected to the extremities of the line are weaker than those directed to the middle parts. As will next be made clear, the rays at the outside of the visual cone are weaker than those at the center.

151.104–106 [911b15–17] ὁρᾶται … λανθάνουσιν αἱ ὄψειϲ: the visual rays are said to "see," which makes sense given that the ray ordinarily returns to the eye with information. Cf. 15.114 [911b23] ὁρῶσι, 15.116–118 [911b25–27].

15.106 [911b17] ἐπιπίπτουσαι (reporting what they see: cf. b23): this verb may indicate either the sending of a visual ray to an object or the return to the eye of the visual ray conveying information about the object seen [cf. Mugler 1964, 157]. In this instance, the latter must be meant.

8 15.106–127 [911b27–34]: the interlocutor/reader is now effectively challenged to integrate his explanation of the curvature of the shape on the ground illuminated by the Sun as it passes through a quadrilateral with our perception of rectilinear objects as curved [15.106–118] and of curved objects/lines as rectilinear [15.118–127].

reater;[9] whereas the visual rays striking at the center [of the shape] persist | because they are crowded together and strong. Thus, when the shape is near, [the visual ays] can also see what is in the angles but, when the same [shape] is far off, they annot.

his is why a curved line appears straight when it is moved away, that is, [why] the Ioon seems to be enclosed by straight lines on the eighth day, since our visual rays rike it not in its breadth but at the | line containing it.[10] That is to say, when the urved line is near, our visual rays can distinguish by how much one part of the curved ne is nearer than another. But, when it is far off, [the eye] does not perceive [this]; ather, [the curved line][11] seems to be at an equal distance. And this is why [the curved ne] appears straight.

9 15.113–114 [911b24] τοῦ ἀποστήματοϲ: the distance here is that from the eye to the object.

10 15.120 [911b28] τῇ ὀγδόῃ: *scil.* at the First Quarter.
15.120–123 [911b29] ἐάν: this introduces the protasis of a present general condition. The point is that the eye and the arc of the shadow on the Moon's surface are in the same plane: cf. Aristarchus, *De mag.* hypoth. 3 with Heath 1913, 353n2.

11 15.127 [911b33] αὐτή: *scil.* ἡ περιφερεία. The subject switches abruptly from αἱ ὄψειϲ (our visual rays) [15.124 / 911b31] to ὄψιϲ (the eye) [15.126 / 911b34] (note: «διαισθά- νεται» and not «διαισθάνονται», as one might expect) to ἡ περιφερεία, the subject of «δοκεῖ» and «φαίνεται». The emendation «αὐτῇ» is cosmetic.

Prob. 15.11 Translation

11 Why is it that during eclipses of the Sun,[12] if one observes [them] through a sieve or through leaves (for example, [the leaves] of a plane tree or another broad-leafed tree) or by joining the fingers of one hand to the other, the beams become lunes on the ground?

Is it because the [Sun's] light is rounded, that is, a cone, | just as [it is] if it shines b
through a small opening with suitable angles?[13] A reason [for this][14] is that there are two cones, the one from the Sun to the opening and the one which is from there to the ground; and they share the same vertex.[15] Thus, with [the

12 15.189–190 [912b11] ἐν ταῖc τοῦ ἡλίου ἐκλείψεσιν: i.e., within the interval from when the leading edge of the lunar disk makes first contact with the leading edge of the solar disk to when the trailing edge of the lunar disk makes last contact with the trailing edge of the solar disk.

13 15.194 [912b15] δι᾽ ὀπῆc. This *problema*, and 15.200–203 [912b21–22] in particular, makes better sense if the opening in question is understood to be small [cf. Hett 1953, 341 "peep-hole"]. This accords with typical usage in which mouse-holes, bodily orifices, a hole in a roof for a chimney or in a wall to receive a beam are instances of «ὀπή».
εὐγωνίου. This adjective is found infrequently in the surviving Greek corpus: apparently, it is only in Galen [Kuhn 1821–1833, 18.2 855.15] that it means "with right angles" or "rectangular" and not just "with regular angles." Here «εὐγώνιοc» simply marks a contrast with the angles of the small openings formed by leaves and fingers. The difficulty is that the adjective «εὐγώνιοc» makes poor sense if it is translated by "regular" or the like. The size of the lune and, hence, its visibility are dependent on the size of the opening—small openings are later specified—as well as on the angle at the corner. If this angle is too oblique, for example, the lune may not be discernible. For this reason, I have rendered «εὐγώνιοc» by "with suitable angles."

14 15.196 [912b16] αἴτιον (a reason): what follows is an explanation of why we see rounded shapes with lunes on the ground when sunlight passes through suitable openings. «αἴτιον» without the definite article is a predicate and so the question is whether it signifies *the* explanation or only *an* explanation, i.e., one to which the questioner is not necessarily committed. I have chosen the latter rendering because, to anyone observing a solar eclipse, the assumption that the Sun is a point-source of light is patently wrong.

15 15.196–198 [912b16–18]: in effect, we have here the unusual thesis that the Sun is a point source of light. It was far more common in such deliberations to start with the palpable fact that the Sun has magnitude. Indeed, the question of its size in relation to those of the Moon and Earth is essential to understanding eclipses. In such instances, the Sun was taken to be the base of a cone with its vertex either on the Earth or beyond it, a configuration implying that solar rays go in every direction from every part of the Sun's surface. In some instances, as Cleomedes notes [*Cael.* 1.7.53–55], "geometers" treated the rays from the Sun as though they are parallel, a thesis that figures in arguments other than just versions of Eratosthenes' demonstration of the size of the Earth [see 2.1.197–209].

ones] disposed in this way, when the upper [cone] is cut off from above by circle, there will be a lune | of light opposite on the ground, since the rays come rom the arc of the lune and since the [openings] in fingers and sieves are, as it vere, *small* openings.[16] This is why [the arc of the lune] is clearer than through large penings.[17]

ut [the two cones] do not arise from the Moon, neither when it is eclipsed nor when t waxes or wanes because the beams from its | extremes are not precisely a formed as a cone] but because [its beams] shine at the middle and the lune has middle that s small.[18]

15.197–198 [912b17] «ὁ ἐντεῦθεν» ("from there") has to mean "from the Sun." Forster's "one between the Sun and the aperture and the other between the aperture and the ground and their apices meet" [1927] makes one wonder how the apices of cones defined in this way are to meet.

15.198 [912b18] συγκόρυφοι: lit. "with vertices together." The verb «συγκορυφόω» means "to bring to one point."

16 15.198 [912b18] ἄνωθεν: *scil.* ὁ ἄνωθεν κῶνοϲ (the upper cone), that is, the cone with its base at the opening.

κύκλῳ: *scil.* the Moon seen as a circular disk. It is assumed that the axes of the cones of solar light are at right angles to their base; otherwise their bases would be elliptical. It is also a consequence of this configuration that all solar eclipses are total and that they occur instantaneously.

17 15.202–204 [912b22]: as the opening increases in size, the lune becomes smaller in relation to the illuminated base of the inner cone (the one that touches the sides of the opening): see Figure 18.5, p. 364.

18 15.204–207 [912b23–25]: the argument would seem to suppose that

a. the Moon is a sphere and reflects the Sun's rays radially.

Thus, it will not send light (in any form) to the Earth when it is in eclipse. It seems to be assumed as well that, accordingly, during the lunar waxing and waning phases,

b. the solar rays reflected by the outer parts of the visible parts of the lunar disk may not reach the Earth and that, if they do, they will not form a proper cone, and that

c. any rays reflected to the Earth mostly come from the Moon's middle region and this middle region is (too) small when the Moon is a crescent.

The size of the middle parts of the lunar crescent matters since the cone is to have the axis perpendicular to the base, which is not possible when the Moon is a crescent if the lunar sphere reflects the solar rays radially. It is worth noting, then, that there is no specific mention of those waxing and waning phases in which the Moon is gibbous.

15.206 [912b24] ἀκριβεῖϲ (precisely formed).

15.206–207 [912a25–26] ὁ δὲ μηνίσκοϲ: the lune here is the waxing or waning lunar crescent, not the lune on the ground.

Appendix C

The Theory of the Solar and Visual Cones in *Problemata* 15.6 and 15.11

(1) Solar rays propagate as straight lines in a cone-shaped formation.

(2) When these solar rays strike a rectilinear opening, they produce a rectilinear area of illumination on the ground, no matter where the opening is placed in the solar cone.

(3) What is seen must be illuminated but what is illuminated need not always be seen, even when it is within our visual field.

(4) Visual rays are also emitted (ἔκπτωσις) as a bundle of straight lines in a cone-shaped formation.

(5) In a successful instance of seeing, these visual rays
   (a) return to the eye with information about the object to which they were emitted (ἐπιπίπτειν) and
   (b) are noticed.

(6) Of the visual rays directed to an object, those nearer the cone's sides are weaker than those nearer its axis.

(7) As the distance to the object increases, the rays to the sides of the visual cone become fewer in number than those towards it axis.

(8) Visual rays may fail
   (a) by not reaching an object or
   (b) by not returning to the eye with information about this object or
   (c) by returning but not being noticed.

Bibliography

Premodern Sources

Anonymous

1501 *Problemata Aristotelis cum duplici* [sic] *translatione antiqua versione et nova s. Theodori Gazae cum expositione Petri Aponi. Tabula secundum magistrum Petrum de Tussignano per alphabetum. Problemata Alexandri Aphrodisei. Problemata Plutarchi.* Venice.

[Aristotle]

Problemata. See Sylburg 1585; Bekker 1831; Bussemaker 1857; Forster 1927; Hett 1953; Louis 1991–1994; Mayhew 2011.

Bartholomaeus de Messana (Bartholomew of Messina)

1482 *Translatio* in Petrus de Ebano 1482.

1501 *Translatio antiqua* in Anonymous 1501.

Erasmus, D.

1550 Ἀριστοτέλους ἅπαντα. vol. 2. Basel: I. Bebelius and M. Isingrinus.

Theodorus Gazae (Theodore of Gaza)

1501 *Problemata Aristotelis.* See *Translatio nova* in Anonymous 1501.

1831 *Problemata.* In ed. Academia Regia Borussica, *Aristotelis opera: 3. Aristoteles latine interpretibus variis.* Berlin: George Reimer.

Petrus de Ebano (Peter of Abano)

1482 *Expositio in libros problematum Aristotelis.* Venice. Transcribed by E. Devière, G. Coucke, and T. Swaenepoel in the Aristoteles Latinus Database (Turnhout: Brepols): http://clt.brepo-lis.net.myaccess.library.utoronto.ca/ald/pages/Toc.aspx.

Sylburg, F. (ed.)

1585 *Aristotelis, Alexandri et Cassii problemata.* Frankfurt: Andreas Wechelus.

Modern Sources

Bekker, I. (ed.)

1831 *Aristoteles graece.* vol. 2. Berlin: G. Reimer.

Bowen, A.C

2013 *Simplicius on the Planets and Their Motions: In Defense of a Heresy.* Philosophia Antiqua 133. Leiden: Brill.

2015 '*Problemata* 15: Its Title and Agenda.' Pp. 214–225 in *The Aristotelian* Problemata Physica, ed. R. Mayhew. Leiden: Brill.

Bussemaker, U.C.

1857 *Aristotelis opera omnia graece et latine cum indice nominum et rerum*. vol. 4. Paris: Firmin Didot.

Cadden, J.

2006 "Preliminary Observations on the Place of the *Problemata* in Medieval Learning." See De Leemans and Goyens 2006, 1–19.

De Leemans, P. and M. Goyens

2006 *Aristotle's* Problemata *in Different Times and Tongues*. Leuven: Leuven University Press.

Filius, L.S.

1999 *The* Problemata Physica *Attributed to Aristotle: The Arabic Version of Ḥunain ibn Ishāq and the Hebrew Version of Moses ibn Tibbon*. Leiden: Brill.

Forster, E.S. (trans.)

1927 *Problemata*. In vol. 7, *The Works of Aristotle Translated into English*, ed. W.D. Ross. Oxford: Clarendon Press.

Heath, T.L.

1913 *Aristarchus of Samos: The Ancient Copernicus*. Oxford: Clarendon Press.

Hett, W.S. (trans.)

1953 *Aristotle:* Problems. vol. 1. London: William Heinemann; Cambridge, MA: Harvard University Press. Repr. 1970.

Kühn, C.G. (ed.)

1821–33 *Galeni opera omnia*. 22 vols. Leipzig: C. Knobloch.

Louis, P.

1986 'Les manuscrits parisiens des "*Problèmes*" d'Aristote.' *Revue de Philologie, de Littérature et d'Histoire Anciennes* 60:197–204.

Louis, P. (ed. and trans.)

1991–94 *Aristote.* Problèmes. 3 vols. Paris: Les Belles Lettres.

Mayhew, R.

2011 *Aristotle:* Problems. 2 vols. Cambridge, MA: Harvard University Press.

Mugler, C.

1964 *Dictionnaire historique de la terminologie optique des grecs*. Paris: C. Klincksieck.

Ross, H.E. and C. Plug

2002 *The Mystery of the Moon Illusion: Exploring Size Perception*. Oxford: Oxford University Press.

Ruelle, C.-É.

1922 *Aristotelis quae feruntur problemata physica. Recognovit Hermannus Knoellinger. Editionem post utriusque mortem curavit praefatione ornavit Iosephus Klek*. Leipzig: B.G. Teubner.

Smyth, H.W.

1971 *Greek Grammar. Revised by Gordon M. Messing*. Cambridge, MA: Harvard University Press.

CHAPTER 19

Chaldeans on the Nile: Two Egyptian Astronomical Procedure Texts with Babylonian Systems A_1 and A_2 for Mercury

Mathieu Ossendrijver and Andreas Winkler

Introduction

Mesopotamian influence on the culture of ancient Egypt is well attested or has been suggested in areas such as literature,[1] religious beliefs,[2] divination,[3] mathematics,[4] and astral science.[5] Our contribution concerns primarily the latter aspect. We present two Egyptian Demotic texts, henceforth referred to as Texts 1 and 2, which are written on two ostraca, now housed in the Ashmolean Museum, Oxford.[6] Both are astronomical procedure texts; that is, they contain instructions for computing astronomical phenomena. In total six procedures are completely or partly preserved, five of them well enough to establish that

* We would like to express our gratitude to Liam McNamara for providing us with assistance and giving us the publication rights to the texts presented in this paper, to Alexander Jones for comments, and to Edward Love for correcting our English text.

1 E.g. Collombert and Coulon 2000; Schneider 2003.

2 E.g. Beck 2015.

3 Copies of what may be understood as an Egyptian version of the Mesopotamian celestial omen series Enūma Anu Enlil were present in Egypt from at least the New Kingdom and throughout the Graeco-Roman period (Parker 1959a; Roccati 1984: 350–353; Andrews 1992: 13–14; id. 1994: 31–32; Collombert 2014). There are also plenty of attestations for other techniques, such as lecanomancy (e.g. Demichelis 2002). See also Derchain 1966; Stockhusen 2012.

4 For possible examples of a transfer of mathematical knowledge see Friberg 2005.

5 E.g. the so-called Stobart Tablets (Neugebauer 1942: 210–242; van der Waerden 1960; Neugebauer and Parker 1969: 225–240; van der Waerden 1972); P.Carlsb. 32 (Neugebauer and Parker 1969: 240–241; *ed. princ.* Parker 1962: 146–147); *P.Ryl.* I 27 (Neugebauer 1949; van der Waerden 1958; Jones 1983: 14–16); *P.Oxy.* LXI 4152–4161 (Jones 1999a); Jones 2007a: 312. An example of a pre-Graeco-Roman period text from Egypt that makes use of Babylonian methods is an inscription from Tanis in the Delta that describes the length of the day throughout the year (Hoffmann 2017: 135–153).

6 Text 1 = O.Ashm.Dem. 483; Text 2 = O.Ashm.Dem. 525+732+763.

 | DOI 10.1163/9789004363380_020

they deal with the well-known Babylonian Systems A_1 and A_2 for Mercury.[7] As will be argued, their ultimate purpose is to compute Mercury's first and last evening or morning visibilities, but the texts do not preserve the practical instructions for achieving this. Instead, they deal with core aspects of systems A_1 and A_2, namely period relations and elementary steps. In particular, Texts 1 and 2 offer the first evidence that ancient scholars employed a version of these systems that is based on elementary steps.

Provenance and Dating of the Ostraca

On palaeographical grounds the ostraca can be dated to the Roman period. Although it is difficult to narrow down with greater accuracy the period when the texts were written, a date in the early first century CE seems most plausible. Both ostraca may come from the same storage jar, given that the same type of black pitch is found on the reverse side of both (Figs. 19.2–19.5). Text 1 seems to be the shoulder of a larger storage vessel, perhaps a so-called torpedo shaped amphora. Text 2 is the neck and shoulder of a similar, or the same, container, but the text is written upside down with respect to the orientation of the jar; the top of the text, which ought to have been broader than the bottom, is the shoulder of the vessel, while the lower part constitutes its neck. Looking at the breakage of Text 1, the bottom part seems to have been broken off more recently than the other sides, which is indicated by its lighter colour. It is possible, therefore, that this piece was originally larger and included another similar or related text. As for Text 2, the same observation can be made. Three individual pieces have been identified as belonging together. Upon discovery, the large ostracon was apparently smashed into at least six pieces, with at least one piece missing near the bottom and at least two from the top.

Unfortunately there is little or no information on the find spot of the pieces in question. The museum inventory suggests that they were donated to the Ashmolean Museum by F. Ll. Griffith after he had received them from W.M.F. Petrie, who had supposedly bought them in Egypt by the late 19th century. Most of these pieces seem to come from the Theban area, though some originate from other Upper Egyptian locations, such as Elephantine. Nevertheless, the Thebaid seems to be the most reasonable assumption for their origin—not least because most pieces from the same acquisition come from that locality.

7 For Mercury Systems A_1 and A_2 see Neugebauer 1955: 288–299; van der Waerden 1965: 199–201; Neugebauer 1975: 466–473; Ossendrijver 2012: 69–73.

Orthographic reasons also favour such a conclusion. However, their provenance needs to remain tentative until further evidence is available.

Some clues about the date, provenance and scribal milieu of the texts might be gleaned from the orthography of the zodiacal signs, which will be described in detail in the section "Zodiacal Signs." The scribe hardly ever spelled out the names of the zodiacal signs in full;[8] often he left out the definite article, usually an integral part of the Egyptian designations. Cancer is the only zodiacal sign that he did not write in one ligature or sign, but also this one was never written out completely. Cancer is also the only zodiacal sign that is written with a determinative (Text 1). Though such practices are known from the horoscopic ostraca, primarily from Narmouthis,[9] it is rarely the case in Theban horoscopes.[10] The Theban text whose orthographic conventions mostly resemble those found in our ostraca is known as the Stobart Tablets,[11] a sign entry almanac from the second century CE, probably about a century later than the ostraca presented here. However, the writing in the Stobart Tablets is, as a rule, less cursive.

Assuming that the texts are of Theban origin, the abbreviated zodiacal signs may be taken as reflecting different conventions for writing astronomical and astrological texts. Though the abbreviated forms do find their way into horoscopes, the astrological texts mostly use the full writing. Scribes of the astronomical texts, on the other hand, hardly ever make use of them and appear to have favoured the abbreviated or symbolic writings. Nevertheless, this does not suggest that the astronomers were different individuals from the astrologers[12] but that different rules governed the two types of texts. The most cogent explanation as to why abbreviations would have been used more frequently in purely astronomical texts is that the shortened writings or symbols enable

8 General discussions of the Egyptian zodiacal signs can be found in e.g. von Lieven 1999: 126; Neugebauer 1943: pl. 2 and 3; Neugebauer and Parker 1969: 203–212 and 218–219; Parker 1983: 143; Ross 2006: 153–155; Spiegelberg 1911: 147–148 and pl. 4.

9 E.g. Parker 1983; Ross 2006; id. 2007; id. 2009; id. 2011.

10 Cf. *O.Neugeb.* 1–5 (Neugebauer 1943). Cf. also *O.Leid.Dem.* 333 and *O.Brookl.Dem.* 145. The latter text is surely of Upper Egyptian origin, although it cannot be established with certainty that it comes from the Thebaid. The same applies to the texts published in Neugebauer and Parker 1968 (O.Ashm.Dem. 633 and O.Berl.Dem 6152), though it is plausible that they stem from that area. Also O.Glasg.Dem. 1925.96 (Quack 2008/2009) is from the Theban area.

11 See n. 5.

12 Cf. Winkler 2016: 269–278.

a more economic use of space, perhaps even facilitating the reading of such texts by making them less cluttered.[13]

Although the ostraca are probably to be dated to the first half of the first century CE, the texts must be older. The originals were probably translated into Egyptian from a Babylonian or, less likely, as we shall argue below (**Practical and Institutional Context of Texts 1 and 2**), a Greek original, during the Ptolemaic era. Their usage of the zodiac[14] and the fact that similar procedure texts are not attested in Babylonia before the 4th century BCE sets a *terminus post quem* for the time when these texts could have reached Egypt. Note that the extant Babylonian tablets with mathematical astronomy date to the period 400–50 BCE.[15] Even though this does not give us a definitive *terminus ante quem*, it at least suggests that the transmission of these texts to Egypt can be expected to have occurred before 50 BCE.

The *terminus post quem* also fits with the Egyptian evidence. This can be established on the basis of the nomenclature for the zodiacal signs. It is generally accepted that the Egyptian lexicon was, by and large, borrowed from the Greek rather than being a direct translation or adaptation from Babylonian texts.[16] According to Neugebauer, the Egyptian terminology indicates that the zodiac was known no later than by the mid fourth century BCE.[17] However, it may be possible to date the introduction of the zodiac into Egypt more precisely. O.Strasb.Dem. 521,[18] which correlates zodiacal signs with months, mentions that Scorpio corresponds to the first month of the Egyptian year, Thoth, which, due to the wandering nature of the Egyptian year, was true only for the mid fourth century BCE. At the same time Libra is said to be rising heliacally.[19] Since the name of this zodiacal sign in Egyptian literally mean "Horizon", it has been suggested that, in this period, it received its name from its position at the beginning of that year.[20] However, the oldest preserved document containing

13 Cf. Neugebauer, Parker, and Zauzich 1981: 312.

14 The earliest possible date for the genesis of the current text is the end of the fifth century BCE, when the zodiacal signs were introduced as sections of 30° of the ecliptic (Britton 2010). See also Rochberg 2010: 13 and Ross 2014: 293–294.

15 Ossendrijver 2012: 6.

16 Neugebauer 1943 122–123; Ross 2014: 299–301; but see van der Waerden 1952/1953: 218 and 228–229.

17 Neugebauer 1942: 246; id. 1943: 122–123.

18 Neugebauer 1943: 121–122; Derchain 1966 153–154; *ed.princ.* Spiegelberg 1902: 6–9 (with image); cf. also Müller 1902: 135–136.

19 Neugebauer 1943: 122–123 mentions the third century, but in Neugebauer 1942: 246 and esp. n. 132 he provides the correct calculation, 370 to 250 BC, while admitting that the period covered by the text could extend as far as ca. 130 BCE. Such a late date is implausible.

20 Neugebauer 1943: 122–123.

an explicit mention of the zodiacal signs, P.Berl.Dem. 13146–7, probably dates to the first century BCE.[21]

Editions[22]

The following editions of Texts 1 and 2 include transliterations, translations, and philological notes that are specific to each text. General philological comments are presented below in a separate section following a discussion of the templates underlying the procedures.

Text 1 (O.Ashm.Dem. 483): Mercury System A₁ Evening First

Measurements: 10 cm (Height); 18 cm (Width); 3 cm (Curvature).
Obverse: inscribed; Reverse: blank.

Transliteration

1 *Sbk*
2 *pꜣy=f ḫꜥ imnṱ 1513*
3 *tꜣy=f s.t ꜥꜣ.t 480 pꜣ (sic) ẖe*
4 *n pꜣy=f mꜣꜥ ḫꜥ imnṱ 1513 vac.*
5 *(n)-ṯꜣy Pꜣ-Gn⟨ḥ⟩ḏ 6 šꜥ Iḫ.t 26 r swsw 110 ỉw=f mšꜥ*
6 *tny ỉwṱ 20 r mꜣꜥ ḫꜥ 3 r swsw 1 ỉw=k ir 3 r {100} sp 110 r 330*
7 *(n)-ṯꜣy Iḫ.t 26 šꜥ ⌜Tbṱ.w⌝ 10 r swsw 134 ỉw=f mšꜥ tny ỉwṱ ⟨13 20 r mꜣꜥ ḫꜥ⟩*
⌜4 1/2⌝
8 *r swsw 1 ỉw=k 4 ½ r sp 134 r 603 vac.*
9 *(n)-ṯꜣy Tbṱ.w 10 šꜥ Pꜣ-Gn⟨ḥ⟩ḏ 6 r swsw 116 ỉw=f mšꜥ tny*
10 *ỉwṱ 12 r mꜣꜥ ḫꜥ 5 r swsw 1 ỉw=k ir 5 r sp 116 r 580*
11 *ỉw=k wꜣḥ nkt r pꜣy=f iry n.im=w r mꜣꜥ ḫꜥ ⌜1513⌝ ꜥn*

Translation

1 Mercury:
2 its western rising: 1513,

21 Parker and Zauzich 1981; Neugebauer, Parker, and Zauzich 1981.

22 [ ] = reconstructed section; ⌜ ⌝ = partially obliterated word; { } = scribal mistakes; ⟨ ⟩ = scribal omission; ( ) = commonly omitted grammatical elements; / \ = supralinearly written word; vac. = vacat; ° is used in the translation when degrees are implied but not specified as such.

3 its great place: 480. The method of
4 its part of western rising: 1513.
5 From Cancer 6° until Libra 26° makes 110 degrees, proceeding
6 at the rate of 0;20°, as the part of rising, 3 (of them) make 1 degree. You shall multiply 3 times 110, it makes 330.
7 From Libra 26° until Pisces 10° makes 134 degrees, proceeding at the rate of 0⟨;13,20°, as the part of rising⟩, 4 ½ (of them)
8 make 1 degree. You shall multiply 4 ½ times 134, it makes 603.
9 From Pisces 10° until Cancer 6° makes 116 degrees, proceeding at the rate of
10 0;12°, as the part of rising, 5 (of them) make 1 degree. You shall multiply 5 times 116, it makes 580.
11 You shall add (the) number(s) to one another, it makes the part of rising ⌜1513⌝ again.

Philological Notes

1. : *Sbk*, "Mercury" (Erichsen 1954: 418; Neugebauer 1943: pl. 4; Neugebauer and Parker 1969: 180 and pl. 63; Ross 2009: 63), written without a final *ꜣ* or vowel is also attested in some of the Theban materials such as *O.Neugb.* 1 and 3 and the Stobart Tablets, which also, however, contain writings with a final vowel. Cf. the writing of the same planet in Text 2.21: . In the Fayumic manuscripts, a phonetically more correct rendering *Swgꜣ* is often encountered, corresponding to the Old Coptic ⲥⲟⲩⲕⲉ (Crum 1939: 368b; cf. Vycichl 1984: 200).
2. Cf. the writing of *imnṱ* "west(ern)" (Erichsen 1954: 31), , with the abbreviated form in l. 4 [].

3. : *ẖe*, "way, manner" (Erichsen 1954: 375). It seems probable that the scribe conflated the gender of the word, since he wrote the masc. def. article *pꜣ* in front of it instead of the expected fem. def. article *tꜣ* (see Text 2.21). The flesh determinative is unusual in this word, but to take it as "copy" (Erichsen 1954: 374), i.e. "copy of (the procedure for establishing) 4| the western part of rising" does not seem plausible; the same word appears in Text 2.21 in front of another procedure, but then with the correct def. article. Cf. e.g. P.Carlsb. 13 b.II.14; 14 a.1 and f.16 (Volten 1942); P.Jena 1209 (Zauzich 1980: 96–98) for a comparable usage of *ẖe*.

6a. : *iwṱ*. The word literally means "without, nonexistent" (Erichsen 1954: 25). It is employed here to indicate the absence of a digit (see below).

6b. : *ḫꜥ*, "appearance, rising" (Erichsen 1954: 350–351), is written in an abbreviated form compared to the writing found in ll. 3 and 11: .

6c. : {*100*} *sp*. Our scribe committed a mistake while writing down the present operation. He wrote the number 100 after *r*, indicated with grey in the facsimile. He must immediately have seen his mistake and corrected it by writing *sp*, "times" (Erichsen 1954: 425), over the numeral, whereupon he inserted the correct figures.

***Text** 2 (**O.Ashm.Dem.** 525+732+763)*

Measurements: 21 cm (Height); 17 cm (Width); 8.5 cm (Curvature).
Obverse: inscribed; Reverse: blank.

Transliteration

A (Mercury System A_1 Morning First)

1 ---] ⸢...⸣ [---

2 ---] *r mꜣꜥ* ⸢*ḫꜥ*⸣ [---

3 ---] *(n)-ṯꜣy Pꜣ-Kꜣ 30 šꜥ Mꜣy* [*1 r swsw 61 iw=f mšꜥ tny*]

4 *iwṯ 6 40 r mꜣꜥ*] ⸢*ḫꜥ*⸣ *9 r swsw 1 iw=k ir 9 r* ⸢*sp*⸣ [*61 r*] ⸢*5*⸣[*4*]⸢*9*⸣

5 [*iw=k*] ⸢*wꜣḥ*⸣ *nkt r pꜣy=f iry n.im=w r mꜣꜥ ḫꜥ 2673* ⸢*ꜥn*⸣

B (Mercury System A_1 Morning First)

6 ⸢*pꜣ*⸣ *wn n pꜣy=f mꜣꜥ n ḫꜥ vac.*

7 *pꜣ Mꜣy 241 vac. Tꜣ-Rpy.t Iḫ.t Ḏlꜣ Pꜣ-nty-ꜣtḥ* <*240*> *Pꜣ-(Ḥr)-ꜥnḫ 212 vac. Pꜣ-Mw Tbṯ.w Isw kꜣ 180*

8 *Ḥtry Gn(ḥḏ) 270 vac.*

C (Mercury System A_2 Morning Last)

9 *pꜣy=f ḥtp iꜣbṯ 1223 tꜣy=f s.t ꜥꜣ.t 388 vac.*

10 *(n)-ṯꜣy Isw 5 šꜥ Ḥtry 30 r swsw 85 iw=f* {*mꜣꜥ*} *mšꜥ tny iwṯ 20*

11 *r mꜣꜥ ḥtp 3 r swsw 1 iw=k ir 3 r sp 85 r 255*

12 *(n)-ṯꜣy Ḥtry 30 šꜥ Rpy.t 30 r swsw 90 iw=f mšꜥ tny iwṯ 16 40*

13 *r mꜣꜥ ḥtp 3 ½ 1/10 r swsw 1 iw=k ir 3 ½ 1/10 r sp 90 r*

14 *324 (n)-ṯꜣy Rpy.t 30 šꜥ ꜥnḫ 6 r swsw 96 iw=f mšꜥ*

15 *tny iwṯ 20 r mꜣꜥ ḥtp 3 r swsw 1 iw=k ir 3 r sp 96*

16 *r 288 (n)-ṯꜣy ꜥnḫ 6 šꜥ Isw 5 r swsw 89 iw=f mšꜥ*

17 *tny iwṯ 15 r mꜣꜥ ḥtp 4 r swsw 1 iw=k 4 r sp 89 r 356*

18 *iw=k wꜣḥ nkt r pꜣy=f iry n.im=w r pꜣy=f mꜣꜥ n ḥtp 1223 ꜥn*

D (Mercury System A_2 Morning Last)

19 *pꜣ wn*/ *(n) pꜣy=f mꜣꜥ n ḥtp Isw 95 Pꜣ-Kꜣ Ḥtry 90 Gn(ḥḏ) Mꜣy Rpy.t 108 Iḫ.t Ḏlꜣ Pꜣ-nty-ꜣtḥ 90*

20 [*Pꜣ-(Ḥr)-ꜥnḫ*] *114 Mw Tbṯ.w* 1⸢2⸣[0]

E

21 ---] ... *pꜣ gy n* ... ⸢ ... ... *tꜣ*⸣ *h̠e* (*n*) *Sbk i͗w⸗k i͗p pꜣy⸗f wš* (*n*) *swsw*
22 ---] ⸢...⸣ (*n*)-*tr.t pꜣy⸗f* ⸢... ... *i͗w*⸣⸗*k* ⸢*wꜣḥ*⸣ (?) *n⸗f* ⸢*nꜣ s.wt*⸣ (?) *n*
23 ---] *n pꜣ* ⸢ ... ... *i͗r*⸣ *pꜣy⸗f ḥtp* -

Translation

A (Mercury System A_1, Morning First)

1 ---] ⸢...⸣ [---
2 ---] as the part of ⸢rising⸣ [---
3 ---] From Taurus 30° until Leo [1° makes 61 degrees, proceeding at the rate of]
4 [0;6,40, as the part of] ⸢rising⸣, 9 (of them) make 1 degree. You shall multiply 9 ⸢times⸣ [61, it makes 549.]
5 [You shall] ⸢add⸣ (the) numbers to one another, it makes the part of rising 2673 again.

B (Mercury System A_1, Morning First)

6 The specification of its part of rising *vac.*
7 [Leo 241 *vac.* Virgo, Libra, Scorpio and Sagittarius , Capricorn] 212 *vac*, Aquarius, Pisces, Aries and Taurus 180,
8 Gemini and Cancer 270.

C (Mercury System A_2, Morning Last)

9 Its eastern setting: 1223; its great place: 388 *vac.*
10 From Aries 5° until Gemini 30° makes 85 degrees, proceeding at the rate of 0;20°,
11 as the part of setting, 3 (of them) make 1 degree. You shall multiply 3 times 85, it makes 255 *vac.*
12 From Gemini 30° until Virgo 30° makes 90 degrees, proceeding at the rate of 0;16,40,
13 as the part of setting, 3 ½ 1/10 (of them) make 1 degree. You shall multiply 3 ½ 1/10 times 90,
14 it makes 324. From Virgo 30° until] Capricorn 6 makes 96 degrees, proceeding ⸢at the rate of⸣
15 0;20, as the part of] setting, 3 (of them) make 1 degree. You shall multiply 3 times 96,
16 it makes 288. From Capricorn 6° until Aries 5° makes 89 degrees, proceeding
17 at the rate of 0;15, as the part of setting, 4 (of them) make 1 degree. You shall multiply 4 times 89, it makes 356.

18 You shall add the numbers to one another, it makes the part of setting 1223 again.

D (Mercury System A_2 Morning Last)

19 The specification of its part of setting: Aries 95, T]⸢aurus⸣ and Gemini 90, Cancer, Leo and Virgo 108, Libra, Scorpio and Sagittarius 90,

20 [Capricorn 114, Aquarius and Pisces] ⸢120⸣ *vac.*

E

21 ---] ⸢... the⸣ method of Mercury. You shall calculate its space of degrees

22 ---] ⸢... ...⸣ you ⸢shall⸣ add to it ⸢the places⸣ of

23 ---] ⸢... ... ...⸣ its setting -

Philological Notes

7. It is apparent that the scribe employed a *vacat* between each cluster of zodiacal signs that contain the same number of elementary steps (δ_j; see the commentary). He, however, forgot to write the number 240 following the sequence Virgo, Libra, Scorpio, and Sagittarius. Instead he continued with Capricorn, providing the number of elementary steps, then again introducing a *vacat* before writing the next sequence, Aquarius, Pisces, Aries, and Taurus.

10. . Instead of writing *mšʿ*, , the scribe first wrote *mꜣʿ*, , clearly visible below the correct verb; both words begin with the letter *m*. Nevertheless, he probably immediately corrected his mistake into the proper word, because he seems not to have completed the last sign in the group read as *mꜣʿ*, the house determinative, which usually looks like  or similar.

16. Note the unusual form of *Isw* (see below, Zodiacal signs).

19. The scribe at first forgot to write *pꜣ wn*, "The specification", but later added it supralinearly above *mꜣʿ*.

21a. *gy* : "way, method" (Erichsen 1954: 571–572). The "heading" indicates that a new procedure is introduced. It is commonly found before instructions (see e.g. Winkler 2009: 369). After *gy* an infinitival form of a verb is expected, but we have not been able to resolve the following group: .

21b. For *ẖe*, "way, manner", see Text 1.3.

21c.  is best read as *wš*, "emptiness, gap", a term that could also designate a "space" (Erichsen 1954: 101). The latter translation seems to be the most sensible one, considering that the word is qualified by *swsw* "degree(s)" (see below).

22. The line is difficult to decipher; the writing is heavily abraded and smeared out. Its interpretation therefore remains tentative. In the beginning, the compound preposition *n-tr.t* (Erichsen 1954: 664) can be read. Thereafter, only the reading ⌜*iw*⌝⸗*k* and *n⸗f* appear to be secure. It is tempting to read the group after *iw⸗f* as *wꜣḥ*, "add", but it appears to deviate from the attestations of the same word in 1.11 and 2.5 and 18; the stroke in the middle seems to be shorter and more slanting. The word after the dative is uncertain due to the poor condition of the surface. If the reading of *wꜣḥ* is accepted, it has to be a noun. The last word is badly rubbed but may perhaps be read as the plural of *s.t* without a house determinative (cf. 1.3 and 2.9); the left "ear" of the ligature is almost lost, but the original seems to reveal vague trances of ink that fit such a reading. What it specifically designates here is unclear.

Notes on the Technical Terminology of Texts 1 and 2

Synodic Phenomena, Other Astronomical, and Mathematical Terms

Elementary steps are called *mꜣꜥ (n) ḫꜥ|ḥtp* (e.g. 1.4; ⟨7⟩; 11 and 2.5; 6; 11), literally "Part of rising|setting." The terms *ḫꜥ*, "rising, appearance" (Erichsen 1954: 350–351), and *ḥtp*, "setting" (Erichsen 1954: 340), together with *mꜣꜥ*, "place, part (or similar)" (Erichsen 1954: 149), should be understood as the technical term for elementary step (see Elementary Steps).

The four synodic phenomena dealt with or implied in Texts 1 and 2 are *ḫꜥ imnṯ*, "western rising" = Evening First, *ḥtp imnṯ*, "western setting" = Evening Last, *ḫꜥ iꜣbṯ*, "eastern rising" = Morning First, *ḥtp iꜣbṯ*, "eastern setting" = Morning Last. All four amount to literal translations of the corresponding Babylonian terms (see Ossendrijver 2012: 57). For the astronomical meaning of these terms see below (Comparison with Babylonian and Graeco-Roman Evidence for Mercury's Systems A_1 and A_2).

The period Z (see table 19.2) is called 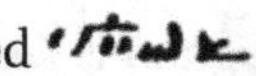 (1.3);  (2.9): *s.t-ꜥꜣ.t*, literally "Great place." As a location it is usually interpreted as a particular place within the temple precinct, as attested in, for instance, *O.Narm.Dem.* II 52 and 82 (Gallo 1997: 131). In the latter case, it is a place where astrological computations were carried out, while in *O.Narm.Dem.* II 52 a light service is to be performed at the "Great place." More specifically, the term designates a sacred area in the sanctuary that was tightly connected with divine burial, e.g. in the Osirian cult and the festival of Khoiak (cf. e.g. *P.Harkness* cols. V.17 and VI.18–19 [Smith 2005: 228; see also Leclère and Coulon 1998: 655–659]). In some sense, this term refers to cyclical processes, which may explain its usage in Texts 1 and 2 (see Templates and Terminology of the Procedures in Texts 1 and 2).

Degrees are called  (1.5);  (2.4);  (2.14), *swsw* (Winkler 2009: 368 n. 28). The reading of one star is *s(j)w* (Erichsen 1954: 413); evidence that two stars are to be read as *swsw* is provided by some of the astrological papyri from the Tebtunis Temple Library, which contain phonetic writings of the word, e.g. P.Carlsb. 71 (unpubl.): *šʿ-ip=k n-ṯꜣy pꜣ swsw* (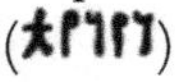) *mḥ-20 n Tꜣ-Iḫ.t* ...: "You always calculate from the twentieth degree of Libra ....." The same manuscript also contains a similar sentence with the writing  of the same word, which demonstrates that we are dealing with the identical term in all instances. Another instance derives from a transcribed passage in the astrological text preserved on P.Carlsb. 66 (Chauveau 1994: 104), but note that the author translates *swsw* erroneously as "day" instead of "degree."

The term is not found in any earlier stage of the Egyptian language and does not appear explicitly in Coptic (cf. Crum 1939). Since the word is only attested in an astrological or astronomical framework it is possible that its usage was limited to this particular context. It has been suggested, however, that the word may derive from the word *sw*, "time" (cf. Erichsen 1954: 461), which is related to the older term *sw* "day, time" (Erman and Grapow 1954: 58); the Egyptian year had 360 "normal" days, which corresponds to the total number of degrees in the zodiac (Winkler 2009: 368 n. 28). It may therefore be tempting to connect the Demotic term *swsw* with the Coptic ⲥⲟⲩⲥⲟⲩ "moment, point" (Crum 1939: 371a). It has been hypothesised that the Coptic term derives from *sw*, "time", as has been proposed for the Demotic one (Osing 1976: 306; Vycichl 1984: 197; Westendorf 1977: 197), but there are other alternatives (Černý 1973: 169), which, if correct, prohibit such an equation.

There may be another solution for the etymology of the word, however, which, in the light of the background of this text, provides a more cogent explanation. In Late Babylonian astronomical texts, the word for degree is written with the logogram UŠ (Neugebauer 1955: 495; Rochberg 2010: 168–169), the phonetic value of which has not yet been established beyond doubt. One well-attested value of the logogram is *šūšu*, meaning "60."[23] It is possible that the Akkadian term for degree was pronounced in a similar way. Hitherto overlooked supporting evidence for such a reading is that the Syriac word for degree, *ssʾ* (Payne-Smith 1903: 383), can most plausibly be interpreted as

23 See the entry *šūši*, "sixty," in the *Chicago Assyrian Dictionary* (Vol. Š III: 380–382). For a discussion of this and various other suggested readings of UŠ, "degree," see *Reallexikon der Assyriologie*, Vol. M, *Masse und Gewichte*: 465–466.

a loanword from Akkadian. It seems reasonable to assume that the Demotic term, *swsw*, is also a loanword from that language. The particular orthography that was adopted by the Egyptian scribes, consisting of two signs representing a star, can perhaps be explained by the astronomical context in which it appears.

See Menchetti 2009: 233 for the use of *mšʿ* to describe the motion of heavenly bodies, scil. planets. The third pers. pron. *=f* thus refers to the planet whose orbit is calculated, that is, Mercury. The sentence *iw=f mšʿ*, lit. "it proceeds", must be taken as a circumstantial present I to fit in grammatically (cf. Johnson 1976: 25). See Erichsen 1954: 635 for the meaning of *tny* with the preposition *r* in the meaning "at the rate of."

The verbal form of *iw=k* plus verb ought to be understood as a future III (Johnson 1976: 100–103), i.e. "You shall (do so and so)", since the text is instructional. See Parker 1972: 7 for *ir* X *r sp* Y r XY, "make X times Y, it makes XY", to indicate multiplication (cf. also *id.* 1959b).

wꜣḥ: "add" (Erichsen 1954: 76), together with the preposition *r* is the verb commonly used in additions. The sum is usually indicated with *r* , i.e. *wꜣḥ* X *r* Y *r* X+Y: "Add X to Y, it makes X+Y" (Parker 1972: 6).

The word for "number",  (1.11); 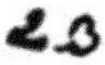 (2.18), *nkt* (Parker 1959b: 278; *ibid.* 1972: 65), literally means "things, possessions, property" (Erichsen 1954: 229–230), but can acquire the suggested translation in mathematical texts.

Zodiacal Signs

In this section we compile the terms for the zodiacal signs attested in Texts 1 and 2 (Table 19.1), and present a detailed discussion of their orthography. As mentioned above (**Introduction**), this provides several clues about the date, provenance and scribal milieu of the texts. In general, both texts employ abbreviated writings; e.g. the definite article is often omitted.

Pꜣ-Isw: Aries. This sign representing Aries is the hieratic version of the animal hide determinative. This sign is on occasion read as *Sr* (e.g. Ross 2006: 153), reflecting a pseudohistorical pronunciation (Erichsen 1954: 441; Vycichl 1984: 195), but there is no reason to do so. The ligature is instead to be understood as an abbreviated spelling of the full writing, which is always spelled out as transliterated above.

TABLE 19.1: *Zodiacal Signs in Texts 1 and 2*

Latin name	Egyptian name	Facsimiles from Texts 1 and 2
Aries	*Pꜣ-Isw*	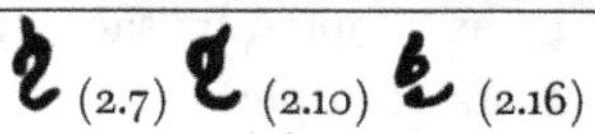 (2.7) (2.10) (2.16)
Taurus	*Pꜣ-Kꜣ*	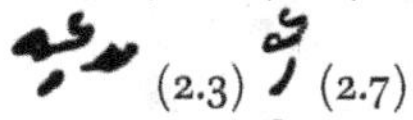 (2.3) (2.7)
Gemini	*Nꜣ-Ḥtry*	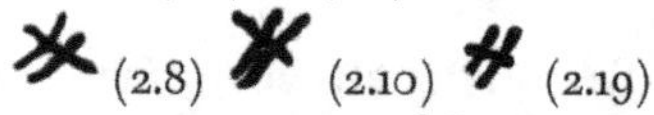 (2.8) (2.10) (2.19)
Cancer	*Pꜣ-Gnḥd̠*	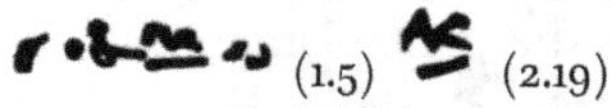 (1.5) (2.19)
Leo	*Pꜣ-Mꜣy*	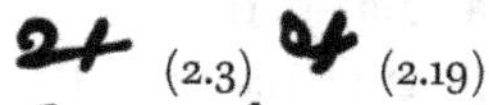 (2.3) (2.19)
Virgo	*Tꜣ-Rpy.t*	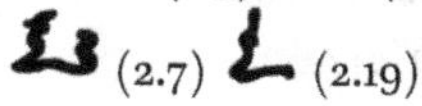 (2.7) (2.19)
Libra	*Tꜣ-Iḫ.t*	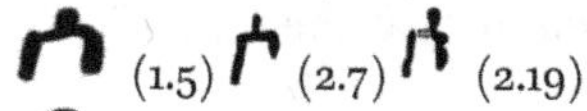 (1.5) (2.7) (2.19)
Scorpio	*Tꜣ-D̠lꜣ(.t)*	 (2.7) (2.19)
Sagittarius	*Pꜣ-nty-ꜣtḥ*	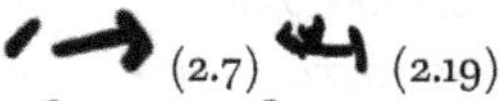 (2.7) (2.19)
Capricorn	*Pꜣ-(Ḥr)-ꜥnḫ*	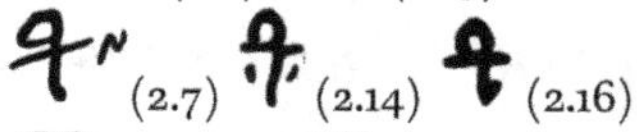 (2.7) (2.14) (2.16)
Aquarius	*Pꜣ-Mw*	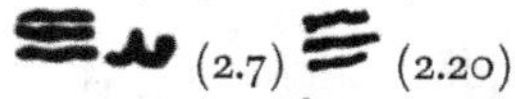 (2.7) (2.20)
Pisces	*Nꜣ-Tbṯ.w*	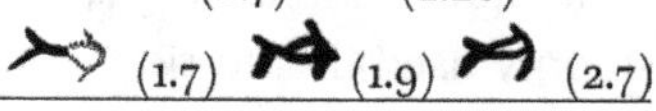 (1.7) (1.9) (2.7)

Pꜣ-kꜣ: Taurus. The orthography of this zodiacal sign is unusual but seems to be based on the hieratic writing (Möller 1936²: 12) rather than deriving from its typical Demotic counterpart (e.g. Erichsen 1954: 555–556).

Pꜣ-Gnḥd̠: Cancer. The scribe always abbreviated the writing of this sign. He wrote either *Pꜣ-Gnd̠*, omitting *ḥ*, or simply *Gn*, an orthography that is also common in the horoscopic ostraca from Narmuthis (Ross 2006: 153). Even more truncated writings exist; in P.Berl.Dem. 8279 the writing *G(nḥd̠)*  is found as an abbreviation for the same sign. The sun determinative at the end of the word in Text 1 is to be noted, because a star usually has this function. The Egyptian term does not mean "crab," but is probably the name for a kind of beetle (Spiegelberg 1911: 147).

Tꜣ-Rpy.t: Virgo. The Egyptian name literally means "The Noble Lady" (Erichsen 1954: 244–245; Spiegelberg 1911: 146), and is also a common designation for a statue of a female deity.

Tꜣ-Iḫ.t: Libra. The Egyptian word is to be translated literally as "The Horizon" (Neugebauer 1943: 122–123), though there have been some unconvincing attempts to interpret the name and the sign used to write this zodiacal sign as a variant writing of a word meaning "scales" or similar (Müller 1903: 8–9; Ross 2006: 154, but cf. Spiegelberg 1911: 146 n. 6). The current ligature may be understood as a reduced writing of the horizon sign (cf. Neugebauer 1943: pl. 3; Neugebauer and Parker 1969: 219), although the two flaps on each side may in fact be interpreted as the cords holding scale pans. If this were the case, it would still not change the basic interpretation of the nomenclature of this sign, which was often artistically represented as a pair of scales (Neugebauer and Parker 1969: 210 and e.g. pls. 40; 42; 47; 50), i.e. the writing would in that case be logographic.

Tꜣ-Ḏlꜣ(.t): Scorpio. The transcription of the sign in question is secure (cf. Ross 2006: 154); the full writing appears in many of the Theban horoscopic ostraca (e.g. *O.Neugeb.* 2; Spiegelberg 1911: 147).

Pꜣ-nty-ꜣtḫ: Sagittarius. The arrow stands as an abbreviation of the full writing. A similar orthography is encountered in, for example, the Stobart Tablets. The full writing is present in several ostraca from the Theban region (*O.Neugeb.* 1, 4, and 5; Spiegelberg 1911: pl. 4), which demonstrates also that the arrow is to be read in the same way. The arrow alone is encountered in several documents from the Fayum, e.g. P.Berl. 8279, and is also common in the horoscopic ostraca from Narmouthis. Ross (2006: 154) attempts an alternative reading, however, proposing that the sign is to be read as *Sty*. This interpretation is probably incorrect. It is argued that such a reading would correspond to the Coptic ⲥⲟⲧⲉ "arrow" (Crum 1939: 361b). Yet this word does not correspond to the Greek τοξότης, but to τόξευμα. Ross further suggests that a word in *O.Narm.Dem.* II 72.4 may offer a paleographic justification for this reading. However, this word, which he reads as *sty*, consists not of two signs (i.e. an arrow and a final vowel), but of three signs—*š* written above *t* and a final vowel. The word is thus read as *šty* following the *ed.princ.* (Gallo 1997: 71 and 136), which means "dough", cf. Coptic ϣⲱⲧⲉ (Crum 1939: 595a). In addition, it would be awkward if two different designations for one and the same zodiacal sign would be used within one locality.

Pꜣ-Ḥr-ꜥnḫ: Capricorn. This zodiacal sign is here written with the common abbreviation *Pꜣ-ꜥnḫ* "The Goat." The full writing is not infrequent in Theban material (e.g., *O.Neugeb.* 1 and 3; O.Strasb.Dem. 521 [Neugebauer 1943: 121–122; *ed. princ.* Spiegelberg 1902]), and is literally to be understood as "The Goat Faced

(One)." The name probably derives from the fact that the sign in question often was represented as a composite creature whose forepart was a goat with a fishtail. Such a creature obviously had no Egyptian name and therefore the locution describes the artistic representation of this sign (Spiegelberg 1911: 148).

Pꜣ-(twtw)-pꜣ-Mw: Aquarius. This zodiacal sign is to be translated literally as "The Water." This writing, however, is apparently the abbreviated form of the Egyptian rendering of this zodiacal sign. The reason is that there are several documents where the sign is referred to as *Pꜣ-twtw-(n)-pꜣ-Mw* "The zodiacal sign of the water", for instance, in O.Strasb.Dem. 521.12 (cf. Spiegelberg 1911: 148 and pl. 4). This example, however, is effaced, and has been misread in earlier attempts at deciphering the writing. It also occurs in, for instance, P.Berl.Dem. 13146–7ro, cols. I.15 and II.6.9.16 (cf. Neugebauer, Parker, and Zauzich 1981: 218). That the term *twtw* "image" (Erichsen 1954: 616) is a generic term for a zodiacal sign can be seen in P.Carlsb. 9. col. I.6 (Winkler 2009: 370 n. 49; id. 2016: 256). See also *P.Zauz*. 50 published by Quack (2004).

Astronomical and Mathematical Commentary

Number Notation

Texts 1 and 2 employ a mixed number notation. The integer part of a number is expressed in a decimal, non-positional notation, while the fractional part is written either in a sexagesimal, positional notation, or as a sum of unit fractions. The non-positional nature of the integer parts is apparent from the use of distinct signs for 10, 100, etc.[24] A more or less equivalent Greek version of the number notation used in Texts 1 and 2 is found in the astronomical papyri from Oxyrhynchus (Egypt) and in the Almagest of Claudius Ptolemy, who was based in Alexandria. However, in the Greek sources, sums of unit fractions, which reflect a traditional Egyptian notation, are only attested in non-technical contexts.[25] Both the Demotic and the Greek conventions contrast with the Babylonian number notation, which is fully sexagesimal and fully positional (see Table 19.3). An interesting feature of the number notation in Texts 1 and 2 is that the sign *ỉwṯ*, which indicates vanishing digits (0), is also systematically used for indicating vanishing initial digits, e.g. in 0;12 (Text 1.10), which stands for 12/60. This usage of the 0 is very rare in the Babylonian texts.[26]

24 In a positional notation, a set of ten signs for the digits 0–9 would suffice for writing arbitrary numbers. Note that the non-positional nature of the notation for whole numbers is not apparent in the transcriptions and translations, in which the modern positional notation is used.

25 See Jones 1999a: 1:9; Toomer 1998: 6–7.

26 Note that *ỉwṯ* was used for writing vanishing digits, but not the *number* 0 (i.e., the outcome of 1–1). See Hoffmann 2004/2005: 44–52 for an outline of the Egyptian tradition for

Comparison with Babylonian and Graeco-Roman Evidence for Mercury's Systems A_1 and A_2

As an aid to the interpretation of Texts 1 and 2, and in order to determine how they could have been used for computing Mercury's synodic phenomena, we recall how the Babylonian astronomers implemented Mercury Systems A_1 and A_2. In the Babylonian context, they function as alternative systems for computing zodiacal positions and times of the four synodic phenomena Evening First (EF), Evening Last (EL), Morning First (MF), and Morning Last (ML). EF and EL denote the planet's first and last visible settings as an evening star just after sunset; MF and ML refer to its first and last visible risings as a morning star, just before sunrise. A complete cycle of Mercury's synodic phenomena, from EF through EL, MF and ML to the next instance of EF, lasts about 116 days (0.32 year).[27]

System A_1 is attested on seventeen Babylonian tablets: nine synodic tables, two procedure texts, one tablet inscribed with a synodic table and procedures, and five auxiliary tables.[28] For the synodic tables an approximate date follows from their initial dates, which cover the years 194–88 BC, whilst for most of the other tablets only a rough date 350–50 BCE can be indicated. The synodic tables contain up to four groups of columns with times and zodiacal positions (modern symbol: *B*) of EF, EL, MF, and ML. In each column, successive rows pertain to successive instances of the same phenomenon. The four phenomena are computed with two different methods. The positions of EF and MF are each updated independently, by adding the distance travelled along the zodiac, also known as synodic arc (modern symbol: σ), i.e.[29]

$$B_{i+1} = B_i + \sigma.$$

Here *i* labels successive instances of the same synodic phenomenon, corresponding to the rows of a synodic table. By contrast, the positions of EL are

signifying non-existent entities. See also Jones 1999a: 1:61–62; Stockhusen 2012: 88. For 0 in Late Babylonian astronomy, see Neugebauer 1941 and Ossendrijver 2012: 18–19.

27 For a more detailed explanation of Mercury's synodic phenomena see Ossendrijver 2012: 55–58 and 68.

28 Synodic tables: Neugebauer 1955: Nos. 300, 302–305, 303a, b; BM 48147 (Aaboe 1977); A 3405 (Steele 2000); synodic table combined with procedures: BM 45980 (Neugebauer 1955: No. 301; Ossendrijver 2012: No. 1); auxiliary tables, all concerned with pushes: Neugebauer 1955: Nos. 800a–e.

29 The same symbol is used in Ossendrijver 2012. According to another frequently used convention, the synodic arc is represented by $\Delta\lambda$.

computed from those of the immediately preceding parent phenomenon EF. This was achieved by applying so-called pushes, which are additive or subtractive distances, to the position of the parent phenomenon. By the same token, the Babylonian scholars computed ML as a satellite of its parent phenomenon MF. Since the method based on pushes is not attested in Texts 1 and 2 it will not be discussed any further here.[30]

System A_2 is attested on seven Babylonian tablets: three synodic tables and four fragmentary procedure texts.[31] The initial dates of the tables cover the years 297–291 BCE. They preserve up to four sets of columns with times and zodiacal positions of the same four phenomena, which are tabulated in the order EL, MF, ML, and EF. System A_2 is implemented using the opposite scenario compared to System A_1: the positions of ML and EL are updated independently by adding the synodic arc, while MF and EF are computed as their satellites. In other words, the positions of MF are obtained by applying pushes to those of the immediately preceding instances of EL, and the same for EF and ML.

In comparison with the cuneiform sources, Texts 1 and 2 only offer a subset of the Babylonian algorithms of Mercury Systems A_1 and A_2. First, they do not preserve any instructions for computing times. Second, as already mentioned, Texts 1 and 2 do not contain evidence of pushes. In particular, the procedures for System A_1 only concern EF and MF, while those for System A_2 only deal with ML and EL. In each respective system, these are precisely the phenomena whose positions Babylonian astronomers computed by means of the synodic arc.

One might argue that this could be an accident of preservation and that instructions for the method based on pushes were written on the missing parts of Texts 1 and 2 or on other ostraca. Nonetheless, it seems more likely that the Egyptian astronomers applied Systems A_1 and A_2 in a different manner than their Babylonian counterparts, namely by using System A_1 only for the positions of EF and MF, and System A_2 only for the positions of EL and ML. It is currently impossible to confirm or refute this on the basis of Demotic sources, since Texts 1 and 2 are silent on this matter and no Demotic tables computed with Mercury Systems A_1 and A_2 have been found. However, tables with positions of Mercury's EF, MF, EL, and ML, computed in accordance with Systems

30 For Mercury's pushes see Neugebauer 1955: 293–299; Neugebauer 1975: 403; Ossendrijver 2012: 70–72.

31 Synodic tables: Neugebauer 1955: Nos. 300a, b; BM 37179 (unpubl.); procedure texts: Ossendrijver 2012: Nos. 2, 3, 4; BM 36950 (unpubl.).

A_1 and A_2, are also attested on a number of Greek papyri from Oxyrhynchus dating from about 100–300 AD.[32] As pointed out by Jones,[33] the astronomers who computed these tables used System A_1 for the positions of EF and MF and System A_2 for the positions of EL and ML. This strongly suggests that the Egyptian astronomers who wrote Texts 1 and 2 combined Systems A_1 and A_2 in exactly the same manner as their colleagues from Oxyrhynchus writing in Greek. Judging from the age of the manuscripts in question and their probable transmission history (see 6 below), it is plausible that the latter group took over this procedure from astronomers writing in Egyptian.

The Babylonian Algorithms of System A_1 and A_2 for Updating the Position of Mercury

Unlike Texts 1 and 2, the Babylonian procedures for updating the positions of EF and MF with System A_1 and those for updating the positions of ML and EL with System A_2 employ step functions for the synodic arc. Nevertheless, the algorithms underlying Texts 1 and 2 are, mathematically speaking, fully equivalent to the Babylonian algorithms. In order to see this, we shall briefly discuss the Babylonian step functions and explain their connection to the procedures in Texts 1 and 2.

The Babylonian step function for the synodic arc involves a division of the zodiac into several zones—three in System A_1 and four in System A_2. As will become apparent, exactly the same divisions of the zodiac underlie Texts 1 and 2. Throughout this paper the index j is used for labeling the zones of a step function. Each zone has a left boundary and a right boundary, which are denoted as b_j and b_{j+1}, respectively. In each zone j, the synodic arc assumes a different constant value σ_j. These and other quantities pertaining to the zones are listed in Table 19.2. Their numerical values for Mercury Systems A_1 and A_2 are compiled in Table 19.3. The step function algorithm as it is formulated in the Babylonian procedure texts involves two steps.[34] As an illustrative example we consider the following procedure for updating the position of Mercury's MF in accordance with system A_1:[35]

32 *P.Oxy.* LXI 4152–4156c (Jones 1999a: 1:123–134); *PSI inv.1* (Jones 2007b).

33 Jones 1999a: 1:114–115; 1:123; 1:128–130.

34 Ossendrijver 2012: 47–50.

35 The tablet originates from Babylon and was written near 177 BC. For the template underlying this procedure see Ossendrijver 2012: 48 (STEP.A.1).

TABLE 19.2 *Modern symbols for the defining parameters of a step function for the synodic arc*

b_j	left boundary of zone *j*
α_j	length of zone *j* expressed in degrees.
σ_j	synodic arc in zone *j* expressed in degrees
r_j	coefficient for crossing from zone *j* into zone *j*+1
δ_j	length of elementary step in zone *j* expressed in degrees
ν_j	number of elementary steps contained in zone *j*
Π	whole number of instances of a synodic phenomenon after which the planet returns to exactly the same position for the first time
Z	corresponding whole number of times that the planet travels through the zodiac

TABLE 19.3 *Parameters of Mercury's Systems A_1 and A_2. Note that System A_2 for Evening Last is not preserved in Texts 1 and 2*

j	zone ($b_j - b_{j+1}$)	α_j	σ_j		r_j	δ_j	$1/\delta_j$	ν_j
System A_1, Morning First: Π=2673, Z=848								
1	1° Leo – 16° Cap	165°	1,46°	=106°	1;20	0;7,30°	8	1320
2	16° Cap – 30° Tau	134°	2,21;20°	=141;20°	0;40	0;10°	6	804
3	30° Tau – 1° Leo	61°	1,34;13,20°	=94;13,20°	1;7,30	0;6,40°	9	549
System A_1, Evening First: Π=1513, Z=480								
1	6° Cnc – 26° Lib	110°	2,40°	=160°	0;40	0;20°	3	330
2	26° Lib – 10° Psc	134°	1,46;40°	=106;40°	0;54	0;13,20°	4;30	603
3	10° Psc – 6° Cnc	116°	1,36°	=96°	1;40	0;12°	5	580
System A_2, Morning Last: Π=1223, Z=388								
1	30° Gem – 30° Vir	90°	1,47;46,40°	=107;46,40°	1;12	0;16,40°	3;36	324
2	30° Vir – 6° Cap	96°	2,9;20°	=129;20°	0;45	0;20°	3	288
3	6° Cap – 5° Ari	89°	1,37°	=97°	1;20	0;15°	4	356
4	5° Ari – 30° Gem	85°	2,9;20°	=129;20°	0;50	0;20°	3	255
System A_2, Evening Last: Π=684, Z=217								
1	30° Gem – 30° Sgr	180°	1,48;30°	=108;30°	1;6,40	0;30°	2	360
2	30° Sgr – 30° Aqr	60°	2,0;33,20°	=120;33,20°	0;54	0;33,20°	1;48	108
3	30° Aqr – 30° Ari	60°	1,48;30°	=108;30°	1;15	0;30°	2	120
4	30° Ari – 30° Gem	60°	2,15;37,30°	=135;37,30°	0;48	0;37,30°	1;36	96

Ossendrijver 2012, Text 1 (BM 45980), *P1.a:*

Transliteration

1 *e-pe-šu*$_2$ *ša*$_2$ IGI.MEŠ *an-ne*$_2$*-e-tu*$_4$ IGI *ša*$_2$ KUR *ana* IGI *ša*$_2$ KUR TA ⸢1 A EN 16 MAŠ$_2$ 1.46 TAB⸣ *ša*$_2$ ⸢*al*⸣ 16 ⸢MAŠ$_2$⸣ D[IRI GAM 1.20 DU TA 16 MAŠ$_2$ EN 30 MUL$_2$.MUL$_2$]

2 2.21.20 TAB *ša*$_2$ *al* 30 MUL$_2$.MUL$_2$ DIRI GAM 40 DU TA 30 MUL$_2$.⸢MUL$_2$ EN⸣ [1 A 1.34.13].⸢20⸣ [TAB *ša*$_2$ *al* 1 A DIRI GAM 1.7.30 DU]

Translation

[1]| Procedure for these appearances. Eastern appearance (MF) to eastern appearance.
From 1 Leo until 16 Capricorn you add 1,46.
(The amount) by which it exceeds 16 Capricorn [you multiply by 1;20.
From 16 Capricorn until 30 Taurus] [2]| you add 2,21;20.
(The amount) by which it exceeds 30 Taurus you multiply by 0;40.
From 30 Taurus until [1 Leo you add 1,34;13],20.
[(The amount) by which it exceeds 1 Leo you multiply by 1;7,30.]

After an introductory phrase, three analogously formulated rules are given, one for each zone of the step function. Each rule consists of two steps. In step 1, a preliminary update of Mercury's position is achieved by adding a value of σ_j to Mercury's previous position (B_i). The value of σ_j to be used is determined by the zone in which Mercury is located. The updated position thus obtained (B_{i+1}) is final if it is in the same zone (j) as the previous position. If the updated position ends up in the next zone (j+1) then step 2 is invoked, which can be summarised as follows: the distance by which the position obtained in step 1 exceeds the boundary to the next zone (b_{j+1}) is multiplied by a coefficient (r_j). The product is added to b_{j+1}, and the outcome becomes the updated position of Mercury (B_{i+1}). The coefficients r_j that are used always equal σ_{j+1}/σ_j. They are therefore not independent parameters of the step function algorithm, but fully determined by the synodic arcs, which are defined in step 1.

Even though the Babylonian procedure texts do not explain why step 2 is executed, the reason is clearly connected to the fact that step 2, if implemented with coefficients defined as $r_j = \sigma_{j+1}/\sigma_j$, can be shown to result in sequences of positions that strictly repeat after a certain whole number, Π, of instances of the synodic phenomenon (i.e. rows of a synodic table). Along with the number period Π there is a second characteristic period, Z, representing the number of revolutions of the synodic phenomenon around the zodiac. Furthermore, the Z revolutions of the phenomenon also correspond to a whole number of years, or revolutions of the Sun; in the case of Mercury this number is also Z. The values of Π and Z were known to the Babylonian astronomers and are quite often mentioned in the procedure texts. For a proper understanding of the importance of the number period Π, note that it has two interconnected meanings: it is the number of repetitions after which the planet returns to exactly the same position in the zodiac and, secondly, the total number of different positions that can be assumed by the planet. Both meanings were fully understood by the Babylonian astronomers and, as we shall see, also by their Egyptian colleagues. For a given step function for the synodic arc, we can compute Π and Z by determining the smallest whole numbers whose ratio satisfies the following modern expression:[36]

$$\frac{\Pi}{Z} = \sum_j \frac{\alpha_j}{\sigma_j}.$$

For instance, for Mercury's MF as computed with System A_1 we obtain 165/106 + 134/141;20 + 61/94;13,20 = 2673/848, i.e. Π = 2673 and Z = 848 (Table 19.3). Note that there is no evidence that the ancient astronomers derived Π and Z in this manner. However, Texts 1 and 2 provide clues about the ancient understanding of these parameters. In order to see how it is necessary to introduce the concept of elementary steps.

Elementary Steps

As was first proposed by van der Waerden,[37] the step function algorithm for the synodic arc, as defined by steps 1 and 2 of the Babylonian procedures, can be conveniently reformulated and summarised in a single rule. This is achieved by introducing the auxiliary concept of elementary steps, to be defined as follows:

$$\delta_j = \frac{\sigma_j}{Z}.$$

36 Neugebauer 1975: 377; Ossendrijver 2012: 51.

37 See van der Waerden 1957. Subsequently Aaboe 1964 arrived at a similar result. Note that Aaboe 1964 quotes neither van der Waerden 1957, nor van der Waerden 1960, which also deals with elementary steps, nor does he reveal any awareness of van der Waerden's work on this topic. Subsequently van der Waerden discussed elementary steps in 1965 in his well-known monograph, *Erwachende Wissenschaft II. Die Anfänge der Astronomie*, 187–188, of which an English edition, *Science Awakening II. The Birth of Astronomy*, appeared in 1974. See also Neugebauer 1975: 421–426. As in Ossendrijver 2012: 51, the term "elementary step" is used here instead of "step" in order to distinguish it more clearly from "step function."

In other words, the synodic arc, σ_j, is viewed as consisting of Z tiny intervals of length δ_j, henceforth referred to as elementary steps. Like σ_j, they have a different constant length in each zone j (Table 19.3). As an aid to the reader, Fig. 19.1 shows a modern visualisation of the distribution of elementary steps along the zodiac for the example of System A_1 for Evening First. With the help of elementary steps, the step function algorithm can be expressed as follows: from one to the next instance of a synodic phenomenon the planet always moves by Z elementary steps, irrespective of its starting position. Their added length defines the synodic arc σ. Therefore, if the planet sets out from a position in zone j and all Z elementary steps fall within that zone, σ reduces to $Z{\cdot}\delta_j{=}\sigma_j$, in accordance with step 1. If, however, the planet leaves zone j before finishing the Z steps, then a number of steps, say N, are situated in the next zone, j+1, and the resulting synodic arc becomes $\sigma = (Z{-}N){\cdot}\delta_j + N{\cdot}\delta_{j+1}$. Accordingly, it assumes a value in between σ_j and σ_{j+1}. This leads to exactly the same result obtained by executing steps 1 and 2 of the Babylonian procedure texts.[38] Further below, the updating of Mercury's position by means of elementary steps will be illustrated for System A_1 for Evening First.

Van der Waerden and Aaboe, recognizing that the usage of elementary steps result in a simpler and more elegant formulation of the step function algorithm for the synodic arc, proposed that they represent a core concept of the Babylonian step function algorithms. Yet the elementary steps are, thus far, not directly attested in Babylonian texts, all of which formulate the computation of the synodic arc in terms of the steps 1 and 2 mentioned earlier.[39] A most surprising aspect of Texts 1 and 2 is that they are formulated completely in terms of elementary steps, thereby offering the first explicit evidence of this elusive concept in an ancient text. Furthermore, the Egyptian scholars who wrote Texts 1 and 2 turn out to have understood and used elementary steps in essentially the same manner as proposed by van der Waerden and Aaboe. In the following section we shall present an analysis of the procedures of Texts 1 and 2.

---

38 This can be seen by noting that the part of σ that lies in zone j+1, i.e. $N{\cdot}\delta_{j+1}$, can be expressed as $r_j{\cdot}N{\cdot}\delta_j$, in exact agreement with the transition rule (step 2) of the Babylonian procedures.

39 As pointed out by van der Waerden 1960: 117; van der Waerden 1965: 187–188, elementary steps are indirectly attested in the Babylonian procedure text BM 33801 (Ossendrijver 2012: No. 44; pp. 88–89; 320–322). Procedure 3' of this tablet lists numbers that are (nearly) identical with the values of δ_j for Mars System A. However, they actually represent the planet's net displacement along the ecliptic after 37 cycles, corresponding to 79 years, which happens to equal δ_j in this system. Hence this procedure does not, by itself, prove that Babylonian astronomers computed with elementary steps.

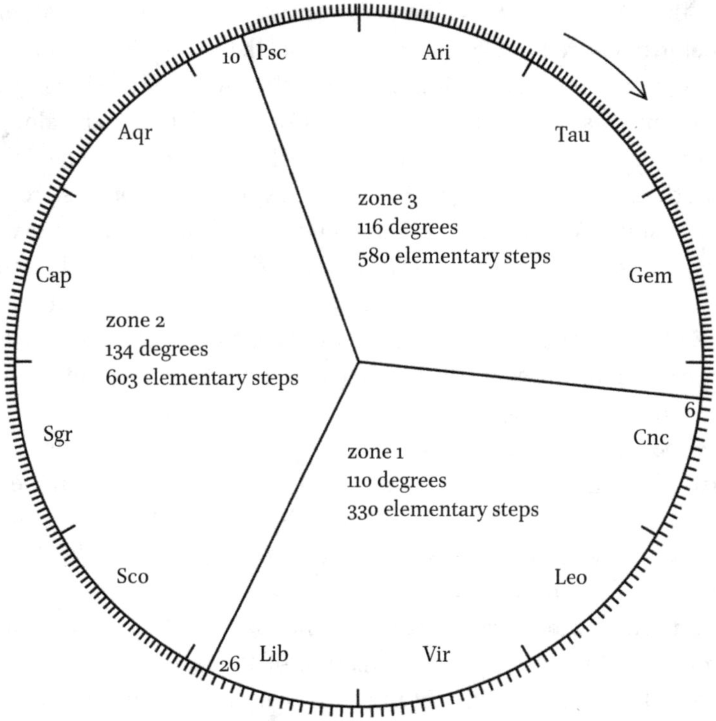

FIGURE 19.1 *Modern visualization of the elementary steps for System A_1 for* EF. *For practical reasons each dash corresponds to five elementary steps.*

Templates and Terminology of the Procedures in Texts 1 and 2

The six (partly) preserved procedures in Texts 1 and 2 belong to three different templates. Three of the procedures—Text 1 and Text 2 A, C—follow a very similar template consisting of eight phrases, here labeled **a–h** (Table 19.4).[40] Each procedure begins, ideally, with phrases **a–d**. Phrase **a** provides the name of the planet and is only preserved in Text 1. We assume that Text 2 A also began with phrases **a** and **b**, but the upper part of the text is lost. Phrases **b** and **c** state the purpose of the procedure and they mention the two characteristic periods of the algorithm, Π and *Z*, respectively. The introductory phrase **d** is only found in Text 1 and can be understood as a reference to the method of computation. It is possible that Text 2 A also once contained a similar notation.

40 Phenomena particular to one text are discussed in the Philological Notes following the editions of Texts 1 and 2.

Phrases **e-g** are provided separately for every zone (j). Phrase **e** mentions the zodiacal positions of the boundaries of the zone, b_j and b_{j+1}, and the resulting length (α_j) of the zone expressed in degrees. Phrase **f** mentions the length of the elementary step (δ_j) in degrees and the number of elementary steps contained in 1°, i.e. $1/\delta_j$. In phrase **g**, the latter number is multiplied by α_j, which yields the number of elementary steps contained in each entire zone ($\nu_j=\alpha_j/\delta_j$). After providing these parameters for all zones, the template ends with phrase **h**, which states that the total number of steps summed over all zones equals Π, the number that was already mentioned in phrase **b**, i.e.

$$\sum_j \nu_j = \Pi.$$

In other words, the entire zodiac contains Π elementary steps. Note that this expression for Π is mathematically equivalent to the one mentioned above (5.3), as can be seen by inserting into the latter $\sigma_j=Z\cdot\delta_j$ and $\alpha_j/\delta_j=\nu_j$.

In Text 2, both examples of this template (2 A, C) are followed by lists (2 B, D), which define a second template. Each list begins with an introductory phrase, *pꜣ wn*[41] *n pꜣy=f mꜣꜥ (n) ḫꜥ|ḥtp*, "The specification of its part of

TABLE 19.4 *Template of Text 1 and Text 2 A, C*

a	(PLANET)	(PLANET)
b	*pꜣy=f ḫꜥ|ḥtp i͗ꜣbṱ|i͗mnṱ* Π	Its eastern|western rising|setting: Π.
c	*tꜣy=f s.t ꜥꜣ.t* Z	Its great place: Z.
d	(*tꜣ ẖe n pꜣy=f mꜣꜥ (n) ḫꜥ|ḥtp i͗ꜣbṱ|i͗mnṱ* Π)	(The method of its part of eastern|western rising| setting: Π.)
e	*(n)-ṯꜣy* b_j *šꜥ* b_{j+1} *r swsw* α_j	From b_j until b_{j+1} makes α_j degrees,
f	*i͗w=f mšꜥ tny* δ_j *r mꜣꜥ ḫꜥ|ḥtp* $1/\delta_j$ *r swsw* 1	proceeding at the rate of δ_j, as the part of rising|setting, $1/\delta_j$ (of them) make 1 degree.
g	*i͗w=k i͗r* $1/\delta_j$ *r sp* α_j *r* ν_j	You shall multiply $1/\delta_j$ times α_j, it makes ν_j.
h	*i͗w=k w ḥ nkt r pꜣy=f i͗ry n.i͗m=w r mꜣꜥ ḫꜥ|ḥtp* Π *ꜥn*	You shall add (the) number(s) to one another, it makes the part of rising|setting Π again.

41 Only preserved in Text 2.6 and 21.

rising|setting."[42] In this phrase, "part" is, again, the technical term for elementary step. What follows are the numbers of elementary steps contained in each zodiacal sign. Both lists are arranged according to the structure of the zones, beginning with the first sign in the first zone and terminating with the ultimate sign in the last zone. Note that in a cluster of signs sharing the same number of elementary steps this number is only written out once (e.g. Text 2.4). The number of elementary steps per zodiacal sign, say *M*, is easily computed using the ingredients already provided in the preceding procedure. Taking into account that some zodiacal signs contain a boundary between two zones, so that one segment of the sign, of length *m* degrees, is in zone *j* while the remaining segment, of length 30 – *m* degrees, is in zone *j*+1, the general expression for *M* becomes

$$M = \frac{m}{\delta_j} + \frac{30 - m}{\delta_{j+1}}.$$

This formula reproduces all values of *M* preserved in Text 2 (setting *m*=30 if a sign is entirely within zone *j*). A complete set of values of *M* for all four systems is compiled in Table 19.5. Note that all *M* are whole numbers, which strongly suggests that the parameters of Systems A_1 and A_2, e.g. the boundaries of the zones, were deliberately tuned in order to achieve this. It can also be noted that for each system the sum of the numbers *M* over the twelve zodiacal signs again equals Π, as it should.

Finally, the procedure 2 E is too strongly damaged to identify what is being computed, but enough is preserved to establish that it deals with Mercury's position and that its formulation deviates from the other procedures. The term *wš (n) swsw*, provisionally translated as "space of degrees", might refer to a deficit, i.e. a subtractive number of degrees. It may be noted that this is suggestive of the topic of net displacements along the ecliptic for multiple repetitions of

42 Cf. the phrasing of the contemporary O.Strasb.Dem. 521.1 and 6–7. The text contains two lists of celestial bodies. The first one identifies the planets, arranged according to the standard Mesopotamian layout, with various deities. The second list correlates the zodiacal signs with the 12 months of the year. The first line of the text reads: *pꜣ wn pꜣ 5 syw ꜥnḫ*: "The specification of the five living stars (planets)," whereupon each planet is attributed to a specific Egyptian deity, but following a Babylonian line of thought, i.e. "Saturn is the star of Re (i.e. Night Sun)," "Mercury is the star of Thoth (i.e. Nabû)," or "Jupiter is the star of Amun (i.e. Marduk [cf. Goebs 1995: 218–221 for the identification between the Egyptian and Mesopotamian god])." Line 6 reads: *pꜣ wn pꜣ syw nty sr* ⌝| *pꜣ ỉbd* 12 {*dmḏ*}: "The specification of the stars (i.e. the zodiacal signs), which are distributed over the 12 months ...." Then follows a list of which zodiacal sign belongs to which specific month.

TABLE 19.5 *Number of elementary steps contained in each zodiacal sign*

system	Ari	Tau	Gem	Cnc	Leo	Vir	Lib	Sco	Sgr	Cap	Aqr	Psc
A_1 MF	180	180	270	270	241	240	240	240	240	212	180	180
A_1 EF	150	150	150	102	90	90	96	135	135	135	135	145
A_2 ML	95	90	90	108	108	108	90	90	90	114	120	120
A_2 EL	60	48	48	60	60	60	60	60	60	54	54	60

a synodic phenomenon, which is often encountered in the Babylonian procedures.[43] However, a definite interpretation of the procedure and a reconstruction of its template cannot be established at this point.

Updating the Position of Mercury by Counting Elementary Steps

As mentioned above, the practical rules for updating the zodiacal positions of Mercury's synodic phenomena are lacking from Texts 1 and 2, but they are easily reconstructed with the help of the procedures and the lists contained in Texts 1 and 2. This may be illustrated by an example taken from the verso of *P.Oxy.* LXI 4152 (Jones 1999a: 1:123), a Greek papyrus with positions of Mercury's EF computed according to System A_1. The first preserved entry of *P.Oxy.* LXI 4152 is 12;36° of Pisces. The date to which it corresponds is unknown. As can be seen in Table 19.3, this position is in zone 3, at a distance of 2;36° beyond the boundary with zone 2. In order to locate the next instance of EF, we need to proceed by Z=480 elementary steps. We may begin by counting how many steps are contained between 0° Pisces and 12;36° Pisces; the answer (using Table 19.3) is $10/\delta_2 + 2;36/\delta_3 = 10 \cdot 4;30 + 2;36 \cdot 5 = 58$. From here we must count the 480 steps. The remaining part of Pisces contains 145–58 = 87 steps, which leaves 480–87 = 393 elementary steps to be taken beyond Pisces. The next two signs, Aries-Taurus, contain 2·150 = 300 elementary steps, which leaves 393–300 = 93 for the final stretch in Gemini. That sign is entirely in zone 3, where δ_3 = 0;12° (see Table 19.3 and Fig. 19.1). Hence the 93 steps yield the position 93·0;12° = 18;36° Gemini, which is indeed written in line 2 on the verso of *P.Oxy.* LXI 4152. It seems reasonable to assume that this updating procedure

43 For a general treatment of this topic see Ossendrijver (2012: 52–53). For Mercury, subtractive net displacements are obtained by considering, for instance, the distance covered after 3 or 63 synodic cycles, corresponding to 1 or 20 years, respectively (Ossendrijver 2012: 71, 73).

explains why Text 2 includes lists with the number of elementary steps per zodiacal sign.

Practical and Institutional Context of Texts 1 and 2

The content, date and mere fact that the texts are written in Egyptian would suggest that the texts belonged to an astronomer connected with, or having access to, a temple. Only these institutions would have educated astronomers of the calibre required for understanding the present texts.[44] If the texts indeed are Theban, they are not isolated compositions; several astronomical and astrological texts are known from that region throughout the Roman period. Evidence for such activities include astronomical texts, e.g. the so-called Stobart Tablets,[45] several horoscopes,[46] funeral monuments,[47] and magical papyri containing spells relating to astrological practices.[48] The fact that the texts are written on ostraca and contain some blunt scribal mistakes may suggest that they were not written for a temple library but rather for an astronomer's private usage. Perhaps they were copied from an original written on papyrus or on a wooden board, similar to the Stobart Tablets. Nevertheless, this does not suggest that the astronomer necessarily worked from an independent workshop, nor does it exclude the possibility that he used the texts for computing tables within the temple.

The immediate purpose of these texts, beyond a purely astronomical interest, escapes a definite interpretation. The positions of Mercury's synodic phenomena that one would obtain would not really be useful for horoscopic astrology; the procedures do not account for the daily motion of the planet, which would have been required for that.[49] However, together with an interpolation algorithm for computing daily positions in between synodic phenomena, the present texts could be employed for such a purpose.[50] If we look for an alternative motivation for knowing the first or last visibility of Mercury,

---

44 Jones 2007a: 311. See also e.g. Quack 2002: 168–169; Dieleman 2003; Winkler 2016: 269–272.

45 See n. 10 above. Greek astronomical texts that probably stem from the same region include *O.Bodl.* II 2176.

46 See n. 9 above.

47 E.g. Neugebauer and Parker 1969: 93–95 and pl. 50 (cf. also Töpfer 2014: 155–168); Riggs 2005: 182–205; id. 2006.

48 E.g. Gundel 1968; Gieseler-Greenbaum 2016: 195. See e.g. Dieleman 2003: 283–287; id. 2005: 123–126.

49 Jones 1999a; id. 2007a: 209–214.

50 Tables of this kind, with second-order interpolation schemes for computing daily positions in between first and last appearance of Mercury, are in fact attested in Oxyrhynchus (Jones 1999a). We would like to thank A. Jones for pointing this out.

tables produced by the methods found in the two texts could possibly be used for another form of judicial astrology, namely forecasts based on the appearance of Mercury. Omens and other astrological forecasts involving Mercury are well known from Mesopotamia.[51] A Babylonian procedure text from Seleucid Uruk (3rd century BC) concerned with weather prediction begins as follows: "If at the beginning of the cold, in Ulūlu (month VI) or Tašrītu (VII), Mercury or Venus rises in the east, there will be rain in that year. If at the beginning of the heat, in Addaru (month XII) or Nisannu (I), Mercury or Venus rises in the west: there will be flooding in that year."[52] Throughout that composition, Mercury's synodic phenomena feature prominently in weather prediction, which might partly explain why they were computed. Not unlike their Mesopotamian counterparts, the Egyptian forecasts concern the political situation of the country, its relation to neighbouring regions, climate and weather, and agricultural affairs. Predictions derived from the appearance of Mercury quite similar to those in the quoted Babylonian text are contained in, for instance, P.Brit.Mus. EA10661,[53] e.g.: "[i]f Mercury (*Swgꜣ*) is high, the inundation in Egypt will be great in that year." The text continues with a statement about the appearance of winds at midday (*mtry*) and its implication for the Chief of the Medes (*wr Mty*),[54] who will die if that is the case. From the subsequent forecasts, we learn that the fields will be prosperous in that year, though a great wind will come and rain will fall. The text also contains a forecast relating to the sixth lunar day, which, if it occurs simultaneously with the appearance of

51 Compared to other planetary omens from Mesopotamia, relatively little is known about the Mercury omens. It is unknown which tablets of the main celestial omen series, *Enūma Anu Enlil*, were dedicated to this planet. Four fragments with descriptions of Mercury's synodic cycle, cast in the form of omens, were edited by Pingree and Reiner 1975. The catalogue of astrological fragments published by Reiner 1996 mentions two more, unpublished fragments with Mercury omens (K. 12602, K. 12769). However, many other fragments lack a description and it can be assumed that some of them deal with Mercury. Furthermore, Mercury features as an accompanying phenomenon in other omens (e.g. of Venus: Reiner and Pingree 1998: 43, 59, 70, 85, 95, 133, 222, 247, 249). Apart from traditional omen divination, the phenomena of Mercury, along with those of other planets, are used for predicting weather and market prices in Late Babylonian astrological procedures (see next footnote).

52 TU 19, edited in Hunger 1976: 246–260.

53 Andrews 1992: 13–14; id. 1994: 31–32.

54 It is often assumed that the Medes in prophetic and divinatory texts, as well as other types of literature, can function as a circumlocution for the Seleucid Empire (e.g. Thiers 2007: 2007; Thissen 1998: 1048–1049; id. 2002: 122–124), although this view has met some resistance, e.g., Quack 2011: 103–131.

Mercury, signifies that the Chief of the Medes will die.[55] Such forecasts might be Egyptian versions of what is found in the Mesopotamian sources. In fact, these practices can be viewed as building on a Egyptian traditions too, since it is clear that the priesthood, for whatever reason, engaged in making forecasts based on the appearance of Sothis in the beginning of the year.[56]

Texts 1 and 2 and the Transfer of Babylonian Astral Science to Egypt

The Demotic texts presented here offer the first evidence that the hitherto elusive concept of elementary steps was known to Egyptian scholars. We can only speculate as to whether Mercury Systems A_1 and A_2 reached Egypt in a version with or without elementary steps. The former seems more likely, even though no such version is attested in Babylonia. Texts 1 and 2 may also provide new clues about the question of whether the new types of Babylonian astral science that emerged after about 400 BCE reached the Egyptians directly or through Greek sources or mediators.[57] One indication for the latter path of transfer is the planetary order most often found in Egyptian texts from the period after Alexander's conquest, which, by and large, follows a Hellenic template.[58] Also, as mentioned above, the zodiacal signs are believed to have reached Egypt through the Greeks, and not through direct contact with Babylonian astronomers. It is nevertheless implausible, in our view, that the present texts were translated from the Greek language, although it cannot be excluded. That they may rather be direct translations of Babylonian tablets is supported by the fact that several other Egyptian texts were most likely originally written

55 Ibid. 2011: 130.

56 For examples see also e.g. *P.Oxy.* LXV 4471–4473 and comments *ad loc.* See also Hughes 1951; Sijpesteijn and Weherli 1985. Among other scholars, Kákosy 1976: 46 and von Lieven 1999: 103–104 argue that these texts may date back to the New Kingdom, though there are in fact no concrete examples of such texts prior to the Late or Graeco-Roman period (cf. Krauss 1985: 37–38). It is possible that *P.Oxy.* LXI 4126 relates to a festival that included divinatory practices connected with the celebration of the appearance of Sothis.

57 The texts mentioned in n. 5 above clearly indicate that there was a continuous flow of divinatory and perhaps astronomical text into Egypt from Mesopotamia from the end of the second millennium BC until the Ptolemaic period. And even if the majority of the divinatory manuals of Mesopotamian origin known in Egypt are Ramesside, some Amarna Tablets may indicate that such influences date back to at least to 18th Dyn. See e.g. EA 358 (Izre'el 1997: 62–65), which contains a story about the king visiting a diviner.

58 Neugebauer and Parker 1969: 175. O.Strasb.Dem. 521 suggests that the Babylonian order also had found its way into Egypt (Derchain 1966: 154; cf. n. 33 above), though it did not gain any prominence.

in the Babylonian language.[59] Furthermore, our identification of the Egyptian word for degree as a probable Babylonian loanword (see 4.1.C above) also supports this assertion. This implies that the concept of degrees was known to the Egyptians through direct contact with Babylonian materials and not through a Greek intermediary.[60] Nevertheless, it cannot be excluded that the idea of degrees was known to the Egyptians prior to the translation of these texts, and that the Egyptian texts relied on Greek sources. But we deem such a scenario as less credible. Instead, we would like to suggest the possibility of an alternative scenario, according to which the Demotic texts were translated into Greek and later on disseminated to environments outside the Egyptian temple milieu. Of course, we cannot exclude the possibility that the procedure texts with which, for instance, *P.Oxy.* LXI 4152–4161 were computed, were independently translated into Greek from a Babylonian original, if they indeed were Greek.

FIGURE 19.2 *Image of Text 1.*

59 Cf. n. 41 above.

60 The most common Greek term seems to be μοῖρα, literally meaning "part" (see e.g. Ross 2014: 298).

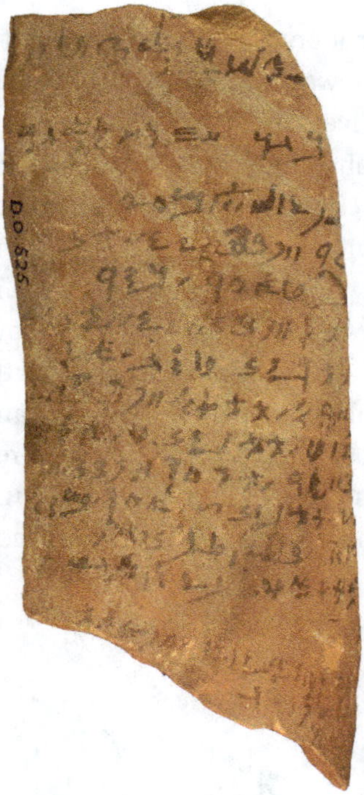

FIGURE 19.3 *Image of Text 2 (over view).*

FIGURE 19.4 *Image of Text 2 (left half).*

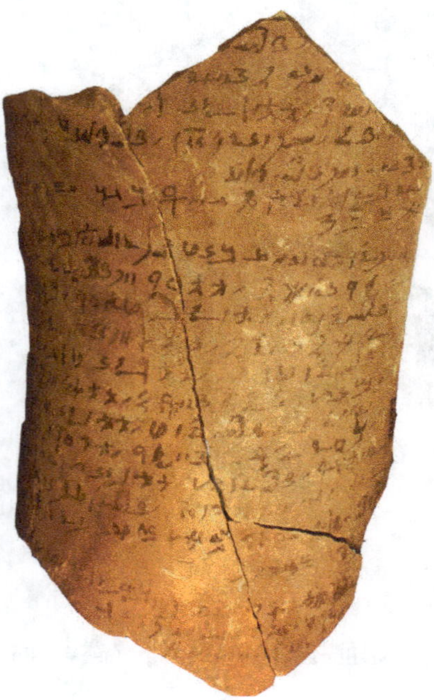

FIGURE 19.5 *Image of Text 2 (right half).*

Bibliography

Aaboe, Asger

1964 "On period relations in Babylonian astronomy." *Centaurus* 10: 213–231.

1977 "A Computed Cuneiform Text for Mercury from Babylon: BM 48147." Pp. 1–8 in *PRISMATA. Naturwissenschaftliche Studien. Festschrift für Willy Hartner*, eds. Y. Maeyama, W.G. Saltzer. Wiesbaden: Steiner Verlag.

Andrews, Carol A.R.

1992 "Unpublished Demotic Texts in the British Museum." Pp. 9–14 in *Life in a Multilingual Society: Egypt from Cambyses to Constantine and Beyond*, ed. J. Johnson. SAOC 51. Chicago: University of Chicago Press.

1994 "Unpublished Demotic Papyri in the British Museum." *EVO* 17: 29–37.

Beck, Susanne

2015 *Sāmānu: ein vorderasiatischer Dämon in Ägypten*. ÄuAT 83. Münster: Ugarit-Verlag.

Britton, John P.

2010 "Studies in the Babylonian Lunar Theory, Part 3: The Introduction of the Uniform Zodiac." *AHES* 64: 617–663.

Černý Jaroslav

1973 *Coptic Etymological Dictionary*. Cambridge: Cambridge University Press.

Chauveau, Michel

1992 "Un traité d'astrologie en écriture démotique." *CRIPEL* 14: 101–105.

Collombert, Philippe

2014 "Omnia brontoscopiques et pluies de grenouilles." Pp. 15–26 in A*cts of the Tenth International Congress of Demotic Studies: Leuven, 26–30 August 2008*, eds. Y. Broux and M. Depauw. OLA 231. Leuven: Peeters.

Collombert, Philippe and Coulon, L.

2000 "Les dieux contre le mer. Le début du «papyrus d'Astarté»." *BIFAO* 100: 192–242.

Crum, Walter E.

1939 *A Coptic Dictionary*. Oxford: Oxford University Press.

Demichelis, Sara

2002 "La divination par l'huile a l'époque ramesside." Pp. 149–165 in *La magie en Égypte: à la recherche d'une définition. Actes du colloque organisé par le Musée du Louvre les 29 et 30 septembre 2000*, ed. Y. Koenig. Louvre, conférences et colloques. Paris: La Documentation Française.

Derchain, Philippe

1966 "Essai de classement chronologique des influences babyloniennes et hellénistiques sur l'astrologie égyptienne des documents démotiques." Pp. 147–158 in *La divination en Mésopotamie ancienne et dans les régions voisines: 14e Rencontre assyriologique internationale, Strasbourg, 2–6 juillet 1965*, ed. Anon. Paris: Presses universitaires de France.

Dieleman, Jacco

2003 "Claiming the stars: Egyptian priests facing the sky." Pp. 277–289 in *Basel Egyptology Prize* I, eds. S. Bickel and A. Loprieno. Aeg.Hel. 17, Basel: Schwabe Verlag.

2005 *Priests, Tongues, and Rites: The London-Leiden Magical Manuscripts and Translation in Egyptian Ritual (100–300 CE)*. RGRW 153. Leiden: Brill.

Erichsen, Wolja

1954 *Demotisches Glossar*. Copenhagen: Einar Munksgaard.

Erman Adolf and Grapow, H.

1953 *Wörterbuch der Ägyptische Sprache* IV. Berlin-Leipzig: J.C. Hinrich'sche Buchhandlung.

Friberg, Jöran

2005 *Unexpected Links between Egyptian and Babylonian Mathematics*. Singapore: World Scientific.

Gallo, Paolo

1997 *Ostraca demotici e ieratici dall'archivio bilingue de Narmouthis II (nn. 34–99)*. Quaderni di Medinet Madi 3. Pisa: Editice ETS.

Gieseler-Greenbaum, Dorian

2016 *The Daimon in Hellenistic Astrology: Origins and Influence*. AMD 11. Leiden: Brill.

Goebs, Katja

1995 "'Horus der Kaufmann' als Name des Planeten Jupiter." *Enchoria* 22: 218–221.

Gundel, Hans, G.

1968 *Weltbild und Astrologie in den griechischen Zauberpapyri*. MBPR 53. Munich: C.H. Beck.

Hoffmann, Friedhelm

2004/2005 "Astronomische und astrologische Kleinigkeiten IV: Ein Zeichen für ‚Null' im P. Carlsberg 32?" *Enchoria* 29: 44–52.

2017 "Astronomische und astrologische Kleinigkeiten VII: Die Inschrift zu Tages- und Nachtlängen aus Tanis." Pp. 135–153 in *Iluminating Osiris: Egyptological Studies in Honor of Mark Smith*, eds. R. Jasnow and Gh. Widmer. Atlanta: Lockwood Press.

Hughes, George R.

1951 "A Demotic Astrological Text." *JNES* 10: 259–264.

Hunger, Hermann

1976 "Astrologische Wettervorhersagen." *ZA* 66: 234–260.

Izre'el, Shlomo

1997 *The Amarna Scholarly Tablets*. CM 9. Groningen: Styx.

Johnson, Janet

1976 *The Demotic Verbal System*. SAOC 38. Chicago: University of Chicago Press.

Jones, Alexander

1983 "The Development and Transmission of 248-Day Schemes for Lunar Motion in Ancient Astronomy." *AHES* 29: 1–36.

1999a *Astronomical papyri from Oxyrhynchus* (P. Oxy. 4133–4300a). MAPS 233. Philadelphia: American Philosophical Society.

1999b "A Classification of Astronomical Tables on Papyrus." Pp. 299–340 in *Ancient Astronomy and Celestial Divination*, ed. N.M. Swerdlow. Dibner Institute in the History of Science and Technology. Cambridge, MA; London: The MIT Press.

2007a "Astrologers and their Astronomy." Pp. 307–314 in *Oxyrhynchus—A City and its Texts*, eds. A.K. Bowman et al. GRM 93. London: Egypt Exploration Society.

2007b "Dai Papiri della Società Italiana: Two Astronomical Tables from Oxyrhynchus Based on Babylonian Planetary Theory." *Comunicazioni dell'Istituto Papirologico «G. Vitelli»* 7, 1–8 + Tavola I.

Kákosy, Lázló

1976 "Die mannweibliche Natur des Sirius in Ägypten." *Stud.Aeg.* 2: 41–46.

Krauss, Rolf

1985 *Sothis- und Monddaten: Studien zur astronomischen und technischen Chronologie Altägyptens*. HÄB 20. Hildesheim: Gerstenberg Verlag.

Leclère, François and Coulon, L.

1998 "La nécropole osirienne de la « Grande Place » à Karnak. Fouilles dans le secteur nord-est du temple d'Amon." Pp. 649–659 in *Proceedings of the Seventh International Congress of Egyptologists, Cambridge, 3–9 September 1995*, ed. Ch. Eyre. OLA 82. Leuven: Peeters.

von Lieven, Alexandra

1999 "Divination in Ägypten." *AoF* 26: 77–126.

Menchetti, Angiolo

2009 "Un aperçu des textes astrologiques de Médinet Madi." Pp. 223–238 in *Actes du IXe Congrès international des études démotiques: Paris, 31 août—3 septembre 2005*, eds. Devauchelle and G. Widmer. BdE 147. Cairo: Institut Français d'archéologie oriental.

Möller, Georg

1936[2] *Die Aegyptische Buchschrift in ihrer Entwicklung von der fünften Dynastie bis zur römischen Kaiserzeit, Band 3: von der Zweiundzwangzigsten Dynastie bis zum dritten Jahrhundert nach Chr.* Berlin-Leipzig: J.C. Hinrich'sche Buchhandlung.

Neugebauer, Otto

1941 "On a Special Use of the Sign 'Zero' in Cuneiform Astronomical Texts." *JAOS* 61: 213–215.

1942 "Egyptian Planetary Texts." *TAPA* N.S. 32.2: 209–250.

1943 "Demotic Horoscopes." *JAOS* 63: 115–127.

1949 *The Astronomical Treatise P.Ryl. 27*. KDVS Hist.-Fil. Med. 32.2. Copenhagen: Einar Munksgaard.

1955 *Astronomical Cuneiform Texts*. London: Lund Humphries.

1975 *A History of Ancient Mathematical Astronomy*. New York: Springer.

Neugebauer, Otto and Parker, R.A.

1968 "Two Demotic Horoscopes." *JEg.Arch.* 54: 231–235.

1969 *Egyptian Astronomical Texts, III: Decans, Planets, Constellations and Zodiacs*, BEStud. 6. Providence, RI: Brown University Press.

Neugebauer, Otto, Parker, R.A., and Zauzich, K.-Th.

1981 "A Demotic Lunar Eclipse Text of the First Century B.C." *PAPS* 125: 312–327.

Osing, Jürgen

1976 *Die Nominalbildung des Ägyptischen*. Mainz am Rhein: Philip von Zabern Verlag.

Ossendrijver, Mathieu

2012 *Babylonian Mathematical Astronomy: Procedure Texts*. New York: Springer.

Parker, Richard A.

1959a *A Vienna Demotic Papyrus on Eclipse- and Lunar Omina*, BEStud. 2. Providence, RI: Brown University Press.

1959b "A Demotic mathematical papyrus fragment." *JNES* 18: 275–279.

1962 "Two Demotic Astronomical Papyri in the Carlsberg Collection." *AcOr* 26: 143–147.

1972 *Demotic Mathematical Texts*. BEStud. 7. Providence, RI: Brown University Press.

1983 "A Horoscopic Text in Triplicate." Pp. 141–143 in *Grammata Demotica: Festschrift für Erich Lüddeckens zum 15. Juni 1983*, eds. H.-J. Thissen and K.-Th. Zauzich. Sommerhausen: Gisela Zauzich Verlag.

Parker, Richard A. and Zauzich, K.-Th.

1981 "The Seasons in the First Century BC." Pp. 472–479, in *Studies Presented to Hans Jacob Polotsky*, ed. D.W. Young. East Gloucester, MA: Pirtle and Polson.

Payne-Smith, Jesse (Mrs. Margoliouth)

1903 *A Compendious Syriac Dictionary founded upon the Thesaurus Syriacus of R. Payne-Smith*. Oxford: Clarendon Press.

Pingree, David and Reiner, Erica

1975 "Observational Texts Concerning the Planet Mercury." *RA* 69: 175–180.

Quack, Joachim, F.

2002 "Die Dienstanweisung des Oberlehrers aus dem Buch vom Tempel." Pp. 159–179, in *5. ägyptologische Tempeltagung, Würzburg 23.–26. September 1999, Akten der ägyptologischen Tempeltagung*, ed. H. Beinlich. ÄuAT 33.3. Wiesbaden: Harrassowitz.

2004 "Fragmente memphitischer Religion und Astronomie in semidemotischer Schrift (pBerlin 14402 + pCarlsberg 651 + PSI Inv. D 23)." Pp. 467–496 in *Res severa verum gaudium: Festschrift für Karl-Theodor Zauzich zum 65. Geburtstag am 8. Juni 2004*, eds F. Hoffmann and H.-J. Thissen. StudDem 6. Leuven: Peeters.

2008/2009 "Ein astrologisches Ostrakon der frühen Kaiserzeit (oGlasgow D 1925.96)." *Enchoria* 31: 104–112.

2011 "Ist der Meder an allem Schuld? Zur Frage des realhistorischen Hintergrundes der gräkoägyptischen prophetischen Literatur." Pp. 103–131 in *Ägypten zwischen innerem Zwist und äußerem Druck: Die Zeit Ptolemaios' VI. Bis VIII.*, eds A. Jördens and J.F. Quack. Philippika 45. Wiesbaden: Harrassowitz.

Reiner, Erica

1998 "Celestial Omen Tablets and Fragments in the British Museum." Pp. 215–302 in *Festschrift für Rykle Borger zu seinem 65. Geburtstag am 24. Mai 1994*, ed. S. Maul. CM 10. Groningen: Styx.

Reiner, Erica and Pingree, David

1998 *Babylonian Planetary Omens: Part Three*. CM 11. Groningen: Styx.

Riggs, Christina

2005 *The Beautiful Burial in Roman Egypt: Art, Identity, and Funeral Religion*. Oxford Studies in Ancient Culture and Representation. Oxford: Oxford University Press.

2006 "Archaism and Artistic Sources in Roman Egypt: the Coffins of the Soter Family and the Temple of Deir el-Medina." *BIFAO* 106: 315–332.

Roccati, Antonio

1984 "Lessico meteorologico." Pp. 343–354 in *Studien zu Sprache und Religion Ägyptens: zu Ehren von Wolfhart Westendorf, überreicht von seinen Freunden and Schülern*, ed. F. Junge. Göttingen: Seminar für Ägyptologie und Koptologie.

Rochberg, Francesca

2010 *In the Path of the Moon: Babylonian Celestial Divination and its Legacy*. AMD 6. Leiden: Brill.

Ross, Micah T.

2006 "An Introduction to the Horoscopic Ostraca of Medînet Mâdi." *EVO* 29: 147–180.

2007 "A Continuation of the Horoscopic Ostraca of Medînet Mâdi." *EVO* 30: 153–171.

2009 "Further Horoscopic Ostraca from Medînet Mâdi." *EVO* 32: 61–96.

2011 "A Provisional Conclusion to the Horoscopics Ostraca from Medînet Mâdi." *EVO* 34: 47–80.

2014 "The Role of Alexander in the Transmisson of the Zodiac." Pp. 287–305 in *Alexander the Great and Egypt: History, Art, Tradition*, ed. V. Grieb et al. Philippika 74. Wiesbaden: Harrassowitz.

Schneider, Thomas

2003 "Foreign Egypt: Egyptology and the Concept of Cultural Appropriation." *ÄL* 13: 155–166.

Sijpesteijn, Pieter, C. and Weherli, C.

1985 "Prédictions astrologiques." *Mus.Helv*. 42: 328–333.

Smith, Mark

2005 *Papyrus Harkness (MMA 31.9.7)*. Oxford: Griffith Institute.

Spiegelberg, Wilhelm

1902 "Ein aegyptisches Verzeichnis der Planeten und Tierkreisbilder." *OLZ* 5: 6–9.

1911 "Die ägyptischen Namen und Zeichen der Tierkreisbilder in demotischer Schrift." *ZÄS* 49: 146–151.

Stockhusen, Marco

2012 "Babylonische Astralwissenschaften im römerzeitlichen Ägypten: das Beispiel Medînet Mâdi." *WdO* 42: 85–109.

Thiers, Christophe

2007 "*Ptolémée Philadelphe et les prêtres d'Atoum de Tjékou: Nouvelle éditon commentée de la «stèle de Pithom» (CGC 22183)*." Orientalia Monspeliensia 17. Montpellier: Presses universitaires de la Méduterranée.

Thissen, Heinz-Joseph

1998 "»Apocalypse Now« Anmerkungen zum Lamm des Bochoris." Pp. 1044–1065 in *Egyptian Religion the Last Thousand Years, II: Studies Dedicated to the Memory of Jan Quaegebeur*, eds. W. Clarysse, A. Schoors, and H. Willems. OLA 85. Leuven: Peeters.

2002 "Das Lamm des Bochoris." Pp. 113–138 in *Apokalyptik und Ägypten: Eine kritische Analyse der relevanten Texte aus dem griechisch-römischen Ägypten*, eds. A. Blasius and B.U. Schipper. OLA 107. Leuven; Paris; Sterling, VA: Peeters.

Toomer, Gerald J.

1998 *Ptolemy's Almagest*. Princeton: Princeton University Press

Töpfer, Susanne

2014 "Heter—Ein Thebanischer Priester mit drei Funerärtexten und einem Sarg." Pp. 155–168 in *Persönlichkeiten aus dem alten Ägypten im Neuen Museum*, ed. V.M. Lepper. Berlin: Ägyptisches Museum und Papyrussammlung, Staatliche Museen zu Berlin.

van der Waerden, Bartel L.

1952/1953 "The History of the Zodiac." *AfO* 16: 216–230.

1957 "Babylonische Planetenrechnung." *Vierteljahresschr. Naturforsch. Ges. Zürich* 102: 39–60.

1958 "The Astronomical Papyrus Ryland 27." *Centaurus* 5: 177–191.

1960 "Babylonische Methoden in ägyptischen Planetentafeln." *Vierteljahresschr. Naturforsch. Ges. Zürich* 105: 97–144.

1965 Erwachende Wissenschaft II. Die Anfänge der Astronomie. Groningen: Noordhoff. [revised English edition: 1974, Science Awakening II. The Birth of Astronomy. Dordrecht: Reidel; Groningen: Noordhoff; New York: Oxford University Press].

1972 "Ägyptische Planetenrechnung." *Centaurus* 16: 65–91.

Volten, Axel

1942 *Demotische Traumdeutung* (Pap. Carlsberg XIII und XIV verso). AnAe 3. Copenhagen: Einar Munksgaard.

Vycichl, Werner

1984 *Dictionnaire étymologique de la langue copte*. Leuven: Peeters.

Westendorf, Wolfhart

1977 *Koptisches Handwörterbuch*: Heidelberg: Karl Winter Verlag.

Winkler, Andreas

2009 "On the Astrological Papyri from the Tebtunis Temple Library." Pp. 361–375 in *Actes du IXe Congrès International des Études Démotiques: Paris, 31 août—3 septembre 2005*, eds. D. Devauchelle and G. Widmer. BdÉ 146. Cairo: Institut Français d'Archéologie Orientale.

2016 "Some Astrologers and Their Handbooks in Demotic Egyptian." Pp. 245–286 in *Circulation of Astronomical Knowledge in the Ancient World*, ed. J.M. Steele. Time, Astronomy, and Calendars 6. Leiden: Brill.

Zauzich, Karl-Theodor

1980 "Aus zwei demotischen Traumbüchern." *Arch.Pap.* 27: 91–98.

CHAPTER 20

Clever Machines and the Gods Who Make Them

The Antikythera Mechanism and the Ancient Imagination

Daryn Lehoux

After sitting, cryptically, silently, in the Archaeological Museum in Athens for nearly a century, the Antikythera mechanism only really began to yield up its secrets in the latter part of the twentieth century.[1] To be sure, there had been some speculation that the thing was some kind of planetarium or orrery from within a few years of its discovery, but it was hard for most scholars to really believe this was possible in the cold light of day. Not until Derek de la Solla Price managed to obtain images of the thing using gamma rays in the 1970s did we finally realize the complexity of its internal gearing, and even then there were plenty of surprises in store for us as imaging technology improved over the next 30 years. Publishing in 1974, Price was now certain that the device was some kind of "calendrical Sun and Moon computing mechanism," and he was quick to try and establish a line between the mechanism and several ancient reports of "planetarium devices" associated with Archimedes by Cicero and others. These reports, too, have had a curious history. In the wake of our new-found (and quite revolutionary) understanding of the inner workings of the mechanism in the last ten years, it is now very easy to see many of the ancient reports of wonderful planetary machines as fairly unambiguous references to complex geared Antikythera-type planetaria. But before the last generation or so, most professional historians of astronomy resisted committing themselves to such a conclusion, wondering whether the stories in Cicero and his contemporaries were fancified accounts of much simpler instruments such as armillary spheres or decorative globes or something similar. Non-historians-of-astronomy, editors and translators of Cicero, for example, were happier to refer to such descriptions as referring to "planetaria" and "orreries" even if professionals were more sceptical that devices of such complexity could have existed as early as the first century BCE. To put the complexity of the mechanism in context: no geared device of this sophistication appears in the European historical record prior to the *fourteenth century CE*, which is to say, until the late-medieval astronomical clocks of Givanni de Dondi and Richard of Wallingford.

1 See, e.g., Freeth and Jones 2012; Freeth, Jones, Steele, and Bitsakis 2008; Wright, 2007; Freeth *et al.* 2006; Price, 1974; See also Field and Wright 1985.

 | DOI 10.1163/9789004363380_021

And for their part, historians of horology had often remarked on how complex the de Dondi and Wallingford clocks seemed to be, given that there was no apparent prior tradition from which they may have developed.

What Does the Antikythera Mechanism Do?

The mechanism, found by sponge divers excavating a Mediterranean shipwreck for treasure at the very beginning of the twentieth century, is badly decayed and missing much of its original material. Nevertheless, recent imaging work has revealed it to be a complex astronomical-calendrical computer that uses intricate gearing to turn inputted motion, in the form of a cranked handle, into sophisticated output using a number of pointers on several dials. Apart from being hand-driven, the mechanics of the mechanism are very much analogous to those of a modern clock, but with more faces, and much more complex motions on its dials. The device itself was originally housed in a wooden box, perhaps 30 cm high by 18 cm wide by 12 cm thick. Hinged doors covered each of the two display sides, opening up to reveal clock-like dials on both faces, two main dials on what we are now calling the rear face of the machine, and a single dial with multiple pointers and scales on the front. Rather than telling the time, as a modern clock does, the mechanism told the Egyptian-calendar date on the front dial, as well as the month in what we now believe to be the calendar of Epirus (including intercalary information) on one of the rear dials.[2] It also had a pointer for cyclical games (Olympia, Isthmia, etc.) and dials for eclipse prediction on the rear face. The front face contained a dial for the motion of the moon through the zodiac (noted also for the sun), and a clever rotating ball mechanism showing the moon's phase. Most scholars now believe that the front face also originally had pointers for the motion of the planets through the zodiac—which is no mean feat.

Republics

Now that we know this thing exists, we can look to literary descriptions of *sphaerae* in ancient sources with more appreciation of detail than we had done previously, since we now have a better idea of what the possibilities were. The earliest such description and in many ways the archetype of all later ones comes from Cicero's *Republic.*

---

2 The identification of Epirus is suggested by Freeth, Jones, Steele, and Bitsakis, 2008.

Although written in the late 50s BCE, the dialogue of the *Republic* is set a couple of generations earlier, in the turbulent aftermath of Tiberius Gracchus' disastrous land-reform proposals, and specifically in the year in which Scipio Aemilianus—a prominent character in the *Republic*—would be found dead under what many took to be suspicious circumstances. Early in the dialogue, after a bit of give and take on the relative importance of then-recent reports that two suns had been seen simultaneously (an ominous sign for ominous times), the character Philus makes a plea for the importance of our knowledge of the cosmos to our understanding of the solutions to the ills of the Roman republic. This appeal is one of Cicero's most cherished philosophical ideas (shared with the Stoics, among others), that human justice, and ethics in general, is rooted in an understanding of nature. And so for Cicero, right reason is grounded in an understanding of the divine governance of the cosmos. And this turning to nature to see how the gods govern the world has immediate political import for Cicero, since, as he says very early on in the *Republic,*

> neque enim est ulla res, in qua propius ad deorum numen virtus accedat humana, quam civitatis aut condere novas aut conservare iam conditas.[3]

> There is no other area in which human virtue comes closer to the majesty of the gods, than in building new states, or in maintaining existing ones.

Given the strife rocking the state when Cicero wrote the *Republic,* this is clearly an important claim. Furthermore, we see from a number of related passages that the cosmos as a divine artifact plays heavily into his account of how the best state should be constituted. He goes so far as to frame the entire *Republic,* at its beginning and end, with cosmological discussions, as well as to pepper cosmological terminology throughout his political passages.[4]

And it is in this context that Cicero invokes the earliest reference we have to an Antikythera-type mechanism. Having stepped back from his own day to 129 BCE for the main body of the dialogue, Cicero now has his character Philus briefly step us back one more generation, to 166 BCE, when a similar report of a double sun had been circulating. Philus was at that time, he tells us, visiting Marcus Marcellus, grandson to the Roman conqueror of Syracuse, in the company of the younger Marcellus' co-consul Gaius Sulpicius Gallus.

3 Cicero, *Rep.* 1.12.

4 See Gallagher 2001.

nam memoria teneo C. Sulpicium Gallum, doctissimum ut scitis hominem, cum idem hoc visum diceretur et esset casu apud M. Marcellum, qui cum eo consul fuerat, sphaeram quam M. Marcelli avus captis Syracusis ex urbe locupletissima atque ornatissima sustulisset, cum aliud nihil ex tanta praeda domum suam deportavisset, iussisse proferri; cuius ego sphaerae cum persaepe propter Archimedi gloriam nomen audissem, speciem ipsam non sum tanto opere admiratus; erat enim illa venustior et nobilior in volgus, quam ab eodem Archimede factam posuerat in templo Virtutis Marcellus idem. sed posteaquam coepit rationem huius operis scientissime Gallus exponere, plus in illo Siculo ingenii quam videretur natura humana ferre potuisse iudicabam fuisse. dicebat enim Gallus sphaerae illius alterius solidae atque plenae vetus esse inventum, et eam a Thalete Milesio primum esse tornatam, post autem ab Eudoxo Cnidio, discipulo ut ferebat Platonis, eandem illam astris quae caelo inhaererent esse descriptam; cuius omnem ornatum et descriptionem sumptam ab Eudoxo multis annis post non astrologiae scientia sed poetica quadam facultate versibus Aratum extulisse. hoc autem sphaerae genus, in quo solis et lunae motus inessent et earum quinque stellarum quae errantes et quasi vagae nominarentur, in illa sphaera solida non potuisse finiri, atque in eo admirandum esse inventum Archimedi, quod excogitasset quem ad modum in dissimillimis motibus inaequabiles et varios cursus servaret una conversio. hanc sphaeram Gallus cum moveret, fiebat ut soli luna totidem conversionibus in aere illo quot diebus in ipso caelo succederet, ex quo et in caelo <et in> sphaera solis fieret eadem illa defectio, et incideret luna tum in eam metam quae esset umbra terrae, cum sol e regione … <lacuna> …[5]

I am reminded of C. Sulpicius Gallus, a most learned man, as you know. At a time when a similar phenomenon was reported and he happened to be at the house of Marcus Marcellus, his colleague in the consulship, he ordered the sphere to be brought out which the grandfather of Marcellus had carried off from Syracuse when that very rich and beautiful city was taken (though he took home with him nothing else out of the great store of booty that was captured). Though I had heard this sphere mentioned quite frequently on account of the fame of Archimedes, when I actually saw it I did not particularly admire it; for that other one, also constructed by Archimedes, which the same Marcellus placed in the temple of Virtue, is more beautiful as well as more renowned among the people. But

5 Cicero, *Rep.* 1.21–22. Translation modified from Keyes.

> when Gallus began to give a very learned explanation of the device, I concluded that the famous Sicilian had been endowed with greater genius than one would imagine it possible for a human being to possess. For Gallus told us that the other kind of sphere, which was solid and full, was an old invention, the first one of that kind having been lathe-turned (*tornatam*) by Thales of Miletus, and later marked with the constellations which are fixed in the sky by Eudoxus of Cnidus (a disciple of Plato, it was claimed). He also said that many years later, Aratus, borrowing this whole arrangement and plan from Eudoxus, had described it in verse, without any knowledge of astronomy but with considerable poetic talent. But this newer kind of sphere, he said, in which were delineated the motions of the sun and moon and of those five stars that are called planets, or as we might say, wanderers, contained more than could be shown on the solid globe, and the invention of Archimedes deserved special admiration because he had thought out a way that a single turn saved the anomalous and changing courses in [their] completely different movements.[6] And when Gallus moved the sphere, it happened that the moon advanced on the sun by as many turns in the bronze as it did by days in the sky itself. Thus the same eclipse of the sun happened in both the sphere and the heavens, and the moon came to the cone where the shadow of the earth would have been, at the very time when the sun … from the region …

The text unfortunately breaks up here and we have a lacuna of (probably) a few pages in the original before the dialogue picks up again with the character Scipio recounting a story about the time that Gallus talked his troops down from panic at the sight of a lunar eclipse by explaining the physics of the earth's shadow to them.[7]

In examining Cicero's account of this "Archimedean sphere" I do not wish to worry overmuch about the plausibility of the device's survival (or its functionality) from the siege of Syracuse in 212, to the date of Philus' purported encounter with the device in 166–46 years is a long time in the life of a delicate and difficult-to-clean bronze instrument. Nor am I particularly committed to the idea that Archimedes himself built such a sphere. We do not need to take any of these claims as historical just so long as Cicero can be expected to be describing a *class* of device that really existed in his day, and that he did not think would be wholly unfamiliar to his readers.

6 I thank Michael Wright for suggestions on the translation.

7 On this story, see Bowen 2002.

As for the sphere itself, Cicero says that it modeled all five planets plus the sun and moon, and he seems to indicate that the geometry of it had the sun and moon pointers configured in such a way that one could talk of knowing where the shadow of the earth (presumably represented at the centre of the dials) would be: *et incideret luna tum in eam metam quae esset umbra terrae,* "the moon [pointer] came to the cone where the shadow of the earth would have been." The use of the odd verb just before this, *servo,* for the relationship between the display and the heavens—the device is said to have "saved the anomalous and changing courses"—seems to have been an attempt to translate the sense of the Greek σῴζει τὰ φαινόμενα, "save the phenomena," meaning that it modeled the phenomena precisely. The phrase *in dissimillimis motibus,* "in [their] completely different movements," with its superlative of *dissimilis,* is a favourite of Cicero's for describing planetary motion (and it turns up, so far as I can find, rarely outside of Cicero).[8] I take it here to be describing the surprisingly complex independent motions of the various pointers, one of the most striking features of the working Antikythera mechanism.

There are several features of Cicero's account that we will see prominently recurring in later references to mechanized *sphaerae.* First is the association with Archimedes, which, although not universally accepted by modern scholars, nevertheless proved to be a long-lasting ancient tradition. Second, we see Archimedes, insofar as he is the purported inventor of this astounding device, being lauded as having a transcendent, almost super-human intelligence for having come up with the thing: "I concluded that the famous Sicilian had been endowed with greater genius than one would imagine it possible for a human being to possess." Indeed, the standard invocation of the mechanized sphere sees it as the very pinnacle of human ingenuity, instantiated usually in the person of Archimedes. Although he is not quite called "divine" here, the impressiveness, complexity, and sheer cleverness of the device is such that it is scarce believable that it could have been the product of human hands. Indeed, this is still a common reaction to reconstructions of the Antikythera mechanism: *How,* everyone seems to ask, was it possible for the ancients to do that, way back then?

The other important theme is less marked here than it will be in later sources, but we will see that it will come to be central to later accounts as it gets more fully elaborated. This is the idea that the machine somehow instantiates the heavens themselves in miniature. We have already seen his use of the verb

8 Compare *ND* II.119 where the planets themselves are said to show that *tantus est concentus ex dissimillimis motibus.* He uses the superlative again in *Tusc.* I.63 to describe the motions in a *sphaera.*

servo, which may be taken as a reference to modeling rather than instantiating, but the issue becomes more complex when Philus comes to describe his amazement at the combined motions of the sun and moon. Here the wording is interesting: the sun and moon make the "same" motions in the device as they make in the heavens: *et in caelo <et in> sphaera solis fieret eadem illa defectio,* the very same eclipse happens in both the sphere and the heavens.[9]

This doubling is a curious feature of the entire passage, recurring in multiple, layered ways. The story is set in the past within a dialogue set in the past; it is reported by the character Philus, who is himself reporting an experience of a second person, Gallus. The mechanized sphere is second to the more famous Archimedean sphere in the temple of Virtue. The *sphaera* invented by Thales was later marked with constellations *described by* Eudoxus, presumably in his book called the *Phaenomena,* which prose version was itself copied into verse by Aratus, as Cicero has Philus tell us here.

It will be worth spending a moment on the—for our purposes—second sphere in this account. The fixed kind, turned on a lathe, we are (plausibly) told is the older. Marcellus is said to have taken as booty for himself only the mechanical sphere from Syracuse, but he did also, on this story, donate a fixed *sphaera* of Archimedes in the temple of Virtue at Rome. Scholars have often suspected that this is in fact a reliable account of a real sphere, and perhaps even the same sphere mentioned by Ovid as hanging in the temple of Vesta. As Ovid describes it, there is an image of the earth "suspended by Syracusan art" and apparently still extant in his day, hung up in the temple of Vesta at Rome:

> terra pilae similis nullo fulcimine nixa,
> aere subiecto tam grave pendet onus.
> ipsa volubilitas libratum sustinet orbem,
> quique premat partes, angulus omnis abest
> cumque sit in media rerum regione locata
> et tangat nullum plusve minusve latus,
> ni convexa foret, parti vicinior esset,
> nec medium terram mundus haberet onus.
> arte Syracosia suspensus in aere clauso
> stat globus, immensi parva figura poli,
> et quantum a summis, tantum secessit ab imis
> terra; quod ut fiat, forma rotunda facit.[10]

9 The addition of *et in* here follows Dobree. Ziegler prefers to excise *caelo* instead, leaving the comparator only implied.

10 Ovid, *Fasti* 6.269–280.

> The earth is like a ball, supported by no pillar; so heavy a burden weighs on the air beneath. This roundness[11] sustains the orb, balanced. It has no angle to press on any part. It is always situated in the middle of everything, and touches no side more and on no side less; were it not round, it would be closer in one part and the world would not have the earth as central mass. There stands a globe, suspended in enclosed air by Syracusan art, a tiny form of immense heaven, and the earth stands distant by the same amount from the top as from the bottom, caused as it is by the round shape.

Clearly this is no mechanical *sphaera,* as it shows only the position of the earth in the cosmos and Ovid makes no mention of the planets or of moving parts. In fact, it sounds very much like a device described in Hero of Alexandria (*Pneum.* 2.7) that has often puzzled modern commentators (it is perhaps the least complex or interesting of the devices in Hero's *pneumatica,* and its purpose or significance remains entirely unclear). If so, one wonders why the Archimedean globe attracted any renown at all. Cicero calls it *venusta,* charming, and *nobilis,* renowned, and Ovid's account seems to point toward a degree of subtlety in its design when he says that it is "suspended in enclosed air by Syracusan art." It is hard to imagine why, but it would seem that there must have been some kind of clever suspension system keeping the earth balanced in the middle, or some kind of *trompe l'oeil* effect hiding the strings, as it were, to make it worthy of the literary attention it gets.

Just Like God Did

Now, if mechanized planetaria—made by clever craftsmen—can be seen as modeling or even in a sense bringing the cosmos into the hands and homes of people on earth, the metaphor of the cosmos itself as a product of craftsmanship cannot be far off, putting the human craftsman into a direct conversation with the cosmic creator.[12] Indeed, even the simpler type of globe mentioned

11 One often sees *volubilitas* translated as "rotation" or some such, but this is anachronistic. Ovid means nothing more than roundness, and this for two reasons: (a) *very* few ancients ever even floated the idea that the earth could be rotating, and (b) the physics he is appealing to here, going back to something like Anaximander's principle of indifference, requires roundness, not motion, as Ovid himself points out repeatedly in the subsequent lines. Furthermore, I know of no ancient physical theory that claims that rotation could keep something in equilibrium.

12 I thank Eduardo Escobar for this suggestion.

by Ovid and Cicero could be used to draw larger analogies about the construction of the whole cosmos. This becomes clear in a curious little passage from a little-known work attributed (albeit insecurely) to Apuleius (2nd c. CE). Apuleius is now perhaps most famous as the author of the celebrated novel *The Golden Ass.* He also, if the present work is in fact his, took enough of an interest in cosmology to translate into Latin the short Pseudo-Aristotelian text called *De mundo.* This translation shows some interesting departures from the original. One passage concerns the shape of the world, where Pseudo-Aristotle highlights the simple fact that, because the cosmos is a rotating sphere, it must have two unmoving poles around which it is turning:

> τοῦ δὲ σύμπαντος οὐρανοῦ τε καὶ κόσμου σφαιροειδοῦς ὄντος καὶ κινουμένου, καθάπερ εἶπον, ἐνδελεχῶς, δύο ἐξ ἀνάγκης ἀκίνητά ἐστι σημεῖα, καταντικρὺ ἀλλήλων, καθάπερ τῆς ἐν τόρνῳ κυκλοφορουμένης σφαίρας, στερεὰ μένοντα καὶ συνέχοντα τὴν σφαῖραν, περὶ ἃ ὁ πᾶς ὄγκος κύκλῳ στρέφεται· καλοῦνται δὲ οὗτοι πόλοι.[13]

> The whole heaven and cosmos being spherical and moving, as I said, perpetually, there are two necessarily unmoving points opposite to each other, just as on a ball being turned on a lathe, that remain still and contain the sphere, around which the whole body of the circle rotates. These are called poles.

But if we turn to the Latin of (Pseudo?) Apuleius, we find that the lathe analogy has been expanded in an interesting way. The Latin no longer talks of a ball simply "being turned" in the passive voice. Instead, it inverts the grammar to more clearly focus on the craftsman who is doing the turning (the craftsman was not even mentioned in the original), and indeed, adds a craftsman to the manufacture of the world itself:

> sed cum omne caelum ita revolvatur ut sphaera, eam tamen radicibus oportet teneri, quas divina machinatio verticibus adfixit, ut in tornando artifex solet forcipe materiam comprehensam reciproco volumine rotundare: eos polos dicimus.

> Because the whole heaven is turned as a sphere, nevertheless it is necessary that it remain still at the extremities, which divine artifice has fixed

13 Ps-Aristotle, *De mundo,* 391b19–392a2. Latin translation in Apul. De Mundo 1.290.

> at the points, just as in lathe-turning the craftsman rotates the wood held by pincers in a repeating rotation. These we call poles.

That is, the two fixed points of the cosmos have been established (*adfixit*) by divine "artifice" or "machinery" (*machinatio*). The lathe comparison then comes in not just as a simple, passing observation ("just like something you have seen"), as it did in the Greek, but as a direct comparison to how god built the universe.

God as lathe operator can also be found in the fourth-century Christian writer Diodorus of Tarsus. In his commentary on verse 3 of Psalm 8, ὅτι ὄψομαι τοὺς οὐρανούς, ἔργα τῶν δακτύλων σου ("Because I shall see the heavens, the works of your fingers"), he says:

> ἀλλ' οὐδέν, φησί, θαυμαστὸν εἰ τῆς σῆς τέχνης ἔργα ἦν ταῦτα, τό τε ἀνοιγῆναι νηπίων στόματα καὶ τὸ καταλυθῆναι τοσούτων ἐχθρῶν εἰρωνείαν. ἐκεῖνοι γὰρ ταῦτα ἐνήργουν, φησίν, οἱ δάκτυλοι, οἱ καὶ τὸν οὐρανὸν παρὰ τὴν ἀρχὴν εὐρύθμως ἀποτορνεύσαντες.
>
> It is no surprise, he says, if this was the work of your *techné,* the mouths of infants being opened and the false modesty of your enemies being destroyed. For those fingers, he says, that did these things also lathe-turned the heavens in beautiful proportion in the beginning.

God as craftsman: employing *techné,* and in particular the lathe, to fashion the world.

One is immediately reminded in these two passages of Plato's *Timaeus,* where the demiurge is explicitly said to "turn" the heavens as on a lathe: κυκλοτερὲς αὐτὸ ἐτορνεύσατο.[14] And the god of the *Timaeus* finds himself in fullest play in yet another passage from Cicero, this time from his *Tusculan Disputations,* where he invokes the full complexity of an Antikythera-like mechanism as an analogy for how the whole cosmos was put together by the demiurge in the beginning. As in his earlier *Republic,* when Cicero again mentions Archimedes in the *Tusculan Disputations,* we see the praise of his near-superhuman genius at the forefront, but this time Cicero takes it much further, using that genius as the keystone to an argument about the very immortality of the human soul. This is a long way from just invoking lathes and spheres as analogies.

The *Tusculan Disputations,* written in 45 BCE, opens with a discussion of death and what happens to the soul afterwards. Hearkening back to the *Dream*

14 *Tim.* 33B.

of Scipio, the character "M." (Marcus himself)[15] wonders at one point what the view will be like when, after death, the soul will be able to rise up to the heavens and look down on the whole earth. The conversation then moves on to how memory might work in the soul, and, ultimately for our purposes, to that other great ability of the soul: invention. A list of great inventions follows, starting with the most important crafts, music and astronomy:

> nam et auribus oblectatio magna parta est inventa et temperata varietate et natura sonorum, et astra suspeximus cum ea quae sunt infixa certis locis, tum illa non re sed vocabulo errantia, quorum conversiones omnesque motus qui vidit, is docuit similem animum suum eius esse, qui ea fabricatus esset in caelo. nam cum Archimedes lunae, solis, quinque errantium motus in sphaeram illigavit, effecit idem quod ille, qui in Timaeo mundum aedificavit, Platonis deus, ut tarditate et celeritate dissimillimos motus una regeret conversio. quod si in hoc mundo fieri sine deo non potest, ne in sphaera quidem eosdem motus Archimedes sine divino ingenio potuisset imitari.[16]

> For through them our ears have gained keen delight from the discovery of the due combinations of musical sounds of diverse quality, and we have looked up at the stars, both those that are fixed in certain spots and those that by name are "wandering" (though they are not really so), and he who has seen their revolutions and all their movements has taught that his soul resembles his who had fashioned them in the heavens. For when Archimedes fastened the movements of the moon, the sun, and the five planets on a sphere, just like Plato's god who built the world in the in the *Timaeus,* he made one revolution of the sphere control several movements completely different in slowness and speed. Now if in this world of ours phenomena cannot take place without the act of god, neither could Archimedes have imitated the same movements upon a sphere without divine genius.

Reflecting upon this genius of discovery, the puzzles of how memory might work, and some quiet amazement even at simple perception, Cicero goes on to offer us a strikingly novel account of the soul. Virtually all Greek and Roman philosophers, so far as I am aware, thought the soul to be a physical entity, made up of the same stuff as the rest of the cosmos, whatever that may have

15 It has also been suggested that the M may be for *Magister,* but I think the internal cross-references to Cicero's other works as his own suggest Marcus very strongly.

16 Cicero, *Tusc.* 1.62-3. Translation modified from King.

been taken to be. Cicero tells us that he had realized, while grieving for his recently deceased daughter Tullia, that the elements of which the earth and cosmos are made could *in no way* be responsible for memory, thought, deliberation, or perception—earth, air, water, and fire simply do not have properties that could produce reason and consciousness. "Whatever it is," he says, quoting his now-lost *Consolatio,* "that perceives, that is wise, that lives, that flourishes, must be heavenly and divine and for that reason eternal."[17] Cicero now moves us up from the divinity of the human mind to the divine governance of the universe as a whole, in a version of the same argument from design that we will see again shortly in the *De natura deorum.* In one extremely long, breathlessly flourishing sentence of 260-odd words, he takes us from our observation of the brilliance and wonder of the sky to the necessity of a divine creator or governor. Harking back to the *Dream of Scipio,* he says that the sun is (astrologically) called the *temperatio* and *moderator* of the planets. Picking up on Stoic themes from the *De natura deorum,* the natural order sees animals near-miraculously well suited to our every need (they serve as food and as draft animals, for example), and in the capstone statement of these dual themes, humans exist specifically *quasi contemplatorem caeli ac terrarum cultorem,* "to contemplate the heavens, as it were, and to till the earth."[18]

Where Cicero had held back from calling Archimedes "divine" in the *Republic,* here he shows no such reticence, and in fact the marvel of the sphere itself is central to the whole claim: "if in this world of ours phenomena cannot take place without the act of god, neither could Archimedes have imitated the same movements upon a sphere without divine genius." Archimedes's genius is divine, and it is identified with the divine soul of the creator of the cosmos. But—and this is curious—the analogy runs unexpectedly backwards to the expected argument from design: it is not that the cosmos must have a creator and operator since it is like the sphere, which clearly has a creator and operator. It is instead that since the cosmos has a creator and operator who is divine, the imitation must also have a creator and operator, Archimedes, who is likewise divine. On the face of it, it is perhaps not a very compelling argument, until we supply the intermediary assumption that this divinity is implied in both cases by the difficulty and cleverness of the created artifact. The marvel is so wondrous in both cases, that it could only have been devised by *divinum ingenium,* "divine genius." Furthermore, the identity of the two makers is further reinforced when Cicero says that Archimedes "fastened" the parts of

17 *ita quidquid est illud, quod sentit, quod sapit, quod vivit, quod viget, caeleste et divinum ob eamque rem aeternum sit necesse est. Tusc,* 1.66.

18 *Tusc.* 1.69.

the mechanism together, where the verb for to fasten, *illigo,* is the same word used by Cicero in his translation of Plato's *Timaeus* to describe how the gods fixed soul and body together to form living things, and, in case we might miss this subtle cross-reference, he then explicitly tells us that Archimedes' construction is "just like Plato's god who built the world in the *Timaeus.*"

This elevation of Archimedes to divine status gets further developed and brought almost full circle in a charming short poem by Claudian (4th century), *On the Sphere of Archimedes,* where Jupiter himself marvels at the cleverness of the inventor:

> Iuppiter in parvo cum cerneret aethera vitro
> risit et ad superos talia dicta dedit:
> hucine mortalis progressa potentia curae?
> iam meus in fragili luditur orbe labor?
> iura poli rerumque fidem legesque deorum
> ecce Syracusius transtulit arte senex.
> inclusus variis famulatur spiritus astris
> et vivum certis motibus urget opus.
> percurrit proprium mentitus Signifer annum,
> et simulata novo Cynthia mense redit,
> iamque suum volvens audax industria mundum
> gaudet et humana sidera mente regit.
> quid falso insontem tonitru Salmonea miror?
> aemula naturae parva reperta manus.[19]

> When Jupiter saw the firmament in tiny glass, he laughed and put these words to the gods: "Has the power of mortal thought come so far? Is my work mimicked in a fragile globe? Behold! An old Syracusan has by handicraft translated the ordinances of the heavens, the truth of things, and the laws of the gods. An enclosed soul is servant to the various stars and drives the living work in precise motions. A counterfeit zodiac traverses its year, and an imitation moon returns at the beginning of the month. Now daring diligence rejoices at moving its own world, and conducting the stars by human mind. Why am I surprised at guiltless Salmoneus with his pretend thunder? His tiny hand is shown to be the rival of nature."

One of the more curious features of this poem is Claudian's use of "soul" as a description for the inner mechanism of the device. He is being very subtle here, simultaneously drawing the reader's eye away from the messy gearwork

19 Claudian, *Shorter poems* LI.

inside the device while at the same time elevating its mystery by identifying it with the invisible and ineffable *spiritus* that drives the heavens.[20] The machine thus transcends itself and becomes a living thing, and this at the hand of a *human* creator. But Jupiter is not jealous; he is amused. Nevertheless, Claudian reminds the reader of how close a call this all was for Archimedes when he compares the "old Syracusan" to Salmoneus, the mythological king of Elis, who had demanded that he be called *Zeus* by his subjects and then went so far as to outfit his chariot with special effects so that it made the sound of thunder while it rode and shot out lit torches to imitate lightning.[21] Zeus accordingly struck him down with (what else?) a thunderbolt. Here Claudian may also be making a subtle reference to Manilius, who in his great astrological poem describes Salmoneus as *qui caelum imitatus in orbe,* one who imitated the heavens *in orbe,* or on a globe (one presumes Manilius means "on earth," although Claudian's playful reference reimagines the passage completely).[22] That Jupiter could have seen Archimedes, with his "tiny hand," as overstepping human bounds is reinforced repeatedly throughout the poem: the heavens were "mimicked," *luditur,* by Archimedes, which word (in Latin as in English) can refer to mockery as well as to sport; Archimedes is the rival of nature, where the semantic overlap between neutral emulation and outright hostility in *aemulus* (rivalling, imitating) plays out; the pejorative sense of *mentitus Signifer,* "counterfeit zodiac," is hard to miss. Nevertheless, Archimedes is said to be "blameless," and Jupiter to be highly amused. Perhaps the threatening talk was just the god's jest.

It is interesting, too, that Archimedes does not in the first instance *copy* the heavens, but instead "translates" them. And here legal metaphors abound: what Archimedes has translated are the judgments or laws of the cosmos, the laws of the gods, and the *fidem rerum,* literally the promise or binding trust of things.

Just Like Archimedes Did

The trope of Archimedes imitating or emulating the gods in his sphere gets turned entirely on its head by the 5th century writer Martianus Capella. His encyclopedic *Marriage of Philology and Mercury* is a survey of the seven liberal arts, in the elaborate conceit of a wedding feast, where each branch of learning

20 On Claudian's fascination with hidden causes, see also his poems on the electric eel and the magnet, respectively.

21 See e.g., Vergil, *Aen.* 6.585 f.

22 See also Manilius 4.267-8: *sidereas caelumque novum versabit in orbem.*

is introduced as a goddess, coming as a bridesmaid to the wedding of Philology and Mercury, and giving an account of her branch of learning. It is a difficult and often unnerving text, a mixture of florid prose and highly allegorical poetry. C.S. Lewis once famously said of it "this universe, which has produced the bee-orchid and the giraffe, has produced nothing stranger than Martianus Capella."[23]

In the *Marriage,* two of the characters are Geometry and Astronomy, Geometry entering after Grammar, Dialectic, and Rhetoric. She is followed by Arithmetic and then Astronomy. Harmony comes last to close the book. In these proceedings, we find another reference to mechanized sphere, again associated with Archimedes, but it is surprisingly not in the hands of Astronomy. Instead it is Geometry who comes in bearing it. What is even more surprising about the sphere Geometry holds, perhaps, is that it turns our whole previous narrative on its head, with the goddess now bearing what must be a divinely made sphere that imitates the one Archimedes is supposed to have made in imitation of the heavens.

When she makes her appearance, the character of Geometry enters only after two of her attendants had preceded her. They were bearing that most basic tool of geometry, an *abacus,* that is, an erasable mathematical drawing-tablet, on which, we are told, the shapes of lines, the curves of circles, or the angles of triangles may be etched.[24] Note here the slightly curious wording. We are not drawing mathematical objects—lines, circles, triangles—in the first instance. We are drawing, to use Martianus' wording "linear shapes" (*lineares ductus*), "circular curves" (*circulares flexus*), and "triangular bends" (*triangulares anfractus*). He continues by telling us that the abacus can show the *facies,* the image, of the cosmos as a whole—and note here the idea of representation starting to creep in, and the perhaps surprising appearance of the cosmos as the ultimate object of Geometry's drawing. That we are dealing with a high degree of abstraction is emphasized by Martianus' summation of the *abacus* as a tool for depicting (*videbis istic depingi*) "whatever cannot be explained in words" (*quicquid verbis non valeas explicare*). That imitation is one of Geometry's paramount skills gets further emphasized when Martianus next compares her to three, and only three, legendary figures, saying that her skill surpasses them all. These are Apelles, the painter, Polyclitus, the sculptor, and Daedalus, who was famous in fable for having made statues that could move about on their own.

23 Lewis 1936: 78.

24 6.579.

A further peculiarity of Geometry, especially considering that the character of Astronomy is waiting in the wings for her own grand appearance just two books away, is that she is so thouroughly adorned with astronomical imagery. Indeed, apart from the mention of circles and lines and triangles in connection with the abacus, nearly all of Geometry's attributes in these early passages are astronomical. She comes in bearing the *sphaera* and wearing a mantle adorned with images of celestial magnitudes, planetary bodies and their orbits, as well as a picture of the earth's shadow and of a gnomon. If we contrast this to Astronomy's later entrance (book 8), we see again an emphasis on Geometry as modeler. Where Geometry comes in with a material sphere in hand, Astronomy simply materializes out of a magically appearing ball of what sounds like actual aether:

> et ecce globus quidam lucis aetheriae et concava perspicui ignis aggestio, ut apparebat intra se quandam virginem claudens, miti vertigine sensim volutus inlabitur. quo candore luminis propinquantis plures irradiati refulsere divi, fatalesque maxime, quorum etiam habitus motusque et quicquid in his ignotum credebatur emicuit.[25]

> And behold: there appeared a sphere of aetherial light, a concave filled with transparent fire, enclosing within it a maiden, and turning with the slightest perceptible motion. A number of gods, especially the ones who determine fates, were bathed to the point of reflecting the brightness of the outpouring light, their habits and motions and other such things that had been concealed, shone forth.

Astronomy is carrying a book in one hand, and in the other, a *cubitalem fulgentemque mensuram,* "a shining cubit-measure" (probably something like a dioptra).[26] She proceeds to describe, and indeed to verbally construct before the audience, the various circles in the heavens. Here she nods to Geometry several times, but in contrast to the account given by Geometry, for Astronomy all the constructions are presented as taking place in the real world: circles, that is, are drawn in the heavens and not on models. She then offers us a detailed list of paranatellonta and rising-times for the zodiacal constellations, and an account of the eccentricity of the sun and the motions of the planets and moon.

---

25 Martianus Capella, *De nuptiis,* 8.810.

26 Martianus Capella, *De nuptiis,* 8.811.

Turning back to Geometry's entrance, we note that in the description offered of the Archimedean sphere she carries, there is the familiar note of human genius emulating that of the gods:

> ipsa etiam laeva sphaera fulgebat honora,
>   assimilis mundo sideribusque fuit;
> nam globus et circi zonaeque ac fulgida signa
>   nexa recurrebant arte locata pari.
> tellus, quae rapidum consistens suscipit orbem,
>   puncti instar medio haeserat ima loco.
> hanc tener et vitreis circumvolitabilis auris
>   aer complectens imbrificabat aquis.
> quae tamen immenso, quo cingitur illa, profundo
>   interrivata marmore tellus erat.
> texerat exterior qui fulget circulus orbis
>   aetheris astrifico lumina multa peplo.
> hinc nitidus rutilum Titan succenderat orbem,
>   moxque imitata pium lactea Luna diem.
> sic igitur furvae oculos splendescere noctis
>   cernas ex auro ut sacra flamma micat.
> sic Cypris Oceani perfuso sidere lymphis
>   Lucifero annuerat lumina tota suo.
> omnia compar habet paribus sub legibus ordo,
>   nec minus haec mira est quam domus alta deum.
> hanc mundo assimilem stupuit Trinacria tellus
>   Archimedea astrificante manu.
> o felix cura, et mentis prudentia maior,
>   corpore sub nostro aequiperasse Iovem.[27]

The glorious sphere shone at her left hand; it was very like the world and the stars. And the sphere and the circles and the zones and the shining fixed stars orbited, put skillfully in place. The earth, which, unmoving, holds up the fast orb, hangs in the middle, the lowest place, like a point. A soft atmosphere holding it in, blowing round in gentle breezes, waters it with rains. The earth was riven by an immense and deep sea which belted it. The outer circle was covered by a very bright mantle that shone with aetherial stars. Next the gleaming sun kindled his golden orb, and then the milky moon imitating true daylight. Thus you could see these sights shining in the dark of night, as a sacred flame flashes from gold. So

27 Martianus Capella, *De nuptiis*, 6.583-5.

> Venus gave her morning star full light when it was still immersed in the waters of the ocean. A similar order maintains all under equal laws, and this (sphere) is no less astonishing than the lofty home of the gods. Sicily was amazed by this copy of the cosmos made by an Archimedean star-making hand. What fruitful skill, and great intelligence of mind to make oneself equal to Jupiter with subterranean matter.

Notice that this is an entirely fictional globe, invented by Martianus for the story. It has properties no earthly model could possess: the winds and rain, the glowing of the sun. So, too, it is envisaged as an actual sphere in shape, entirely unlike the Antikythera box. It is self-lit, and is somehow shrouded in an outer shell of fixed stars that nevertheless is transparent enough to see through.

As in our earlier discussions of mechanized spheres, we see here the amazement at the mechanism: "this sphere is no less astonishing than the lofty home of the gods." And note again the ubiquitous reference to Archimedes. Martianus credits Archimedes with manufacturing "this copy of the cosmos" but surely he is speaking loosely or at best allegorically, given the impossibility of the many features already mentioned. Finally, Archimedes transcends mortality in the machine, making himself equal to Jupiter.

Divine Watchmaker

We have seen mechanized sphaera used as proofs of the divinity of Archimedes' genius, as well as its deployment in Cicero's proof of the immortality of the soul. Sphaera have been the go-to example in our authors of the intellectual heights attainable by human beings—look what we can do!—one that impresses even the gods themselves. It also gets invoked in another context, one hinted at already but so far not fully developed. This is in a version of what theologians know as the "divine watchmaker" argument, usually attributed to the English philosopher and theologian William Paley (1743–1805).[28] On Paley's formulation, the finding of a rock lying unattended upon a heath should occasion no surprise and no pressing question of how it came to be—one could easily suppose it had simply always been there. The finding of a working watch, on the other hand, would not lead to the supposition that it had always been there. Instead, given its complexity, utility, and obvious signs of deliberate design, one would suppose that it had in the first instance been manufactured by an intelligent being for some purpose. Paley then argues for an analogous

28 See for example, Shapiro 2015.

complexity and purposiveness to the world as a whole, and draws the conclusion that the world must therefore have had a designer just as surely as the watch did, one as far above the human watchmaker as the world is above the watch in majesty.

But the argument is a very old one, and we find several versions of it in antiquity, only, instead of a watch as the exemplar of complex, deliberate design, it is a mechanized sphere. The earliest of these arguments appears, once again, in Cicero.

In the *De natura deorum,* Cicero has his Stoic character, Balbus, argue at length for the rationality and divinity of the cosmos as a whole. In the course of laying out the case for the governance of the world by divine reason, Balbus argues that the perfect regularity shown by the motions of the planets shows not only that their motions are not due to chance, but further, that their motions are due to deliberate, divine design:

> quodsi in Scythiam aut in Britanniam sphaeram aliquis tulerit hanc quam nuper familiaris noster effecit Posidonius, cuius singulae conversiones idem efficiunt in sole et in luna et in quinque stellis errantibus quod efficitur in caelo singulis diebus et noctibus, quis in illa barbaria dubitet quin ea sphaera sit perfecta ratione? hi autem dubitant de mundo ex quo et oriuntur et fiunt omnia, casune ipse sit effectus aut necessitate aliqua an ratione ac mente divina, et Archimedem arbitrantur plus valuisse in imitandis sphaerae conversionibus quam naturam in efficiendis, praesertim cum multis partibus sint illa perfecta quam haec simulata sollertius.[29]
>
> Suppose someone had brought the sphere that our friend Posidonius recently made into Scythia or Britain, on which the individual motions pertaining to the sun and the moon and the five wandering stars are effected just as it happens in the sky on specific days and nights. Who in that foreign land would doubt that the sphere was completed by reason? But these [opponents of mine] waver on the subject of the world, from which everything arises and is made, whether it happens by chance, or whether by necessity, or by divine reason and will. They judge Archimedes to have been better for imitating the motions of the sphere than nature for causing them, even though nature's construction is many times more skillful than the imitation.

For the first time, we see the instantiation of a real sphere in the room, rather than a story about an offstage encounter. Moreover, the *sphaera* is one recently

29 Cicero, *ND* 2.88.

made by "our friend," Posidonius. Where all of our Archimedian spheres have up to this point been essentially hearsay, I see no reason to doubt that Cicero has actually seen a Posidonian one. Nevertheless, Archimedes is still not very far away: it is not Posidonius, this sphere's maker, who gets the final credit for his clever device, but the old Syracusan, presumably for designing the plans followed by Posidonius. (For what it is worth, we do know that Archimedes is supposed to have written a book, now lost, on the art of *Sphaeropoiia,* and the tenacity of the tradition associating him with Antikythera-type displays may be on the grounds of this book rather than on the preservation and display of any particular globe he may have made).

And notice here how Cicero turns the familiar idea of Archimedes' genius against his opponents: "they judge Archimedes to have been better for imitating the motions of the sphere than nature for causing them." For all that he has been quick to praise Archimedes in his other works, here he berates his interlocutors for failing to recognize how much better than Archimedes' little sphere the model from which it was copied really is.

Curiously, we find in the Christian author Lactantius, a direct attack on Balbus's larger position in the *De natura deorum,* but one that nevertheless seems to come to almost the same conclusion as Balbus had, with the exception that instead of the regularity of the motions of the stars proving their own divinity, it instead proves the divinity of the one who created the stars.

> an Archimedes Siculus concavo aere similitudinem mundi ac figuram potuit machinari, in quo ita solem lunamque composuit, ut inaequales motus et caelestibus similes conversionibus singulis quasi diebus efficerent et non modo accessus solis ac recessus vel incrementa deminutionesque lunae, verum etiam stellarum vel inerrantium vel vagarum dispares cursus orbis ille dum vertitur exhiberet, deus ergo illa vera non potuit machinari et efficere quae potuit sollertia hominis imitatione simulare?[30]

> Could Archimedes of Sicily have devised in hollow bronze a model and figure of the world, in which he has arranged the sun and the moon such that they would effect unequal motions and similar ones to the celestial rotations almost day by day, and the sphere, when rotated, show not only the rising of the sun and its setting and the waxing and waning of the moon, but also the disparate courses of the stars, both fixed and wandering, and yet God himself could not devise and bring about what human ingenuity is able to copy by imitation?

30 Lactantius, *Divinarum institutionum,* 2.5.

The argument in which this is embedded wanders considerably, and relies to an extent on a certain amount of sleight-of-hand on Lactantius's part, as indeed the key move in this passage relies on the reader's prior acceptance of the idea that anything Archimedes could do (the Christian) God could do better, but we get the idea: if a man could make something displaying complex regularity like the sphere does, so God could make the stars that move that way too.

Finally, the Pyrrhonian sceptical philosopher Sextus Empiricus had also earlier elaborated a version of the Stoic argument from design, one very similar to that in Cicero's *De natura deorum*. Sextus begins by noting that the cosmos moves with an order that precludes what Democritus had called necessity (ἀνάγκη) or accident, and requires (so the Stoics say) a divine mover. Sextus then compares the situation to that where someone looks at an Archimedian *sphaera:*

> τά γε μὴν αὐτομάτως κινούμενα τῶν κατασκευασμάτων θαυμαστότερά ἐστι τῶν μὴ τοιούτων. τὴν γοῦν Ἀρχιμήδειον σφαῖραν σφόδρα θεωροῦντες ἐκπληττόμεθα, ἐν ᾗ ἥλιός τε καὶ σελήνη κινεῖται καὶ τὰ λοιπὰ τῶν ἀστέρων, οὐ μὰ Δία ἐπὶ τοῖς ξύλοις οὐδ' ἐπὶ τῇ κινήσει τούτων τεθηπότες, ἀλλ' ἐπὶ τῷ τεχνίτῃ καὶ ταῖς κινούσαις αἰτίαις.[31]
>
> Devices that move by themselves are more marvellous than those that do not. Thus we are especially astounded on seeing an Archimedean sphere, in which the sun and the moon and the rest of the stars move—not, by Zeus, marveling at the materials or their motions, but at the artificer and at how the motions are brought about.

This he reports as proof for the Stoics that there must be a divine intelligence guiding the complex motions of the cosmos as a whole.

Realist Objections

Notice in Sextus the implied reference to the inner workings of the machine: we are more impressed by *how the motions are brought about* than by the materials or the motions themselves. It is the gearing and the construction, opening the thing up and looking inside it that really get us excited, he says. And this is true. On seeing a modern reconstruction of an ancient *sphaera,* such as Michael Wright's remarkable reconstruction of the Antikythera mechanism,

31 Sextus, *Adv. math.* 9.115.

the display is very impressive,[32] but for most people that is nothing compared to seeing the elegance, complexity—and above all the non-obviousness—of the inner workings.

But we have one author, believe it or not, who was unimpressed, or at least unhappy with how the motions were brought about. Claudius Ptolemy, the greatest astronomer and one of the greatest mathematicians of the whole ancient period: he thought he had a good idea of how the motions in the heavens were brought about, and in his mind, that didn't look anything like the insides of an Archimedean sphere. Being a good scientific realist, this bothered Ptolemy.

Before turning to Ptolemy, though, let us briefly look back at something both Claudian and Martianus Capella said: "a similar order maintains all under equal laws, and this (sphere) is no less astonishing than the lofty home of the gods" (Martianus), and "an old Syracusan has translated the ordinances of the heavens, the truth of things, and the laws of the gods" (Claudian). Both of them seem to be saying that the Archimedean sphere follows the same laws as the cosmos, which may be to say that the sphere can tell us something about the mechanism of the heavens themselves. Ptolemy is having none of it:

> ἐνταῦθα δὲ προήχθημεν αὐτὸ μόνον ἐκθέσθαι κεφαλαιωδῶς καὶ ὡς ἂν μάλιστα προχειρότερον κατανοηθεῖεν ὑπό τε ἡμῶν αὐτῶν καὶ τῶν εἰς ὀργανοποιίαν ἐκτάσσειν αὐτὰ προαιρουμένων, ἐάν τε γυμνότερον διὰ χειρὸς ἑκάστης τῶν κινήσεων ἐπὶ τὰς οἰκείας ἐποχὰς ἀποκαθισταμένης τοῦτο δρῶσιν, ἐάν τε διὰ τῶν μηχανικῶν ἐφόδων συνάπτωσιν αὐτὰς ἀλλήλαις τε καὶ τῇ τῶν ὅλων. οὐ μὴν ὃν εἰώθασι τρόπον σφαιροποιεῖν· ὁ γὰρ τοιοῦτος καὶ χωρὶς τοῦ διημαρτῆσθαι τὰς ὑποθέσεις τὸ φαινόμενον παρίστησι μόνον καὶ οὐ τὸ ὑποκείμενον, ὥστε τῆς τέχνης καὶ μὴ τῶν ὑποθέσεων γίνεσθαι τὴν ἔνδειξιν.[33]
>
> In this book, we go on to set [these hypotheses] out in summary so that they can be most easily understood both by ourselves and by those proposing to arrange them for the purposes of instrument-making, whether they do this in the more stripped-down manner, returning each of the motions to its proper epoch by hand, or by a mechanical method, joining the motions to each other and to the motion of the whole. But this is not the usual way they do sphere-making, for there—apart from betraying the hypotheses, they represent the appearances only and not the cause (lit: "the underlying"), such that the show is about the craftsmanship, not about the hypotheses.

32 See, for example, Yirka 2011.

33 Ptolemy, *Plan. Hyp.* 1.70.11-23.

What Ptolemy is objecting to here is that the mechanical spheres made by the sphere-makers, although they model the phenomena, do not show the real workings of the cosmos. In a perfect world, he seems to be arguing, the sphere-makers would make their *sphaerae* with a mechanics identical to that set out in the *Planetary Hypotheses:* spheres turning within spheres within spheres. The mechanical sphere, then, would for once be a true model of the cosmos, mirroring not only the appearances of the planetary motions, but also demonstrating in their interlocking moving parts, the actual physical causes of those heavenly motions.

Turning it the other way around, Geminus of Rhodes, the first-century-BCE author of an introductory textbook of astronomy, seems to think that in fact the *sphaerae* can tell us about how the universe is constructed. He goes so far as to tell us that the cosmos is itself a *sphaeropoiia.* Evans and Berggren have found in this claim a strong realist stance on Geminus' part (one with which I agree), and they cite three passages where Geminus describes the cosmos as a *sphaeropoiia,* plus one where they take *sphaeropoiia* to mean the spherical theory of the world more broadly.[34] I will focus on two of these passages, leaving aside the "spherical theory of the world" one, and one of their instances of *sphaeropoiia* as the actual spherical construction of the world (at 12.23) because I think those instances may just possibly refer to *sphaera*-making in the more traditional sense.[35] At 14.9, Geminus is discussing stars that rise and set: some never rise or set, and some do one or the other but not both on any given night. He says:

> διὰ δὴ τὴν τοιαύτην σφαιροποιίαν, οὐ πάντα τὰ ἄστρα ἀνατέλλει καὶ δύνει καθ' ἑκάστην νύκτα, ἀλλὰ τινὰ μὲν ἀνατέλλει καὶ δύνει, τινὰ δὲ ἀνατέλλει μέν, οὐ δύνει δέ, ἃ δὲ οὔτε ἀνατέλλει οὔτε δύνει.
>
> Because of this spherical construction [of the cosmos], not every star rises and sets every night. Instead, some rise and set, some rise but do not set, and some neither rise nor set.

Again, at 16.19, he wants to explain how he knows that there are southern inhabitable zones on the earth, where the antipodes live. He spells out that he has not had traveller's reports from those regions. Instead, it is ἕνεκεν τῆς ὅλης σπαιροποιίας καὶ τοῦ σχήματος τῆς γῆς καὶ τῆς παρόδου τοῦ ἡλίου τῆς μεταξὺ τῶν

34 Evans and Berggren 2006: 52–53.

35 They may well mean what Evans and Berggren take them to mean, but they are just ambiguous enough (as I think Evans and Berggren also see) that I will stick to the two clear-cut instances.

τροπικῶν γινομένης, "because of the spherical construction of the whole and the shape of the earth and the path the sun takes between the tropics" that we can know three to be an inhabitable temperate zone on the opposite side of the earth.

Here the cosmos itself has become the product of the sphere-making craft. Just as in the simpler lathe-turning analogies we saw much earlier, only this time with more explicit reference to those craftsmen who make astronomical *sphaerae* (whether mechanized or not, however, he does not say). The analogy now runs from the model to the original, and the manufacture of the original is somehow modeled on the model. Just as for Ptolemy, Geminus is here betraying a commitment to the idea that the models we make should really capture the structure of the world, and he believes this so immediately that he sees no need to even explain himself. He takes it for granted that he can just call the universe a manufactured *sphaera* and that his readers will accept the move unproblematically. The implications of a divine creator, Geminus does not draw out (this would not be the place in any case), but if our previous authors are any indication, a heavenly creator cannot have been too far in the wings.

Conclusion

The ancients, just as we do now, marveled at the incredible ingenuity on display in their geared astronomical models. Indeed, in many contexts, such devices stood as the very embodiment of the pinnacle of human intelligence, and in many cases elevated human genius on a par—dare they say it?—with the divine. Fully conscious that such hyperbole may have been not entirely theologically sound, Cicero and Claudian gird themselves, Claudian by neutralizing it through playful storytelling, and Cicero by actually embracing the consequences and running them home, holding them up as the very proof of the immortality of the soul. Human creators are modeled on divine ones, just as the human-made mechanisms are modeled on the created cosmos, and the complexity of the task sometimes even has even the gods standing in admiration. So, too, the models could tell us something about the complexity and origin of the cosmos, and for the Stoics, and Cicero siding with them the mechanism's self-implied proof of design furnishes an argument for our earliest version of the theological argument from design.

But the very difficult question of the extent to which such mechanisms can really tell us something about the heavens was not lost either (and here the philosophical literature of modeling and metaphor becomes relevant). After all, if we are to use any model to think about the original thing that it models, we need to first know which features of the model are meant to be representative of the real thing, and which are compromises inherent in the act of modeling.[36] Ptolemy, whose philosophical reflections on the methodology of science have been getting increasing attention in recent years, worried very clearly about which features of the model were, to use the modern parlance, positive analogies (which ones really copied the details of the original, and so could be used heuristically), and which were negative analogies (non-identical and therefore non-heuristic). For him, the negatives seem to have outweighed the positives, and he seems to have been wishing for a better type of mechanism.

But Ptolemy's higher-order question about the nature of the analogy between mechanism and sky seems to have been spoken in his voice alone. Across the board, from Christians to late pagans and all the way back, the idea of a craftsman building something so marvellous struck a chord in the ancient imagination, and told them, so they thought, some very important truths about themselves, the world they lived in, and their place in it.

Bibliography

Bowen, A.C.

2002 "The Art of the Commander and the Emergence of Predictive Astronomy." Pp. 76–111 in *Science and Mathematics in Ancient Greek Culture,* ed. C.J. Tuplin and T.E. Rihll. Oxford: Oxford University Press.

Field, J.V. and M.T. Wright

1985 "Gears from the Byzantines." *Annals of Science* 42: 87–138.

Freeth, T. and Jones, A.

2012 "The Cosmos in the Antikythera Mechanism." *ISAW Papers* 4 (February).

Freeth, T.; Jones, A.; Steele, J.M., and Bitsakis, Y.

36 I am here thinking of what Weisberg 2012 calls "concrete" models, physical objects that are meant to model systems in the world. Such models are intended to capture as many features of the modelled system as possible, but without being exact copies (exact in scale, exact in material, exact in function). They inherently introduce what I am calling "compromises," where they model does things differently than the modelled system (it is smaller or larger, it is made of plastic or metal, it uses strings and rods instead of fundamental forces). This is not necessarily a problem so long as all interested parties are aware of what compromises are being made and which features of the model are meant to be taken as what we might call more literally representative, but it does introduce epistemic issues that are nontrivial. See recently, e.g., Weisberg 2012; Suárez 2009; Giere 2004; Sismondo and Gissis 1999; Morgan and Morrison 1999.

2008 "Calendars with Olympiad Display and Eclipse Prediction on the Antikythera Mechanism." *Nature* 454: 614–617.

Freeth, T. *et al.*

2006 "Decoding the Ancient Greek Astronomical Calculator Known as the Antikythera Mechanism." *Nature* 444: 587–591.

Gallagher, R.

2001 "Metaphor in Cicero's 'De Re Publica.'" *Classical Quarterly* 51: 509–519.

Giere, R.N.

2004 "How Models Are Used to Represent Reality." *Philosophy of Science* 71: 742–752.

Lewis, C.S.

1936 *The Allegory of Love.* Oxford: Oxford University Press.

Morgan, M.S. and Morrison, M. (eds.)

1999 *Models as mediators.* Cambridge: Cambridge University Press.

Price, D.J. de Solla

1974 "Gears from the Greeks." *TAPS,* New Series 64: 1–70.

Shapiro, B.

2015 "A Failed Metaphor for Intelligent Design." *The Atlantic,* Feb. 12, 2015.

Sismondo, S. and Gissis, S. (eds.)

1999 *Practices of Modeling and Simulation,* Special issue of *Science in Context* 12.

Suárez, M., ed.

2009 *Fictions in Science.* New York: Routledge

Weisberg, M.

2012 *Simulation and Similarity.* Oxford: Oxford University Press

Wright, M.T.

2007 "The Antikythera Mechanism Reconsidered." Interdisciplinary Science Reviews 32: 27–43.

Yirka, B.

2011 "Ancient Greek Calculating Device Continues to Reveal Secrets." *Phys.org,* April 4.

Index of Subjects

Index of Ancient Texts and Words

Akkadian/Sumerian
Publication/Titles

www.ingramcontent.com/pod-product-compliance
Lightning Source LLC
LaVergne TN
LVHW010638300926
849161LV00005B/975